Volume 39

Ammonia Plant Safety
& RELATED FACILITIES

"A Continuation of the Series" Safety in Air and Ammonia Plants

a Technical Manual
Published by the American Institute of Chemical Engineers
3 Park Avenue, New York, NY 10016-5991

Copyright 1999

American Institute of Chemical Engineers
3 Park Avenue, New York, NY 10016-5991

Library of Congress Card Number: LC72-625346
ISSN Number: 0149-3701
ISBN Number: 0-8169-0803-6

Authorization to photocopy items for internal or personal use, or the internal or personal use of specific clients is granted by AIChE for libraries and other users registered with the Copyright Clearance Center (CCC) Transactional Reporting Service, provided that the $2.00 fee per copy is paid to CCC, 21 Congress St., Salem, MA 01970. This consent does not extend to copying for general distribution for advertising or promotional purposes for inclusion in a publication or for resale.

Articles published before 1978 are subject to the same copyright conditions, and the fee is $2.00 for each article. Technical Manual fee code: 0149-3701-99/$2.00.

INTRODUCTION

The 43rd annual Ammonia Safety Symposium followed the tradition set by previous symposia, providing a forum for discussions on all aspects related to safe operation of ammonia plants and similar installations. The symposium was attended by close to 500 participants from 38 countries and thus confirmed the status of this series of symposia as the most important international forum for exchange of information within the relevant industries.

The symposium was organized by the AIChE Ammonia Safety Committee. This committee has, until recently, consisted of members from North America and Europe only. However, it has now been decided also to invite members to the committee from other parts of the world, and the first Indian committee member has already been nominated. In recognition of this and of the size and importance of the Indian fertilizer industry, the Managing Director of Indian Farmers Fertilizer Cooperative Ltd. (IFFCO), Mr. U.S. Awasthi, who is also president of IFA (International Fertilizer Association) and former president of FAI (Fertilizer Association of India), was invited to give the key note address at the Symposium. Mr. Awasthi accepted the invitation and prepared the presentation. Unfortunately, unforeseen and unavoidable circumstances prevented his participation, and in his place Mr. I. J. Ohri, Executive Director (N) of IFFCO, presented the address dealing with the status of the fertilizer industry in India. India is the third largest producer and consumer of fertilizers in the world, and the industry is based on the most modern technology ensuring safe, reliable, efficient, and environmentally friendly production.

The success of the symposium is due to the authors and speakers who present the papers, and the individual companies who support them and of course to the participants. The papers presented at the 43rd Ammonia Symposium in Charleston are presented in this publication, along with the subsequent discussions, so that the entire industry can benefit from the proceedings.

On behalf of the ammonia industry and the Ammonia Safety Committee, I would like to thank AIChE for their continued support. Special thanks go to Courtney Bunting and the meetings staff, the very excellent audio-visual crew, and to publications staff of AIChE.

Ib Dybkjaer

AIChE Symposium: Charleston, SC, 1998
Program Chairman, Ib Dybkjaer; Program Vice Chairman, Richard Johnson

Committee Membership: September, 1998
Committee Chairman, Gunnar Schulstok; Committee Secretary, R.W. Clark

Ian Welch	Louis Frey	Gerald Williams	Kevan Vick
Jerry Davis	Theo Huurdeman	J. R. Le Blanc	
William Delboy	Jim Richardson	Ashok Gupta	

AIChE Publications Staff
Managing Editor: Haeja Han
Assistant Editor: Arthur Baulch
Editorial Assistant: Jeff Karpinos
Cover Design: Armand Veneziano

TABLE OF CONTENTS

Safety Performance in Ammonia Plants: Survey VI
Gerald P. Williams 1

Nitrogen Fertilizer Producers Push the Boundaries
Jorge A. Camps 8

Startup and Initial Operation of a 2,400 Metric t/d Methanol Plant
A. Gedde-Dahl and Helge Holm-Larsen 14

Commissioning a New Ammonia Plant in China
Li Wei Chun and I. R. Barton 24

Successful Operation of a Novel Pool Reactor
K. Jonckers, J. Meessen, and W. Lemmen 33

Use of Bimetallic Tubes in Urea Strippers: Technology Improvement
Franco Granelli and Gian Pietro Testa 46

Shroud Failures of Process Air Compressor Turbine
V. Jayaraman and K. Rajagopal 54

Modifications After a Primary Reformer Explosion at a Reforming Plant
Henry de Wet and Rudie O. Minnie 61

Ammonia Process Primary Waste Heat Boiler Shell Failure Experiences
Colin P. Jackson 69

Pressure Relief Valve Piping Failures and Fire: Ammonia Synthesis Loop
Lester E. Sutherland, Michael Holman, and Don Hansen 87

Plugging a Bayonet/Scabbard Tube in an Ammonia Plant Waste-Heat Boiler
Gary G. Osborne 94

Inner Basket Failure of Ammonia Booster Reactor
Kamarudin Zakaria, Lau Nai Tuang, and Reinhard Michel 101

Damage of Electric Motor of Benfield Solution Circulation Pump
Mubashar M. Butt and Mogens Pedersen 108

Vanadium Recovery Solves Soot Recycle Problems
W. Soyez 119

A Novel PSA System for Ammonia Recovery from Synthesis Gas
Kent S. Knaebel 126

Liquid/Gas and Liquid/Liquid Coalescers in the Ammonia Industry
Thomas H. Wines and Michel Farcy 135

Furnace Section Optimization Using High Emissivity Ceramic Coatings
(Discussion Section Only) *John C. Hellender* 143

Secondary Reformer Burner Nozzle Design and Operating Experience
 Sajjad Hussain, Abdul Basit, Olav Holm-Christensen, and Henrik Stahl 147

Dimensional Check of Catalyst Tubes to Assess Remaining Life
 T. Shibasaki, T. Mohri, and K. Takemura 156

Atmospheric Ammonia Storage Tank Inspection
 Peter Jaras 164

Reducing Methanol Byproduct Formation over the LTS Converter
 Jack H. Carstensen and Birgette S. Hammershøi 171

Maximizing Ammonia Production
 Keith Wilson, Mukund L. Bhakta, and Michael Crowley 179

Planning and Execution of Major Revamp in a Running Plant
 P. S. Neelakantan, K. V. Swaminathan, V. K. Anil, and L. Chawes 189

Upgrading a 25-Year-Old Ammonia Plant
 J. J. de Wit and A. Riezebos 213

Upgrading a Synloop for Capacity Increase and Energy Savings
 Keith Wilson, Ermanno Filippi, and Jim Gosnell 222

Experiences with Heavy Fuel-Oil Firing in a Steam Reformer
 Catela Pequeno and Manfred Severin 231

Revamping Urea Plants for Large Capacity Increase
 K. Clayton, B. Summerscales, and F. Zardi 241

Acoustic Emission Monitoring of Pressure Vessels during Proof Test
 Etienne Soutif, Catherine Herve, Fan Zhang, and Marc Deschamps 251

Urea Waste Water Purification
 J. J. M. P. Goorden and E. R. Killian 264

Risk Management Plans for Existing Control Rooms
 R. A. McConnell 272

Risk Management Plan Modeling for Ammonia Retailing
 Gale F. Hoffnagle and Steven E. Zell 279

Case Study of CO_2 Removal System Problems/Failures in Ammonia Plant
 V. K. Bali and A. K. Maheshwari 285

Safe Work Procedures for Profitability
 Margaret M. R. Eastman and James R. Sawers 302

Management Tool for Undertaking Quantified Risk Assessment Studies
 Katherine Filippin and Mark Jarman 307

Syngas Purification in Gasification-Based Ammonia/Urea Plants
 John Y. Mak and David Heaven 319

Safety Performance in Ammonia Plants: Survey VI

The worldwide safety performance is shown of 89 ammonia plants operating in 25 countries around the world. A three-year period from 1994–1996 is covered, and safety incidences including fires and explosions are reviewed. This recent safety performance is compared with the performance history from past surveys. A list of noteworthy safety practices recommended by the plants is also presented.

Gerald P. Williams
Plant Surveys International, Inc., Petersburg, VA 23805

Introduction

Major safety incidents include: (1) fatalities; (2) lost-time accidents; (3) explosion/rupture incidents; and (4) fires. The percentage of plants experiencing these incidents over the 3-year period is shown in Figure 1.

Only one of the plants reported a fatality (1%). A welder died from burns received while working inside a steam boiler.

Most plants did not have a lost-time accident. However, two out of every five plants (39%) did have one or more lost-time accidents. A total of 143 incidents occurred in 32 plants with four of these plants having 68 of the lost-time accidents.

Explosions or unconfined pressure releases (defined as ruptures) happened in 14% of the plants, and three out of every five plants had at least one fire.

Explosion and Rupture Incidents

Each of 11 different plants had one explosion or rupture incident. Each incident is described in Table 1. Eight of the incidents involved hydrogen-rich gases (mainly synthesis gas), two involved natural gas, and one was electrical.

Fires

The most common fires in plants are from flange leaks where hydrogen-rich gas catches fire and oil

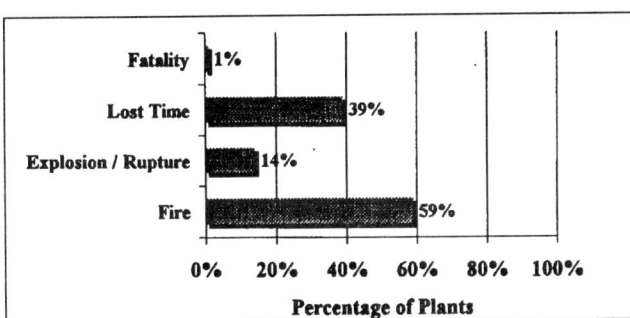

Figure 1. Major safety incidents: plants experiencing incidents.

Table 1. Explosion/Rupture Incidents

- **Transfer line rupture (from secondary reformer to waste heat boiler).**
  Cause: undefined; *Remedy*: damaged repaired.

- **Small explosion inside reformer during startup.**
  Cause: aspirators not turned on before lighting burners; *Remedy*: better communication to workers.

- **Syngas leaking from cryogenic cold box exploded, followed by a fire.**
  Cause: leak inlet expander; *Remedy*: expander replaced by spare unit.

- **Internal explosion in hot potassium carbonate regenerator.**
  Cause: syngas escaped during shutdown while working on sight glass; *Remedy*: (1) explosive gas test should be performed; (2) purge system.

- **CO_2 stripper rupture disc blew and damaged stripper internals.**
  Cause: welder cut bolts without permit; *Remedy*: revised lockout/tagout procedures.

- **Explosion occurred when hole burned through secondary reformer cone section.** *Cause*: burner oxygen vane collapsed directing flame toward the vessel wall; *Remedy*: burner vibration monitoring system installed to detect deviation in flame pattern.

- **Secondary reformer shell ruptured.**
  Cause: misalignment and failure of air sparger which damaged vessel refractory and wall; *Remedy*: developed "fool-proof" method of installing air sparger to assure proper alignment.

- **A minor explosion occurred while lighting the ammonia converter startup heater.**
  Cause: fuel gas leaking through valve; *Remedy*: (1) startup procedure modified; (2) leaking valves were replaced.

- **Potassium carbonate storage tank exploded while draining the CO_2 removal system.**
  Cause: residual hydrogen content in the potassium carbonate solution accumulated. Source of ignition not identified; *Remedy*: (1) communication of shutdown procedures; (2) nitrogen blanketing of tank.

- **Explosion in drain and sewer system.**
  Cause: drain valves from pressure relief system mistakenly not closed allowing syngas to enter sewer system. Probably ignited by hot steam; *Remedy*: improve instructions and use checklist.

- **Explosion at electric power feeder in main substation.**
  Cause: failure of insulation in power control panel causing arcing and flash; *Remedy*: (1) preventive maintenance scheduled; (2) CCTV monitor installed; (3) installation of air sampling fire detection system planned.

Table 2. History of Ammonia Plant Fires

Survey Period	1973–1976	1977–1981	1982–1985	1994–1996
Frequency, months	11.1	14.6	12.2	14.2
Total Number of Fires	125	257	520	215
Plants Having No Fires	2	22	41	35
Plants Having Fires	27	74	95	50
Total Plants Reporting	29	96	136	85

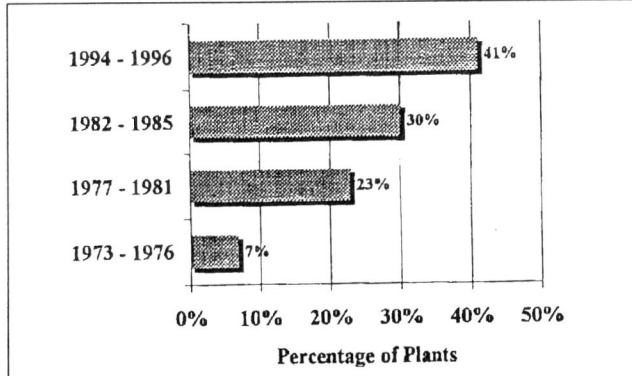

Figure 2. Ammonia plant fires: plants having no fires.

leaks associated with compressor and pump lube oil and seal oil systems.

Other sources are valve packing leaks, piping, electrical, transfer header ruptures, and miscellaneous. Fires within a primary reformer (such as tube, riser, and manifold leaks) are not included in this list of fires.

Fires in ammonia plants have been tracked in previous surveys as shown in Table 2.

The average ammonia plant during the current survey period averaged one fire every 14.2 months. This is better than the previous surveys except for the 1997–1981 period where the frequency was slightly better at 14.6 months. There has been no significant breakthrough in reducing the frequency of fires in the "average" ammonia plant, but there has been steady improvement in the number of plants having no fires, as shown in Figure 2.

In the first survey period (1973–1976), only 7% of the plants had no fires. There has been steady improvement since then where in the latest period 41% of the plants reported having no fires.

Safety: Best Practices

Each plant was asked if they had a safety program which was especially effective in preventing injuries or property damage and "what features of your safety program would you recommend to other ammonia plant operators?" Table 3 is a summary of the responses.

Conclusions

It is evident that there is room for improvement in all aspects of safety. One fatality is one too many, and there is considerable work to be done to eliminate lost-time accidents.

However, it is important to recognize that there are many plants operating without any of these major safety incidents. It is encouraging to note that more plants are operating without any fire incidents.

It is recommended that operators review the *Explosion/Rupture* incidents and assess the exposure in their own plants. There are many excellent *Recommended Safety Practices,* and operators should consider implementing those that can enhance their safety programs.

Table 3. Recommended Safety Practices

As Recommended by Operating Plants, Number of Sites Making Suggestion is shown in Parentheses. Suggestions are in Random Order:

- Full-time safety supervisors assigned to each shift.

- Behavior Accident Prevention Process: Peer to peer observation in "safe" and "at risk" behaviors which could cause accidents. (3)

- Plant modification documentation, critical task analysis, HAZOPs of plant sections, emergency training, personnel protection equipment.

- Unsafe Acts Prevention: Three questions are used: (1) What am I about to do? (2) What could go wrong? (3) How can it be done safely? Everyone is trained as a UAP assessor and is encouraged to help each other maintain safe practices.

- The DuPont Stop/Take Two program is used. Total management support and total employee support and participation is key.

- Genuine management commitment to safety. This is essential for our comprehensive set of safety, health, and environmental policies to be effective. All employees are trained and programs are regularly audited.

- Safety auditing program. Auditing by operators. (2)

- Use STOP program developed by DuPont. Management of Change applied to even the smallest changes of the plant. Quantitative risk analysis to all situations related to the use of chemicals which can cause major accidents. Training and instruction program.

- Conduct emergency simulations.

- Training on-the-job, safety training, safety committee meeting, one-on-one safety tour, job safety analysis, use of work permit.

- Employee safety suggestion system, supervisory conducted safety meeting in addition to formal meetings by safety department, mandatory safety meeting attendance, safety incentive program.

- HAZOP and strong work permit systems.

- Report all abnormalities, no matter how small, strictly enforce work permit system, conduct refresher safety training and involve personnel in fire training exercises, and conduct an annual company safety day to enhance safety awareness.

- See that safety procedures and work permit system is adhered to everyday.

Table 3. Recommended Safety Practices (Continued)

- Conduct on-the-job safety training, maintain plant safety committee.

- Use DuPont safety improvement program (now in stage 3 of 7).

- Safety is top priority of management and every employee, regular plant safety inspections, every incident is discussed within the whole company.

- Implement annual safety improvements plan, every plant change is subject of safety analysis, use *causes tree* to analyze each safety incident.

- Use Process hazard analysis and implement OSHA requirements.

- OSHA Voluntary Protection Program (VPP) Star Plant.

- Use safety suggestion boxes.

- No single program can account for our improvements. Comprehensive safety program used including: (1) Voluntary Protection Program (VPP); (2) use of ILCI/ISRS International Rating System; (3) Responsible Care (CMA); (4) use of OSHA Recordable Special Emphasis Program which focuses on increased employee awareness, involvement, as well as behavior modification.

- Operators have assumed ownership of the safety program. Expect and demand outstanding performance. Set these high standards and hold everyone accountable.

- Always use safety equipment properly and conduct identification-of-hazard analyses.

- Review of working permits, operational hazard analysis, and hazard analysis of specific jobs.

- Committees: discuss safety, health, and environmental items weekly and plantwide six times per year, CTO Committee to review safety of new designs, safety idea reward system, reward system for periods without incidents.

- Use employee participation in developing procedures.

- Implement disaster program and continuous training.

- Root cause analysis, mechanical integrity, behavioral science technology, and basic accident program to detect at-risk behaviors.

- Safety improvement plan is developed each year at plant and factory level. Cross-section of employees involved.

Table 3. Recommended Safety Practices (Continued)

- Each employee is trained with the attitude that "each employee is his/her own safety person" and that they must watch out for themselves.

- Safety committee is comprised of hourly employees from each work group, which meets monthly to discuss safety items that have occurred or need attention. It promotes teamwork in the safety program.

- Maintain comprehensive safety program including: (1) work permit system, (2) emergency evacuation procedure, (3) job working procedure, (4) hazardous materials handling, (5) personnel protection equipment, (6) health, safety, and environment management system.

- Always use safety equipment and check and recheck the condition at the field and control room.

- Performance indexing to drive continuous improvement in safety and "keep yourself safe – take two for safety".

- Extensive on-the-job training which involves cross-training of plant personnel.

- Senior management is involved in safety inspections and audits (demonstrates senior management commitment).

- Plant is committed to a Safety Management Practice System (SMP) which provides a structured employee based approach to protect the safety and health of employees, operations, customers, environment, and the public. Mandatory participation by all employees within the organization.

- UAUC: a self-safety auditing program which encourages anyone to report any noticeable unsafe acts or conditions and take immediate remedial actions.

- Conduct safety refresher training.

- TARGET ZERO: an incentive program to have a zero accident rate. Near miss reporting, formal safety audits, and asset management system (maintain good equipment records).

Survey Basis

The participation base for the 1994–1996 survey was 89 ammonia plants located in 25 countries around the world and operated by 38 different companies. These plants have a maximum annual capacity of 37,000 k metric tons per year. This is slightly more than one-half of the world capacity excluding the Former Soviet Union countries, most of Asia, and portions of Eastern Europe. Regional classification is 28 plants in Europe, 31 plants in North America, and 30 plants in the rest of the world.

Of the 89 plants, seven are reciprocating plants and the remaining 82 plants are large tonnage reciprocating compressor type plants.

Acknowledgment

We appreciate the considerable time and effort spent by the plant operations personnel in reviewing their production records and providing answers to the lengthy questionnaire. The participating companies are to be commended for their willingness to share their experiences with other plant operators and for providing financial support for the benchmarking study.

Literature Cited

Sawyer, J. G., G. P. Williams, and J. W. Clegg, "Causes of Shutdowns in Ammonia Plants," *Ammonia Plant Safety & Related Facilities*, Vol. 14, AIChE, New York, pp. 62–66 (1972).

Williams, G. P., "Worldwide Ammonia Plant Benchmarking Study 1994–1996," Plant Surveys International, Inc. Multi-client Benchmarking Report (Dec. 1997).

Williams, G. P., W. W. Hoehing, and R. G. Byington, "Causes of Ammonia Plant Shutdowns - Survey V," in *Plant/Operations Prog.*, **7**(2), 99 (April 1988).

Williams, G. P., and W. W. Hoehing, "Causes of Ammonia Plant Shutdowns," *Chem. Eng. Prog.*, **79**(3), 11 (Mar. 1983).

Williams, G. P., "Causes of Ammonia Plant Shutdowns," *Chem. Eng. Prog.*, **74**(9), (Sept. 1978).

Williams, G. P., and J. G. Sawyer, "What Causes Ammonia Plant Shutdowns?," *Chem. Eng. Prog.*, **70**(2), 45 (Feb. 1974).

Nitrogen Fertilizer Producers Push the Boundaries

Nitrogen fertilizer producers face the challenge of reducing costs and gaining competitive advantages while pushing technological and other boundaries. Our article discusses nonconfidential details (technology and catalyst innovations, unusual construction conditions, and innovative process and financing schemes) of several current world-scale projects, including the world's largest single-stream plants.

Jorge A. Camps
The Pace Consultants, Miami, FL 33144

Introduction

Nitrogen fertilizer producers have recently embarked on a series of trend-setting projects as global competition increases. Producers have accepted risks that are associated with new technology, the construction of large new plants in emerging areas of the world, and short-cut construction methods in these projects. Novel project financing and project configuration schemes are also being tried.

Worthy of mention are:

• The Farmland MissChem and PCS Nitrogen projects in Trinidad. These are the first applications of the complete M. W. Kellogg advanced ammonia process (KAAP) and the world's largest single-stream grass-roots ammonia plants in operation to date.

• The Profertil project under design and construction in Argentina. These plants are in line to be the world's largest single-stream ammonia and urea plants.

• The Farmland Petroleum coke-to-ammonia and UAN project. This is the first petroleum coke–to-nitrogen fertilizer project in the United States.

• The FertiNitro project in Venezuela. This is the largest capital investment project in the nitrogen fertilizer industry to have successfully achieved financial closure.

In the interest of the various sponsors of these projects and the licensors involved, this presentation will focus on the nonconfidential details of the process technology innovations, the unusual construction conditions, unusual feedstocks, and so forth.

Farmland MissChem Ltd. (FMCL) and PCS Nitrogen Projects in Trinidad

The FMCL and PCS ammonia plants in Trinidad were the first grass-roots projects to be constructed based on the complete Kellogg KAAP process. FMCL is a joint venture of Farmland Industries and Mississippi Chemicals.

The KAAP process utilizes a novel ammonia synthesis catalyst consisting of a precious metal (ruthenium) impregnated on a high-surface-area graphite support. The catalyst is allegedly 10 to 20 times more active than conventional iron catalyst, thus allowing operations at lower synthesis loop pressures. This encourages other innovations, such as the combination of the syngas and refrigeration compression services into a single machine, the construction of potentially less costly synthesis loops, and some potential energy savings. The lower synthesis loop pressure may bring other future benefits as these processes prove to be effective, such as potentially lower maintenance costs and improved onstream time.

The plants were built in a reasonable amount of time with fairly smooth execution of all design and construction phases. Nevertheless, a significant number of challenges occurred along the way.

Some of the most significant challenges were faced early during the construction phase. One of the projects (FMCL) had to be moved to a new location within Trinidad during the early phase of design and construction. The relocation was due to subsoil stability problems encountered at the original site at La Brea (Figure 1). Detailed studies of the site performed before construction revealed numerous underground pitch flow areas. The problem was severe enough to warrant relocation of the project to the Point Lisas area (Figure 2). This relocation occurred approximately 4 months after the commencement of design and was a significant obstacle to on-time completion of the project.

Other obstacles included the preparation of a new industrial park area north of the existing Point Lisas Industrial Complex. The Point Lisas development company had to acquire land, obtain rights of way, demarcate the industrial zone, and construct new facilities for the supply of raw materials and utilities. A new and much longer pipeline and utility corridor (Figure 3) also had to be planned, permitted, and constructed to connect with the existing pier at Point Lisas. A new ammonia loading station was also built at this pier.

In addition to these difficulties, the new location also required the addition of special water-effluent disposal facilities in order to preserve existing mangroves along the shore line. A process flare also had to be added at the request of the local authorities. These two additional requirements were not known until the project was at a fairly advanced stage.

In spite of the above difficulties, the project and its new supporting infrastructure was able to achieve mechanical completion without modifications to the schedule established for the original site. This was primarily due to a fine coordination effort by the project sponsors and the diligence displayed by the various

Figure 1. Map of Trinidad I.

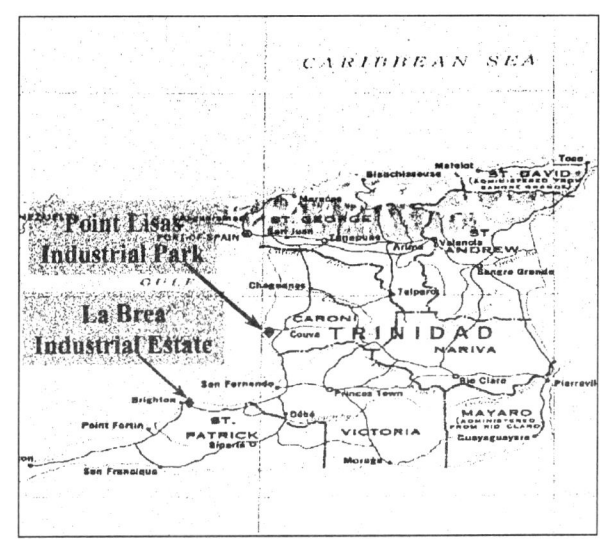

Figure 2. Map of Trinidad II.

government agencies in Trinidad and the EPC contractor.

At the time of this writing, both the FMCL and PCS plants are producing ammonia.

The FMCL project in particular illustrates many of the special situations encountered by producers in the fertilizer industry as they venture to build large projects in emerging areas of the world.

The Profertil Project in Argentina

This project consists of the world's largest single-stream ammonia plant (2,050 metric t/d or 2,260 standard t/d), the world's largest single-stream urea plant (3,250 metric t/d or 3,580 standard t/d) and two conventional Hydro Agri urea granulation plants, each sized for 1,800 metric t/d (1,980 standard t/d). Design and construction were started early this year, and completion of the startup phase is expected to occur in the third quarter of the year 2000.

The design is based on previously proven technologies. The ammonia unit is based on the Haldor Topsøe S-200 converter consisting of two radial beds with an internal heat exchanger. The urea process is the Snamprogetti process.

The ammonia plant scale-up is relatively small. Haldor Topsøe has based its design in plants presently producing 1,800 metric t/d and increased synthesis loop pressures. However, the increased pressures are still within present day conventional designs. In this regard, Haldor Topsøe's approach to ammonia plant design is essentially the opposite of Kellogg's approach.

The urea plant design is also based on judicious scale-up techniques and previous plants that produce up to 2,400 metric t/d. Most of the equipment does not exceed conventional and/or mechanical equipment design parameters. According to Snamprogetti, the maximum size increase for any item of equipment has been on the order of 20%. This does not seem to represent undue risk for the project owners and lenders.

Aside from the technology challenges imposed by size, the Profertil project will face the challenges of having some of its surrounding infrastructure built in parallel with the project. This applies primarily to water and electricity supply. However, this is typical of most large projects built overseas.

The Profertil Project will require improvements to the local supply of industrial water, the local electric power distribution grid, and the regional electrical

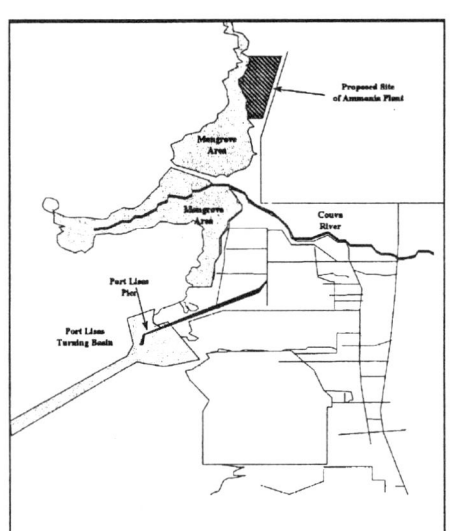

Figure 3. Port Point Lisas, Trinidad.

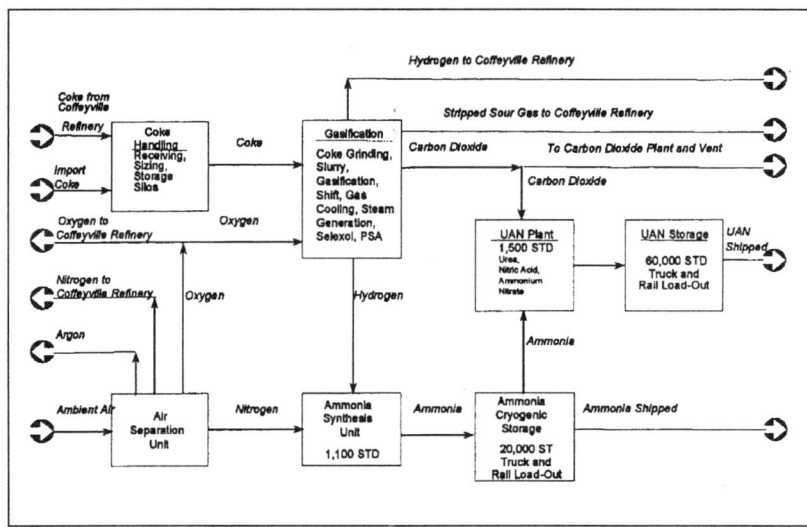

Figure 4. Weatherly urea ammonium nitrate process.

transmission and generation grid. It is to be noted however, that the Bahia Blanca area of Argentina is an established industrial area and the home of various other existing and planned projects.

The Farmland Petroleum Coke-to-Ammonia and UAN Project in the United States

This project consists of the gasification of petroleum coke, the cryogenic separation of nitrogen from air, the shift and clean-up of synthesis gas, ammonia generation in a conventional synthesis loop, a nitric acid plant, and a Weatherly Urea ammonium nitrate process (Figure 4). No novel technologies are being employed. Gasification of petroleum coke in Texaco partial oxidation reactors is presently being done in two other plants. One of these plants produces ammonia at a facility located in Japan. The other plant is in the United States and generates a fuel gas for the generation of electric power and steam.

The most interesting features of this project are the various cost saving techniques employed. The project will reuse solid handling facilities, coke gasifiers, and associated equipment transported from the Coolwater project at Daggett, CA. This will generate very substantial savings in capital costs. Farmland is also assuming a larger project management role to minimize construction management costs. Black and Veatch Pritchard will do most of the construction; however, the construction of several pieces of the project, such as the Weatherly UAN plant and some of the offsites, will be contracted and controlled directly by Farmland.

The project will be located next to Farmland's Coffeyville, KA refinery. In this manner the project will benefit from the synergies created, such as the utilization of a very low-cost feedstock with no transportation costs, the sharing of a significant amount of utilities and other services, and the disposition of byproducts.

Uncoupling nitrogen fertilizer production from the price of natural gas is another of the many advantages of this particular production scheme.

Design and construction commenced early in 1998 and should be completed mechanically in the first quarter of the year 2000.

The FertiNitro Project in Venezuela

The FertiNitro Project has been sponsored by a joint venture between Pequiven, Koch Oil, S.A., Snamprogetti Netherlands B.V., and Polar Uno, C.A..

Pequiven is a government-owned company in Venezuela responsible for the development of chemical projects derived from natural gas. Koch Nitrogen is a subsidiary of Koch Industries in the United States. The Snamprogetti group plays a triple role in this project. Snamprogetti Netherlands BV is one of the project owners. This company is affiliated with Snamprogetti SPA, who perform the dual role of technology supplier and EPC contractor. Polar Uno is part of the Polar group, the biggest food industry in Venezuela.

The project will be located in the Jose Industrial Park along the northeastern shore of Venezuela. This industrial park is presently the home of several other ongoing and potential projects for the upgrading of heavy crudes and the production of petrochemicals (Figure 5).

The FertiNitro project consists of two 1,800 metric t/d (1,984 standard t/d) ammonia plants based on Haldor Topsøe Technology, two 2,200 metric t/d (2,425 standard t/d) urea melt plants based on the Snamprogetti process, and two 2,200 metric t/d granular urea plants based on the Hydro Agri process.

The overall project, valued at approximately $1 billion (United States dollars), is presently the world's largest fertilizer project under construction. The project economics rely heavily on the abundant and inexpensive natural gas in Venezuela and on the plants' strategic location close to the United States and Latin American markets.

The project is unique in having achieved investment-grade ratings, thus allowing the placement of $250 million (United States dollars) of long-term debt in the bond market. The long-term portion of the financing alleviates cash flow requirements during the debt repayment period. The rest of the financing is based on more conventional private placement loans in the banking community.

Challenges faced by the project sponsors include the simultaneous upgrading of natural gas, industrial water and electrical infrastructure in the Jose area, and

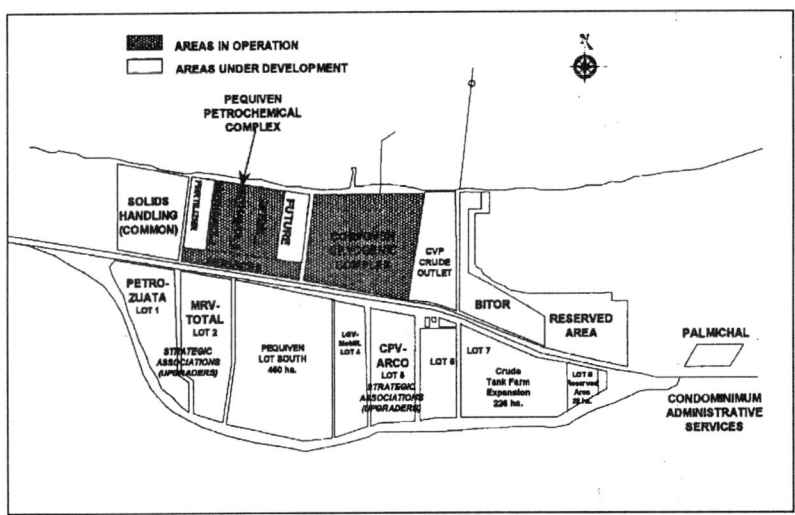

Figure 5. Plot plan of the industrial area.

Figure 6. Map of Oman.

the construction of new port facilities. Other challenges or uncertainties are created by the local inflationary situation in Venezuela; however, sufficient contingencies appear to be in place to see the project through to successful conclusion.

Design commenced in early 1998, and completion of the startup phase is expected in the third quarter of the year 2000.

Other Large Projects

Another sizable project is under serious consideration by a joint venture of Oman Oil and two government-sponsored ammonia–urea producers in India (KRIBHCO and RCF). Oman Oil is a company created for the strategic development of energy-based industries in Oman. KRIBHCO is a multistate cooperative society with a 9% share of the nitrogen fertilizer market in India and an experienced producer of ammonia and urea. RCF supplies nearly 10% of India's current urea requirements and is also an experienced producer of nitrogen fertilizers.

The projects consists of two 1,750 metric t/d (1,930 standard t/d) ammonia plants, two 2,200 metric t/d (2,425 standard t/d) urea melt plants and two urea granulation units. This project is similar to FertiNitro since it is based on the same technology licensors and processes (Haldor Topsøe, Snamprogetti, and Hydro Agri). The facilities will be built along the shore of the Gulf of Oman near the town of Sur (Figure 6). This is the site of a new industrial park that is under development. The industrial park is the home of an ongoing LNG project with a total investment of approximately $2 billion (United States dollars).

All of the urea production will be lifted by the Indian sponsor companies for distribution in their home country. The project is strategically important to both Oman and India. Oman will be able to monetize a portion of its abundant natural gas resources, and India will be able to secure a reliable source of low cost urea, thus freeing available natural gas for other uses such as the generation of electricity.

Aside from the supply of natural gas, the Oman–India project will be self-sufficient in all utilities and will have its own port facility. This is expected to be the largest investment in a nitrogen fertilizer project to date.

Financing

The creative financing modes employed in the projects we have mentioned are interesting.

Four of these—Farmland MissChem, FertiNitro, Profertil, and Oman—are new joint venture companies. The new companies seek and obtain financing in the capital markets. In this manner, the parent companies limit their exposure or risk to the level of capital contribution made in the new projects. In addition, the parent companies also avoid burdening their balance sheets with new debt. With this mode of financing (known as "Project Financing," "Off-Balance Sheet Financing," or "Non-Recourse Financing"), the lenders have no recourse to the assets of the parent corporations. For the most part, the lenders have recourse only to the assets of the new project or joint venture company as a guarantee for their loans.

Lenders are given additional protection by the setting of offshore accounts, usually administered by a trustee, where all receipts are deposited and from which all payments are made. Debt repayments are senior to other debts. The lenders receive payments from these accounts on a preferential basis except for the ongoing operating expenses of the joint venture company (personnel costs and the costs of raw materials, utilities, and other supplies). Surplus funds after the satisfaction of these obligations are then passed on by the trustee to the project owners.

Conclusion

It is noteworthy that as the nitrogen fertilizer industry grows and becomes more complex, producers are finding new ways to compete. These include:
- The incorporation of advances in technology such as the KAAP process.
- The selection of novel process and project schemes such as Farmland's Pet Coke to Ammonia Project.
- The increased reliance in very large plants on economies of scale (Profertil, FertiNitro, etc.).
- The introduction of novel financing techniques.
- The assumption of inherent risks with the construction of plants in emerging areas around the world.

These are all signs of a vibrant competitive industry where no single company exerts dominance and no single technology or contractor controls. The best decisions for every project are made after a careful evaluation of all possible factors.

Startup and Initial Operation of a 2,400 Metric t/d Methanol Plant

The startup and initial operation of the 2,400 metric t/d methanol plant at Tjeldbergodden, Norway is described. This plant feat ures a number of new technologies and operating parameters, including a falling film saturator, prereforming, two-step reforming (tubular reforming followed by oxygen-blown secondary reforming) at low steam-to-carbon ratio, and a three-column distillation. These features make the plant one of the most energy efficient in the world.

A. Gedde-Dahl
Statoil A/S, Tjeldbergodden, Norway
Helge Holm-Larsen
Haldor Topsøe A/S, Lyngby, Denmark

Introduction

The Tjeldbergodden methanol plant is based on associated gas from the Heidrun oil field, which was discovered in 1985 during exploration of the upper part of the North Sea. The discovery was made by a consortium consisting of Statoil, Conoco, and two minority partners.

In addition to oil, the Heidrun field contains a significant portion of associated gas. This gas needed to be processed in order to develop the field. For various reasons the gas could not be piped to the existing grid of gas pipes in the southern part of the North Sea, and continuous flaring is prohibited by the Norwegian authorities. Consequently, Statoil had to find alternative ways to utilize the gas, including power generation, reinjection, or methanol production. Power generation and reinjection proved politically and economically unacceptable (because of CO_2 emission and tax). It was decided to take the gas ashore and produce methanol. The capacity of the plant was pegged at 2,400 metric t/d, or 830,000 t/year, making it one of the largest methanol plants in the world.

Conoco and Statoil divided the responsibilities between them. Conoco was responsible for transporting the gas to the shore (Statoil and Conoco, 1995), while Statoil was responsible for the receiving facilities and the methanol plant. Statoil is the main shareholder in the new plant.

The Heidrun field is located in the Haltenbanken sector just south of the Arctic Circle, approximately 200 km from the Norwegian coast and approximately 600 km northeast of the closest point of the existing North Sea gas pipeline grid. In consideration of this location, it was decided to take the gas ashore at Tjeldbergodden, close to Trondheim, which houses the centre of Statoil's R&D activities.

A location near the Norwegian coast close to the

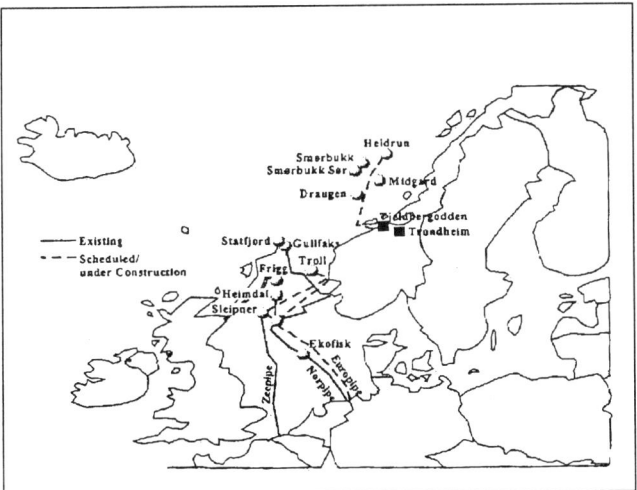

Figure 1. Norwegian gas transport systems and location of plant.

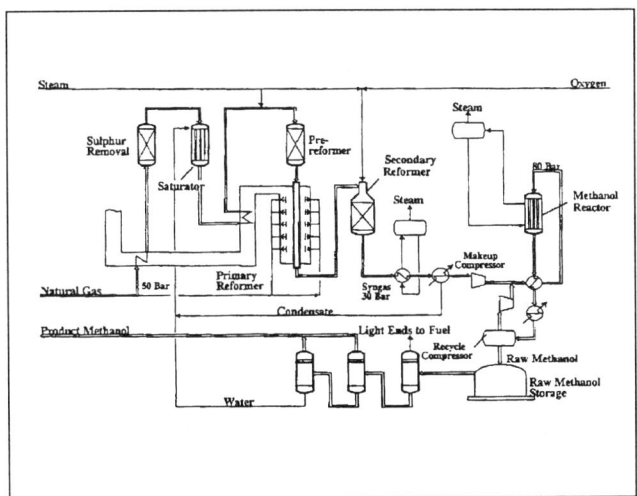

Figure 2. Production process at Tjeldbergodden.

Arctic Circle (Figure 1) may seem too remote and exposed for a petrochemical plant, but, due to the Gulf Stream, the coast is free of ice. Furthermore, the harbor at Tjeldbergodden enjoys natural protection by islands, and the Trondheim fjord is deep, providing ample draught for large ships.

There is no local consumption of methanol in the region, but the cost of shipping the product to Rotterdam is very low compared to the cost from competing plants in Russia, Libya, and the Middle East. The time required to sail from Tjeldbergodden to Rotterdam is less than 24 h.

Strategy for Technology Selection

In 1990 Statoil and Conoco decided to initiate a study of the available methanol technologies. Haldor Topsøe A/S and a number of other companies with methanol expertise were retained to perform various studies. Statoil commissioned Topsøe to make a generic comparison between available technologies.

In addition to this technology evaluation, Topsøe performed a number of sensitivity analyses, illustrating (among other things) the consequences of transporting gas from the other fields on the Haltenbanken sector in the same pipeline.

It was concluded from the technology evaluation that the most attractive synthesis gas generation was the two-step reforming technology, which proved to have both the lowest investment and the highest energy efficiency. In fact, this was the only process layout that met Statoil's requirement for a maximum CO^2 emission corresponding to an overall energy consumption of no more than 30 GJ/metric t of product (7.18 Gcal/metric t). This energy consumption level, which was fixed during the study phase, was a key condition in the plant approval from the Norwegian parliament.

With respect to the methanol synthesis, the conclusion of the technology evaluation was less clear, since one reactor concept (boiling water reactors) had the highest efficiency, while another reactor concept (three adiabatic reactors in series) presented the lowest investment. Statoil and Conoco chose the boiling water reactor concept, primarily because DuPont, the owner of Conoco, had favorable experience with this reactor concept at an existing methanol plant on the United States Gulf Coast, but also because efficiency had a very high priority.

The gas composition varies because the pipeline is prepared for transport of gas from the other fields in the Haltenbanken sector. To enhance the flexibility towards these variations, a prereformer was installed upstream of the tubular reformer. Consequently, the Tjeldbergodden methanol plant is based on synthesis gas generation by prereforming followed by two-step reforming, methanol synthesis in boiling water reactors, and purification in a three-column distillation unit (Figure 2). The synthesis gas and distillation trains are designed by Haldor Topsøe A/S, while the methanol synthesis train is designed by Lurgi AG.

Statoil R&D undertook extensive testing of commercially available catalysts to select the most suitable methanol synthesis catalyst. Topsøe's MK101 catalyst was selected on the basis of this screening, and Topsøe subsequently supplied all catalysts for the entire plant.

Process Description

Gas treatment and reforming

The natural gas enters the plant through a receiver terminal, where pressure is reduced from about 150 to 50 bar, any possible slugs of heavy hydrocarbon are removed, and the temperature is adjusted to about 40°C before the gas is admitted to the methanol plant proper.

In the methanol plant the gas is preheated in the wasteheat recovery section of the primary reformer and sulfur (H_2S) is removed at 400°C in a sulfur absorber containing Topsøe HTZ-3 catalyst. After the desulfurization, the natural gas is saturated with process condensate in a saturator, which is heated by the steam from the methanol synthesis reactors. In the synthesis section, steam is produced at about 35 bar, whereas the pressure upstream the reforming section is in excess of 40 bar. This pressure difference is overcome in the saturator, which is a falling-film type reusing process condensate without pretreatment. The saturated steam from the methanol synthesis reactor is condensed on the shell side of the saturator and returned without contamination in a closed loop to the methanol synthesis reactors. On the tube side of the saturator, the process condensate evaporates into the gas. All process condensate and some of the excess water from the distillation section are reused this way, and the requirement for demineralized water is minimized.

The saturated natural gas from the saturator is mixed with a small amount of medium-pressure steam to allow exact control of the steam-to-carbon ratio at 1.8 and then preheated to about 500°C before being admitted to the prereformer.

The prereformer converts the higher hydrocarbons so that the tubular reformer is fed with a more uniform gas composition containing no hydrocarbon heavier than methane. This has several advantages (Nielsen and Dybkjaer, 1996): the steam-to-carbon ratio can be reduced; the size of the tubular reformer, one of the high-cost items, is reduced; and the lifetime of the tubular reformer catalyst is extended. Purge from the methanol synthesis is recycled to upstream of the prereformer to provide hydrogen that protects the catalyst and to permit adjustment of the hydrogen/carbon ratio of the reformer feed.

After the gas exits the prereformer it is further preheated in a coil in the primary reformer wasteheat section and sent to the tubular reformer. The heat provided in this coil helps reduce the size of the tubular reformer radiant section. Because of the secondary reformer downstream of the tubular reformer, the duty of the tubular reformer is further reduced by 2/3, and significant simplifications can be made in the reformer outlet system because of the low outlet temperature of about 750°C. Only 210 reformer tubes are used for the production of 2,400 metric t/d methanol.

The partially reformed gas from the tubular reformer is transferred to the secondary reformer, where the reforming is completed by catalytic partial oxidation. The oxygen is admitted to the secondary reformer through the CTS burner in the reformer neck.

The most critical item in the secondary reformer is the oxygen-fired burner (Figure 3). This burner must operate in an aggressive atmosphere, under pressure, and at temperatures well in excess of 1000°C. In the past, burners for such applications demanded extensive maintenance and, on occasion, burned through the refractory insulation and pressure vessel, causing significant damage.

Topsøe has developed a new generation of burners based on the CTS burner design, which has eliminated the problems encountered with earlier generations (Christensen et al., 1994). These burners are designed using computer-based flow modeling, which was developed jointly by independent experts on computer modeling and by Topsøe, and verified in physical models in Topsøe's workshops. The first burner of this type was installed in a plant in France in 1992 and has now operated flawlessly for six years, which is much longer than any previous burners.

In the secondary reformer hydrogen is consumed during partial oxidation. In a tubular reformer a surplus of hydrogen is generated. By combining these two process steps, a balanced synthesis gas can be manufactured, and the water formed by combustion is contained within the system. The balanced synthesis

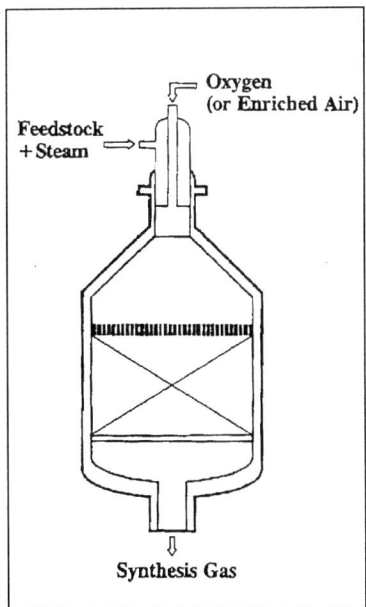

Figure 3. Typical layout of an oxygen-fired reformer.

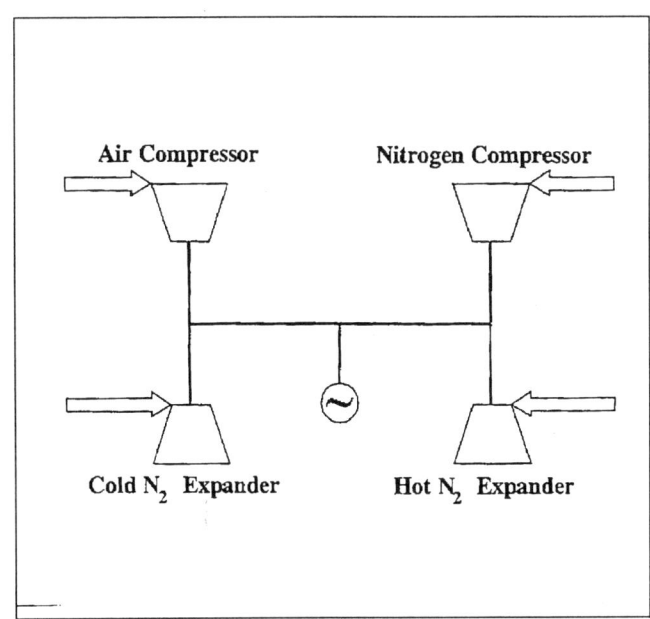

Figure 4. "Compander."

gas and the retention of combustion water are key elements for the process efficiency.

The oxygen for the secondary reformer is produced in a cryogenic air separation unit based on British Oxygen (BOC) technology. The oxygen plant produces liquefied oxygen rather than gaseous oxygen, and a 2,000 m³ buffer storage for liquefied oxygen is installed, helping to maintain a higher onstream factor for the methanol plant compared to the conventional gaseous oxygen.

The oxygen requires a lot of power. Because of the inexpensive and reliable electricity supply available in Norway it was decided to separate the manufacture of oxygen from the process plant proper and form a stand-alone, dedicated oxygen plant. Consequently, all high-pressure steam from the methanol plant is converted to electric power in a turbine-driven turbo generator, and electricity is used in the motor-driven air compressor and booster of the oxygen plant. Due to this configuration it is possible to startup and cool down the oxygen plant independently of the process plant, that is, it is not necessary to wait for sufficient amounts of steam to be generated in the plant or in the auxiliary boiler.

A new feature of the oxygen plant is that an air compressor, a nitrogen compressor, and the hot and cold nitrogen expanders are connected in a so-called "compander" configuration (Figure 4). In this compander, all the rotating equipment is connected by shafts and gears and a motor is provided for startup. The compander is a proprietary concept of BOC.

The stand-alone oxygen plant was contracted separately with the Swedish gas supplier AGA, who took equity participation in the plant. Statoil operates the plant under a long-term contract. In addition to the oxygen and nitrogen required for the methanol plant, excess oxygen, nitrogen and argon is produced for export (Statoil, AGA, and Conoco, 1996).

The synthesis gas is cooled down after it leaves the reformer section in a boiler-superheater-BFW preheater train, where high-pressure steam is raised. Due to the small size of the reformer, the flue gas only holds sufficient heat for process preheat duties, and thus all steam superheat is supplied in the synthesis gas wasteheat recovery. Therefore, controlling the temperature between the wasteheat boiler and the superheater is essential. This control is obtained by applying a new "cooled bypass" boiler layout that was developed by Topsøe (Stahl and Thomsen, 1996).

In this layout (Figure 5) both the bypass and the main synthesis gas stream are cooled to temperatures safely below the kinetic limit for metal dusting corro-

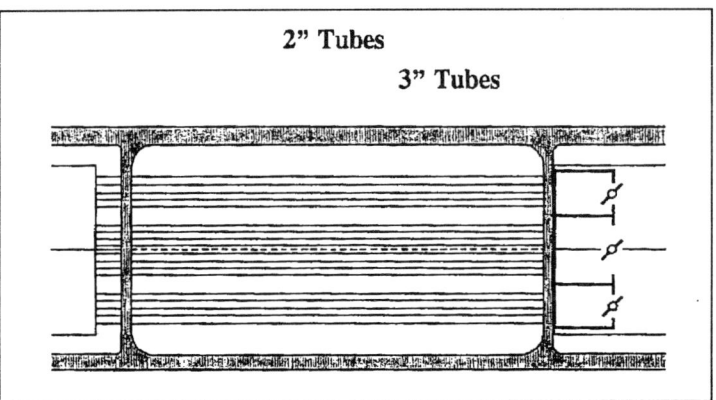

Figure 5. Double-tube bundle wasteheat boiler.

sion, and the risk for metal dusting is completely eliminated, also in the gas mixing zone. The cooling is obtained in one pressure shell in two concentric tube bundles with different specific heat-transfer characteristics. The gas flow distribution between the two tube bundles is controlled on the cold side of the wasteheat boiler.

After the boiler-superheater-BFW preheater section, the synthesis gas is cooled further in the reboilers in the distillation. This helps increase the energy efficiency of the plant.

Compression and methanol synthesis

The synthesis gas compression is obtained in a single-casing, single-stage synthesis gas compressor from Nuovo Pignone. The synthesis gas compressor and the recirculator are on the same shaft. The methanol synthesis occurs in a conventional loop with two parallel boiling-water reactors loaded with Topsøe MK-101 methanol synthesis catalyst.

Methanol distillation

The raw methanol produced contains dissolved gases, water, and higher boiling impurities. These are removed in the distillation section.

The distillation is a three-tower distillation layout with a stabilizer and two concentration columns operating at different pressures. Dissolved gases are removed in the stabilizer, and water and higher boiling impurities are removed in the two concentration columns. The first concentration column is operated at an elevated pressure, permitting the overhead condenser to work as a reboiler for the second concentration column, thereby using the reboiler energy twice over and achieving a very low overall energy consumption amounting to only two-thirds that of the conventional two-column layout.

The Tjeldbergodden distillation section is the largest three-column methanol distillation section ever constructed.

Utilities and control

Cooling water for the plant comes from the nearby fjord, 70–75 m below the surface, and is pumped by three parallel 50% pumps. This structure ensures a reliable flow of a cooling medium with a constant low temperature and no biological activity, providing for excellent efficiency on the condensing turbines. Closed-loop fresh cooling water is used for all other cooling duties. There is no air cooling in the plant.

The plant is controlled using a Bailey digital control system. All signals are transmitted to the control room, and the plant is exclusively controlled from this point. The control room (Figure 6) features four terminals from which the plant can be controlled. Two wall-sized screens are available in addition to the terminals to display video signals from surveillance cameras, plant log, alarm lists, and so forth.

The basement below the control room has a static and dynamic simulator of the plant. The kernel of this simulator is a dedicated version of proprietary Topsøe

Figure 6. Control room.

software, including Topsøe kinetics and thermodynamic data. The simulator runs on a Bailey station identical to the stations in the control room and is used in the training of new operators.

Implementation of the Project

The technology evaluation was completed in 1991 when the process license and engineering agreements were signed with Topsøe and Lurgi (Statoil and Conoco, 1994, 1995).

Statoil had limited experience manufacturing methanol at this point in time, while DuPont had considerable engineering expertise. In order to benefit from this expertise, the project was initiated with a "front-end loading" phase that was inspired by DuPont's project implementation philosophy. In effect, this phase was a conceptual preengineering phase with the objective of ensuring that the design basis for the project was well defined and that any design problems were identified at an early stage, where changes would be less costly. The front-end loading was performed in DuPont's offices in Wilmington, DE, with participation of engineers from Topsøe and Statoil.

Before front-end loading (and to a certain extent in tandem with it), basic engineering was carried out by Lurgi and Topsøe. In addition, Topsøe carried out

Figure 7. Two views of tubular reformer under transport at site.

detailed engineering on the critical parts of the plant, in particular the tubular reformer and other parts exposed to high pressures and temperatures.

General detailed engineering, procurement, construction, and supervision were outsourced under an EPCS contract, which was won in January 1994 by Fluor Daniel Ltd.. Fluor performed the detailed engineering in their London offices in 1994 and 1995, with assistance from key personnel from their office in Canada and engineers from Topsøe and Lurgi.

The construction of the methanol plant and the gas pipeline was complete in the first quarter of 1997. Construction was followed by a commissioning and testing period of about three months. Full production ensued in June 1997. Oil production from the Heidrun field began in 1995. In the interim period, the associated gas was reinjected.

It was necessary to significantly improve the local infrastructure. A new pier capable of servicing ships weighing up to 40,000 t was constructed and the plant site was cleared, blasting away more than 2,000,000 t of rock.

Due to the remote location and the high cost of field labor, it was decided to modularize the plant and to minimize on-site installation work. Construction of equipment and pipe rack modules commenced at several Norwegian yards during work on the foundation, while heavy equipment was purchased from various European vendors.

The modules and equipment were received, installed, and hooked up beginning in September 1995 and continuing through most of 1996.

The tubular reformer (Figure 7) was supplied pre-assembled in three modules. The largest of these, the radiant section, weighed about 1,400 metric t, and the physical dimensions were 15 × 30 × 35 m. The reformer was imported from Kirchner of Venice, Italy. Upon arrival in Tjeldbergodden, the reformer modules were jacked up and rolled off to their final destination on a 232-wheel, 58-shaft transporter (Figure 6). The refractory lining of both reformer and collector was installed before shipment.

The cold box (Figure 8) for the oxygen plant, the distillation columns, and other major vessels were transported to the site and erected fully dressed. This project was the first where a cold box (height 71 m, weight 433 t) was installed in one module. The low-pressure concentration column was the heaviest at 460 t (Figure 9).

The modular construction was a positive experience. The modules were generally delivered on time and according to specifications, and even the largest and heaviest modules could be handled without troubles.

In addition to process licensing and basic engineering, Topsøe's scope of supply comprised catalysts, training, supervision, startup assistance, and supply of certain critical parts, including burners for the secondary oxygen-fired reformer. The catalysts and the burners were delivered in the fourth quarter of 1996.

At the start of the methanol project, Statoil established a firm policy that prioritized safety in design and working practices.

The safety objectives were reflected in all contracts and closely monitored by the project team. Good performance was honored and an improvement system was put in place to register and learn from all accidents and near-accidents. The result has been a very satisfying safety performance with no serious accidents or permanently injured personnel. At project completion in April 1997, only 15 accidents causing loss of worktime had been reported out of close to 6 million accumulated man-hours.

Plant Startup and Initial Operation

Mechanical completion and handover of the plant took place from November 1996 to April 1997 and was closely integrated with the commissioning and start-up activities. The reformer dryout was initiated in late April 1997 and continued for 10 days. Catalyst was loaded and activated in May 1997. Natural gas was first introduced to the plant on May 30, 1997.

The methanol plant was officially inaugurated on June 5, 1997, and the first methanol was produced on the same day. Startup was smooth and fast, and the plant was operating at 85–90% of design capacity within three days. The plant was kept at this level for a period of about 2 months. During this time, a number of bugs in the instrumentation and the control system were corrected, and a bottleneck was identified in the oxygen plant.

The problem in the oxygen plant originated in the compander, where the cross shaft started to vibrate at around 90% of required load. This cross shaft was originally 50 mm in diameter, and it was replaced by BOC at short notice with a 60-mm cross shaft. This replacement permits the oxygen plant to operate at 100–105% of required load without vibrations.

By August 6, 1997, the bottleneck was removed, and the plant reached its design capacity of 2,400 metric t/d. A stable production of 103% of design capacity was then demonstrated and the performance test run was performed from November 11 to November 14, 1998.

After the initial operation period, the new equipment was evaluated as follows.

Saturator. The saturator has operated according to design without causing any problems. In fact, it proved to have a stabilizing effect on the plant because pressure excursions on the medium-pressure steam

Figure 8. Cold box.

Figure 9. Low-pressure column.

could not spread to the process during startup.

The medium-pressure steam pressure is 44 bar, whereas the pressure at the steam addition point is about 42 bar. Consequently, if steam were added directly, the pressure would only have to drop marginally before steam addition would be in jeopardy. Instead, in the saturator, a steam pressure of 28–30 bar is sufficient, because the water on the process side is evaporating against its partial pressure only.

Thus, the saturator has provided extra convenience and safety, particularly during startup of the plant.

Prereformer. The prereformer has operated according to design. C_2+ was reduced from 16.9% to below 100 ppm. It is the prereformer that permits operation at a steam-to-carbon ratio as low as 1.8.

Primary reformer. The primary reformer was inspected after dryout, and it looked fine. The reformer is operating at exit temperatures slightly lower than design, indicating catalyst activity in excess of design. The reformer tubes are inspected visually at regular intervals and show no sign of hot banding or other temperature maldistribution. No detrimental effects of the low steam-to-carbon ratio have been identified. The pressure drop over the tubes has remained constant at 3.6 bar.

Refractory. After the reformer was transported, minor damage was found on the inner layer of castable refractor. This damage was deemed to not compromise the thermal insulation. Except for cleaning, no attempts were made to fix the refractory before the startup. The soundness of the refractory has been proven after startup, since no hot spots have been found in connection with this minor damage. Later, the transfer line was inspected with a video camera, and no changes have been found in the condition of the refractory lining.

In the field weldings between the reformer sections, a few hot spots have developed. Metal temperatures have been found to be within the acceptable range, and the refractory in connection with these hot spots will not be fixed until the next turnaround. Metal temperatures continue to be monitored.

Secondary reformer. The secondary reformer has performed according to design and both oxygen consumption and residual methane concentration in the synthesis gas are as expected.

In accordance with Topsøe design philosophy, the secondary reformer is not water-jacketed, but rather is open to continuous visual inspection. In particular, the conical part of reactor has been surveyed because it harbors the flame with the highest temperatures. No signs of hot spot are found on the conical part or any other part of the secondary reformer, except for a small area around the field welding on the incoming transfer line.

During the first days of operation, the level control in the liquid oxygen storage was not calibrated, and the storage was allowed to overfill during a brief absence of the operator. The overfill resulted in liquid oxygen in the oxygen header. The liquid oxygen evaporated into the oxidant stream and reduced the temperature, compromising the flow control and showing false low flow of oxygen. The plant tripped on high-temperature oxygen exiting the secondary reformer. The overflow in the liquid oxygen storage was eliminated after the trip, and the plant was started up without any problems. Subsequently, the level control was calibrated, and the flow controller was changed into a temperature-compensating type.

Air separation unit. Certain problems have been identified in the oxygen plant. In addition to the vibrating cross shaft that was rectified, the oxygen-flow control valve is too small. In order to obtain the required flow, the plant is operated with a fully open valve bypass, while the control valve is about 75% open. This flow control valve will be replaced with a higher capacity valve during the next turnaround.

Wasteheat boiler. The new layout of the wasteheat boiler with concentric tube bundles has been demonstrated without problems. The outlet temperature is controlled according to design. The only irregular incident in connection with the wasteheat boiler was that the thermowell for the TIC controlling the flow distribution between the tube bundles had been made in alloy 800 because of a fabrication error. This thermowell succumbed quickly to metal dusting and started to leak. When the leak was discovered, the thermowell was replaced with an alloy 601 thermowell in accordance with design, and this well has shown no signs of metal dusting.

Distillation. The three-column distillation section proved easy to operate. The columns are started up against closed valves and with full recycle, and when temperature and pressure profiles are right, the valves are slowly opened. The procedure to startup from cold standby takes about 12 h.

Grade AA quality was obtained from the very start. IMPCA specifications and more restrictive specifications are produced for certain clients.

Catalysts

All catalysts performed according to or better than design. In particular, the methanol synthesis catalyst performed well, obtaining a carbon efficiency in the methanol synthesis of 1.3% above design.

Performance test run

Excellent performance was achieved, with higher production and lower natural gas and utility consumption than was planned for. Details are given in Table 1.

Present and future operation

The plant is presently operating consistently between 100 and 105% of nameplate capacity. During the next plant turnaround, minor repairs on the refractory will be made and the oxygen supply valve will be replaced with a higher-capacity valve.

Table 1. Performance Run Test Results

Guarantee	Guaranteed Value	Realized Value	Deviation
Production (t/h)	100	103.6	+ 3.6%
Quality	Grade AA "+"	Passed	NA
Natural Gas Consumption, (GJ/metric t) (Gcal/metric t)	29.64 (7.09)	28.74 (6.88)	− 3.0%
Total works cost (globalized utilities), Index	100	80	− 20%

Recently, a 105% campaign was conducted for 5 days without meeting bottlenecks, and pending the analysis of collected operating data the plant will continue operating at 105% of nameplate capacity.

After the turnaround, a new campaign is planned, with the aim of further increasing production.

Conclusion

During the startup of the plant and the following operation, all equipment and catalysts in the methanol plant have operated satisfactorily. Statoil and Conoco are both satisfied with the plant.

Certain minor problems originating from the utilities rather than the methanol process were encountered and solved during startup and initial operation. In fact, the reforming, synthesis, and distillation sections have been minor concerns during the initial operation period.

With the methanol plant at Tjeldbergodden, Statoil has found a value adding application for part of the associated gas in the North Sea. The methanol plant is the first major industrial facility in the region, which has so far been dominated by fishing and farming. An estimated 100 jobs have been created for operation and maintenance of the oxygen and methanol plants, and a large number of jobs are likely to be created outside the plants providing various services.

For Topsøe, the cooperation with Statoil has been outstanding, in particular because Statoil has been positive towards the application of new technology. This was already apparent from the start of the technology evaluation study, where technical insight and understanding on Statoil's part typically carried more importance than the number of industrial references.

The willingness of Statoil and Conoco to embrace new technology has created a unique facility that is not only the most efficient, but also the most environmentally friendly methanol plant in the world today.

Literature Cited

The Air Separation Plant at Tjeldbergodden, Statoil, AGA & Conoco, February 1996.

Christensen T. S., I. Dybkjaer, L. Hansen, and I. I. Primdahl, "Burners for Secondary and Autothermal Reforming—Design and Industrial Performance," *Ammonia Plant Safety & Related Facilities*, Vol. 35, AIChE, New York (1995).

Construction of the Plant, Statoil & Conoco, August 1994.

The Construction Period, Statoil & Conoco, May 1995.

Haltenpipe—A Pipeline for Gas Transportation and Reception, Statoil & Conoco, May 1995.

Holm-Larsen, H. "Selection of Technology for Large Scale Methanol Plants," presented at the World Methanol Conference (1994).

Nielsen, S. E., and I. Dybkjaer, "Use of Adiabatic Prereforming in Ammonia Plants," *Ammonia Plant Safety & Related Facilities*, Vol. 37, AIChE, New York (1997).

Stahl, H., and S. G. Thompson, "Survey of Worldwide Experience with Metal Dusting," *Ammonia Plant Safety and Related Facilities*, Vol. 36, AIChE, New York (1996).

DISCUSSION

Inderjit Ohri, *Indian Farmers Fertiliser Cooperative Ltd.*: Have you experienced any metal dusting problem, because the steam to carbon is quite low at the inlet to the reformer? What was the material of construction of the waste heat boiler tubes?

Holm-Larsen: We have not found any signs of metal dusting except for the thermowell in the waste heat boiler. Of course, I could possibly tell you more after the turnaround, but there have been no signs of metal dusting. I'm not sure that I'm at liberty to tell you about the material of the waste heat boiler.

Richard Strait, *M. W. Kellogg Co.*: Can you share with us the CO to CO_2 ratio inlet to this boiler which did not metal dust?

Holm-Larsen: I cannot give you the exact ratio. It's what you would expect from a steam to carbon ratio of 1.8 and 1,000°C. It's close to 20% CO and about 7-8% CO_2, approximately.

Strait: One other question. You didn't share with us the metallurgy of the tubes, but it looks like you used 601 to replace 800 as a solution to metal dusting. Is that correct?

Holm-Larsen: That is correct, yes.

Strait: Thank you.

Commissioning a New Ammonia Plant in China

In the early 1980s ICI developed a new ammonia process to address the need for low energy consumption in the face of rising gas prices. This process, the AMV Ammonia Process, was successfully commissioned in a factory in Canada in 1985 and has subsequently been used for two new fertilizer factories in China. The AMV Ammonia Process and its adaptation for the recent application in China are described.

Li Wei Chun
Hainan Fudao Chemical Co. Ltd., Dong Fang, Hainan Province, 572600, Republic of China
I. R. Barton
ICI Katalco, Billingham TS23 1LB, United Kingdom

Features of the ICI Katalco AMV Ammonia Process

The ICI Katalco AMV Ammonia Process is based on a natural gas feed and combines the traditional steps of gas purification, steam reforming, CO shift, CO_2 removal, methanation, and ammonia synthesis.

The process was developed in the early 1980s to decrease energy consumption in response to rapidly increasing gas prices for the fertilizer industry (Livingstone and Pinto, 1983). The aim of the design was to simplify the process and create milder processing conditions to improve reliability and lower capital cost. The objectives of reliability and low capital cost were equally important to the target of lowering energy consumption.

The ICI Katalco AMV Ammonia Process can be characterized by three main features that differentiate it from other ammonia processes:

• An excess of air is added to the secondary reformer, increasing the proportion of reforming that is done in the secondary reformer, decreasing the load on the primary reformer, and ultimately yielding a smaller, less expensive primary reformer.
• A low ammonia synthesis pressure. This allows both significant savings in the main compressor power and a simpler, less expensive machine configuration. The simpler machine is easier to maintain and improves plant reliability.
• Hydrogen recovery within the synthesis loop. This allows hydrogen to be recovered and inerts to be rejected without the need to recompress the recovered hydrogen through the syngas compressor.

These three features are elaborated on in the following sections.

Use of excess air in the secondary reformer

The amount of air in the secondary reformer is typically 20% above the requirement that would give a synthesis gas with a stoichiometric $H_2:N_2$ ratio of 3.0. This leads to a minor increase in the power of the air compressor, but this is of the order of only 1 MW for a plant with a capacity of around 1,000 metric t/d. The excess of air generates more heat from combustion in the secondary reformer. This heat can be effectively used to drive the endothermic reforming reaction, allowing the primary reformer duty to be decreased and resulting in significant cost savings.

The lower gas conversion requirement for the primary reformer means that the exit temperature from the primary reformer is substantially lower than in a more conventional ammonia plant. This lower temperature leads to easier design requirements for the catalyst tubes in the furnace and allows the primary reformer pressure to be increased without unusually thick reformer tubes. A higher reformer pressure is normally beneficial in the ammonia process because it decreases the pressure ratio for the main syngas compressor.

A comparison of reforming conditions for the ICI Katalco AMV Ammonia Process and a more traditional ammonia process is given in Table 1.

The decreased primary reforming requirement also allows the steam:carbon ratio to be lowered. This saves steam and further lowers the heat duty in the primary reformer furnace, again contributing to lower capital cost and improved efficiency.

In the AMV Ammonia Process the optimum slip of methane from the secondary reformer is approximately 1%. This is higher than a traditional ammonia process because the unreacted methane can be easily removed in the ammonia synthesis loop. In the traditional ammonia process it is important to drive the methane slip to the lowest attainable value to produce as much hydrogen as possible and to prevent a build-up of inerts in the synthesis loop. This leads to high exit temperatures from the secondary reformer and severe design conditions for air pre-heat systems and waste-heat boilers. The use of an excess of air in the AMV process avoids these harsh equipment design conditions, thus improving reliability and lowering capital and maintenance cost.

Low synthesis pressure

The synthesis pressure in the AMV Ammonia Process is normally between 80 and 110 bar (1,160–1,600 psig). This low synthesis loop pressure allows use of a simpler main syngas compressor and significant reduction of energy consumption. The lower synthesis loop pressure does make separation of ammonia in the loop a little more difficult because of the lower condensation temperature, however, the reduction in power in the main compressor more than offsets the increase in power required by the refrigeration compressor.

The low synthesis loop pressure is made possible by a modified catalyst that is optimized for operation at low pressure. This catalyst, known as ICI Katalco

Table 1. Milder Reforming Conditions in the ICI Katalco AMV Ammonia Process

	Conventional Ammonia Process	AMV Ammonia Process
Primary Reformer:		
steam:carbon ratio	3.75	2.75–3.0
outlet temperature (°C)	800–820	780
(°F)	1,470–1,510	1,440
outlet methane (% dry basis)	11	16
Secondary Reformer:		
outlet temperature (°C)	1,000	950
(°F)	1,830	1,740
outlet methane (% dry basis)	0.3	1.0

Catalyst 74-1, is based on reliable, conventional multi-promoted iron technology. Operation at low pressure is achieved by adjusting the catalyst promoters and improving the catalyst manufacturing process. This catalyst is very active, particularly at low pressure, and can operate over a wide range of conditions. In commercial operation it has produced ammonia at an inlet temperature of only 300°C (570°F) and at part-load operation it is stable and effective at 60 bar (870 psig). It has also been operated at H:N ratios as low as 1.1 in periods of unusual plant constraints. The catalyst is also very stable and has shown less deactivation during an operating life of 10 years than is normally expected from traditional high-pressure catalysts.

Recovery of hydrogen from the purge gas

In the AMV synthesis loop a cryogenic separator is used to recover hydrogen from the purge gas. This separation is done within the synthesis loop and the recovered hydrogen can be recycled at full loop pressure without inefficient recompression in the syngas machine. The cryogenic separator also provides an efficient means to reject the excess nitrogen introduced in the process via the secondary reformer. Figure 1 shows the configuration of the purge gas separation. This diagram shows how the pressure rise across the loop circulator is used as the driving force for the cryogenic system. The system therefore is very simple, requiring only heat exchange, a let-down valve, and a separator. A turbo expander is not needed for the process, thus keeping the process simple and improving reliability.

The cryogenic purge recovery system provides a convenient means of balancing the nitrogen and methane in the plant. Swings in process conditions in the front end of the plant, such as a change in methane slip from the secondary reformer or in nitrogen from the process air compressor, can be smoothed by the purge-gas recovery system; this allows the synthesis loop to continue operating effectively. The purge-gas recovery system also frees the traditional constraint of operating the synthesis loop at a hydrogen:nitrogen ratio that is very close to 3. As reaction over the ammonia synthesis catalyst is limited by the rate of nitrogen adsorption, the reaction rates can be improved by operating the synthesis loop somewhat below the conventional ratio of 3. This can be readily achieved with the purge-gas recovery system in the AMV process.

Extensive optimization of the AMV Ammonia Process has been used to identify the most appropriate values for key process variables. This optimization takes into account capital and operating costs and leads to slightly different results for projects with varying balances between energy and capital costs and plants with different feed compositions and ambient conditions.

The results of the optimization are shown in Table 2.

In summary, the AMV Ammonia Process produces ammonia with a system that can be characterized as follows:

- Stress is reduced at key parts of the plant by the use of lower operating pressures and temperatures compared to older processes. For example, the reformer exit temperature and heat duty are low, as are the secondary reformer exit temperature and synthesis loop pressure, thus easing design requirements, lowering cost and improving reliability.

- Simple equipment is used. The lower synthesis loop pressure makes it possible to simplify the compressor configuration and the cryogenic purge gas recovery can be achieved without an expander.

- Reduced sensitivity to fluctuations caused by changes in front-end conditions. This reduction can be a consequence of ambient conditions, operator error, equipment problems, or catalyst damage. The process will work effectively with a methane content as high as 2% in the fresh synthesis gas to the loop, making the process less sensitive to secondary reformer problems or aging low-temperature shift catalyst. The lessened sensitivity of the synthesis loop to changes in the front-end of the plant enables smoother operation of the plant.

Application of the AMV Ammonia Process

The ICI Katalco AMV Ammonia Process has been

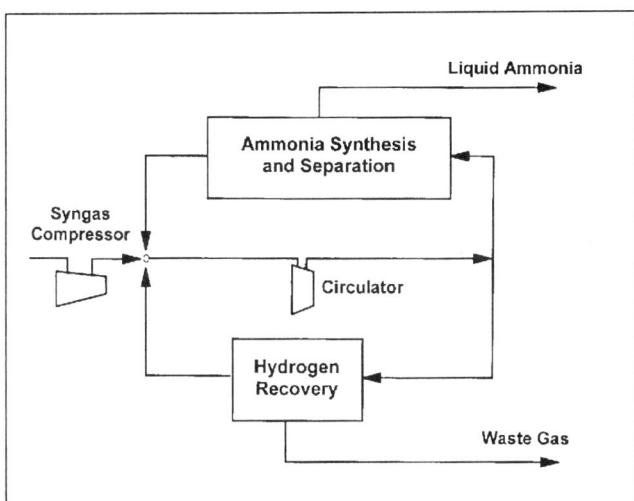

Figure 1. The purge-gas recovery system in the AMV Ammonia Synthesis Loop.

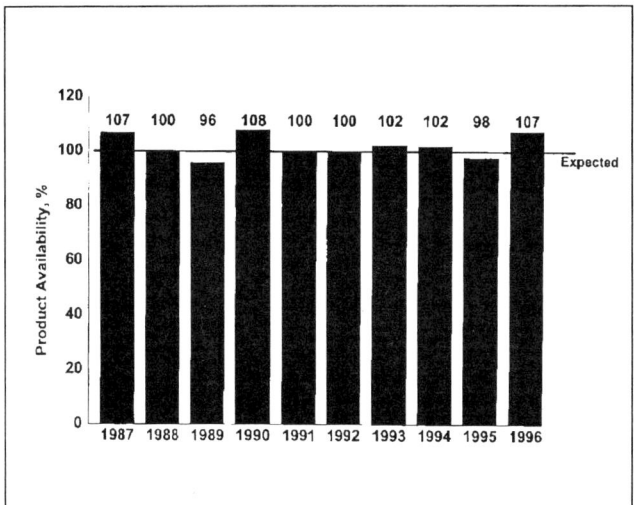

Figure 2. Product availability of the AMV plant at Terra, Canada.

Table 2. Optimal Values of Key Variables in the AMV Ammonia Process

Primary Reformer Outlet Temp. (°C/°F)	770–790/1,420–1,450
Steam:Carbon Ratio	2.75–3.0
$H_2:N_2$ Ratio in the Fresh Syngas	2.4–2.6
$H_2:N_2$ Ratio in the Synthesis Loop	2.0–2.5
Ammonia Separator Temp. (°C/°F)	Approx. –10/–15
Synthesis Loop Pressure Bar (psig)	80–110 (1,160–1,600)

applied to three large ammonia plants. Details of these applications are given in the subsections below.

Terra International, Canada

The first application of the ICI Katalco AMV Ammonia Process was at Courtright in Canada (Taylor and Pinto, 1986). This plant was originally owned by ICI through its CIL subsidiary, but has since been purchased by Terra International. The plant was designed for 1,120 metric t/d and began operation in 1985. The plant has exceeded all of the original targets for reliability and low energy and continues to provide efficient production today at rates of up to 1,380 metric t/d.

Figure 2 illustrates the reliable long-term performance of the plant. This graph shows the availability of the plant relative to the design figure of 330 days per year. The plant has normally exceeded this expected performance and achieved availability up to 355 days per year.

The gas consumption of the plant has been very low (Figure 3). This graph shows the total annual gas consumption for the plant divided by the annual ammonia production. This performance measure, which includes all of the gas wasted during a plant trip or plant restart, normally gives a figure substantially higher than the design flowsheet figure which naturally corresponds to steady operation at the design rate. The difference between annualized efficiency and flowsheet efficiency is often 5–10%. The steady, stable operation of the AMV plant shows that the annual energy consumption is less than expected: in fact, the plant has achieved annual average gas consumption of less than the flowsheet figure of 6.82 GCal/MT (24.6 mm BTU/t).

The application of the AMV Ammonia Process at the Terra plant has also demonstrated exceptionally good environmental performance (Elkins et al., 1993). This is due to the careful recycling of process conden-

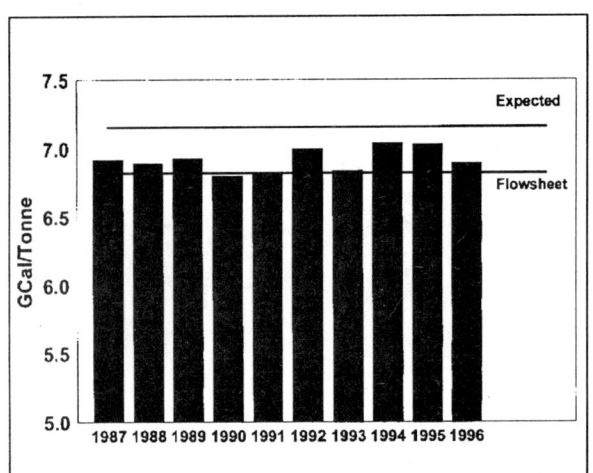

Figure 3. Annualized gas consumption of the AMV ammonia plant at Terra, Canada.

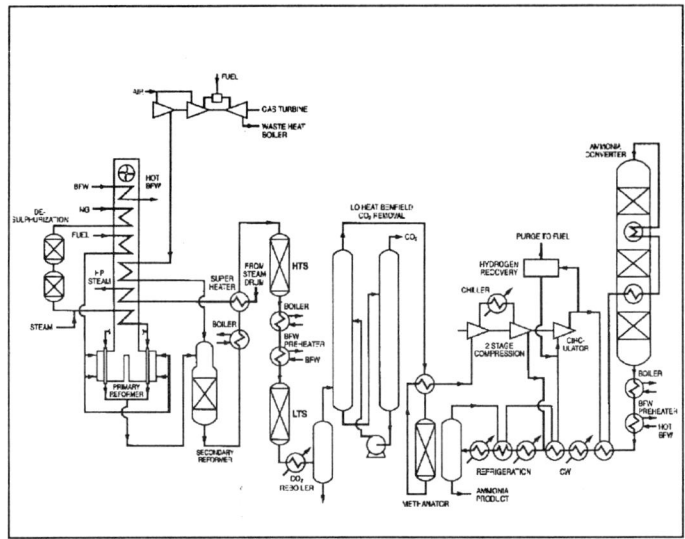

Figure 4. Process flow for the Hainan AMV Ammonia Plant.

sate and the lower gaseous effluents resulting from the reduced primary reformer size.

Zhongyuan Fertilizer Factory, China

The second application of the ICI Katalco AMV Ammonia Process was for Zhongyuan Fertilizer Factory at Puyang in China. The plant was the first large low-energy ammonia plant in China. The plant construction was started in 1986, and full-scale operation began in 1990.

The design energy consumption for this plant was 7.0 GCal/MT (25.2 mm BTU/t) and it represented a dramatic improvement over the previous 13 large ammonia plants in China, which typically had a design energy consumption of 9.8 GCal/metric t (35.3 mm BTU/t).

The Zhongyuan plant was integrated with a matched downstream urea plant, unlike the original Terra AMV plant. This led to a number of process adjustments to adapt to the specific project conditions. These included the use of a gas turbine to drive the air compressor, with the exhaust being used for combustion air in the primary reformer. Although this provides a very efficient configuration, it results in a close integration between the primary reformer and the gas turbine. In this scenario, a trip in the gas turbine will cause a major upset to the whole process.

Hainan Natural Gas Fertilizer Factory, China

The Hainan Natural Gas Fertilizer Factory was established to produce urea for the domestic Chinese market. The factory was enabled by the recent discovery of natural gas off the southwest coast of Hainan Island, which is situated off the southern coast of mainland China. The ammonia plant size was established at 1,000 metric t/d to enable 1,765 metric t/d of urea to be produced at the factory.

After international competitive bidding, Chiyoda Corporation of Japan was selected to engineer and procure the ammonia plant based on ICI Katalco's AMV Ammonia Process. This combination met the requirement of achieving low energy consumption for the lowest capital cost. The design net energy consumption was only 7.06 GCal/metric t (25.4 mm BTU/t), which is unusually low for a plant in a hot and humid climate.

Although three ammonia plants are in operation with the AMV Ammonia Process, the Hainan Fertilizer Factory is more advanced and reliable due to a number of process changes and key equipment selection. These items are elaborated on below.

Process aspects

Figure 4 shows the overall flow for the Hainan plant.
Reliable Combination of Gas Turbine and Boiler. The

process air compressor is driven by a gas turbine. The waste heat in the exhaust gas is effectively recovered by cogeneration of high-pressure steam in a gas turbine waste-heat boiler. This boiler can be operated when the gas turbine is not in service and has two operating modes. The first mode takes the gas turbine exhaust as the combustion air and the second uses air from a blower for combustion. The boiler can automatically switch between these modes, giving the boiler independence from the gas turbine and improving reliability. This arrangement decouples the gas turbine from the primary reformer and provides a fundamentally more reliable design than that used at Zhongyuan.

Two steam drums provide the whole factory with high-pressure steam. One is the gas turbine waste-heat boiler steam drum and the other is shared by boilers located after the secondary reformer, after the high temperature shift reactor and after the ammonia synthesis reactor. The two steam drums produce high-pressure steam at 125 bar (1,800 psig) that is superheated and distributed in a single system between the ammonia and urea plants.

The Hainan Natural Gas Fertilizer Factory uses natural gas with a pressure of 25 bar (360 psig), allowing the gas turbine to be tested and commissioned before startup of the natural gas compressor ignition of the boiler. Similarly, during normal operation the boiler can continue to provide steam if the gas turbine trips. This configuration improves the overall steam system integrity and safety. If the boiler trips due to a loss of waste gas the gas turbine can continue to operate and the primary reformer is protected by steam for an allowed period of time to enable reignition of the boiler and quick resumption of production.

CO_2 Removal System. The plant uses the Benfield LoHeat CO_2 removal system, including a four-stage steam ejector system and power recovery by hydraulic turbine, to improve the efficiency of CO_2 removal. This arrangement significantly lowers the energy required for regeneration of the CO_2 removal solution and improves the heat integration of the plant.

Nitrogen Circulation. The addition of a nitrogen circulation system decreases the time required for plant restart, particularly after a period of maintenance. The system allows circulation of nitrogen for warming up parts of the plant and minimizes the startup time.

Selection of equipment

Hainan Natural Gas Fertilizer Factory's choice of key equipment also differs somewhat from the other two AMV ammonia plants.

Primary Reformer. The Terra plant and the Zhongyuan factory both use top-fired primary reformers. In practice this configuration gives little flexibility for controlling reformer tube temperatures, the pattern of heat input into the catalyst tubes, and the flame length. This configuration also tends to have a long convection section. After careful evaluation, the Hainan plant selected the ICI-Chiyoda side-fired reformer configuration. This design is very compact with the convection section situated directly above the radiant section. The multiple rows of gas burners give the operators flexibility to adjust the pattern of heat input into the catalyst tubes, thus giving some scope for continuing optimization.

Ammonia Converter. For the ammonia converter it was determined that the axial–radial reactor internals from Ammonia Casale would provide the optimum solution for this plant. A configuration with three catalyst beds and two interchangers was used as this allows a high exit ammonia concentration to be achieved, even at the low operating pressure. The axial–radial design minimizes the reactor pressure drop, thus saving power in the circulator and enabling the use of small catalyst particles (1.5–3 mm size), which have improved effectiveness.

Ammonia Converter Catalyst. Because of the low synthesis loop pressure, the design conditions for the ammonia converter are relatively easy, with a thinner wall reactor compared to higher pressure processes. Therefore, the economic optimum size for the ammonia converter is somewhat larger than on traditional plants and a larger catalyst volume can be effectively utilized. During reduction of the ammonia synthesis catalyst it is important to establish uniform gas flow conditions to prevent the possibility of water (produced by the catalyst reduction reaction) poisoning catalyst previously reduced. During the reduction of the ammonia synthesis catalyst in the Zhongyuan ammonia plant, relatively large temperature differences were observed in the first bed of the radial flow converter. It was concluded that this was caused by uneven gas flow and the onset of the exothermic ammonia synthesis reaction at different times as the catalyst reduced unevenly. Although the subsequent operation of the catalyst at Zhongyuan has been satisfactory, improvements were identified for the Hainan project. Distribution could have been improved by increasing the flow during reduction, but this would have resulted in a size increase in the startup heater. It was decided instead to use prereduced ICI Katalco Catalyst 74-1 in the first bed of the converter. This catalyst, which is stabilized with a small amount of surface oxidation, can be activated very quickly, avoiding the difficulties of large temperature differentials.

Gas Turbine. The Hainan plant chose the PGT-10 model from Nuovo Pignone. It is characterized by high efficiency, an advanced control system, and a spray injection system for power augmentation.

Double-wall Ammonia Storage Tanks. The cold ammonia produced at Hainan Natural Gas Fertilizer Factory is stored in two ammonia storage tanks with a capacity of 2,500 metric t each. A double wall configuration was adopted to minimize heat absorption in the tropical climate. Expanded pearlite was added between the inner and outer walls. This gap is filled with nitrogen at a small positive pressure to prevent wet air from entering.

Instruments and automatic control system

The Hainan Natural Gas Fertilizer Factory has incorporated modern instrumentation and control systems throughout the plant. These include:

On-line Analysis. The on-line analysis system carries out analysis on 22 compositions. This reduces manual analysis and number of analytical staff and provides quick and accurate results for the plant operators.

Advanced Control. This is used for five key process areas: steam:carbon ratio (adjustment to account for changes in gas composition); primary reformer exit temperature control; hydrogen:nitrogen ratio control; temperature control in the ammonia converter; and steam system load control.

The advanced control in these areas provides more reliable and stable operation, improving output and efficiency and preventing maloperation.

Choice of Advanced Control System. The distributed control system (DCS) for the Hainan Natural Gas Fertilizer Factory adopted the Centrum-CS system from Yokogawa. To improve the operation of the large compressors the TDM self-diagnosing system is also provided. The DCS system is linked to an upper supervisory computer to allow for the future option of on-line process optimization.

Project execution

The companies responsible for assembling the Hainan Natural Gas Fertilizer Factory were as follows: ammonia process—ICI Katalco Basic; design and equipment procurement—Chiyoda Corporation Detail; design—Chengdu Design Institute; construction—The Ninth Construction Corporation of China National Engineering; commissioning—Hainan Natural Gas Fertilizer Corporation.

The construction of the ammonia plant began December 1994. The first batch of ammonia was produced in October 1996. After initial commissioning the plant entered normal operation very quickly.

The finished plant is shown in the photograph in Figure 5.

Initial operation

The plant operated very smoothly at first, with gradual increase of the production rate over the first few months of operation. Figure 6 shows this attainment of design production. Early in 1997 the market conditions dictated that the plant rate be reduced, so the plant operated for around 3 months at about half the design capacity. This production cut-back, which delayed the formal performance test, is shown in Figure 6. Even with the reduction in capacity due to the limitation of the local urea market, the plant still established the best "first year" production figures for a new ammonia plant in China.

A formal performance test was conducted after several months of operation to establish whether the plant had achieved design conditions. The performance test was conducted over a 3-day period in July 1997. Tables 3 and 4 contain the results of the performance test.

Figure 5. The Hainan AMV Ammonia Plant.

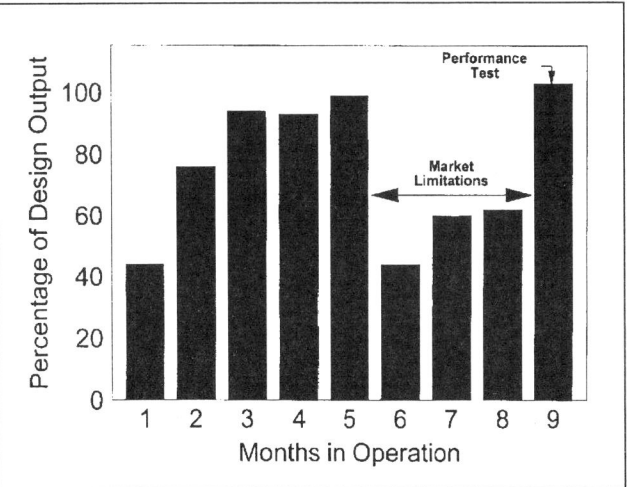

Figure 6. Initial production Hainan AMV Ammonia Plant.

The plant exceeded the daily design production of 1,000 metric t/d (Table 3). Immediately before the performance test the plant had been operating at reduced load and the gradual increase in the production rate during the test was due to the operators gradually retuning the plant for the higher rate operation.

The energy consumption during the performance test also showed that the design performance had been attained. The net energy consumption was 6.92 GCal/ MT (24.9 mm BTU/t) and the Overall Energy Consumption (including utilities such as cooling water and demin water) was 7.06 GCal/MT (25.4 mm BTU/t).

Conclusions

The ICI Katalco AMV Ammonia Process provides a number of advantages for the ammonia plant operator.

• *Reliable, Stable Long-Term Operation.* This has been demonstrated during 12 years of good performance at the Terra plant.

• *Low Energy Consumption.* The energy consumption is only 6.9 GCal/MT (24.8 mm BTU/t) in a tropical climate. In the cooler Canadian environment the Terra plant has achieved annual average gas consumption of 6.8 GCal/MT (24.5 mm BTU/t).

• *Easy Operation.* The milder conditions of the AMV process and the flexibility inherent in the purge

Table 3. Production During the Performance Test

Date	Production (metric t)
July 10, 1997	1,010
July 11, 1997	1,024
July 12, 1997	1,031

Table 4. Energy Consumption During the Performance Test

	GCal/met	mm BTU/t
Natural Gas	8.298	29.87
Electricity	0.094	0.34
Steam Export	−1.470	−5.29
Net Energy Consumption	6.922	24.92
Demin Water	0.018	0.06
Cooling Water	0.120	0.43
Overall Energy Consumption	7.060	25.42

gas recovery system provide easier operation and a plant that is more forgiving in the event of process upsets.

• *Less Investment*. In the AMV process the major items of equipment, such as the primary reformer and the syngas compressor, are smaller and simpler than in traditional ammonia processes, thus saving capital cost.

The above advantages have been demonstrated in commercial operation, not only in a Canadian ammonia plant, but by 8 years of experience in the first large low-energy ammonia plant in China at the Zhongyuan Fertilizer Factory. They have been further confirmed with the recent successful commissioning of the Hainan Natural Gas Fertilizer Factory, also in the Chinese fertilizer industry.

Literature Cited

Elkins K. J., D. Kitchen, and A. Pinto, "Environmental Performance of Modern Ammonia Technology," *Ammonia Plant Safety & Related Facilities*, Vol. 33, AIChE, New York (1993).

Livingstone and Pinto, "New Ammonia Process Reduces Costs," *Ammonia Plant Safety & Related Facilities*, Vol. 23, AIChE, New York (1983).

Taylor and Pinto, "Commissioning of CIL's Ammonia II Plant," *Ammonia Plant Safety & Related Facilities*, Vol. 27, AIChE, New York (1987).

Successful Operation of a Novel Pool Reactor

In this article, Stamicarbon's current highly competitive CO_2 stripping process is described. We then discuss a completely new urea process design. The advantages of the pool reactor in this new process design include lower investment, fewer high-pressure equipment items, less piping, and higher plant flexibility.

K. Jonckers, J. Meessen, and W. Lemmen
Stamicarbon BV, 6160 AB Geleen, The Netherlands

Introduction

Established in 1947, Stamicarbon is the licensing subsidiary of the Dutch chemicals and materials group DSM. Stamicarbon licenses proprietary processes, know-how, and expertise developed and commercially proven by its parent company, DSM.

Stamicarbon has succeeded in vastly improving its CO_2 stripping urea process. The feedstock consumption figures are almost equal to the stoichiometric values for ammonia and carbon dioxide. Steam consumption has also been reduced to a competitive level, and any further reduction would require investment in relatively complex heat-exchange techniques. Electricity consumption is lower than in any other urea process.

The Stamicarbon CO_2 stripping process has extremely low effluent and emission figures (close to zero), leaving no scope for any significant improvement.

The new Urea 2000plus process includes a number of significant improvements that ultimately result in lower investment, lower operating costs, and improved reliability. The first improvement was the introduction of the pool condenser, which was installed in a plant operated by KAFCO of Bangladesh. It has been running excellently since late 1994. The latest improvement, the pool reactor, is now commercially used at DSM's new urea plant at Geleen and has been running without difficulties since March 1998.

Because the pool reactor forms part of the urea synthesis section, this article deals with urea synthesis only.

History of the CO_2 Stripping Process

The CO_2 stripping urea process was developed in order to reduce the steam consumption of the conventional urea process (about 1,600 kg/t). These efforts resulted in steam consumption of 800 to 1,000 kg/t for the stripping process. The process has been a success and more than 100 units have been built. Another breakthrough in the late 1960s was the introduction of type 25-22-2 steel for the stripper tubes. This material lived up to expectations and is still successfully used.

Features of the CO_2 stripping process include:
- A reactor pressure of approximately 140 bar.
- An ammonia/carbon dioxide ratio in the reactor of approximately 3.
- Recycling of unconverted NH_3 and CO_2 at reaction pressure.
- Only a single 4-bar recirculation step is needed.

The evolution of the CO_2 stripping process and the development of the Urea 2000plus process can be summarized as follows:
- Small-scale plants with helicoil carbamate condenser above reactor (total gravity flow) (type A).
- Large-scale plants with vertical falling-film condenser (total gravity flow) (type B).
- Large-scale plants with vertical falling film condenser and low elevation for high-pressure scrubber (partial gravity flow) (type C).
- Large-scale plants with pool condenser (partial gravity flow) (type D).
- Large-scale plants with pool reactor (total gravity flow) (type E).

Flow sheets of the last three types of synthesis sections are shown in Figures 1–3. Figure 4 shows the height reduction in Stamicarbon CO2 stripping plants over the years, another consequence of the stripping-process evolution and a factor in improving safety.

Pool Condenser

Pool condensation has always intrigued us. For example, we found that the inherently unstable operation of the helicoil-type high-pressure carbamate condenser could be avoided by flooding the helicoil. Heat fluxes in pool-condensing high-pressure scrubbers have also been reported to be very high. Thus, it has been a challenge for us to use this technique in the synthesis section. Due to the inverse response of the traditional CO_2 stripping synthesis section, it has been difficult to implement complete and simple automatic process control of the synthesis section.

The differences in performance between a vertical falling-film condenser and a pool condenser can be explained by condensation phenomena and by comparing the situations where either the gas phase or the liquid phase is continuous.

The constraints on condensation in a multicomponent mixture are well known and are extensively described in the literature. Concepts like "concentration gradients," "boundary layer," "temperature gradients," and "diffusional transport rates" are used to describe the fundamental constraints. Considering the vast amount of knowledge available on this subject, it is surprising that the advantages of pool condensation over film condensation have received relatively little attention in the process industry.

Let us try to describe these advantages in plain language:

Gas as the continuous phase: In a vertical falling-film condenser, the composition of the gas varies as it condenses. Other than ammonia and carbon dioxide, the condensing gas contains components such as water vapor and gases, which are considered inert to the process and are not condensable.

First, the aqueous solutions are knocked out. Liquid carbamate can only form if the NH_3/CO_2 molar ratio is within a rather narrow range. The condensed portion rapidly flows down the tubes.

During this condensation process the remaining gases are depleted of the desired components and are hesitant to condense any further (for want of any interested partners, so to speak), resulting in synthesis pressure that is higher than normal and a lower steam pressure. Gas diffusion constraints are important in condensation. The distance of the tube radius (one half of a tube diameter) to the cold cooling wall is very large on a molecular scale. Condensation takes place if heat is removed at the right temperature at the prevailing pressure, the distances to the cooling wall are small, a particular mixture of condensable molecules is available, and the concentration of noncontributing molecules is low.

Liquid as the continuous phase: A pool of previously formed carbamate, possibly already containing urea and water as diluents, circulates from the top of the compartment towards the sparger; it is in equilibrium with the off-gases, which contain some inert gas (approximately 10% on a molecular basis). The circulating liquid therefore is basically subcooled when it is contacted first with fresh ammonia and then with the stripping gases entering at the lower part of the shell. A large proportion of the gas entering immediately

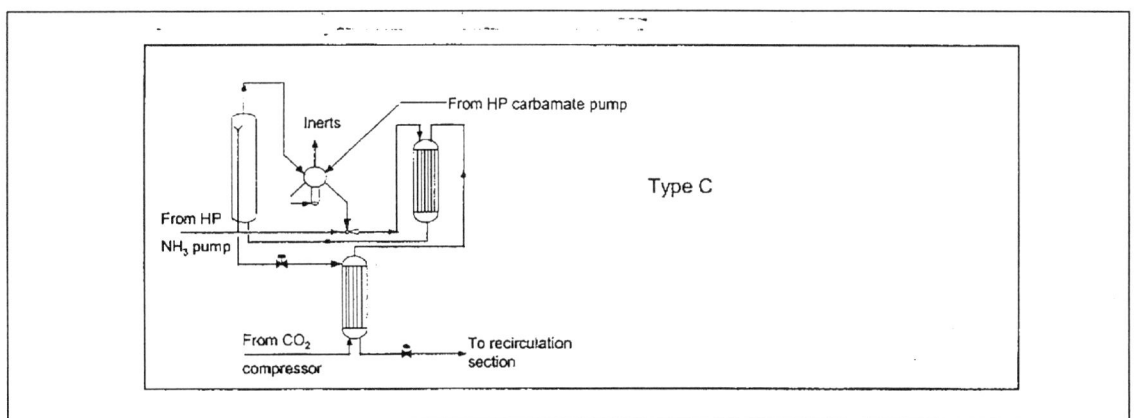

Figure 1. Falling-film HP carbamate condenser.

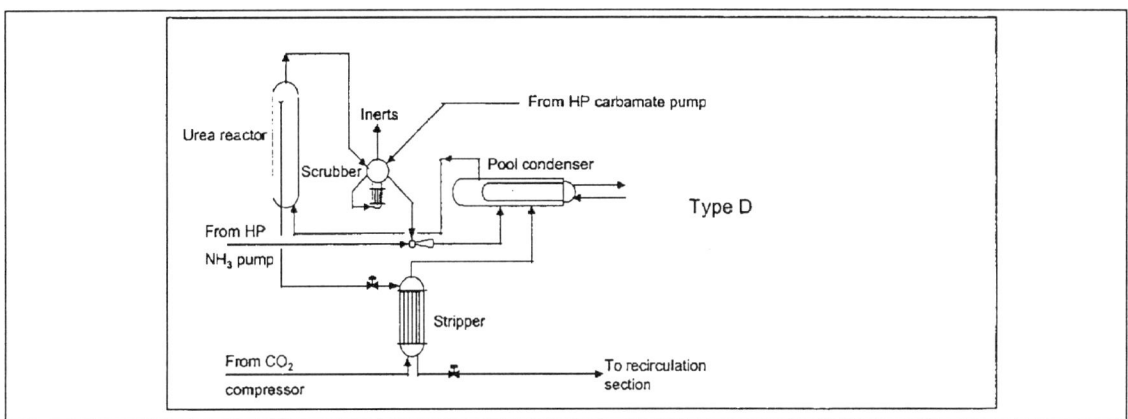

Figure 2. Pool condenser synthesis section.

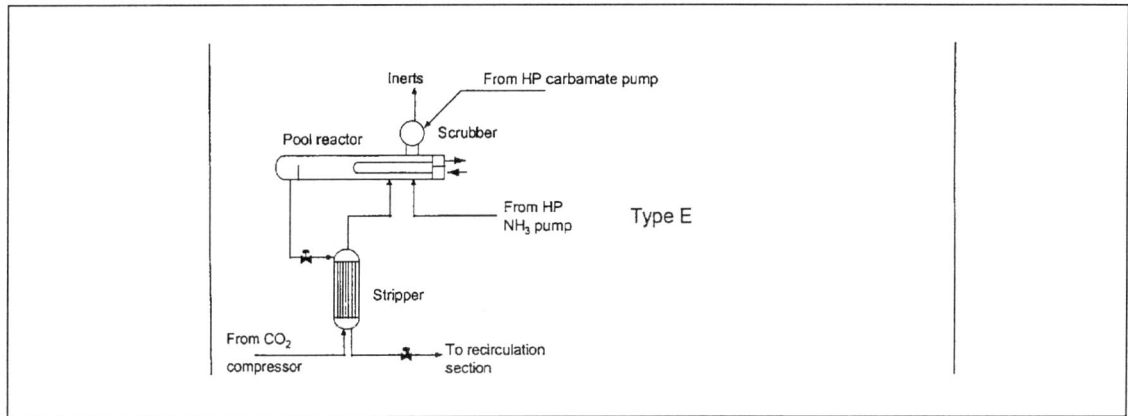

Figure 3. Pool reactor synthesis section.

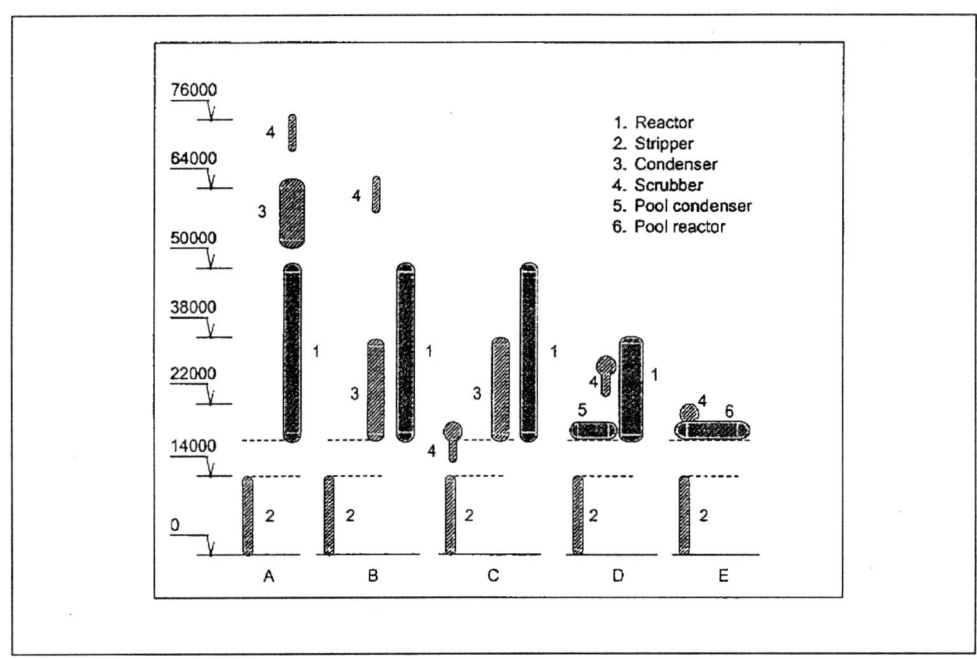

Figure 4. Height reduction in Stamicarbon stripping plants.

condenses in the circulating pool.

The gas bubbles collapse instantaneously while reheating the subcooled liquid to its boiling point. The tube bundle then cools the liquid further, aided by agitation of the remaining gas, which still contains the inert gas.

The size of the gas bubbles is determined by the apertures in the gas distribution system, reducing the influence of gas diffusion constraints. The small gas bubbles are contacted with the liquid phase, which contains condensed, subcooled molecules required for condensation.

Operating principle

Stripping gases are condensed in a pool of liquid on the shell side of the pool condenser (Figure 5), with the heat of condensation being removed by a U-bundle to generate low-pressure steam on the tube side.

The shell side, with substantial residence time, serves as a reactor. Approximately 60% chemical equilibrium is reached at this step. At the same time, the boiling point of the liquid is increased at the prevailing pressure.

The pressure drop across this horizontal vessel is low, and turbulence of the remaining gas gives a very high heat-transfer coefficient.

The characteristics of pool condensation include the following:

• The pool condenser operates as a bubble-agitated liquid cooler, thus requiring less cooling area.

• The temperature in the pool condenser is about 5°C higher due to the residence time.

• The pool condenser, seen as part of the reactor, has a low pressure drop.

• The pool condenser is not sensitive to deviating NH_3/CO_2 ratios.

• The pool condenser is free from inverse response phenomena.

Over the years, we have applied pool condensation in the following process sections: the low-pressure carbamate condenser, the reflux condenser, the atmospheric flash tank condenser, the medium-pressure carbamate condenser, the medium-pressure shell side of first-stage evaporation, the high-pressure scrubber, the pool condenser, and the pool reactor.

Design of the pool condenser

The process design of the pool condenser calls for an operating temperature of 175°C, shell-side 140-bar synthesis pressure, tube-side 4.5 bar steam/BFW, and a gas dividing system with flow deflector plates and baffles.

The mechanical design includes a carbon-steel vessel, stainless steel internals and liner/overlay, a U-tube bundle, internal bore welding, and tube supports.

With internal bore welding there are no crevices between the tubes and the tube sheet, eliminating the risk of stress corrosion cracking that can result from chlorides buildup.

Operational experience

The KAFCO plant has been running above its design capacity of 1,725 metric t/d ever since it first came onstream in May 1995. A corrosion inspection in December 1995 revealed no corrosion in the pool condenser.

In April 1997 ammonia in the parts per million (ppm) range was detected in the low-pressure steam produced. Although all internal bore welds had been checked during fabrication, we were afraid that a leak had developed. However, testing indicated that all welds were satisfactory. We did find a leak that was caused by a manufacturing flaw in the tube.

Because ammonia ingress into the steam system had been very minor, we did not believe that there was any corrosion in the carbon-steel low-pressure steam system. The owner decided to restart the plant and vent excess low-pressure steam.

During the October 1997 turnaround, the tube was plugged at both ends to permanently isolate it. The conductivity of the BFW in the steam drum decreased, indicating that the tube had been leaking since initial startup.

Inspection during this turnaround again revealed no corrosion, so our advice to KAFCO was to have the next corrosion inspection after another four years.

Table 1 shows the production levels of the KAFCO plant.

Pool Reactor

Implementation of benefits

The very moment the pool condenser showed its excellent performance and benefits we realized that the next step was further integration of a pool condenser and a reactor.

One question had to be answered first: Do we know enough about the behavior of the stripping gases pass-

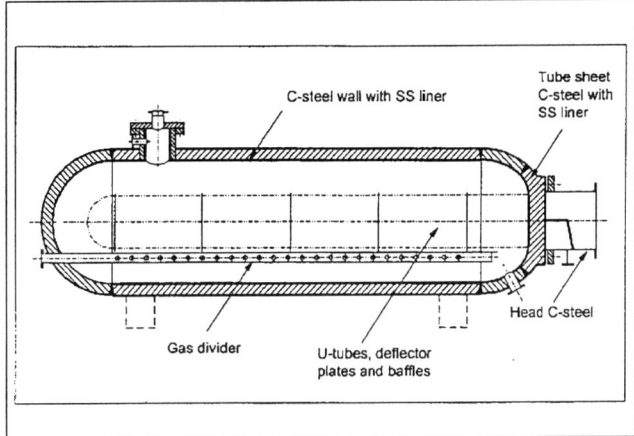

Figure 5. Pool condenser.

Table 1. Production Levels of the KAFCO Plant

Month	Avg. Prod. Urea Unit	No Raw Mat.	Urea Plant Limitations
Aug. 96	102%	1.4	0
Sept. 96	96%	4.3	0
Oct. 96	100%	6.7	0
Nov. 96	102%	7.2	0
Dec. 96	102%	4.0	0
Jan. 97	104%	0	0
Feb. 97	104%	0.8	0
Mar. 97	104%	1.0	0
Apr. 97	104%	7.8	4.5
May 97	101%	7.0	0
June 97	106%	10	0
July 97	106%	0	0
Aug 97	106%	0	0
Sept. 97	106%	0	0
Oct. 97	Turnaround	31	0
Nov. 97	Turnaround	13	0
Dec. 97	110%	1	0
Jan. 98	103%	5	0
Feb. 98	110%	0	0
Mar. 98	106%	0	0

Note: Mat. = materials. Raw materials and plant limitations are measured in days.

ing through the final compartment of a reactor and the influence of these gases on urea conversion?

The answer, which is affirmative, may be explained as follows. In the high-elevation stripping plants of the mid 1970s, all stripping gases passed through the entire reactor (Figure 4, plant type A) to condense in the horizontal helicoil-type high-pressure carbamate condenser. At the time, a serious drawback was the rather nasty inverse response of the stripping gases to the level in the reactor. An advantage, however, was the low partial pressure of the inert gas in them, giving a synthesis pressure of 130 bar at the same reactor effluent temperature of 183°C and the same composition. Because a proportion of the stripping gases is precondensed in the falling-film high-pressure carbamate condenser (Figure 4, plant types B, C, and D) before entering the reactor, we had to accept a pressure of 140 bar in all medium-elevation stripping plants.

The penalty for lowering the elevation was a synthesis pressure that was 10 bar higher. This was necessary for the scale-up from 200 metric t/d to 1,750 metric t/d. We observed that the amount of gas, the composition, and the change in gas holdup did not have any material effect on the approach to chemical equilibrium.

As the synthesis gases pass the reactor, a proportion condenses in order for the urea reaction to take place and for the solution to heat to its boiling point. The adverse effect of newly condensed carbamate on the conversion is negligible.

This yielded the method for increasing the final product temperature at the prevailing pressure and composition. Adding sufficient gas to the last reactor compartments allows enough gas to escape from each compartment to keep the molar percentage of inert low and the temperature high. The off-gas from a vertical reactor of types B, C, and D in Figure 4 contains about 8% inert, while the off-gas from the final compartment of the pool reactor (Figure 6) contains only about 2%.

Another beneficial operational aspect of the horizontal pool reactor is the possibility of monitoring the level increase in the whole reactor by installing a bypass line with valves between the reaction compartments and the level compartment. As the reactor is filled, the stripping process can be started by means of this bypass line, allowing seamless transfer from reactor filling to full production.

The same bypass line makes it possible to heat up the whole synthesis section when the reactor is partly filled with condensate and moderate stripping is applied by adding steam on the shell of the stripper.

Applying the same steam pressure on the low-pressure steam side of the bundle allows heating up of all synthesis equipment simultaneously without live steam, spool pieces, and so forth. The synthesis section can be kept warm during a block-in period by maintaining the steam pressure on the stripper and reactor tube bundle.

Design and operating principles

The design of the pool reactor is similar to that of the pool condenser. Construction, materials, and the leak detection system are based on our experience with reactor fabrication. The pool reactor is basically a combination of the pool condenser and the urea reactor.

The compartments containing the U-bundle serve as a pool condenser, with low-pressure steam being generated in the bundle and gases around the bundle being partially condensed and dehydrated to form urea and water.

In the additional compartments of the reactor (it has more compartments than the pool condenser), further dehydration takes place to achieve 95% chemical equilibrium (Figure 7).

The horizontal condenser-reactor is a gas-agitated vessel with a retention time that enables the condensed carbamate to dehydrate while forming urea and water. The gases condense in a bundle of U-tubes, generating low-pressure steam as in any other urea plant.

Figure 8 shows a cross-section of a typical compartment in the pool reactor, which is much like a pool-condenser compartment. It confirms the results of fluid dynamic calculations and one can tell the location of the gas sparger and its two branches from where the bubbles emerge.

The ascending gas flow induces a circulation pattern from the bottom to the top and down again past the deflector plates flanking the sparger and the tube bundle.

The stripper off-gas, the recycled carbamate solution, and ammonia feedstock are introduced into the reactor. The liquid phase is thoroughly agitated by the gases from the stripper. Condensation heat aids the dehydration reaction and generates steam in the tube bundle.

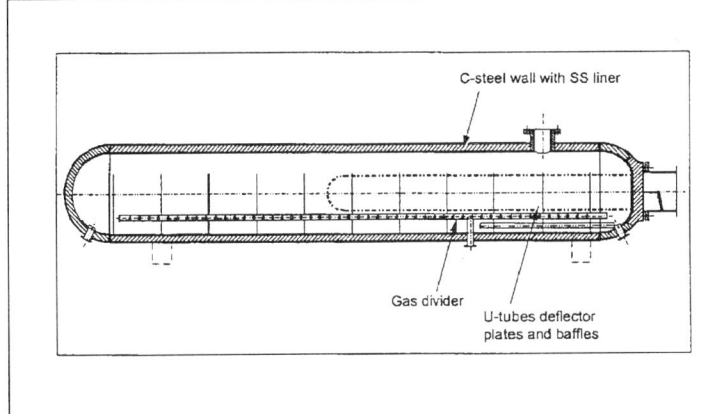

Figure 6. Cross-section of pool reactor.

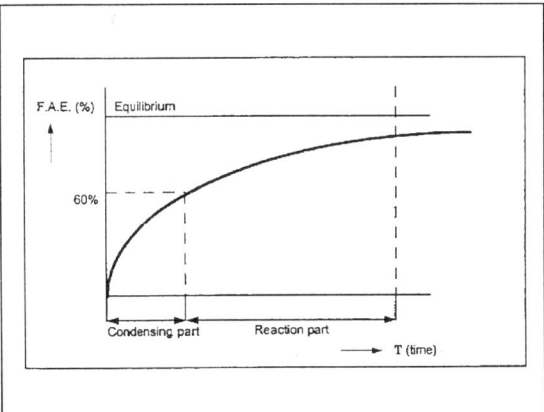

Figure 7. Optimum reactor design.

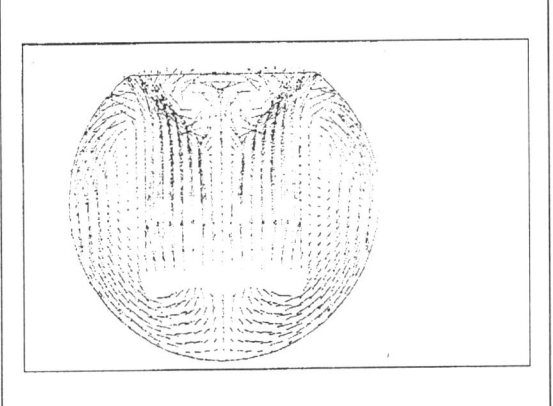

Figure 8. Compartment cross-section with flow pattern.

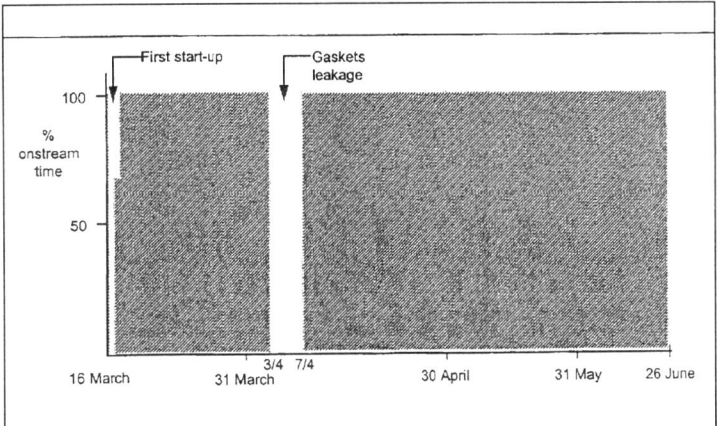

Figure 9. Onstream time since March 18, 1998.

The condensation capacity of the condenser part is determined by the heat-transfer coefficient, the surface area of the tube bundle, and the temperature difference between the process side and the steam side, which is controlled by the pressure of the generated steam.

The process temperature is a function of the rate at which urea is formed, with all other parameters such as pressure, NH_3/CO_2 ratio, initial H_2O content, and retention time remaining constant.

Advantages of the pool reactor

The pool reactor configuration offers the following advantages:
- The amount of heat transferred through the condenser bundle in the reactor is the same as in the configuration of the pool condenser (far better than in falling-film condensers).
- The baffles in the reactor prevent backmixing. This allows the closest approach to equilibrium in the urea reaction and, consequently, the highest conversion and most economical downstream equipment sizing.
- The same operational flexibility as with the pool condenser (i.e., a synthesis that is less sensitive to deviating N/C ratios).
- No inverse response in the synthesis loop.
- Considerably lower investment because of the omission of a costly high-pressure vessel and a high-pressure ejector, far less high-pressure piping in the synthesis section, and lower steel or concrete structure.
- High safety and reliability because of horizontal

layout and fewer flanged joints.

Design of the pool reactor

The process design of the pool reactor includes a 185°C temperature at the reactor outlet, shell-side 140-bar synthesis pressure, tube-side 4.5 bar steam/BFW, and a gas-dividing system with flow deflector plates and baffles.

The mechanical design includes a carbon-steel vessel, stainless steel internals and liner/overlay, a U-tube bundle, internal bore welding, and tube supports.

The first pool reactor, with a design capacity of 1,150 metric t/d, has been installed in the new DSM plant in Geleen, which produces urea melt for melamine production. The conceptual design of this plant commenced on April 1995. The start of basic engineering was during August of that same year, and construction began on August 1996. Urea production started on March 18, 1998.

A consortium of Krupp-UHDE and Raytheon carried out the detailed engineering and construction.

Operational experience with pool reactor

From the initial startup of the synthesis section onwards it became clear that the fundamental design aspects of the reactor were correct.

The temperature readings for each compartment of the pool reactor indicate that the reactor truly acts as a bank of continuously stirred tank reactors (CSTR), and the temperature in each compartment is exactly as predicted.

Heating up of the synthesis section to the required startup temperature is simple and can be done at almost any desired rate. It can be done with either condensate or with a mixture of ammonia water by partly filling the reactor. Synthesis-section heat-up is accomplished by draining a small stream from the reactor through the bypass valve to the stripper, evaporating the drained water with indirect steam on the shell of the stripper, and then condensing it again with the condensate recycling through the bundle.

Starting up a pool condenser might be simple, but starting up a pool reactor is even easier. The steps are as follows:
- Start the absorption circulation loop, preheat the dissociation heater, and start the evaporation section and waste-water treatment system on recycle mode using water or product.
- Prefill the reactor with some water to submerge the sparger, start feeding ammonia while maintaining a constant reactor temperature with steam to the stripper and tube bundle, feed the required amount of CO_2, and open the inert vent valve.
- Supply about 10-bar steam to the shell side of the stripper. Production will begin automatically in 1 h, when the water is expelled from the synthesis section.
- In the case of a total-recycle plant, the water will gradually be replaced by concentrated high-pressure carbamate.
- Change from recycle mode to production mode when the levels in the buffer tanks increase.
- Close the bypass valve when the reactor is full.

The benefits of this are threefold. First, the hectic activities that ensue when the reactor overflows during startup are a thing of the past; second, no adjustment of the molar feed ratio is required; and third, there will be no sudden increases in steam consumption.

Figure 9 shows the results of the initial startup and the onstream time since then. The plant had to be shut down once due to leaking gaskets and production was halted for a few days. Since then, the plant has been operating continuously. The urea melt produced is supplied to the melamine plant, which has been operating continuously since the urea plant's initial startup.

Proprietary equipment

Stamicarbon/DSM have developed the pool condenser and reactor through extensive research efforts, covering both the process aspects and fabrication. Internal bore welding is a specific feature of fabrication. Stamicarbon has extensive experience in high-pressure equipment fabrication. In the interest of our customers, new vessels are fabricated as proprietary equipment by selected manufacturers only.

Availability for licensing

Both alternatives of the Urea 2000plus technology (the pool condenser and the pool reactor) are available for licensing through our licensed con-

Table 2. Typical Consumption Figures for Urea Plants Using Urea 2000plus

Finishing Technique	Prilling	Granulation	Unit of Measure
Feedstocks			
NH_3 (100%)	568	564	kg
CO_2 (100%)	733	730	kg
Utilities			
MP steam (22 bar, 330°C)	855	805	kg
Electricity	14	50	kWh
Cooling Water ($\Delta t = 10°C$)	58	50	t
UF 85	NA	7.5	kg
Utilities Production			
LP steam (4-bar saturated)	370	415	kg
Steam condensate export	235	280	kg
Process condensate export	570	245	kg
Product Characteristics of Prills			wt. %
Nitrogen content		46.4	
Normal biuret content		0.85	
Low biuret content		0.25	
Moisture content		0.25	

Note: Waste-water had 1 ppm wt concentration of both ammonia and urea. The figures include ammonia compression, synthesis section, recirculation section, evaporation section, waste-water treatment section, and finishing section (prilling or granulation). Carbon dioxide compression and storage and bagging facilities are excluded. MP = medium pressure, LP = low pressure.

tractors.

The pool condenser plant in Bangladesh has now commercially operated for more than three years, and the pool reactor plant of DSM is successfully onstream as well. Performance data for these plants appear in Table 2. The Urea 2000plus technology can be used in plant revamps and debottlenecking projects to improve profitability. Old conventional plants with a single 200-bar reactor can benefit enormously from installation of a pool condenser/stripper.

Conclusion

This Urea 2000plus pool reactor process, an innovative updating of mature technology, retains the advantages of the CO_2 stripping process while also being considerably simpler, easier to operate, reliable, and more economical.

Stamicarbon's objective is to remain the world leader in urea technology. With our Urea 2000plus technology, we can license state-of-the-art urea plants with capacities above 2,000 metric t/d for dependable service beyond the year 2000.

DISCUSSION

Dana Baham, *PCS Nitrogen*: You didn't mention passivation. Are you running a conventional passivation process; and, if so, are you using hydrogen destruct units?

Jonckers: Nowadays, passivation goes automatically with heating up. So, you just put in water and you circle around the water and steam by heating up. At the same time, you are passivating the synthesis section.

P. S. Neelakantan, *Madras Fertilizers Ltd.*: You mentioned a figure of 98% approach to equilibrium. Is it the conversion of carbon dioxide?

Jonckers: Yes, that is the approach to equilibrium of the CO_2.

Neelakantan: What is the reaction temperature? Are isothermal conditions existing in the pool reactor?

Jonckers: It's about 184°C.

Neelakantan: What will the pressure be?

Jonckers: The pressure is about 140 bar.

Neelakantan: In comparison to the vertical reactor, what is the length and the height of this horizontal reactor?

Jonckers: The specific volume of this reactor is up to now exactly the same as in our normal reactor.

Neelakantan: Will the ammonia to CO_2 molar ratio be similar to the conventional process or is it different?

Jonckers: The molar ratio is, usually, three. So, as a matter of fact, nothing has changed on the synthesis except that we are using a horizontal reactor and we have a smaller percentage of inert in the final compartment leading to a somewhat higher temperature there. This all leads to a somewhat higher CO_2 conversion. So, if the volume should be the same and the kinetics are higher, the temperature is higher in this reactor through the reactor.

Neelakantan: The final question is: Are there any internals inside the reactor except the compartment partitions?

Jonckers: Well, I have mentioned the gas divider sparger with the branches which leads 80% of the gas to the pool condensing part and the rest to the remainder of the reactor. I have mentioned the ammonia sparger, which divides the ammonia equally over the condensing part of the pool reactor. We see the bundle and the support of the bundle, and we have division plates, which are compartment plates which divide the reactor in, say, 11 compartments. To see to it that the gas will be providing a circulation movement, just like a thermosiphon boiler, we have guidance plates flanking the bundle.

Eiji Sakata, *Toyo Engineering Corp.*: I have one question. In the case of the pool reactor, the mixed gas from the stripper is fed to the condensing part and the reactor part, and the outlet temperature to be controlled is at 185°C. Furthermore, the condensing part temperature will be controlled at 175°C. Is there any temperature profile horizontally?

Jonckers: When I started with the pool condenser, I thought of it as one big CSTR, the whole pool condenser. We did not pay much attention to defining this long, pool condenser as a sequence of CSTRs. However, what happens in reality is that we found out that with having this thermosiphon recirculation, this pool condenser, and also the pool reactor, acts more like a plug-flow reactor. On the pool reactor, we have inserted a thermocouple to measure the temperature in each compartment in the middle of the compartment. Also, we see a temperature profile from compartment to compartment. We can predict a temperature by just calculating locally the residence time, having the kinetics to find out how much urea will be formed, what the inert pressure locally will be, and so find out about the temperature. Then, you see a temperature profile which is, say, 170°, almost horizontal in the condensing part; then, gradually up to 184 or 185°C in the end of the reactor. I had a problem with checking, because I calculated a somewhat higher temperature; however, I then divided my model into not 10 compartments, but 20 CSTRs. Then, I look at the end of the first split up compartment where the thermocouple is, and I see a calculated temperature which is higher than what we see. At the next step, we see a higher temperature. I have the feeling that we have more than just a sequence of CSTRs. I think that we are very close to plug flow. It's hard to prove. However, anyway, in the pool condenser, I assumed 175°C and I got 172° to 178° in KAFCO, which clearly indicated that I have much more than just one CSTR. Again, having a pool reactor and having more inert in the top of the layer in that compartment No. 1, 2, 3, 4, or whatever, where we have the pool, that is, the condensation part, I calculate also a temperature profile which is within 1° or 2°C accurate to what we see. Also, what we then see is not because the model is poor, it is just that our way of interpreting the CSTR is probably wrong. So, there is a temperature profile. You can consider the pool condenser and the pool reactor as a sequence of a number of CSTRs. Does that answer your question?

Sakata: Thank you.

S. Balu, *PCS Nitrogen*: I'd like to have some explanation of all the material and construction of tubes, and the shell and the gasket that you're using in this pool reactor.

Jonckers: We normally use in Stamicarbon, Geleen, all internals that are 2522 material. The lining, the tubes, the supports, and the deflection and compartment plates are made of the same material.

Balu: What about the closing gasket?

Jonckers: I don't know. I'm not particularly familiar with that kind of material, but I don't think that there is a special kind of material. It is just the normal procedure.

Balu: Yes. The only reason I ask is because you had pointed out a gasket leak there.

Jonckers: That's a funny story. It also proves that my calculations are very accurate. When pressurizing the whole synthesis and checking for leakage, a situation began that is comparable to a heart bypass. By accident, too much CO_2 was put in while the sparger was submerged, having 140 bar on the atmosphere of the CO_2 compressor and having five bar in the synthesis with it filled with liquid, including the stripper. However, this is not the stripper gas line and not the discharge line of the CO_2 compressor, which are almost equal in volume. One can imagine that one tips over and there is a quick shutoff valve. The pressure in the discharge stripper line becomes 70 bar, just half of the 140. That can happen so quickly that you have no time to push out the water, which was there blocking all the little holes in the branches of the sparger. At that very moment, the one and only flange in the 20 m long sparger line blew out. Then, a funny phenomenon occurred. I had low temperatures in the condensing part, which was about 6° to 7° low. Also, I had a sudden leak in the compartment No. 5 where the flange was located. I had an increase to the normal temperature, and then everything went on normal again. The funny thing was that this accident never reduced our melamine production. So, like a heart with a bypass, it still operated. It was functioning. And we just stopped it, repaired it, packed it, and made the new packing. Everyone is happy now.

Balu: Thank you sir.

Hussain A. Al-Hajari, *Saudi Arabian Fertilizer Co.*: I'd like to know, during emergency shutdown, how do you drain the whole reactor? Is it different from normal?

Jonckers: It's easier. In a present Stamicarbon plant, you have also an interconnecting line between the liquid line from the carbamate condenser to the urea effluent line from the reactor. And you open that drain line connecting all kinds of compartments of the synthesis and you drain your liquid effluent from the reactor or from the carbamate condenser. You drain that via the stripper to the back end of the plant. Now, here, this is exactly the same. There's no difference.

Al-Hajari: Does it take a longer time or the same time?

Jonckers: No. The actual volume of the reactor is exactly the same. So, overall, the volume should be a bit bigger because I have tubes in it and I have to add that volume to the total volume. However, the net of volume for formation of urea is the same.

Al-Hajari: Can you elaborate on the final product specifications? Are they different from the normal plant?

Jonckers: No. They are a bit better.

Al-Hajari: In terms of biuret and moisture, can you elaborate more figure-wise?

Jonckers: I have no hard evidence yet. The plant is brand new. The problem is that it's our own plant and nobody is interested in a test run so far. So, for me, it is very difficult and very frustrating to get the right evidence that I need for you. What I can give you is all on a theoretical base.

Al-Hajari: What are the design specifications for the final product in terms of biuret and moisture?

Jonckers: I don't have that yet, because it is not important. Everything goes to a melamine plant. However, trust me, with my NC ratio and my HC ratio, everything is normal according to our standard.

Al-Hajari: Based on the practical experience of the melamine technology, if you have an increased biuret level in the urea going to melamine, then you end up having some problems with concentration and the final product of melamine, too. I'm asking this question, because I want to know if this is different from the normal reactor or not. Do you have the final specifications for biuret and moisture? Is it the same or is it different?

Jonckers: It is not different. It goes through the same kind of operation of evaporation and recirculation stage. So, this is just putting in the right biuret. This has nothing to do with the synthesis, which produces exactly the same kind of urea. Also, the HC ratio, the NC ratio, and the temperature are almost the same. So, you cannot expect that the effluent of the synthesis should differ from normal.

Al-Hajari: Concerning the vented gases in the synthesis loop, what are the main components?

Jonckers: Well, there is a difference between what we see normally and what is in this plant. In this plant, we don't have a recirculation; we have a neutralization. Also, we don't bother so much about trying to get everything inside. So, we just have the valve in a certain position and wider than what we are used to at Stamicarbon plants. Here, the off-gases are going to the four bottom absorber. They are absorbed, dissolved, and then neutralized. Also, since there is no need for going so far, and also due to the severe safety regulations which are very strong in Holland, they don't want to go to scrub so deep as we normally do. This is although we have an explosion-proof environment there. However, the four bottom absorber can cope with any position of the vent valve, the pressure control valve, and absorbs it to a level of approximately 5 to 10 PPM in the off-gases, which are very clean.

Al-Hajari: Thank you.

Federico Zardi, *Urea Casale S.A*: I have one question regarding the compartments that you have mentioned in the pool reactor. You mentioned these compartments generate a lot of turbulence and act as CSTR reactors. This looks very similar to what happens in a normal reactor between the trays. So, normal reactors are also divided into different compartments. Is it that in the pool reactor you can fit more compartments and make it better than a normal reactor? What is the difference? Why is the pool reactor better than a normal reactor?

Jonckers: First, I don't have more compartments than in a normal reactor. It is about the same, 11 compartments. Depending on how you construct the trays and how you play with it, and how you play with the division of the gas and the liquid, you can have a poor CSTR. You can have bad spots. You can channel. All these kinds of things can happen in a vertical reactor unless you improve on your tray design. I don't face that kind of problem any longer in a horizontal reactor. I have the strong feeling that channeling is out; it is not there. It is hardly convincing that if you look at extensive fluid dynamic simulation that you might expect dead centers or long residence times. There are certain spots. So, it is more or less reducing the actual active volume of the reactor. This is what I see. It is a high turbulence circle almost like a flying wheel. If I have a flying wheel and everything is going around like mad due to the bubble, why should there be any backmixing or forward mixing? It is like a flying wheel which goes around with the axis having a slant on it. Also, it goes about one meter per second around and it moves in the horizontal direction with about three to four millimeter per second. Also, I see no evidence which proves counterwise that this is almost a perfect plug-flow reactor.

Zardi: Is it easier to get the CSTR condition in a horizontal reactor?

Jonckers: I think a pool condenser is much simpler than a horizontal reactor. The way it has been built is a much simpler way in my belief than having a vertical reactor. In a vertical reactor, you have to see that you have either local turbulence or whatever. Whatever you choose, it is just wrong. It can be terribly wrong. And I have the feeling that this, independent from capacity, just works, and that the thermosiphon reboiling effect is just canceling all these kind of drawbacks of a vertical reactor.

Ahmad Al-Badrani, *Saudi Arabian Fertilizer Co.*: I believe about your reactor that the corrosion rate allowance is higher than the conventional reactor. Is it

true?
Jonckers: No.
Al-Badrani: How much is the corrosion allowance?
Jonckers: I don't know about allowance. I know about the corrosion.
Al-Badrani: However, I think that this is very critical in the reactor.
Jonckers: We don't see any corrosion in our pool condenser.
Al-Badrani: I think that it must be a corrosion allowance there.
Jonckers: Yes, I know that, but I'm not a mechanical man. You have to ask that question to somebody else.
Al-Badrani: I am concerned about the long range, because you have been operating for just four years and you will be operating initially at a lower temperature in your plant. This will contribute, because I believe, as well, it's much more difficult to do an inspection in your reactor than in the conventional reactor. Can you elaborate?

Jonckers: I cannot understand your question.
Al-Badrani: No, I'm saying this because our plant is around 30 years old. So, this is all to your advantage, because your plant has been operating for four years. I think that if you consider the long range and consider inspection difficulties in the reactor rather than the conventional reactor, as well as the corrosion allowance, I think that makes it difficult to make comparisons with conventional reactors.
Jonckers: I think that it's good to compare. However, if we look at results and look at the checks that we did so far in KAFCO, normally, we inspect a synthesis every two years. However, we have postponed that to four years.
Al-Badrani: You are in the initial stages after only four years. My concern is if you go around for a long range.
Jonckers: However, I cannot help you. I cannot speed up time.
Al-Badrani: Thank you.

Use of Bimetallic Tubes in Urea Strippers: Technology Improvement

An up-to-date report confirms the excellent operating performance of the strippers designed using bimetallic tubes and heads lined with stainless steel. Modifications to the original design are also discussed.

Franco Granelli and Gian Pietro Testa
Snamprogetti, Milan, Italy

Introduction

The stripper has always been the most critical piece of equipment in urea processes that use stripping technology. In the urea process, the conditions that cause corrosion are more severe in the stripper than elsewhere in the synthesis section. These unfavorable conditions include high pressure, high temperature, high corrosiveness of the solution, and the presence of both liquid and vapor phases (which is always a potential source of corrosion).

Different licensers, based on their own process conditions and experience, have selected different construction materials for the stripper tubes and the two heads that are in contact with the process solution.

Snamprogetti used titanium for the tubes and for the lining of the two heads in the first industrial plants that were based on its own technology.

A few years ago, Snamprogetti changed the material of the tubes from titanium to bimetallic stainless steel/zirconium, and the lining of the two heads from titanium to stainless steel (Miola et al., 1995).

Profile of Snamprogetti and its Role in the Urea Technology Field

Snamprogetti is the international engineering contractor and technology company of the Italian ENI Group and is engaged worldwide in the development, design, and construction of industrial facilities and associated infrastructures that include pipelines and plants for offshore processing, refining, gas treatment, fertilizers, and chemicals.

Although dedicated to these various activities, Snamprogetti is better known in many countries as a licenser of urea technology. Snamprogetti has attained worldwide success in this field, and there are now 96 urea plants in the world based on its technology. Recently the company has been awarded a contract for the design and construction of the highest-capacity urea plant in the world: 3,250 metric t/d.

The capacity of urea plants has increased tenfold compared with the capacities at the beginning of the 1960s. Figure 1 represents the urea plant capacity increase (in a single line) over the last four decades.

Stripper Technology in Urea Plants

The main concepts relevant to the stripper in the urea plants have been described previously in Miola et al. (1995). These concepts are repeated briefly here.

Urea is normally produced by direct reaction between ammonia and carbon dioxide (Figure 2). The reaction is composed of two distinct steps (Figure 3). The first is the reaction between ammonia and carbon dioxide that leads to an aqueous urea solution at about 70% wt. From this solution, solid urea product is produced by crystallization, granulation, or prilling in the second step.

This reaction (Figure 3) takes place in the urea reactor and appears to be very simple. Unfortunately, the reaction is incomplete and large amounts of unreacted ammonia and carbon dioxide leave the reactor with the

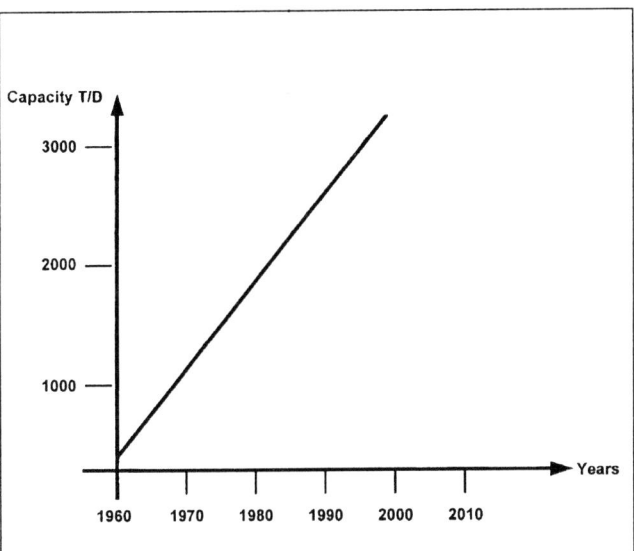

Figure 1. Single-line urea plant capacity.

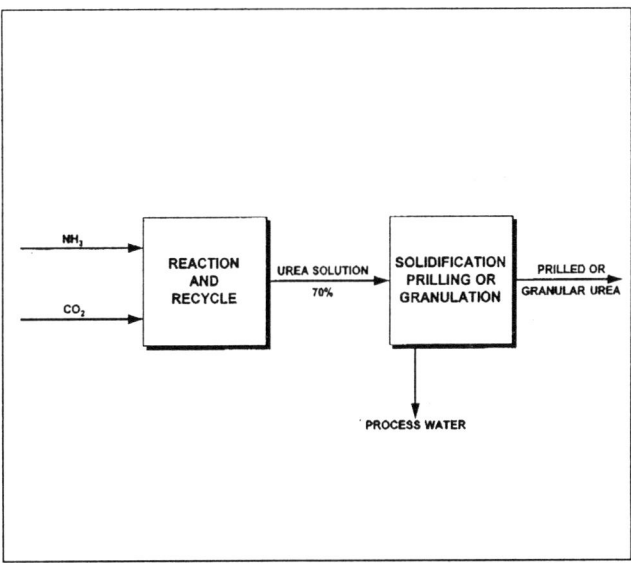

Figure 3. Block diagram of urea production.

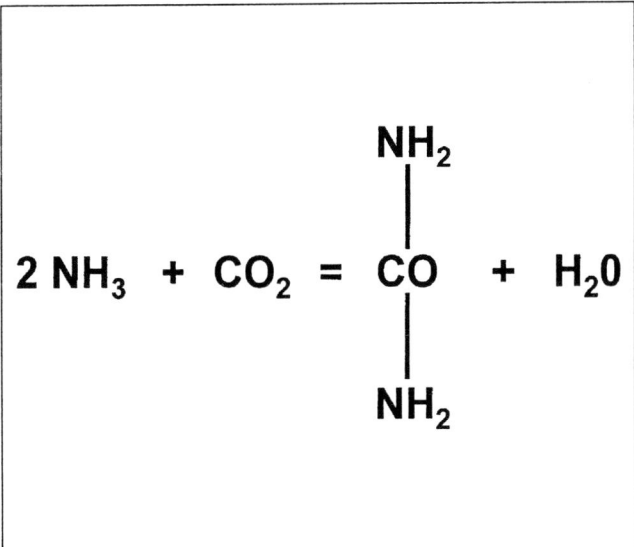

Figure 2. Urea synthesis reaction.

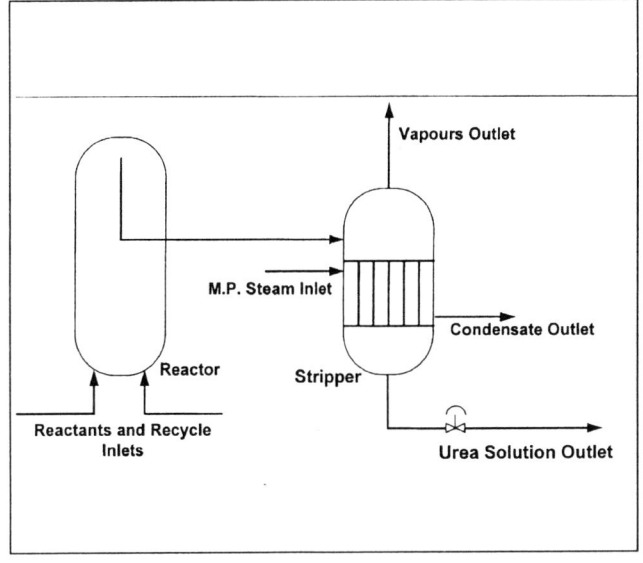

Figure 4. High-pressure reaction and decomposition in Snamprogetti urea stripping process.

urea solution. These components must be separated from the urea solution at the reactor outlet and recycled to the reactor. The separation of these components and the recycle to the reactor take place in subsequent steps and consume a considerable amount of energy. Designers have reduced the energy required in these steps and intend to reduce it further.

The process conditions, that is, temperature, pressure, and particular fluid compositions at various operating steps (with the presence of intermediate compounds), can be highly corrosive. In the past, corrosion was the most critical problem in urea plants. Corrosion is fairly well controlled in today's plants, but careful attention must always be paid to it at every step of the design stage and during plant operation.

The introduction of the stripping technology was a milestone in urea plant design because it drastically reduced energy consumption. The stripper is the primary piece of equipment in the stripping technology. In the stripper, a large part of the unconverted raw materials (ammonia and carbon dioxide) are separated from the urea solution leaving the reactor. This separation is made at practically the same pressure as the reactor pressure (Figure 4).

In the stripper, all conditions causing corrosion are present simultaneously. These include high pressure and temperature, high corrosiveness of the solution, and the presence of both liquid and vapor phases.

Apart from material selection, it is important to pay careful attention during the design stage as well as during stripper fabrication to ensure an appropriate fluid-dynamic, particularly to achieve adequate heat treatment.

The Stripper in the Snamprogetti Stripping Technology

The function and design of the stripper in the Snamprogetti urea technology, as well as both the material of the tubes and the lining of the two heads, are described in Miola et al. (1995). We present a summary here.

The stripper separates ammonia and carbon dioxide from the urea solution leaving the reactor. In this solution ammonia and carbon dioxide are present as free compounds and as an intermediate compound, ammonium carbamate. The heat needed for this separation (mainly for the decomposition of ammonium carbamate) is supplied by medium-pressure steam condensing in the shell side of the stripper. The stripper consists of a vertical tube bundle with the process solution flowing down along the internal walls of the tubes to ensure low residence time (that is, prevent biuret formation) and obtain high heat transfer.

Snamprogetti has used titanium for the tubes and the two head linings since its first industrial urea plant. Today, the majority of urea plants based on Snamprogetti technology (96 plants in operation or under construction) are built with titanium tubes and head linings. The reason for this choice is that titanium has excellent resistance to corrosion from the urea/carbamate solutions and the ability to operate the plant at practically any capacity below design capacity. Unfortunately, due to the softness of titanium, some erosion phenomena have been observed, especially where there is a sudden change of fluid direction, an impingement, or strong evaporation.

Snamprogetti tried several materials in both laboratory and industrial plant tests to solve this problem. Not all of these tests yielded successful results. After testing, bimetallic tubes (zirconium internally and stainless steel externally) were the solution selected by Snamprogetti. This option guaranteed the excellent anticorrosion characteristics of titanium without its inconveniences.

With the introduction of the bimetallic tubes, the material of the lining of the two heads was also changed from titanium to stainless steel (same as the external parts of the tubes). This resulted in a lower total cost for the stripper than with the previous design solution that used titanium for tubes and head linings.

The main test that convinced Snamprogetti of the soundness of the new solution regarding the tubes was the performance of 15 tubes that were installed in 1986 in the stripper of a 350 metric t/d urea plant. Laboratory analysis and a visual inspection of some of these tubes after some time in operation yielded good results. The last two tubes removed for inspection were taken out in August 1991 after about 40,000 h and were in excellent condition.

Based on these good results, three strippers with bimetallic tubes were manufactured (Figure 5). The first one was put into operation in September 1991 in the 1,200 metric t/d urea plant of PCS-Nitrogen (Memphis, TN), the second was put into operation in June 1992 in the 1,300 metric t/d plant of Agrolinz (Austria), and the third was installed in the 600 metric t/d plant of Chimco

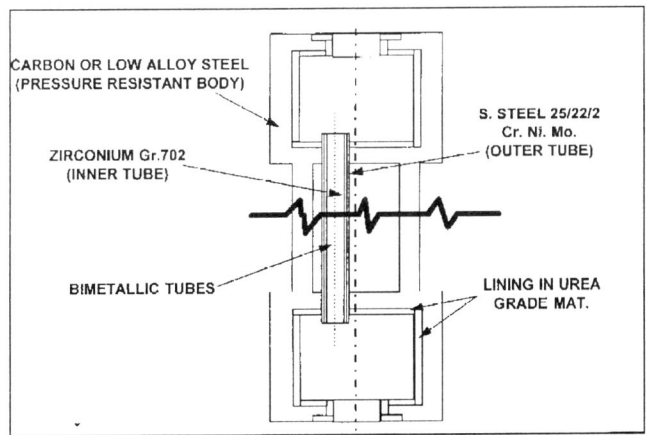

Figure 5. First three bimetallic strippers.

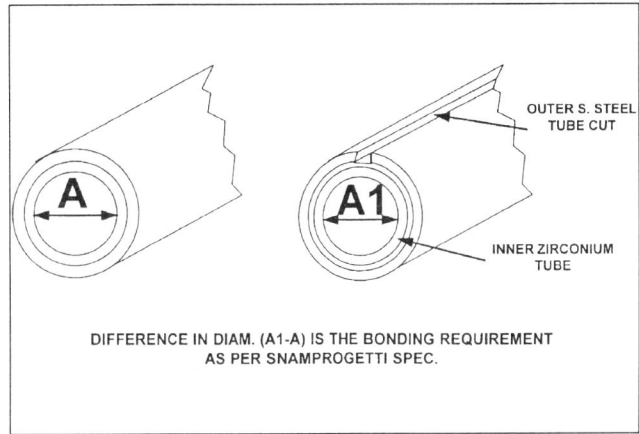

Figure 7. Bimetallic tubes: Snamprogetti bonding evaluation.

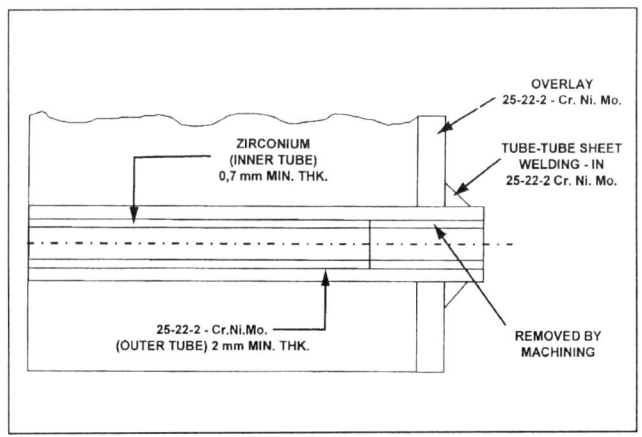

Figure 6. The bimetallic stripper.

Figure 8. Details of the tubesheet.

(Bulgaria) and put into operation in December 1992. No operating problems have been discovered in any of these units. Furthermore, some of the 15 tubes that were installed in the 350 metric t/d plant are still in operation without problems after 12 years.

Many strippers with bimetallic tubes are now in operation or in various stages of fabrication, as indicated in the "Current Situation" section of the article.

Description of the Stripper and the Bimetallic Tubes

Figure 6 represents an overall sketch of the stripper supplied with bimetallic tubes. The article by Miola et al. (1995) describes the drawing, and this description is not repeated here. The bonding between the external tube of stainless steel and the internal tube of zirconi-

AMMONIA TECHNICAL MANUAL 49 1999

Company	Country	Plant Capacity T/D	In Operation Since
PCS Nitrogen	U.S.A.	1200	Sept. 1991
Agro-Linz	Austria	1300	June 1992
Chimco	Bulgaria	600	Dec. 1992
Hydro Agri	Germany	1400	Nov. 1995
Oswald Bindal	India	1400	Dec. 1995
Oswald Bindal	India	1400	Dec. 1995
Fauji Fertilizer Co. Ltd.	Pakistan	2100	Sept. 1996
CNCC-Chishui-Guizhou	China	500	Oct. 1996
IFFCO	India	1400	Dec. 1996
IFFCO	India	1400	Dec. 1996
NFL - Vijalpur	India	1400	March 1997
NFL - Vijaipur	India	1400	March 1997
Nagarjiuna Fertil-Kakinada	India	2100	January 1998
Gulf Petroch. Ind. Corp.	Bahrain	1700	January 1998
CNCCC - Zhanyi - Yunnan	China	400	June 1998
CNCCC - Pingdingshan - Henan	China	400	July 1998

Figure 9. Strippers in service up to August 1998 with bimetallic tubes.

Company	Country	Plant Capacity T/D
CNTIC - Nanjing	China	1760
Fertinitro	Venezuela	2200
Fertinitro	Venezuela	2200
Profertil S.A.	Argentina	3250
Agrium	U.S.A.	250
Kribhco	India	1100
Kribhco	India	1100
Pakarab	Pakistan	300
CNTIC - Qianan - Hebei	China	400
CNCCC - Anyang - Henan	China	600
IFFCO - Phulpur	India	1100
IFFCO - Phulpur	India	1100
CNCCC - huaxian - Shaanxi	China	400
CNCCC - Hechi - Guangxi	China	400
CNCCC - Huainan - Anhui	China	900
Petronas Fert.	Malaysia	1800
CNCCC - Qingzhen - Guizhou	China	400
Agrium	Canada	2150

Figure 10. Strippers in different stages of construction up to August 1998 with bimetallic tubes.

um is purely mechanical and there is no welding between the coaxial tubes. The minimum required bonding is indicated in Figure 7, and the details of the tube sheet are shown in Figure 8.

Current Situation

At present 16 strippers are in operation (Figure 9) and 18 others are in various manufacturing stages (Figure 10).

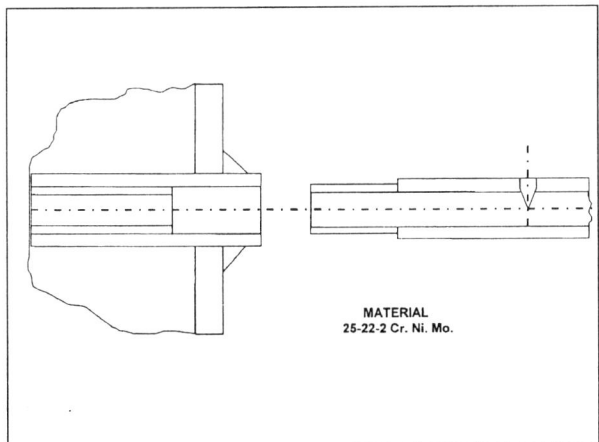

Figure 11. Bimetallic stripper: coupling of ferrules.

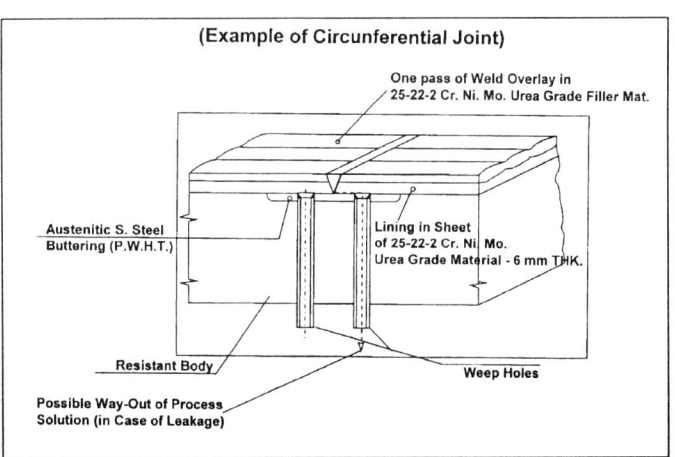

Figure 12. Weld overlay coating on 25-22-2 Cr-Ni-Mo sheets.

The tubes of the first three strippers discussed previously in "The Stripper in the Snamprogetti Stripping Technology" section (Figure 5) have been checked by eddy current method (absolute and relative methods). Their condition is excellent and they look like new tubes, so it may be concluded that further checks are not necessary. For safety reasons it is advisable to check them after 4–5 years.

The bimetallic tubes of the Agrolinz stripper were checked after 8,500 h and 13,000 h of operation. The bimetallic tubes of the Chimco stripper were checked after 13,000 h of operation and the tubes of the PCS-Nitrogen stripper were checked after 27,000 h of operation.

Improvements on the Technology

Some improvements have recently been introduced into the design of this equipment.

We have improved the liquid distribution on top of the tubes even though, from the point of view of corrosion, neither the bimetallic tubes nor the titanium tubes require an even distribution of the process solution. The purpose of this modification is to prevent abnormal hydraulic patterns that could jeopardize the efficiency of the equipment. This modification consists of the introduction of two distribution trays in the top head of the stripper. We consider this important, especially in high-capacity urea plants. Strippers with this improved distribution are already successfully operating.

The design of the ferrules was modified as indicated in Figure 11. This modification, which has already been tested, guarantees a better flow of the solution along the internal wall of the tubes.

Air was introduced at the bottom head in some strippers as additional protection against the aggressively corrosive nature of the urea/carbamate solution when unusual process conditions were present. In strippers without air injection, it has been noted that all of the welding in the bottom head was in much better condition than the lining.

For this reason, we have worked out a new solution for the bottom head that make the injection of passivation air at the bottom of the stripper unnecessary. This solution (Figure 12) consists of making the lining of the bottom head with a welding overlay coating on 25-22-2 Cr-Ni-Mo sheets. A deposit consisting mainly of iron oxides has been noticed on the bimetallic tubes. This deposit does not compromise the heat-transfer coefficient, even after several years of operation. Snamprogetti has also successfully tested a hydroblast method that can fully eliminate the deposit.

Conclusion

In all the urea processes using the stripping technology, the stripper and the tubes in particular are the most critical pieces of equipment.

Based on the excellent results from long operation times in industrial plants, Snamprogetti can affirm that the bimetallic tubes have completely solved the problem of corrosion in the urea strippers.

The solution is so reliable that it is not necessary to make a regular check of the tubes' condition. Some operators may, for safety, inspect after several years of operation.

Literature Cited

Miola, C., F. Granelli, and G. Testa, "Use of Bimetallic Tubes in Urea Strippers," *Ammonia Plant Safety & Related Facilities*, Vol. 36, AIChE, New York, p. 254 (1996).

DISCUSSION

Dana Baham, *PCS Nitrogen*: We've been running your bimetallic tubes since 1991. We have, in fact, extracted tubes and measured zirconium corrosion, and there is absolutely no corrosion of the tubes themselves. However, I caution users or potential clients to be extremely careful of the tube-to-tubesheet weld, to do as many measurements as possible there, and to be cautious of the temperature that you run these strippers at. They cannot run at the temperatures of the old titanium strippers and must be run less than the 400°F. Otherwise, you'll incur corrosion of the 2522 liner and, particularly, the bottom piece of the stripper.

Michael Hyland, *Snamprogetti*: I asked Franco (Granelli) whether that was where the ferrule improvement occurred. Franco says that the basic design does not run at those temperatures. We've reduced the temperature of operation. The improvement here is basically where we're cutting back on the zirconium, so that it's further down below the welding area. That was probably an improvement based on seeing some of what you're talking about.

Jorge Camps, *The Pace-Consultants Inc.*: Earlier this morning, I made a reference to your 3,250 MTPD plant in Argentina. I tried to stay away from any confidential information. However, since you reintroduced the subject, you may want to give some details to the audience about the design of that stripper and what Snamprogetti has done to minimize the scale-up risks in going from, for example, the Baruch plant in India. That plant has your largest tonnage, which is 2,400 tons a day to the 3,250 size.

Hyland: The question is how are we dealing with the scale-up from our largest prior design up to the 3,250 stripper design.

Granelli: We have already in operation a stripper even bigger than the one in Argentina. It is bigger in the number of tubes or very similar. So, we already have in hand how to make the distribution. It's already applied in a plant, having a lower capacity, but with a higher fluid, so the stripper is nearly the same. The most critical equipment is already in operation, having the same design as the one for Argentina.

Camps: Does that also include the same number of tubes?

Hyland: Franco's comment was that he has additional fluid. The number of tubes and size is similar to what's in operation, but has that changed the characteristics in the sense of the fluid-flow rate? I hope that answers your question. If not, we can talk about it.

Camps: No, thank you. All that I needed was for you to perhaps provide additional information regarding the stripper scale-up. Thank you.

Will Lemmen, *Stamicarbon BV*: From the presentation did I hear that there are tubes in operation for more than 100,000 h?

Hyland: That's correct.

Lemmen: Do they look brand new?

Hyland: That's what I'm told.

Lemmen: I assume that's the zirconium part. What about the tube-to-tubesheet welds and the connections after 100,000 h?

Granelli: All the tubesheets are made and lined in stainless steel, and the tubes that are zirconium internal are stainless outside, the same material, and are welded to the tubesheet.

Lemmen: I understand that, Mr Granelli, but you said brand new after 100,000 h. Does the same situation occur with the tube-to-tubesheet connection?

Granelli: Yes, also the tubesheet. There is no problem at all. This plant is still operating in Italy, 350 tons per day. We didn't see any problem.

Lemmen: That brings me to the second question. What is the oxygen content in your CO_2 to prevent corrosion on the tube-to-tubesheet connection?

Granelli: We maintain the same old 0.25% volume in the CO_2. However, in some plants where we have some problem at the bottom, we add a small quantity of air, let us say for 20 cubic meter for 1,500 tons per day. However, with a new solution, with a lining on the bottom, we are not going to use any more air in the bottom.

Baham: In answer to Mr Lemmen's question, I indicated that clients should be careful of the tube-to-tubesheet weld. We have experienced no corrosion of that tube-to-tubesheet weld. What I was referring to is primarily defects in the original tube-to-tubesheet weld. Again, we've seen no corrosion in the tube-to-tubesheet weld itself.

Ali Al-Qahtani, *IBN Al-Baytar Saf, Saudi Arabia*: You mentioned that there is erosion, but you didn't locate the erosion of the titanium steel parts.

Granelli: It was mainly where you have an impingement of a sudden evaporation, and, as regards the stripper, it was located in the upper part of the tubes just below the upper tubesheet.

Shroud Failures of Process Air Compressor Turbine

The steam turbine driving the process air compressor in the Southern Petrochemical Industries Corporation (SPIC) ammonia plant experienced shroud failure during the fourth stage (out of six possible stages). This article describes the failure and its probable causes, modifications carried out to avoid the failure in the future, and finally the redesigning of the fourth stage for lower shroud stress levels and higher efficiency to increase machine availability.

V. Jayaraman and K. Rajagopal
Southern Petrochemical Industries Corporation, Tuticorin, India

Introduction

SPIC operates a fertilizer complex that manufactures 5,12,000 metric t/year of urea and 4,15,000 metric t/year of diammonium phosphate (DAP). Ammonia is produced using the conventional steam naphtha reforming process. Air to the secondary reformer of the ammonia plant is supplied by a two-casing centrifugal compressor driven by a steam turbine (process air compressor turbine).

Steam Turbine

The steam turbine uses expanding steam to turn a shaft. The shaft is mounted with blades and shrouds in one or more stages. The nozzles and diaphragm in a turbine are designed to direct the steam flow into a well-formed, high-speed jet. This jet strikes the moving rows of blades mounted on the rotor. The blade converts kinetic energy of steam into the rotation energy of the rotor.

Construction

In a multistage turbine, the diaphragm separates each stage. The stationary blades can be a part of the diaphragm, which directs the steam so that it hits the next stage of blades.

The shroud band is a segmental rim, which connects the tip of blades in a stage. This rim is composed of several segments, each of which connects seven to eight blades. This is required to improve the rigidity of blades.

To minimize steam leakage across blades, radial sealing fins are provided on casing, which maintains close clearance with the shroud. Details of the PAC steam turbine are shown in Table 1.

The History of the Process Air Compressor (PAC) Turbine

Use of a PAC-driven steam turbine began in 1975. A brief history of the working and spare rotors is as

Table 1. Details of PAC Steam Turbine

Form & Type	Hitachi Impulse-type condensing turbine
Max. Output	9,550 kW (12,993 Hp)
Max. Speed	7,875 rpm
Inlet Steam pr./Temp	12ksc A/278°C (170.6 PSIA/532.4°F)
Exhaust Steam pr.	0.17 ksc A. (2.42 PSIA)
Turbine Stages	Rateau 4 + (1 × 2) double flow
Bearing	Tilting pad journal and tapered land thrust bearings

Note: pr. = pressure.

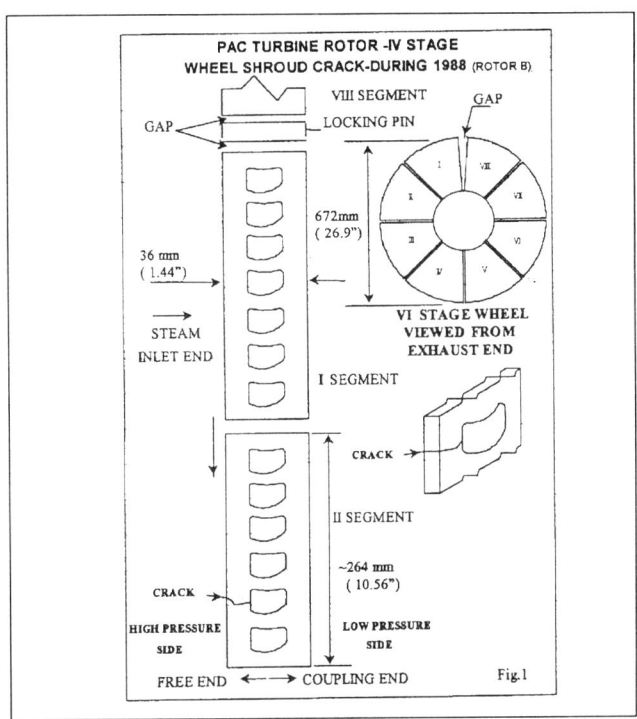

Figure 1. PAC turbine rotor fourth-stage wheel shroud crack.

follows: In the years 1975–1983, 1988–1990, and 1997, rotor A was used, while rotor B was used from 1983–1988, 1990–1995, and from 1997 to the present.

The rotor has been changed six times. Shroud cracks were observed on three of these six occasions. Rotor A had cracks on one occasion and rotor B had cracks on two occasions.

When rotor A was taken out in 1983, three cracks were present on the steam inlet side of the fourth wheel shroud. The shroud was rebladed as per original design and installed in 1988. When rotor B was removed in 1988 the steam inlet side of the fourth wheel shroud had one crack. The location of the crack is shown in Figure 1.

Reducing the stress level on the shrouds was suggested. The original manufacturer of the equipment was consulted, and the shrouds were ultimately modified to achieve this end (Figure 2).

The expected shroud stress was far below the maximum allowable stress level for the shroud material after this modification (Figure 3).

It was believed that shroud cracks occurred because of stress levels that were caused by condensate impingement on the shrouds and exceeded the allowable values. These events could be attributable to inadequate draining of the condensate through the drain holes in the diaphragms and casing. It was decided to enlarge the drain passage in the casing beneath the fourth-stage diaphragm. The original drain hole of 6-mm (0.236-in.) diameter was to be enlarged to 10 mm (0.394 in). To effect proper draining of condensate, removal of the shroud radial sealing fin in the fourth-stage diaphragm on the low-pressure side was also proposed (Figures 4a and 4b).

Rotor B was unchanged from the original design except for the aforementioned modifications. This rotor was installed in 1990. The modification to the casing drain hole was also carried out (this is described later in the article). The fourth-stage diaphragm was removed. Due to space constraints, instead of enlarging the drain hole in the casing, two 25-mm- (1-in.-) wide channels were cut in the bottom casing on either side of the existing drain hole. The second radial fin was also removed (sections b and c in Figure 4b).

The rotor performance was satisfactory until 1995, when vibration started increasing gradually. The vibration pattern was unsteady and phase changes were unpredictable (Figure 5). Vibration levels were watched closely and compared with bearing load so that Babbitt fatigue limits were not reached.

During a plant stoppage in 1995 due to a leak in the reformed gas boiler downstream of the secondary reformer, the turbine was opened for inspection. Rotor B was removed and rotor A was installed. When

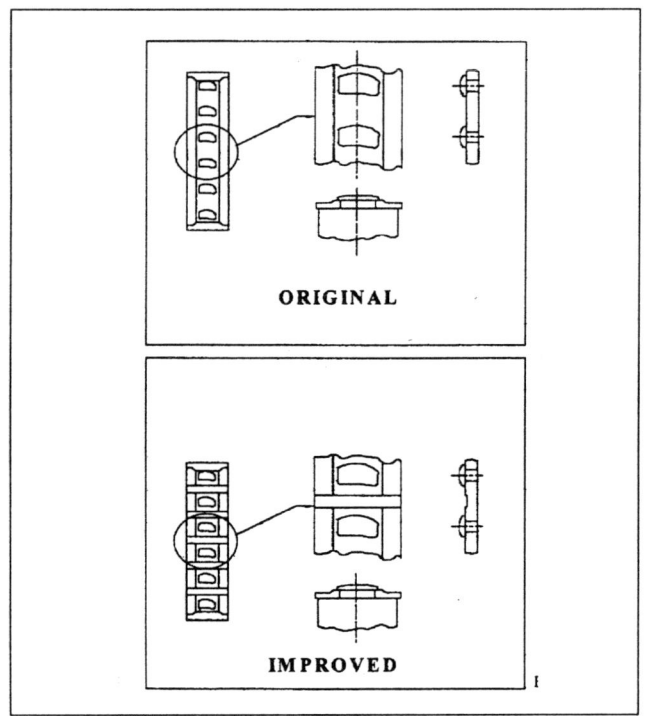

Figure 2. Shape of the fourth-stage shroud.

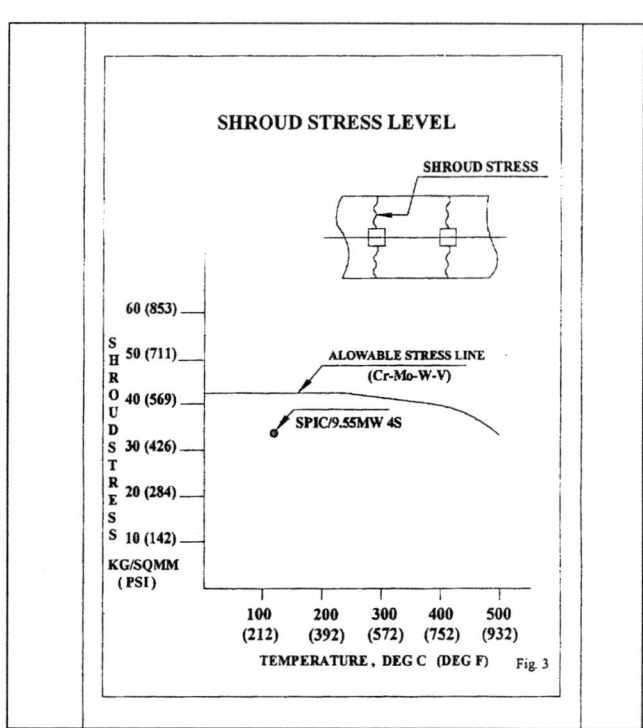

Figure 3. Shroud stress level.

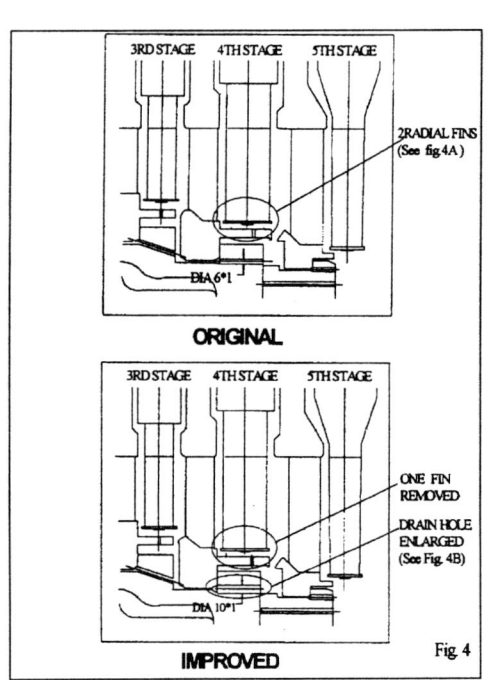

Figure 4a. The fourth-stage diaphragm and casing.

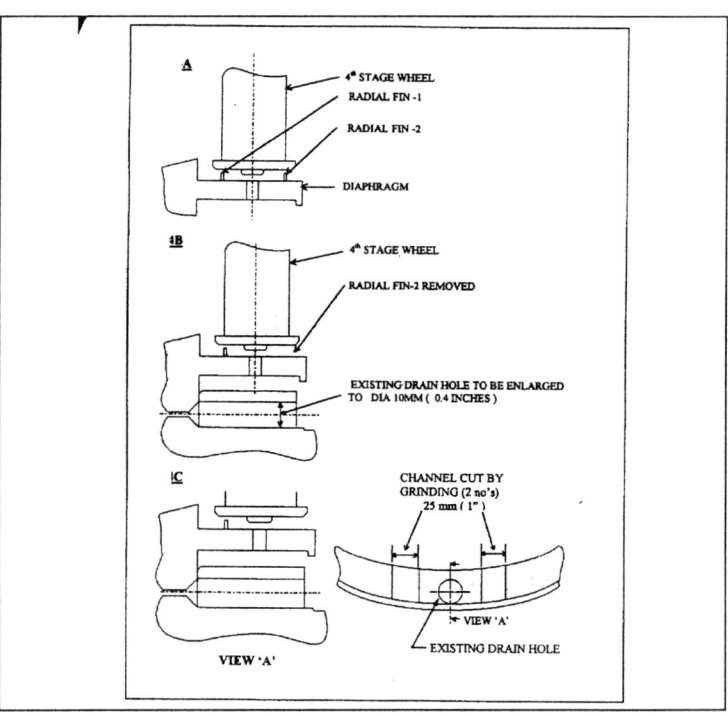

Figure 4b. Details of the fourth-stage diaphragm casing.

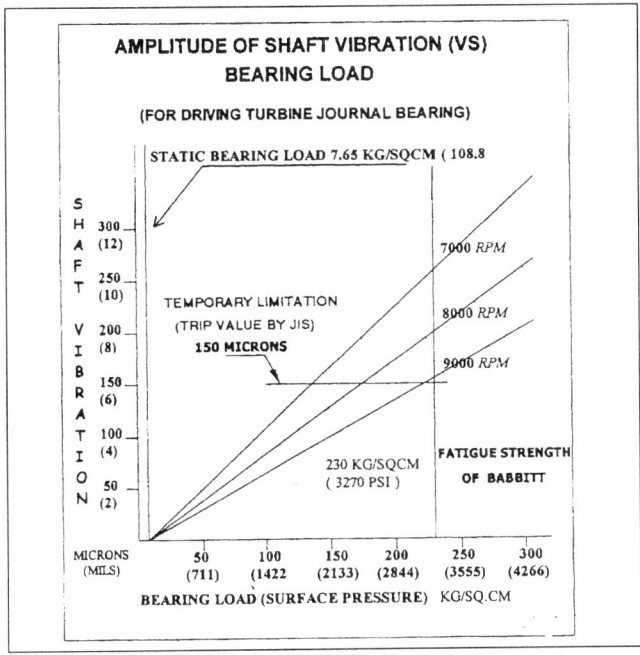

Figure 5. Amplitude of shaft vibration vs. bearing load

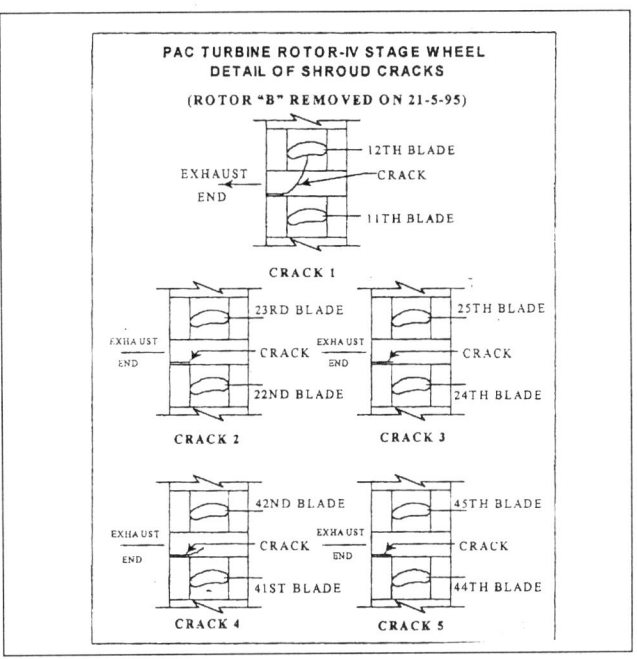

Figure 6. Detail of shroud cracks.

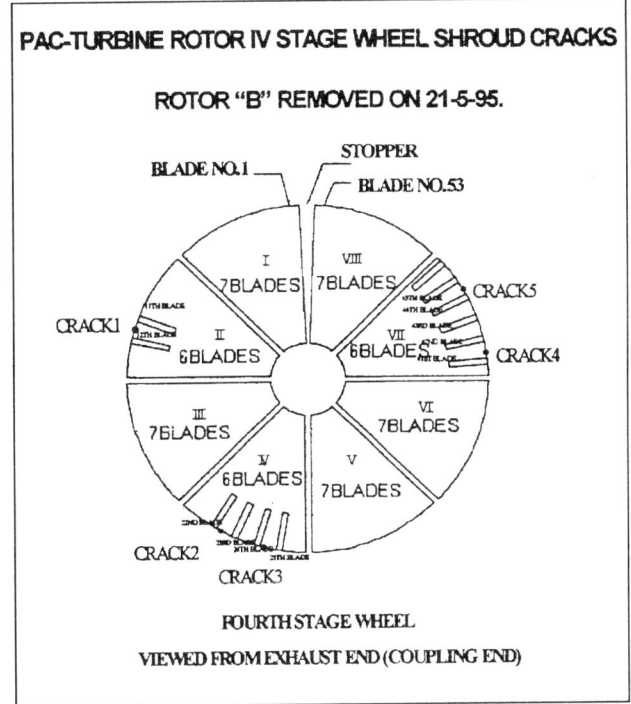

Figure 7. PAC-turbine rotor fourth-stage wheel shroud cracks.

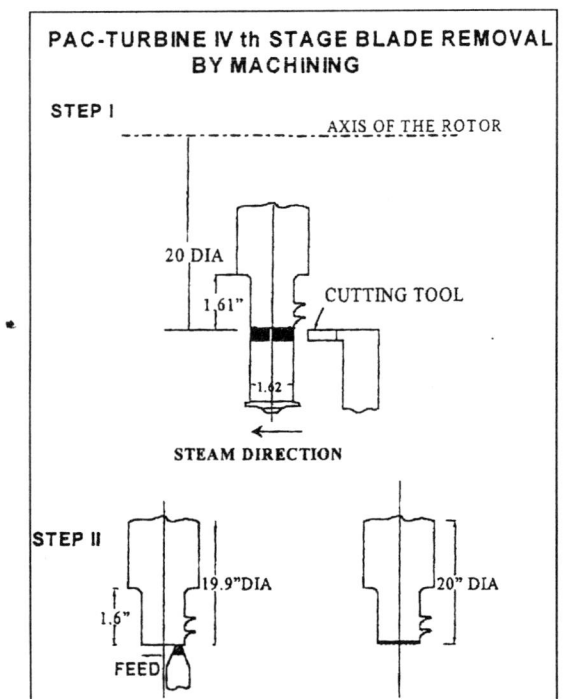

Figure 8. PAC-turbine fourth-stage blade removal by machining.

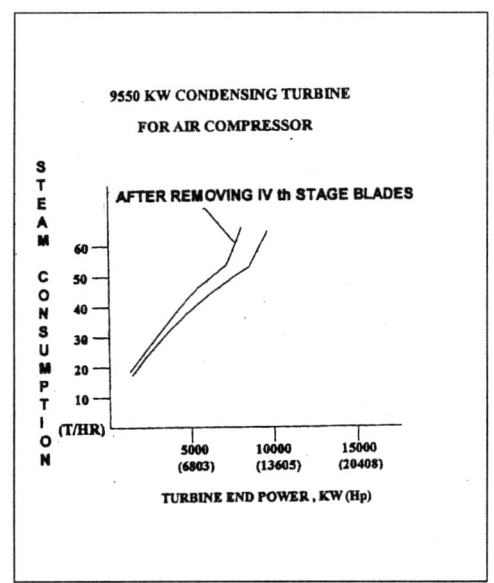

Figure 9. Steam consumption curve.

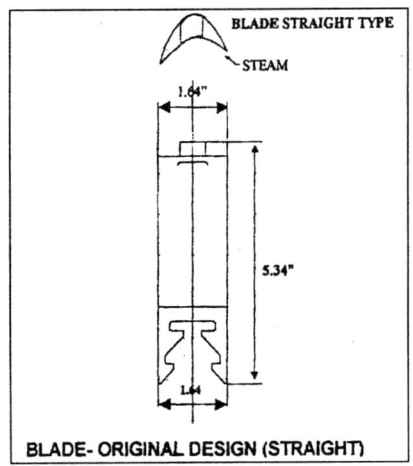

Figure 10. Original blade design.

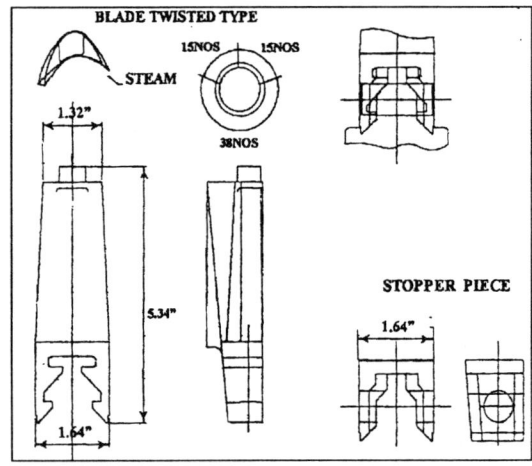

Figure 11. Improved Blade Design

Table 2. Revised Blade Design

SL. No.		Original Design	Improved Design
1	Profile type	Straight Laminar	Balanced Vortex
2	Axial width	Tip/Root: both 40.6 mm (both 1.62 in.)	Tip/root: 33.1/41.0 mm (1.32/1.64 in.)
3	No. of Blades	53 + 1	68 + 1
4	Root Diameter	499 mm (19.96 in.)	Same as original
5	Length	83 mm (3.32 in.)	Same as original
6	Material	12 Cr steel KT5302AS7E TS $\geq$ 689 N/mm^2 YS $\geq$ 482 N/mm^2	12 Cr Nb Steel KT5325BS12 TS $\geq$ 758 N/mm^2 YS $\geq$ 551 N/mm^2
7	Average bending stress on shroud cover	218 N/mm^2 (100%)	121 N/mm^2 (55.5%)
8	Expected stage efficiency	82.5% (Base)	85% (3% higher)
9	Type of blade	Straight type (Figure 10)	Twisted type (Figure 11)
10	Impact on steam consumption	Base	0.5% less

rotor B was inspected, it was found that the shrouds on the fourth-stage wheel had cracked at five different places (Figures 6 and 7). Apparently stress levels on the shroud were still above the limit of the material.

The manufacturer of the equipment was contacted and asked to redesign the fourth stage so that shroud stresses would be low enough to avoid failures. If redesign, procurement of new blades, and reblading of rotor B had been necessary, the plant would not have had a spare rotor for at least one year. It was therefore decided to cut away the fourth-stage blades on rotor B as shown in Figure 8. After dynamic balancing it served as an emergency spare.

The steam consumption is elevated by 20% because of the missing stage (Figure 9). If it had been necessary to install the rotor, reduction of the plant load would have ensued. Fortunately, installation was not necessary.

The fourth-stage blades were redesigned so that a 3%-higher stage efficiency was achieved, and the shrouds were redesigned so that the average bending stress was only 55.5% of that in the original shrouds. The original and modified design are compared in Table 2 and Figures 10 and 11.

The redesigned blades and shrouds were installed in rotor B, and this rotor was installed in 1997 to check the performance. The modified rotor has been performing well. During the inspection in July 1998 the rotor condition was satisfactory, leading to its reinstallation. It is now operating smoothly.

Conclusion

The condition of rotor A, which was removed in 1997, is satisfactory. This rotor is being kept as a standby at present. The previously discussed failures all occurred due to very high stress levels on the shrouds. The response to this was to modify both the shrouds and the blading to lessen the amount of stress. In order to prevent condensate impingement, draining arrangements should be checked for adequacy during the design stage itself.

DISCUSSION

Jack Poole, *Augusta Service Co., GA*: Your figure seven, as shown in your paper, suggests that there may be a natural frequency problem with the six-blade grouping. Instead of machining that row of blades off, did you consider changing the grouping?

Jayaraman: Yes, you can see from my last table of blade design that there was previously about 53 blades plus one. And in the improved design, it is 68 plus one. That means the regrouping is changed.

Poole: Thank you. However, to clarify, the sketch shows you only have cracks in a six-blade grouping, not in the seven. One possible solution would have been to change the six-blade grouping to three-blade grouping by cutting THE SHROUD BAND as an option without machining the row off. Was that something that OEM analyzed, and it was not an option?

Jayaraman: Yes, we have thought of that also, but we felt that it was better to remove the blades instead, as the blades, in other segments, will have the same problem. The whole set of segments of blades have undergone the same stress level during the years of operation. Even if we find a problem in one blade, we are prepared to change the entire lot. That is our philosophy. Based on this philosophy, when one segment has gone, before long, three segments will be having a problem. We thought it better to change the total set of blades. However, at that time, we did not have the blades. As we want to keep the rotor as a standby to meet any eventuality of the plant, we have removed the total fourth stage. We know that we are going to lose 20% efficiency. We have not used it, because we have no opportunity to use that rotor.

Poole: After you removed the fourth stage blading, did you consider the effect that would have on the fifth stage blading?

Jayaraman: No, because the fourth stage is the vulnerable stage where condensate is formed first. In the other stages, as you see, it is open to the condenser.

Dean Damin, *Dupont Engineering*: When you analyzed those cracks in the shroud bands, was there any component of environmental cracking associated with those cracks?

Jayaraman: No, it is purely mechanical. It is nothing else. It starts at the point where this turbine blade is fixed. That is the point where the shroud is highly stressed in the mechanical assembly.

Hussain A. Al-Hajari, *Saudi Arabian Fertilizer Co.*: You mentioned that you have a method to remove the silica deposits on the plate. Can you elaborate on that method?

Jayaraman: Presently, there is one method called grit blasting, using grit for blasting or the turbine blades to remove the silica deposits. Somebody uses shots, and, in our opinion, the grit or shots will slightly damage the blade material, by giving some dents. So, after giving it some thought, if the blasting material is soft, it will be better. Also, we have tried rice with a normal, sandblasting machine. This removes the silica deposit efficiently, and, at the same time, it doesn't damage the rotor blades.

Modifications After a Primary Reformer Explosion at a Reforming Plant

During the restart of a second reformer train in October 1997 after a factory shutdown, a powerful explosion occurred in the primary reformer firebox. Failure of equipment ensued that required total rebuilding of the primary reformer. Normally installed protection and control systems as well as operating procedures for startup conditions of a primary reformer are evaluated, and modifications implemented prior to restarting the remaining two reformer trains are presented.

Henry de Wet and Rudie O. Minnie
Mossgas (Pty.) LTD., Mossel Bay 6500, South Africa

Introduction

The Mossgas plant produces synthetic fuels from synthesis gas. The synthesis gas is produced by reforming natural gas in a three-train combined reforming plant utilizing both primary tubular steam reforming and oxygen-blown autothermal reforming. The front end of the plant layout incorporates a gas reforming unit that was designed according to the principle of the LURGI combined reforming process.

After completion of a scheduled biennial factory shutdown and during restart of a second reformer train in October 1997, a powerful explosion occurred in the firebox of the primary reformer. This resulted in extensive damage to the primary reformer and to the train adjacent to the reformer, which was fully operational at that time.

Gas reforming unit

The LURGI combined reforming process is well documented in the open literature and includes both primary tubular steam reforming and oxygen-blown secondary reforming. The novelty of this patented process is that a portion of the lean natural gas feed is bypassed around the tubular reformer directly to the oxygen-blown secondary reformer. In the case of Mossgas, the design provided for three identical trains that under normal running conditions can comfortably sustain 850,000 m_n^3/h (761 mmscfd) syngas to the synthol plant at a carbon monoxide concentration of 23 mol. % (dry) to the downstream units. This is equivalent to 2,500 metric t/d of methanol per train. Figure 1 shows a simplified diagram of the flow.

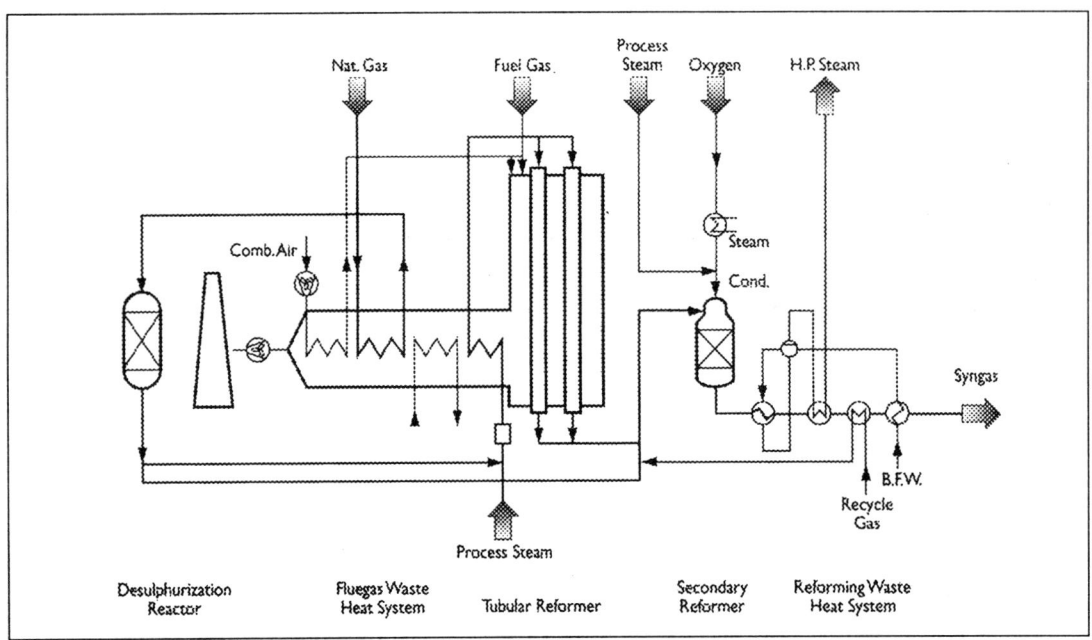

Figure 1. Flow diagram of the LURGI process.

Overview of the explosion

The entire refinery complex was shut down for its biennial maintenance inspection. Reformer train 1 was still in shutdown mode due to the replacement of two new wasteheat boilers as well as a new 110 bar (1,595 psig) steam superheater tube bundle. Reformer train 2 was restarted on October 20, 1997, and was in full operation at design capacity. Reformer train 3 was in the process of initial startup (dry-out of refractory, boil-out of new wasteheat boilers) on October 24, 1997, after completion of the shutdown.

During this startup period a powerful explosion occurred in the primary reformer that caused significant damage to equipment in reformer trains 2 and 3 as well as to the power generation unit. The whole refinery complex had to be shut down as a result of the incident.

The incident occurred in the early hours of the morning with very few people in close proximity of the 3 reformer trains. The two operators that were in the top part ("penthouse") of the primary reformer miraculously escaped serious injury and suffered only minor bruises and cuts. The three reformer trains had been successfully started up more than 100 times since 1992 without any problems. Problems with leaking wasteheat boilers, secondary reformer burners, and refractory problems were the main reason that all of these startups were necessary. (This was discussed in our article at the 1997 AIChE symposium [de Wet et al., 1998].)

Primary reformer fuel gas system

The purpose of this article is to highlight possible shortcomings in a fuel gas system of a primary (tubular) reformer and to share our operational experience, with specific emphasis on the following items:

• Primary reformer fuel gas burner management system.
• Comparison with other gas-firing furnaces.
• Modifications considered and implemented.
• Safety features to be considered for future plants.

Primary Reformer Fuel Gas Burner Management System

General layout of a primary reformer at Mossgas

The general layout of our primary reformer, including the combustion air system and the fuel gas system, is shown in Figure 2.

In summary, a primary reformer contains 84 top-firing burners (seven rows, each with 12 burners), seven flame-registering monitoring systems/infrared flame detectors (one in each row), 228 catalyst tubes per reformer, and both induced-draft as well as forced-draft fans.

Startup procedure of the primary reformer

The following preparations and checks are carried out prior to a startup:
• The main fuel gas isolation valves (UVs and PV) are closed.
• All combustion air dampers at burners are opened to 50%.
• Forced-draft and induced-draft fans are started and negative pressure is established of 3- to 5-mm water column vacuum (WC) with a combustion air flow rate of 40,000 mn³/h (36 mmscfd) minimum.
• After a nitrogen purge to air-free the system, the fuel gas headers downstream of the trip valves (UVs) are tested for tightness (leak test). This is to prevent the opening of the UVs with burner cock valves that are still open after a trip or during a cold startup.
• A "ready for leak test" is obtained from the local panel.
• The bypass around the UVs is opened as soon as a pressure of 50 kPa (7.3 psig) is reached in the fuel gas header, but within one minute the bypass closes.
• The gas pressure downstream of the UVs is monitored and the system fails the leak test if the pressure falls below 30 kPa (4.4 psig) within 1 minute.

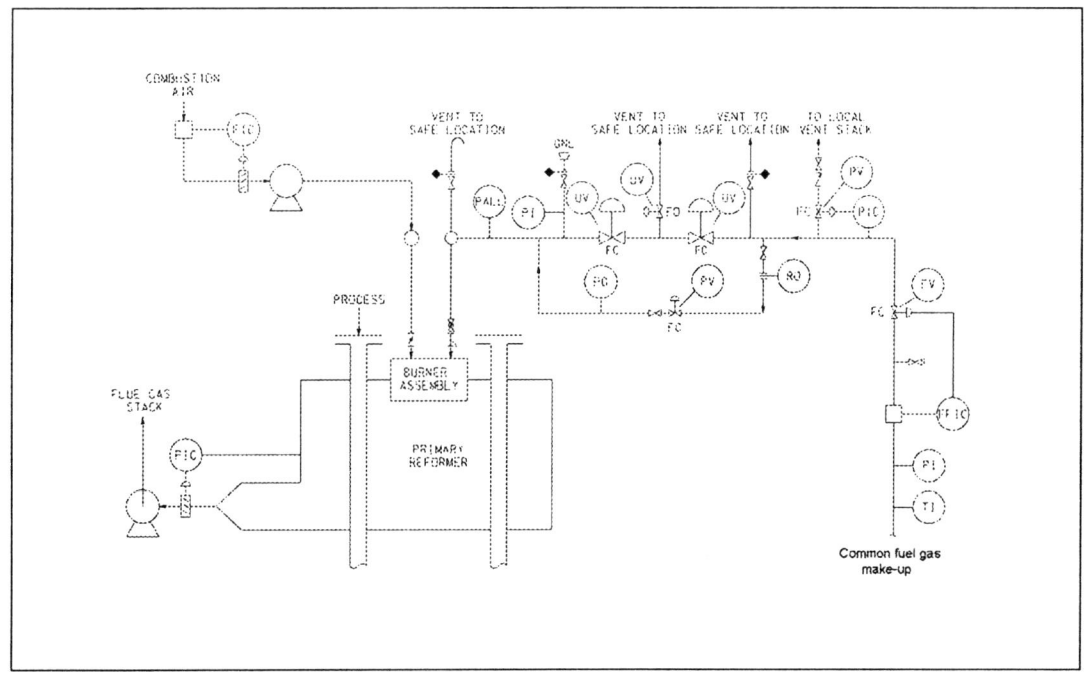

Figure 2. General layout of primary reformer fuel gas system.

- The operator then must recheck the system for tightness and repeat the leak test after 5 minutes.
- When the leak test has been passed the UVs are opened from the local panel.
- The seven burners with the flame detectors are then sequentially ignited with a portable ignitor.
- A four-out-of-seven voting system trips the reformer on loss of flame.

Comparison with Other Gas-Firing Furnaces

Typical layout of a gas-firing furnace

Figure 3 illustrates the general layout of a fuel gas system on a gas-firing refinery furnace. A gas-firing furnace typically includes the following:
- Five to ten bottom-firing burners.
- A pilot burner for each burner.
- Each pilot burner, as well as the common pilot burner fuel gas header, is equipped with a quick shut-off isolation valve.
- Each burner, as well as the common main burner fuel gas header, is equipped with a quick shut-off isolation valve.
- A fixed ignitor is installed at each pilot burner assembly.
- Flame registering monitoring systems and infrared/ionization flame detectors at each burner assembly.
- Natural draft with a stack to atmosphere, or central refinery stack.

Generic startup procedure of a gas-firing furnace

The following preparations and checks are carried out prior to a startup:
- All fuel- and pilot-gas burner isolation valves are closed.
- The stack damper of the furnace is opened.
- A steam purge is introduced into the furnace firebox to ensure that no hydrocarbons are present when the pilot burners are ignited and that there is

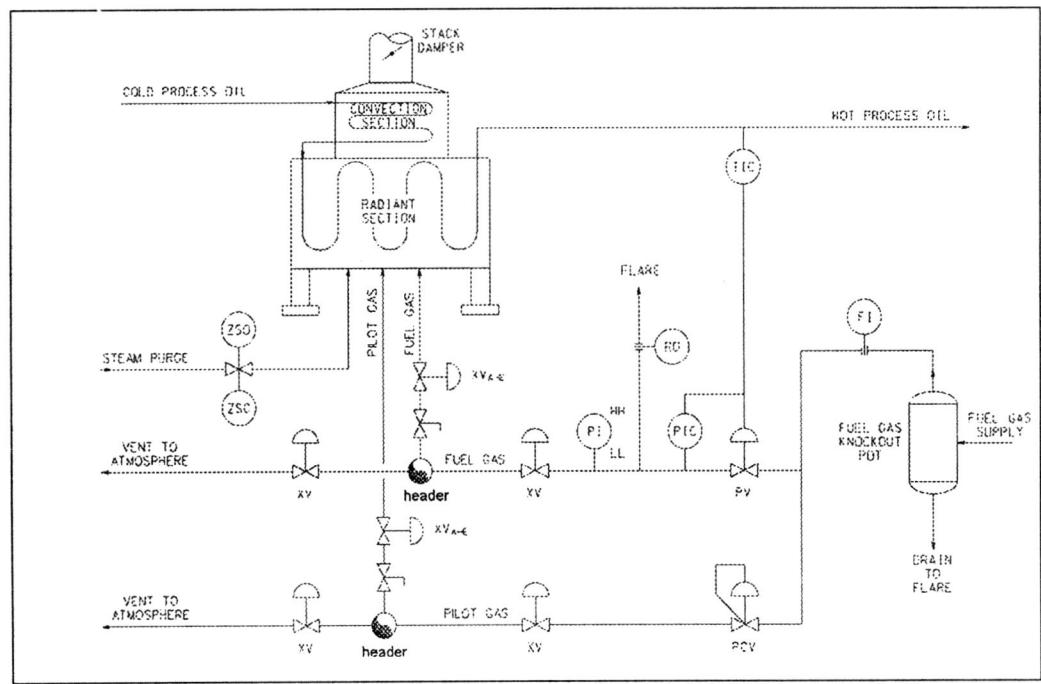

Figure 3. General layout of a fuel gas system on a gas-firing refinery furnace.

a draft through the firebox.
- The firebox is checked for any hydrocarbons with a normal portable gas detector before fuel gas is introduced to the pilot burners.
- After a successful fuel gas purge, the fuel gas piping system downstream of the fuel gas knock-out drum is vented to atmosphere. This is necessary to ensure that the pilot fuel gas header is nitrogen-free and ready to be commissioned.
- The main header quick closing valve of the pilot burners can be opened only after all of these conditions are satisfied.
- Furthermore, a remote panel is used to start the selected pilot burner.
- If the pilot burner flame detector does not detect a flame within five seconds, or if three attempts to ignite the burner fail, the quick closing valve will close and the furnace will trip.
- The furnace has to be restarted again, beginning with the steam purge.
- A predetermined number of pilot burners need to be healthy as a permissive to open the main burner fuel gas header quick-closing valve.

Modifications Implemented Since the Explosion

After investigating the cause of the primary reformer explosion, it was evident that the possibility of mixtures exploding in the firebox needed to be eliminated as much as practically possible. The modifications to the burner management system included instrumentation modifications, changes to the operating procedures, new improved portable ignitors, and additional supervision during commissioning.

Instrumentation modifications in the fuel gas system

Utilize the Fuel Gas Calorific Analyzer to Determine Maximum Fuel-Gas Pressure. The fuel-gas system is designed for several types and mixtures of fuel gas. Under normal operating conditions, fuel gas is supplied by both lean tail gas from the synthol process and off-gas from the PSA hydrogen purification unit. A lean natural gas (LNG) supplement is used to ensure a constant calorific value of at least 16 MJ/m_n^3 (463 Btu/scft). During startup and upset conditions, provision is made for syngas and rich natural gas as fuel gas, with calorific values of 11 and 39 MJ/m_n^3 (304 and 1,111 Btu/scft), respectively. The required fuel gas pressure at the burner therefore depends on the type of fuel gas used, and it is therefore clear that different pressure profiles should be used at the burner for different compositions of fuel gas. The correct pressure for different compositions of fuel gas can only be obtained from the burner performance curves. The normal operating pressure at the burner is 50 to 60 kPa (7.3 to 8.7 psig) (when firing lean tail gas) but during a cold startup with rich natural gas (RNG) this pressure should not exceed 10 kPa (1.5 psig). In order to prevent selection of an overly high fuel gas pressure for LNG/RNG during a cold startup, the aforementioned burner performance curves are utilized to calculate the correct fuel gas pressure, which is then shown on the DCS. This calculated pressure is now utilized by the panel operator for startup to ensure that the high explosive limit is not exceeded in the combustion chamber. A comparison between the range of fuel gases being used is shown in Table 1.

Modifications to the cold startup logic

The installed startup logic was changed extensively. After completion of the fuel gas header leak test, the logic did not provide for a time restriction to allow ignition of the first five burners. (The original design incorporated a time limit of thirty minutes to ignite at least five burners, but was removed during initial commissioning for the dry-out of the new refractory installation.) Furthermore, if the fuel gas system trips, both the induced and forced draft fans are stopped. These two issues were addressed by incorporating some

Table 1. Comparison of Different Fuel-Gases

	Design	LNG	RNG	Syngas
Lower Expl. Limits (Vol. %)	4.5	4.4	4.3	4.8
Higher Expl. Limits (Vol. %)	34.4	14.1	13.8	70.4
Lower Heating Value (MJ/m_n^3)	16.34	38.34	39.24	10.75
(BTU/scft)	*463*	*1,087*	*1,111*	*304*

modifications to the startup logic. The emergency shutdown (ESD) logic allows the operator only ten minutes to start at least one burner (with a flame eye) and thirty minutes to have at least five burners healthy. This timer starts as soon as the main fuel gas supply valves open. The fuel gas system trips if either of these conditions is not satisfied.

To ensure a constant sweeping of the firebox during cold startup, both the induced and forced draft fans are kept running at all times, whereas in the past these fans were also tripped by the logic. When a reformer is fully on-line and a trip occurs, the unit has to be restarted as soon as possible. If the induced- and forced-draft fans are then kept on-line, the firebox will cool down rapidly, creating unnecessary thermal shocks. A secondary reformer inlet temperature below 300°C (573 K/572°F) is regarded by the ESD logic as a "cold" start, and a temperature above 300°C (573 K/572°F) as a "hot" start. During a "hot" trip both fans will be stopped.

Improved manpower and supervision for startup

A single operator performs the following tasks:
• Frees the fuel-gas header from air downstream of the isolation valve with nitrogen by utilizing the utility station in the "penthouse".
• Ensures that all the burner cock valves are closed.
• Performs the fuel-gas header leak test and ignites the first burners.

These tasks would take that operator about two hours to complete. Since the incident, two operators have been used to perform these tasks. The shift supervisor controls the ignition operation.

Safety Features to be Considered for Future Tubular Reformers

This incident has not only resulted in extensive damage to this reforming train, but also has had a significant impact on the production capacity of the whole plant, resulting in substantial financial losses. The incident occurred in the early hours of the morning when very few people were around. Regardless of what is considered as the "norm" for the reforming industry, a lot can be learned from the oil refining industry. Their fuel gas and burner-management systems may appear as "going overboard," but can be considered safer than the "norm" in the traditional tubular reforming industry. A traditionally designed system is regarded as "sufficient for a knowledgeable operator adhering to procedures to safely commission and operate such a unit".

Features to be considered for a new plant design include the following.
• There should be at least one pilot burner per row of burners.
• Each pilot burner should have its own dedicated ignitor and flame detector.
• Each pilot burner should have a dedicated quick shut-off valve.
• A dedicated pilot burner fuel gas header with

common quick shut-off valve.
- Automated purging (air-free) and pressurization of the fuel-gas header with fuel gas (leak test).
- The pilot fuel-gas flow should be kept separate from the explosive mixture in the firebox.
- The characteristics of the different fuel gases should be incorporated into the burner management system.
- The burner cock valve handles positioned to "up" to open.
- Institution of facilities to monitor the explosive limit of flue gas prior to ignition of first burner.

Some of these elements are already included in the current design of the running plant. The intention is to implement the remaining features during the next biennial factory shutdown.

Conclusion

At Mossgas, we have experienced how dangerous tubular reforming operations can be and what dangers are inherent in a fuel gas system. The message that we would like to convey to operations in the reforming industry is to review the safety systems around your tubular reformers. Modify the systems if you are not absolutely convinced of their integrity.

Acknowledgment

The authors express a special word of thanks to their colleagues at Mossgas who have either directly or indirectly contributed to the contents of this article.

Literature Cited

De Wet H., R. Minnie, and A. J. Davids, "Post-Commissioning Operating Experience at the Mossgas Reforming Plant," *Ammonia Plant Safety & Related Facilities*, Vol. 38, AIChE, New York (1998).

Murdoch R., and K. Still, "Elimination of Carbon Formation in Secondary Reformer System," *Ammonia Plant Safety & Related Facilities*, Vol. 36, AIChE, New York (1996).

Shaw G., H. de Wet, and F. Hohmann, "Commissioning of the World's Largest Oxygen Blown Secondary Reformers," *Ammonia Plant Safety & Related Facilities*, Vol. 35, AIChE, New York (1995).

DISCUSSION

Rudy Frey, *M.W. Kellogg*: This is a very interesting paper with very unfortunate results. I'd like to address a few items that you discussed. Number one, with regard to steam purging your furnace boxes, we've eliminated that in all of our refinery designs at least ten years ago due to condensation corrosion problems that this presents. Consequently, we wouldn't find that in any of our fired heaters. The IDFD fan system is far superior. And, if you're timing to a number of air volume changes, I'm sure that you're going to have a safe procedure for lighting off that furnace. The concern that I had with regard to your recommendations is the increased complexity that you're adding to the furnace firing system, and the operator confusion that may result from this. Also, as soon as you have a good pilot system going in a furnace, I don't see where you have any particular hazard. I don't have any difficulty with the pressure compensation for fuel calorific value. However, we found in trying to adjust furnaces for calorific value that you have to ensure that you have correct dynamic response from your measurements. The most unsafe feature that I see in your original design was a bypass around the UVs, which you use for your original testing. Also, I feel that has been a hazard for whoever prefers to test with fuel gas. We always insist that you make this pressure test to ensure the cocks are closed with nitrogen. I did feel that some of these things may be somewhat overstated, and I feel that our furnaces are adequately protected. Thank you.

Minnie: I would like to answer you on the steam

purge of our refinery furnaces. At Mossgas, we had similar problems with wet steam purges, which prolonged the commissioning exercises of the refinery furnaces after a shutdown (wet ignitors, wet pilot burners, etc.). Furthermore, we are now in the process of changing all the steam purges on the refinery furnaces with nitrogen purges. Nitrogen is not a concern on our complex, because we have the biggest air separation units in the world on-site, which means that there is always excess nitrogen available. The second topic that I would like to comment on is the complexity of our proposed pilot fuel-gas system. We know that the pilot fuel-gas system can work, which we have proved on the refinery furnaces. Furthermore, if you consider the initial capital layout for a pilot fuel-gas system and compare it with the production loss incurred if your reformer is out of commission for 400 days, I think it is a small price to pay for a big insurance.

Ammonia Process Primary Waste Heat Boiler Shell Failure Experiences

Three cases of pressure shell failures during 20+ years of operation are described. The results of the failure analysis following each event are also presented, as well as a description of the repair methods employed. Although the nature of the failures were not all similar, some common elements surfaced as a result of the failure investigations.

Colin P. Jackson
Canadian Fertilizers Ltd., Medicine Hat, Alberta, Canada

Introduction

The Primary Waste Heat Boilers in an MW Kellogg ammonia plant were first constructed to the subject design in the mid-1960s. It is estimated that as many as 300 waste heat boilers of similar design may be in operation around the world. Although generally reliable, the operating experience over the ensuing 30 plus years has resulted in a number of failures of varying types. Due to the nature of the process and materials involved, any incident has the potential for serious equipment damage and injury to personnel. The information presented in this paper is largely based on the operating experience of CF Industries Inc. owned and managed facilities in the U.S. and Canada. For the most part, the material is comprised of extracts from the failure investigations conducted by others at the time of the related events.

Background

CF Industries Inc. (CFII) operates two 1,000 T/D (900 tonnes/day) and two 1,150 T/D (1,025 tonnes/day) ammonia trains in Donaldsonville, LA and two 1,150 T/D (1,025 tonnes/day) units at the Canadian Fertilizers Limited (CFL) site in Medicine Hat, Alberta, Canada. All of these units have been substantially debottlenecked in several stages to current capacities, approximately 140% of original design. The primary waste heat boilers were not changed as a part of the debottlenecking projects.

The MW Kellogg Steam / Methane reforming operation and its attendant waste heat recovery boilers are the first processing stage in the preparation of a hydrogen rich synthesis gas for the production of anhydrous ammonia.

Hydrogen is typically produced from natural gas and steam through the reforming reaction in two stages. A tubular fired primary reforming furnace is followed by a secondary reforming reactor vessel. Air, injected at the inlet of the secondary reformer, consumes a portion of the primary reformer product to produce the heat required to reform the balance of the methane to a residual level of about 0.2%. The reformed gas exits the secondary reformer at approximately 1,800°F (982°C) and passes through parallel transfer lines to two vertical bayonet style, waste heat boilers. (Figure 1)

These boilers produce 1,500 psig (10,300 KPa) steam and are a major source of power for the steam turbine driven compression equipment required for process operation. The primary components of the process gas stream at this point are hydrogen, nitrogen, carbon dioxide, carbon monoxide and steam.

Waste Heat Boiler Design

The extreme temperatures and high reactivity of the hot reformed gas stream requires some special consideration of design and materials for reliable operation. Figure 2 illustrates the construction of the pressure containing the shell of the boiler and it's design features to compensate for the extreme operating conditions presented by the process.

The pressure shell is constructed of carbon steel plate (SA 516- Gr. 70) and has additional thickness in the upper and lower portions to provide the necessary reinforcement for the inlet and outlet nozzles. The pressure shell is enclosed in an atmospheric pressure water jacket which was originally provided with level controls and operates with a supply of steam condensate as the cooling medium.

The inside of the vessel shell is lined with 5 in (13 cm) of bubbled alumina refractory. The refractory is shielded from the erosive action of the flowing process gas by stainless steel shrouding. The nozzles are constructed in similar fashion.

The weight of the vessel is supported by a "skirt" style base set on pedestals on a concrete foundation. The downstream vessel is also a waste heat boiler which is supported on spring supports and is connected to the primary waste heat boiler via approximately four feet of welded transfer line. (See Figure 3 for the general structural arrangement.)

Operating Experience/Design Performance

The greatest majority of the operating experience with this design waste heat boiler has been favorable.

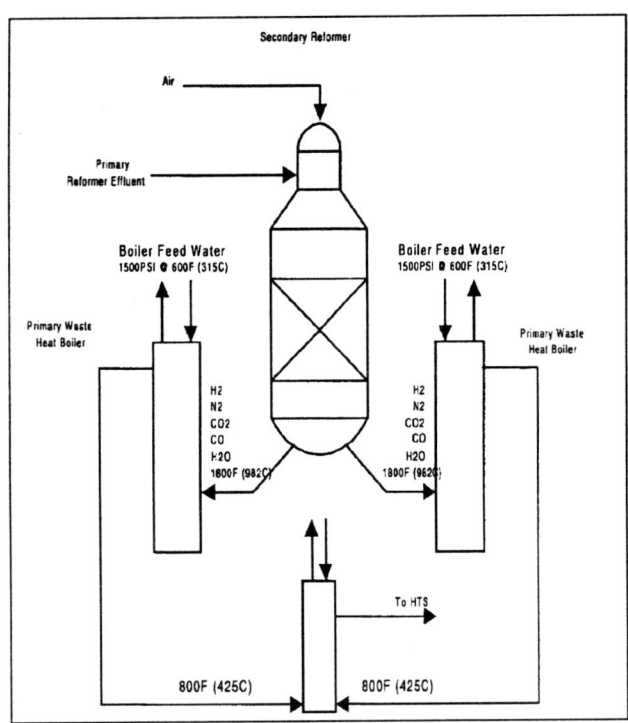

Figure 1. Process flow.

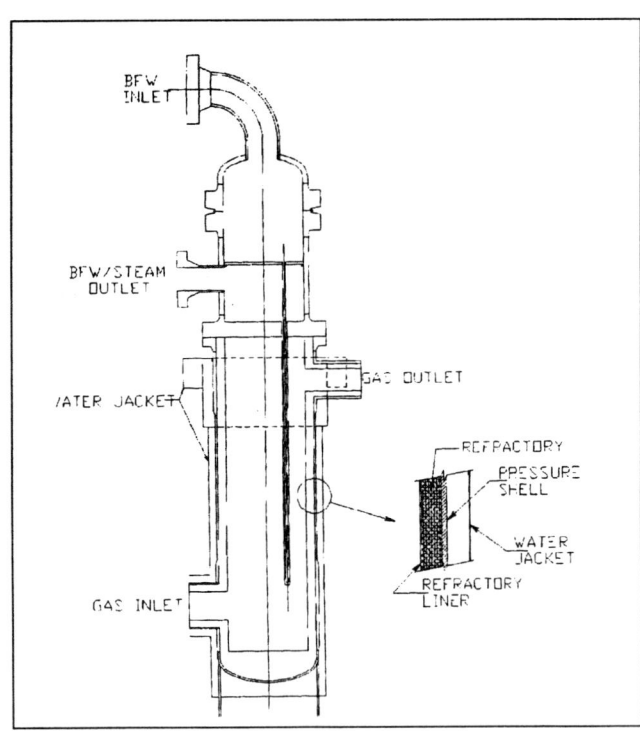

Figure 2. Waste heat boiler shell design features.

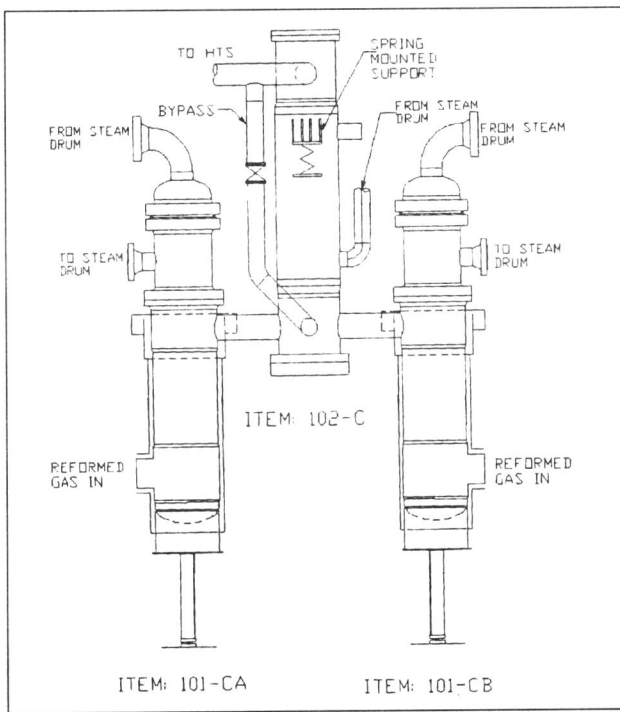

Figure 3. General arrangement of vessel attachments and supports.

Operators recognized early in the life of the plant design that the integrity of the shell temperature protection was critical to the safe operation of the plant. The general reliability of the design is supported by the finding that the majority of operators have had only minor shell failures or none at all.

Early operating experience indicated the water jacket level control system was in need of improvement as the turbulence due to boiling in the jacket was sufficient to produce repeated activation of the jacket water level alarms. A number of solutions have been implemented. At the CFL facility the level control was replaced with a flow control and the jackets are operated with an overflow of approximately 60 gpm (230 lpm) shell. The low level alarm remains in operation as further verification that water remains in the jackets.

Annual scheduled plant maintenance turnarounds were the rule through the mid 1980s. The shell insulation of the waste heat boilers together with their inlet and outlet lines was a priority for inspection during each opportunity. Most often these inspections revealed some distortion of the internal shroud material but only minor deterioration of the visible portion of the refractory. Through the history of these inspections, areas of crumbled refractory and voids as large as 1 cubic ft (1,600 cc) have been infrequently found. Failures of the boiler tubes nearly always resulted in significant thermal distortion of the shroud material. The bundle often became lodged in the shell and its removal resulted in added damage to both liner and the tube bundle itself. In more recent times the reliability of plant equipment has increased run cycle length to two to four years. Although there are currently fewer inspections, recent inspections indicate that the integrity of the shell protection is much as noted earlier. The one notable exception to this observed performance is the frequency of shroud deterioration due to Metal Dusting, (Shibaski et al., 1996). Metal Dusting related deterioration of the shroud material has been noted almost entirely in the upper 10 ft (3 m) of the vessel and has ranged from minor to near complete disappearance of the shroud in this area (Figures 4a and 4b). The refractory has generally remained intact where the shroud loss occurred. The CF facili-

Figure 4a. Complete disappearance of a shroud area due to metal dusting.

Figure 4b. Typical inner shroud deterioration due to metal dusting.

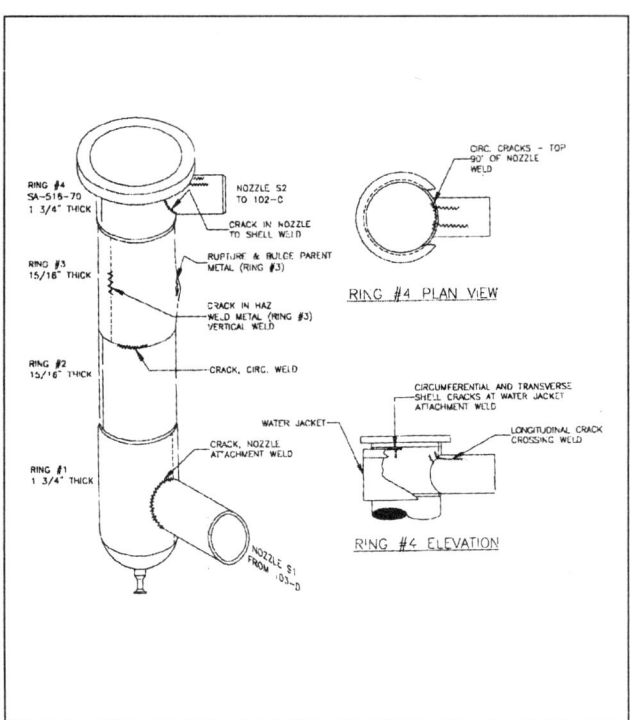

Figure 5. Composite diagram of typical vessel shell failure locations.

ties have installed Inconel 601 liners in some applications. To date this material has performed well.

The liners in the inlet lines and lower portion of the vessel shell deteriorate due to carburization. The service life is generally 5–10 years.

Although the general operating experience with this design has been satisfactory, there have been shell failures. Undetected failures of the shell refractory insulation and/or the failure of the water jacket cooling system together with manufacturing defects, are the most often cited root causes of the failures. In each case the issues of plant safety and that of access to assure suitability for service present an unresolved dilemma.

Failure Experiences

Figure 5 is a composite diagram illustrating the typical locations and types of shell failures that have been experienced. These fit into two general classifications; shell plate cracks and ruptures, and shell nozzle attachment weld cracks. Table 1 contains a table summarizing the experiences of some other North American facilities. The facilities surveyed involve approximately twenty-four applications of similar vessel design. The aggregate of failures consists of nine cases of nozzle weld attachment cracks and six shell cracks.

CF Failure Experiences

(1) CFL No. 2 Plant, 1978, shell failure east vessel, due to a vertical rupture of shell coarse 2 caused by hydrogen accumulation in slag contaminated shell plate. The vessel had been in service 2 years.

(2) CFL No. 2 Plant, 1986, shell failure east vessel, due to vertical rupture shell coarse 3 caused by hydrogen attack (extended metal exposure minimum temperature 650°F (340°C) but below 1,000°F (537°C) as no creep was evident). The vessel had been in service 10 years.

(3) CFL No. 2 Plant, 1996, gas outlet nozzle West and East vessels partially penetrating crack top quadrant due to cyclic mechanical and thermal stress shallow cracking of vessel shell at water jacket. Upper

Table 1. Industry Survey of Waste Heat Boiler Shell Failure Experience

Facility	Material of Construction	Failure Type	Location	Cause	Age at Time of Failure	Plant Capacity	Year Built
1		Cracks	Both nozzles all around	Operations upset. Overheat. Slag in plate and weld defects	11 yr.	1,150	76
2	SA 212 gr. B	Bulge, Crack	15/16 in (2.3 cm) Shell section	Loss of refractory suspected	5 yrs	1000	66
3	N/A	None	-	......			75, 77
4	SA 516-70	Transverse cracks	S2 nozzle weld	Thermal fatigue due to the fluctuating water level in the jacket	15 yrs	1,150	77
4		Circumferential and transverse cracks, partial and full penetration.	Upper flange attachment weld/ joint to jacket. support ring	Thermal fatigue due to the fluctuating water level in the jacket.	15 yrs		77
4	SA 516-70	Circumferential and transverse cracks; Partial penetration.	Upper flange attachment weld/ joint to jacket	Thermal fatigue due to the fluctuating water level in the jacket.	15 yrs	1,150	77
4		No Shell Failures, Inspected by removing the lining at last T/A no			15 yrs	1,000	68
5		Weld crack	Outlet nozzle	Original assembly; Stress Induced			

attachment weld due to thermal cyclic stress. The vessels had been in service 20 years.

(4) CFII No. 1 Plant, 1978, gas outlet nozzle south vessel, a full penetrating crack and several shallow cracks in the nozzle attachment weld. Thermal or mechanical stress was the probable cause. The vessel had been in service 11 years.

(5) CFII No. 1 Plant, 1986, gas outlet nozzle south vessel, full penetration crack initiating near top of pipe at vessel weld, caused by lack of a post weld heat treatment of a prior repair in an area of cyclic stress. Shallow cracking in adjacent vessel wall. The vessel had been in service 20 years.

(6) CFII No. 1 Plant, 1986, outlet gas nozzle north vessel, shallow cracks in vessel wall near top of nozzle; probable cause was cyclic stress as above. The vessel had been in service 20 years.

(7) CFII No. 4 Plant, 1995, inlet nozzle attachment weld failure south vessel, partially penetrating circumferential cracks to the mid wall area in the attachment weld of the inlet nozzle. Cause was attributed to a shroud failure, refractory loss and manufacturing defects in the original weld. The vessel had been in service 19 years.

(8) CFII No. 3 Plant, 1997, shell failure south vessel due to vertical rupture in shell coarse 3, caused by creep due to extended exposure in the temperature range 900–1150°F (480–621°C). No evidence of hydrogen attack.

• Gas outlet nozzle partial through crack due to cyclic mechanical and/or thermal stress.

• The vessel had been in service 21 years.

Case Histories

Case 1: Canadian Fertilizers Ltd, November 1986.

The Incident. At approximately 10:45 p.m. on November 13, 1986, the pressure shell of the east Primary Waste Heat Boiler in the No. 2 Ammonia Plant failed, shutting down plant operation. Operators on shift at the time reported hearing a series of three loud explosion like sounds and observed a large cloud of fire and steam to the south and east of the east primary waste heat boiler location. A loud roar was also

Figure 6a. Elevation looking west at rapture site, extensive insulation & minor piping damage (Wagner, 1986).

noted following the explosions. The feed and fuel gas were immediately isolated from the plant and the normal process vents were opened to depressure the plant via the vent stacks.

An initial inspection, following the depressurization of the plant, revealed a large amount of insulation had been stripped from the piping in the area and all normal access ways were littered with debris. More comprehensive inspection was delayed until the following morning as the area lighting was out of service.

In addition to the shell damage, an electrical conduit tray was severely damaged by the resulting fire. There was minor piping and instrumentation damage, and a large area where the pipe insulation was stripped from the lines. (Figures 6a and 6b.)

The failure occurred in the parent metal of the upper 15/16 in. (2.4 cm) thick section of the pressure shell (ring 3) and was recognized as a vertical thick lip rupture approximately 30 in. (76 cm) long with a 3 1/4 in. (8.25 cm) opening at the widest point. The ruptured material protruded outward approximately 4 in. (10 cm). One end of the split extended to the circumferential weld joining the two 15/16 in. (2.4 cm) courses and although there was a crack in the circumferential weld, the ruptured split did not link up with the circumferential weld crack (Figures 7a, 7b, and 7c)

Figure 6b. Plan view - east of rapture location, damaged electrical conduit and pipe insulation (CFL, 1986).

(Radian, 1987).

The Cause. Samples of the failed vessel shell material were sent to Radian Corporation (now Mechanical and Materials Engineering) for failure analysis.

The shell section showed a severe bulge in the rupture area which initially indicated the shell had been locally overheated. The area surrounding the rupture showed signs of secondary cracking which indicated the possibility of hydrogen damage. An examination was done using straight beam ultrasonic which indicated the damage *was localized to the area surrounding the rupture.*

Figure 8 shows a polished and etched photomicrograph of the secondary cracking near the rupture lip. The cracking is intergranular in nature and most of the pearlite is missing from the microstructure as a result of decarburization. The higher magnification photomicrographs shown in Figure 9 show the microstructure of the shell at the rupture and the intergranular cracking is associated with the pearlite breakdown.

Microcracking and the breakdown of pearlite by decarburization is the classic manifestation of high temperature hydrogen attack. The microcracking is caused when hydrogen diffuses into the steel at temperatures above 440°F (225°C). The hydrogen reacts with the carbon present in solid solution or in the form of iron carbide to form methane. The methane accu-

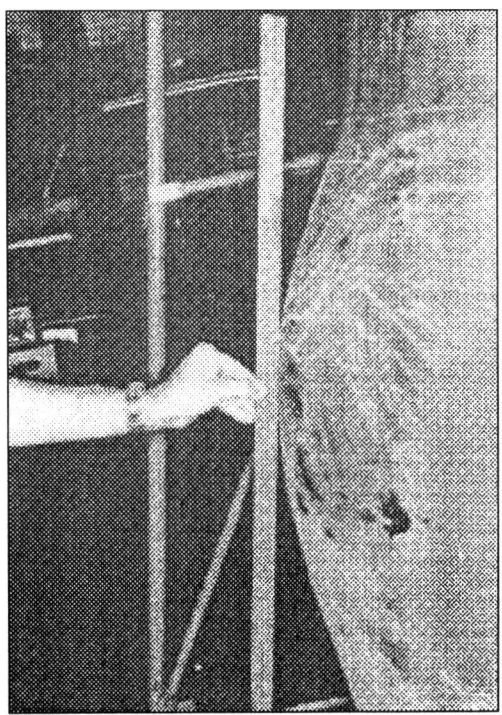

Figure 7a. Bulged shell section, ring 3 (CFL, 1986).

Figure 7b. Rapture damaged shell section (CFL, 1986) 30 in. long x 3 1/4 in. wide x 4 in. (82 cm x 8cm x 10 cm) protrusion.

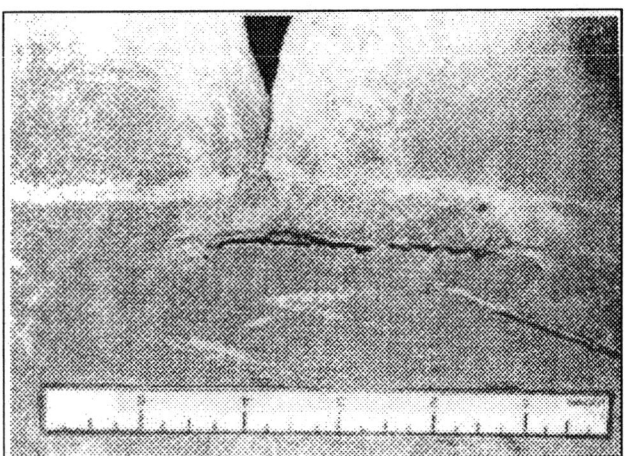

Figure 7c. Lower end of rapture showing crack in circumferential weld (CFL 1986).

mulates in the microstructure until it has reached sufficient pressure to form microcracks. In the advanced stages of attack, the large number of microcracks in the microstructure causes a pronounced reduction in mechanical properties (strength) of the steel. The microcracking also accounts for the indication of microstructural degradation detected ultrasonically.

The vessel operates at approximately its design pressure of 485 psig (3,300 KPa). The partial pressure of 35% hydrogen in the process gas would be approximately 175 psia (1,200 KPa). Figure 10 shows the Nelson curves for the time for incipient attack of carbon steel in hydrogen service. For a hydrogen partial pressure of 175 psia (1,200 KPa), the Nelson curves show exposure to temperatures above 650°F (340°C) for 10,000 h is required for hydrogen attack to occur. The Nelson curves also show at low hydrogen partial pressures, the exposure time necessary to produce hydrogen attack does not decrease with increasing temperatures above 650°F (340°C).

At the design pressure of 485 psig (3,300 KPa), the hoop stress in the shell is approximately 13,200 psi (964.51 Kg/cm^2). If the bulging exhibited by the shell were caused by long term exposure to elevated temperatures only, the temperature necessary to cause the shell to fail in 10,000 h is approximately 900°F (480°C) (Table 2). At a temperature of approximately 830°F (443°C), the shell would be expected to fail after 100,000 h, and at 650°F (340°C) the shell is expected to have an almost infinite life. Given the fact

Figure 8. Secondary cracking near rupture lip is intergranular in nature.
Pearlite is almost completely absent from the microstructure (CFL, 1986). Magnification 50x, etching Nital.

the shell is exposed to both a hydrogen partial pressure of 175 psia (1,200 KPa) and a hoop stress of 13,200 psi (964.51 Kg/cm^2), the shell will suffer hydrogen attack at 650°F (340°C) without any degradation of the microstructure as a result of temperature effects.

The above analysis indicates a loss of the insulating ability or the cooling of the shell, need only produce a hot spot of approximately 650°F (340°C) to produce a similar failure.

Conclusion. The analysis of the bulged and ruptured shell revealed that the rupture was a result of hydrogen attack. The presence of hydrogen attack indicated the shell must have been at a temperature in excess of 650°F (340°C) (but not greater than 900°F {480°C}) for at least 10,000 h.

The hydrogen attack was confined to the area of the rupture. A sufficient failure of the shell thermal protection must have occurred to allow the metal temperature to reach 650°F (340°C). It is unlikely a failure of the water supply *and* low level warning system could

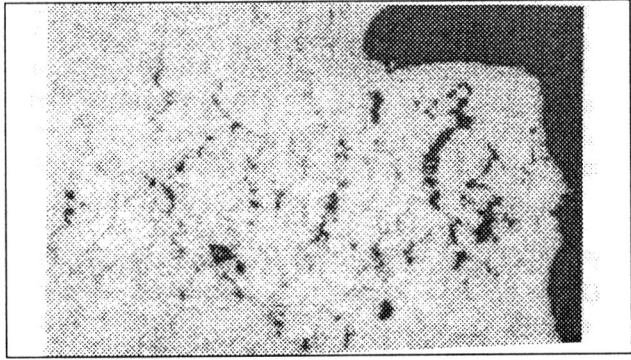

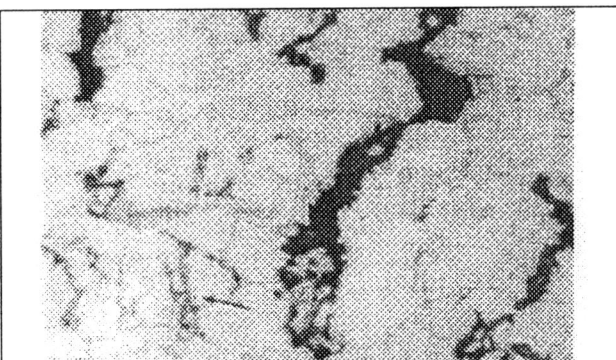

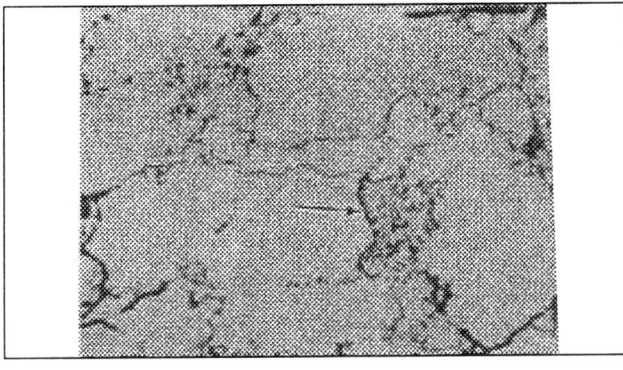

Figure 9. The microcracking associated with the breakdown of pearlite colonies, as indicated by the arrow, is a result of hydrogen attack.
Magnification (a) 100x; (b) 400x; (c) 1,000x.

have persisted for this period. It was therefore further concluded that loss of the refractory in a localized area caused heating of the shell material. Sufficient "steam blanketing" of the heated metal surface must have occurred to allow the metal to achieve the noted temperature, in spite of the continuing presence of water in the jackets. The cooling effect of the water/steam was sufficient to prevent the failure of the shell due to metal creep or thermal softening (900°F+, 480°C+).

The Repair. A previous shell failure was known to have occurred in this vessel in 1978. That failure occurred along a longitudinal weld of shell section 2. It was also attributed to a hydrogen induced failure involving weld defects and a slag contaminated section of shell plate. The affected area was repaired with a "flush patch". There was no detailed information available from this prior incident at this writing.

In light of this previous experience the repair method chosen involved the complete replacement of shell sections two and three (the two 15/16 in. (2.4 cm) sections). The upper portion of the vessel was removed from the plant to a fabrication shop where the two shell sections were replaced with available 1 in. (2.5cm) material of the same composition (SA 516 Gr. 70). The refabricated upper portion of the vessel was returned to the plant, installed and welded to the lower section by plant maintenance personnel. New shroud sections were fabricated offsite and installed in the field, together with the insulating refractory. The refractory was cured and the waste heat boiler was reassembled ready for startup December 5, 1987. The vessel has operated without incident since that time. The repair procedure resulted in 22 days lost production.

Case 2 - CF Industries Inc., November 1986

The Incident. Following the shell failure in Medicine Hat, a survey was conducted of the water jacket vents on the operating vessels at the CFII facility in Louisiana. An explosive gas meter was employed to test for the presence of process gas. Combustible gas was detected above the outlet gas nozzle of the south vessel in the No. 1 Ammonia Plant. An orderly shutdown of the plant was initiated. The plant was ready for repair at 07:00 November 22, 1986.

The Cause. A pneumatic test of the vessel revealed a leak in the nozzle attachment weld at the top of the outlet nozzle. The water jacket in the area of the failure was removed to gain access to the entire nozzle to vessel weld. UT inspection revealed a number of cracks in the upper quadrant (90°) of the nozzle circumference that extended longitudinally into the parent material of the nozzle. Some small cracks were also detected which extended across the weld into the vessel shell material. Two shallow cracks were identified in a similar location in the vessel wall of the adjacent north positioned boiler shell. The nozzle material of construction was A357 (5 Cr. 0.5 Mo.). These vessels had a history of cracking in the upper portion of the nozzle weld area approximately 8 years previously. As there was no record of post weld heat treatment following the previous repair, it was concluded this was a factor contributing to the recurrent cracking in the area of mechanical and thermal cyclic stress.

The Repair. The repair procedure was developed in conjunction with a consulting metallurgist and the Harford Insurance, Authorized Inspector responsible for this facility. The original material of construction was no longer available; it has been replaced by A387. The chosen repair method involved removal of the damaged upper portion of the nozzle and replacement with a section of the A387 material which was prepared to fit. All cutting and welding operations were preceded by a "hydrogen bake out". The repair area was preheated to 350°F (177°C) prior to welding and post weld heat treated following the completion of welding.

The cracks, which extended into the vessel wall, were removed by grinding and repaired by welding with Inco A filler material. The plant was restarted following the repair, with approximately 5 days lost production.

Case 3 - CF Industries Inc. Dec. 1997

The Incident. On December 1, 1997, the pressure shell of the south waste heat boiler in No. 3 Ammonia Plant failed, resulting in a fire in addition to the damage to the boiler shell. The plant appeared to be operating normally and had started up normally after both an extensive September turnaround and a short October outage. The tube bundle had been replaced on the September turnaround. During the September turnaround, a routine inspection found no damage to the internal liner and hence no reason to suspect any refractory failure in this area.

On November 30, 1997, the routine explosive test of the water jacket was negative. On the evening of December 1, the routine explosive test indicated positive. The field operator was called away to another location in the plant just after completing the explosive test. A few minutes later, Operations personnel heard a loud blowing noise in the plant. When the control room door was opened to investigate, fire was observed coming from the gap at the top of the south waste heat boiler water jacket overflow trough. The Operations personnel immediately took the natural gas out of the front end of the plant and shut the unit down. The fire extinguished as the gas was taken out of the plant. The fire had been hot enough to melt the aluminum covering of nearby insulation. However, there was no significant damage to surrounding facilities (Figure 11).

The Cause. Samples were removed from the failure areas (2) and from an area of the same shell section remote from the failure locations and sent for metallurgical analysis to Scientific Testing Laboratories and to Radian International (now Mechanical & Materials Engineering) (Pagendarm et al., 1998).

The damaged shell section contained two failures: a bulged area with a vertical rupture approximately 8 in. (20 cm) in length with an opening of 1/2 in. (1.3 cm) (Figures 12a and 12b) and a 5 in. (15 cm) long partially penetrating

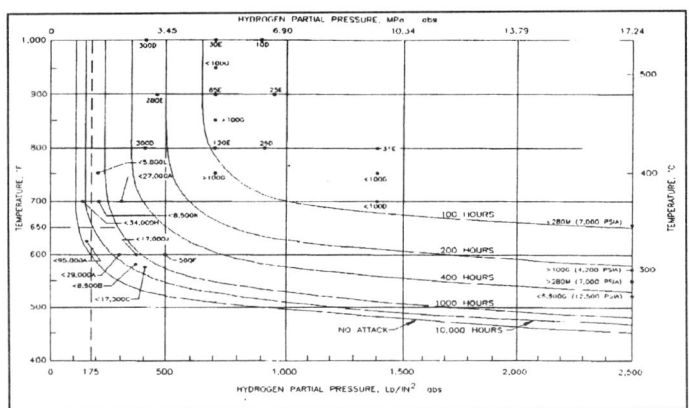

Figure 10. Nelson curve for temperature effect on time for hydrogen attack of carbon steel (ref. 1 Radian, Report 1987).

crack adjacent to a longitudinal weld seam 180° from the rupture (Figure 13). The shell liner shroud was not significantly damaged as found following the shell failure however the refractory was almost entirely missing from behind the shroud in the area of the shell damage.

Samples were removed from the damaged shell section and examined visually and with the aid of a low power stereomicroscope. The rupture was longitudinal and was in the center of a large bulge. The rupture was surrounded by secondary cracking which paralleled the failure. The outside and inside surfaces of the shell were relatively free of deposits and corrosion. A large dent in the OD surface of the shell was visible (Figure 14) where a spacer from the water jacket pressed into the shell when the shell expanded. The section from the vessel area opposite the rupture contained a vertical seam weld. Cracks were visible which ran along the edge of the weld seam (Figure 13). The cracks did not go through wall and began on the OD surface.

Sections were cut from the shell samples and prepared for metallographic examination. The prepared sections were examined using a metallurgical microscope to assess microstructural conditions.

Substantial necking and thinning was present in the area surrounding the failure (Figure 15). The failure lip was square edged and had an oxide layer. The area surrounding the rupture contained creep voids and lacked plastic deformation (Figure 16). The microstructure consisted of spheroidized carbides in a ferrite matrix (Figure 17). These features are typical of thick-lip stress rupture due to long term overheating.

The microstructure of the sample remote from the failures consisted of pearlite in a ferrite matrix. Slight spheroidization was evident along some grain boundaries. This microstructure is typical of a slightly overheated carbon steel and would be considered typical for a boiler application.

Creep fissures were present near the crack in the Heat Affected Zone (HAZ) of the seam weld. The microstructure in this area was similar to that seen at the rupture area. The microstructure was completely spheroidized and creep fissures were evident (White, 1997).

The creep life of a component is inversely proportional to increases in temperature and stress over time.

Standardized calculation methodologies have been developed to characterize the creep life of a component. This process provides an approach toward estimating the period(s) of total time the (exchanger) shell was exposed to excessive temperatures using the Larson-Miller parameter.

The average creep life at various temperatures, and at the nominal stress level of the exchanger shell has been calculated for carbon steel and 1 1/4 Cr. 1/2 Mo. steel. These data are presented in Table 2. The numbers show how rapidly the estimated creep life drops off for the carbon steel with increasing temperature. For example, the heat exchanger shell would not have failed due to creep after its entire service life at 825°F (440°C) however, failure would occur in less than 200 h at a metal temperature of 1,000°F (537°C). The creep calculations also show the 1 1/4 chrome steel offers a considerable increase in expected creep life over the range of temperatures for which the creep life was calculated. At the higher range of temperatures, (probably above the 1,150°F {621°C}) the carbon steel would be subject to tensile failure as its, at-temperature, yield strength would be reduced to the operating stress level.

In addition to the evidence of creep fissures another indication of the temperature exposure of a steel sample can be drawn from an examination of its microstructure. As noted above, some of the metallurgical mounts exhibited spheroidization of the pearlite constituent in the microstructure. Spheroidization occurs due to elevated temperature exposure of steel

Figure 11. Primary waste heat boiler shell failure minor fire damage (Yelverton, 1997).

Figure 12a. Shell "bulge" in failure area (Yelverton, 1997).

Figure 13. Cracks adjacent to weld seam remote from the leak site (Yelverton, 1997).

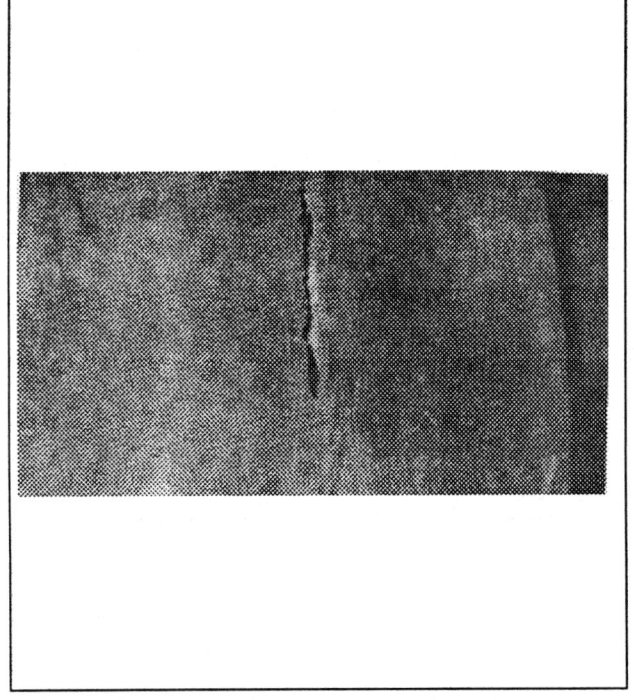

Figure 12b. Bulged shell area and crack (Yelverton, 1997).

Figure 14. Shell dent on OD of bulged surface (Yelverton, 1997).

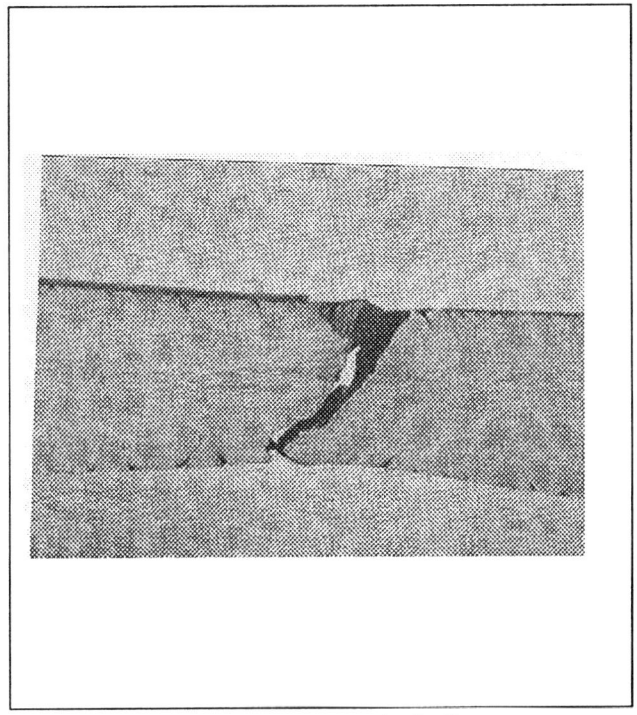
Figure 15. Section through shell rupture.

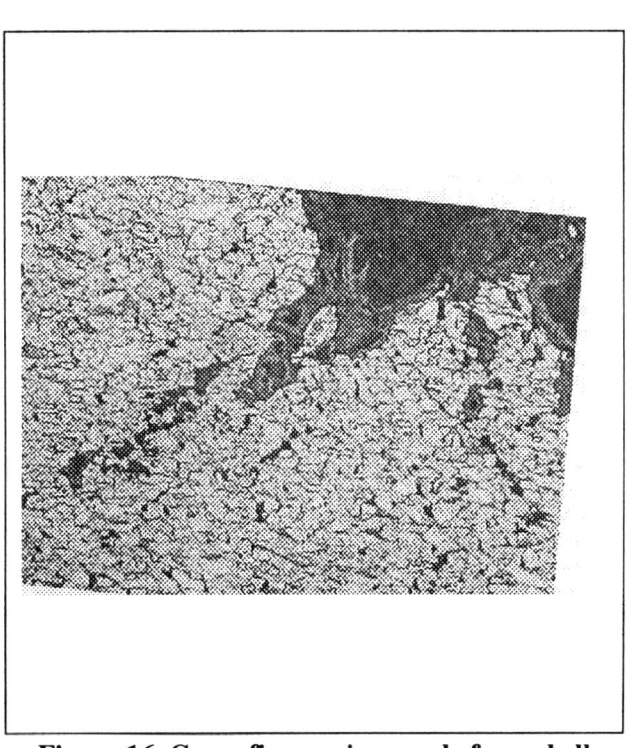
Figure 16. Creep fissures in sample from shell rupture (50x, Nital).

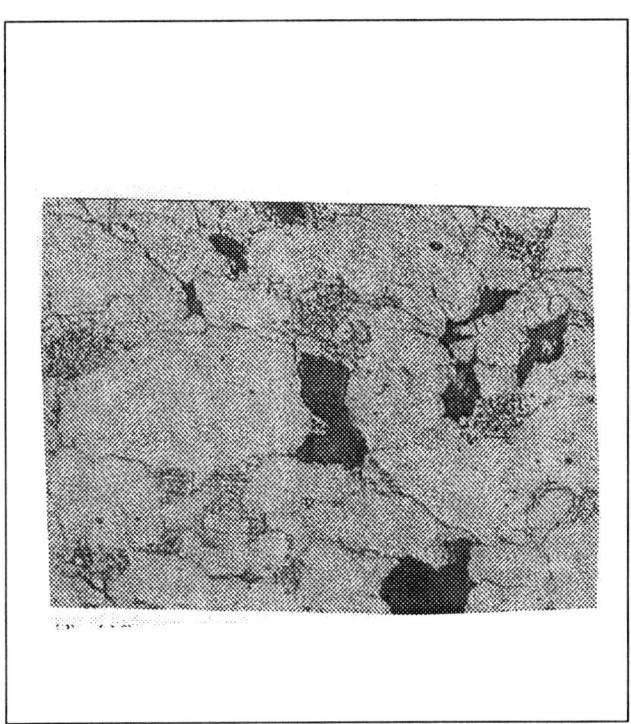
Figure 17. Spheroidized microstructure and creep voids in sample from creep rupture (500x, Nital).

over time. Spheroidization can typically occur above about 800°F (426°C). The pearlite (or iron carbide) phase in the steel changes shape from a plate-type or an unresolved form into a spherical shape. This change is proportional to the temperature and the time at temperature. Above about 1350°F (730°C) the steel is said to transform or re-crystallize. This microstructural change is usually evident and was not observed in the metallurgical samples examined.

The strength of the steel decreases as the exposure temperature increases. Thus when a high enough temperature is reached, such that the stress loading on the component is approximately equal to the yield strength of the steel, a hot tensile rupture failure will occur. The failure zone will typically exhibit ductile thinning. A hot tensile separation is also typically the final event in the creep failure sequence. This occurs because the creep cracking effectively reduces the cross sectional area of the component such that the effective local stress reaches the yield strength of the steel. The metallurgical report indicates that "substantial necking and thinning was present" in the rupture area.

Conclusion. The pressure shell from the South waste heat boiler failed in service due to thick-lip stress rupture (creep).

The damage was due to prolonged exposure of the shell material, in the temperature range of 900–1150°F (480–620°C).

No evidence of hydrogen related damage or cracking was noted in the samples examined. The tensile strength of the remote sample was slightly below ASTM A516, Grade 70 specifications. This was due to slight thermal softening of the material. The sample retained adequate impact properties. The failed sample met the chemical composition requirements for ASTM A516 Grade 70 material.

The presents of creep damaged metallurgy in the area of the shell crack opposite from the failure area indicates a large area of the shell was exposed to high temperature. The degree of carbide spheroidization in the same areas reinforces a conclusion the exposure time was relatively long perhaps, 4,000 - 6,000 h. The extent of the exposure time is limited by the fact that there was no hydrogen damage present as would have been the case had the exposure been 10,000 plus hours. The time and temperature effect on the rate of spheroidization is not known well enough to be precise in this regard, however the rapid reduction in creep life of the material as the temperature rises above 850°F (454°C) leads to the probability the temperature was in the range of 850 to 925°F (454–495°C) (Table 2).

In light of the conclusions from the 1986 Medicine Hat case; it is probable the loss of a large area of the insulating refractory and the resulting steam blanketing effect could have caused this failure. This conclusion is supported by the observation that the jacket water supply is routinely monitored and a low level condition in the water jacket is unlikely to go unnoticed for an extended period of time. The depression in the shell material was likely caused following the initiation of hot gas flow through the crack. The gas would further displace the water in the jacket and produce the high temperature required for the thermal softening that is necessary for the impression noted in Figure 14 to occur.

The Repair. The stainless steel internal shroud and refractory were removed from the vessel. The vessel shell was cut at the location of the 15/16 in. (2.3 cm) shell weld below the failed shell section (ring 3) and at the field weld on the nozzle connection to the downstream vessel. The upper section was sent to a fabrication shop where the 15/16 in. (2.3 cm) section was replaced with new material of the same (SA516 Gr. 70) specification. The entire shell section was "baked out" to remove hydrogen prior to attempting any welding. The refabricated shell section was returned to the field and installed by replacing the welds at the previous cut points. All welds were post weld heat treated. The interior lining, refractory, shroud and water jacket were replaced according to the original design. (Yelverton, 1997)

The vessel was returned to operation on December 16, 1997. The failure repair resulted in a total of 15 days lost production.

Considerations to avoid shell failures

The operators of the MW Kellogg waste heat boilers of this design have become increasingly aware of the potential for failures as the consequences and various

Table 2. Calculated Creep Life (Hours) at Normal Design Stress for Vessel Shell (13,200 psi) 91 n/mm^2

Temperature	C.S.	1^1/$_4$ Cr.
800°F (425°C)	622,000	215,000,000
825°F (440°C)	196,000	60,500,000
850°F (455°C)	64,400	17,900,000
875°F (468°C)	22,100	5,520,000
895°F (480°C)	9,670	2,220,000
900°F (482°C)	7,890	1,780,000
925°F (495°C)	2,920	597,000
945°F (507°C)	1,350	257,000
950°F (510°C)	1,120	209,000
975°F (523°C)	445	75,500
980°F (525°C)	371	61,900
1000°F (537°C)	182	28,300
1005°F (540°C)	153	23,400
1020°F (548°C)	91	13,200
1035°F (557°C)	55	7,580
1050°F (565°C)	33	4,390
1055°F (568°C)	28	3,670
1060°F (570°C)	24	3,070
1070°F (575°C)	17	2,150
1080°F (582°C)	13	1,520
1090°F (587°C)	9	1,080
1100°F (593°C)	7	767
1125°F (607°C)	3	334
1140°F (615°C)	2	205
1150°F (620°C)	2	149
1200°F (648°C)	.4	32

near misses have been experienced. There have been few and in many cases no design changes to the vessels undertaken by the user community. Inspection and early detection, is the most common approach to avoiding potentially serious failures.

The continuing theme surrounding early detection of failures, is the limited access to the affected parts, afforded by the design of the vessel. Frequent or continuous analysis of the vapor space above the water level in the jackets has resulted in the detection of process gas leakage in two cases at CFII. This program has become a standard operating procedure at the CF facilities for this reason. This is not a panacea however, as evidenced by the fact that Case 3 involved a gas check within 24 hours of an event which disabled the plant for 15 days.

The area of the most frequent nozzle failures, at or near the outlet nozzle attachment welds, is located inside the water jacket. These failures have been attributed to mechanical and thermal stress cycles. It has become common practice to remove sections of the water jacket during plant shutdowns in an effort to detect the onset of cracking at an early stage through the use of PT or MT inspection of the sensitive area.

Shell cracks have been associated with loss of the refractory insulation which protects the inner surface from exposure to the process temperature. Inspection to assess the condition of the refractory is hampered by the presents of the protective stainless steel shroud on the inner surface. It can be surmised from the analysis of the two shell failure cases discussed earlier that the loss of the refractory may have gone undetected by conventional inspection methods. The resulting long exposure of the pressure shell to elevated temperatures contributed to the cause of the eventual failures. An inspection technique that assures the presence of refractory is essential. Probing the refractory via a grid of small holes in the shroud has been successfully applied.

Considerations For Design Upgrades

CF Industries has had stress analysis performed which indicates the attachment area of the vessel's outlet nozzle is susceptible to stress cycles which will eventually cause the observed cracking. A redesign of the nozzle attachment and the support system for the connected piping has been suggested to alleviate this situation.

The vessel's designer has developed an alternate lining system which employs two layers of refractory and has no metal internal shroud. This design requires a replacement shell with a larger diameter to provide the necessary space for the additional refractory, but has the advantage of providing the opportunity to verify the condition of the insulation to a much greater degree.

The use of 1 1/4 Cr. 1/2 Mo. or higher alloy shell material would offer the increased margin of protection that may avoid shell failures due to overheating, if the exposure temperatures are below 1000 - 1100°F (535 - 595°C). Both the resistance to hydrogen attack and the creep resistance of 1 1/4 Cr. 1/2 Mo. material in this temperature range are substantially better. In both the temperature related Cases (1 & 3) presented here, the metal temperature was well below this range.

The ability to assure early detection of breaches to the internal insulation of this vessel may only be realized following adoption of a vessel design that provides access, or a monitoring method, which would allow the shell temperature to be indicated during vessel operation.

Final Thoughts

This short summary highlights some industry experiences which indicate the potential for failures in the primary waste heat boilers in MW Kellogg design ammonia plants and the consequences which can result. There remains a need for continued application of present inspection techniques and for innovative alternatives to present equipment design and inspection methods. It is hoped the information presented here will assist other operators to avoid a failure or lessen the consequences as a result of increased awareness of the potential causes of failures and their avoidance.

Acknowledgments

We wish to acknowledge the following Authors for their assistance with this article: Joe Boley, CF

Industries Inc., Donaldsonville, Louisiana - editorial assistance; Larry Cizmar, MW Kellogg - provided historical data, equipment and graphics Figures 2, 3; Len Hodas, Mechanical & Materials Engineering (formerly Radian Corporation) - provided file slides, 1986 Canadian Fertilizers Ltd, Medicine Hat, Alberta, shell failure analysis and valued advice; J. Yelverton, T. Pagendarm, CF Industries Inc.; Donaldsonville, Louisiana - vessel failure reports and supporting information.

Literature Cited

Hansen, D. E., "Stress and Flexibility Analysis of Piping Systems and Vessel Shells (101CA & 101CB)," Piping Analysis Inc. (1998)

Harding, S., "Failure Analysis, Primary Waste Heat Boiler, Canadian Fertilizers Ltd. Medicine Hat Alberta," Radian Corp. (unpublished, 1987)

Hodas, L. J., Failure Analysis, 101CB Waste Heat Exchanger Shell Unit No. 3 Ammonia Plant, CF Industries Inc. Donaldsonville, LA, Radian International (unpublished., Jan. 1998).

Laing, K. (to R. D. Shannon), "Canadian Fertilizers Ltd, Medicine Hat Alberta, No. 2 Ammonia Plant Shell failure report," (unpublished, Dec. 3, 1986).

Lapinskie, R. (to H. Matheson), "Canadian Fertilizers Ltd, Medicine Hat, Alberta 101C Failure and Repair Report" (unpublished, 1987).

Pagendarm, T., et al., "Incident Investigation No. 3 AIChE Ammonia Plant," CF Industries Inc., Donaldsonville, LA (unpublished, April 1998)

Pagendarm, T., et al.,"Incident Investigation No. 3 Ammonia Plant," CF Industries Inc., Donaldsonville, LA,(unpublished, April 1998)

Shibasaki, T., et al., "Experience with Metal Dusting in $H_2/CO/CO_2/H_2O$ Atmosphere, *Ammonia Plant Safety & Related Facilities,* Vol. 36, AIChE, New York (1996).

Stahl, H., and S. G. Thomsen "Survey of World Wide Experience with Metal Dusting," *Ammonia Plant Safety & Related Facilities*, Vol. 36, AIChE, New York (1996)

Sutherland, L., "Failure Repair Report No. 1 Ammonia Plant, 101C" CF Industries Inc., Donaldsonville, LA (unpublished, Nov. 24, 1986).

Wagner, E. (to K. Laing), "Canadian Fertilizers Ltd, Medicine Hat, Alberta No. 2 Ammonia Plant Shutdown Incident Report" (unpublished, Nov. 14, 1986).

White, M. T., "Stress Rupture of Boiler Shell, Scientific Testing Laboratories Inc." (unpublished, Dec. 11, 1997).

Yelverton, J., CF Industries Inc., "Inspection / repair report Donaldsonville, Louisiana / Radian," (unpublished, 1997).

DISCUSSION

P. Khetarpal, *Kribhco*: I would like to know whether you have an inlet gas distributor within the waste heat boiler, and, if so, is there any problem with that?

Jackson: We do not.

Khetarpal: It's not there?

Jackson: It's not there. It was originally provided, but it was removed, probably, within the first year of operation and we haven't had one since.

Khetarpal: We also faced a problem inside the waste heat boiler. The liner inside shell got bulged. It is maybe because of overheating or metal dusting. Liner breaks very easily.

Jackson: Yes.

Khetarpal: Do you suggest some improvement for that?

Jackson: I cannot. Our experience would be that the hot end of the exchangers in the inlet line and in the first 10 ft or so of the shell liner, maybe a little less, suffers from what I would call carburization. It gets brittle and it cracks and what you have typically lasts in excess of five years. We've had some in service maybe ten years. I think our original installation was 310 stainless in that area, and we've continued to replace it with that. I really don't have an offering for, or any experience with, any other material in that application.

Harry Van Praag, *Terra Nitrogen*: You were asking about monitoring of shell temperatures. We found heat sensitive paint to be a very useful tool for finding any minor refractory failures, and also major failures. It tells you right away.

Jackson: How can you apply heat sensitive paint over a water jacket?

Van Praag: I thought you said you wanted to do away with the water jackets.

Jackson: Yes, indeed, heat sensitive paint is a great idea if we can get ourselves a shell that's not jacketed.

Pressure Relief Valve Piping Failures and Fire: Ammonia Synthesis Loop

This article describes a piping failure and fire resulting from inadequate support of an ammonia synthesis loop pressure relief valve, the nondestructive testing and evaluation of fire-impinged equipment, and the pipe failure mode. It also addresses support criteria for relief valves and other piping systems illustrating the importance of evaluating both static and dynamic loads.

Lester E. Sutherland and Michael Holman
CF Industries, Inc., Donaldsonville, LA 70346
Don Hansen
Piping Analysis Inc.

Introduction

On May 1, 1996 at approximately 4:15 PM, while in the process of restarting CFIIs Ammonia Plant 3 following a nonscheduled maintenance shutdown, a fire occurred in the vicinity of the Ammonia Refrigerant Flash Drums. The cause of this fire was later identified as an insufficiency in the piping/support system for the Synthesis Gas Loop Relief Valve (RV-105D).

CFII's Anmonia Plant 3 is a M. W. Kellogg designed 1,000 TPD Stretch unit that was commissioned in 1976. The First, Second and Third Stage Ammonia Refrigerant Flash Drums (112F, 111F, and 110F, respectively) appear to be a single horizontal drum mounted approximately 20 ft above grade. This drum is separated into 3 distinct chambers by 2 internal heads. The vessel has an inside diameter of 7 ft and is approximately 70 ft in length. The material of construction of the flash drum is SA-516 Grade 60 and has a nominal 0.4 in. shell thickness and a 1/16 in. corrosion allowance. The vessel was post weld heat treated after fabrication. The design pressure of the vessel is 100 psig. The 110F operates at 85 psig and 56°F with the 111F and 112F operating at progressively lower pressures and temperatures. The drums contain ammonia liquid and vapor and have been in service for approximately 20 years. The drums were insulated during initial plant construction with foam glass insulation and aluminum weather proofing. A zinc rich epoxy coating was applied to the shell prior to installing the insulation.

A common platform runs the length of the drum which provides access to an array of level, pressure, and temperature instrumentation. Several relief valves are also serviced from this platform.

RV 105D is a pilot operated relief valve that protects the Synthesis Gas Loop from excessive pressure. The 3 in diameter relief valve inlet is connected to a 14 in. diameter line through approximately 15 ft of 3 in. diameter pipe. The pilots, which serve to actuate the relief valve, are connected to the 14 in. diameter line through a separate 1/2 in. diameter remote sensing line. Access to this valve is provided by the above mentioned platform.

While restarting the plant, the RV-105D lifted. It was reported by operations personnel that the tail pipe began to vibrate violently and that the valve was chattering. Apparently, the stress levels in the 90° elbow immediately below the valve exceeded their allowables and the elbow failed. The failure of this elbow allowed the relief valve and tail pipe assembly to lay over into the horizontal plane pointing directly parallel with the axis of the 1lOF, 111F, 112F platform (see Figure 1). Cracks developed on either side of the elbow and the leaking high pressure synthesis gas eventually ignited. As the valve had still not seated properly, synthesis gas continued leaking from the tail pipe. This gas stream eventually ignited. At this point, fire was emanating from the cracks in the 90° elbow as well as from the tailpipe. The fire continued to burn for approximately 1 h as the synthesis gas section depressured.

The shock waves, which followed the two ignitions, damaged a great deal of insulation in the general vicinity. The synthesis loop eventually depressured sufficiently to allow the RV to seat.

This incident appears to be the result of at least the following factors:

The Remote Sensing Line was not Properly Connected to the Relief Valve Pilots. During a routine plant turnaround, RV-105D was removed from the unit for inspection and testing. In order to test the valve, the pilots were temporarily connected to the valve inlet. These connections were not removed from the valve prior to its return to the plant. The technicians that reinstalled the valve in the plant did not reconnect the remote sensing line to the pilots, leaving the test lines in place.

With the pilot sensing line connected to the pressure tap located just below the valve seat, problems developed. As the valve was in the flowing condition, a pressure drop in the line occurred. The pressure immediately upstream of the valve was instantaneously reduced, allowing the valve to seat. The pressure immediately rebuilt to the header pressure and caused the valve to reopen. This situation was repeated many times per second. The cumulative result was a cyclical load being applied to the support system.

The Tail Pipe was not Properly Cut at the Discharge. This situation resulted in the application of a tremendous moment arm and hence extremely high stress levels to the piping system.

The Support System for the RV was not of Sufficient Strength. The tail pipe was restrained by an angle steel bracket welded to a handrail. This support, more likely than not, failed the instant the valve lifted.

Discussion of Subsequent Inspections

The following is a summary of the equipment inspections performed following the incident. The intention of these inspections was to assess the extent of equipment damage which may have occurred as a result of the fire and to assist in the adoption of intelligent remedial measures.

The types of inspections performed on the various pieces of equipment, as well as the acceptance/rejection criteria, were developed by CFII and were based upon input from several outside engineering firms.

110F

The south side of the west head of the 110F was directly exposed to heat from the fire emanating from cracks in the 90° elbow located immediately below the 105D RV. Insulation was absent from this area, apparently the victim of one of the shock waves which followed the ignition of the leaking synthesis gas. All of the instrumentation and associated small bore piping in the immediate area were also found to be without insulation and to have been exposed to intense heat.

The following measures were taken to assess the extent of damage to the 110F and assure its fitness for continued service.

- Hardness readings were taken at those areas which were found to be without insulation. This included the west head, as well as the south west side of the 110F shell. The purpose of this test was to determine whether or not the properties of the material had been altered by exposure to excessive heat. All of the shell plate readings were found to be within acceptable limits. All of the head readings were at or slightly below acceptable limits for SA 516-60 material. Ultrasonic thickness measurements taken on this head yielded a minimum thickness of 0.537 in. Using the yield

Figure 1. Photograph of damaged 110F, 111F, and 112F.

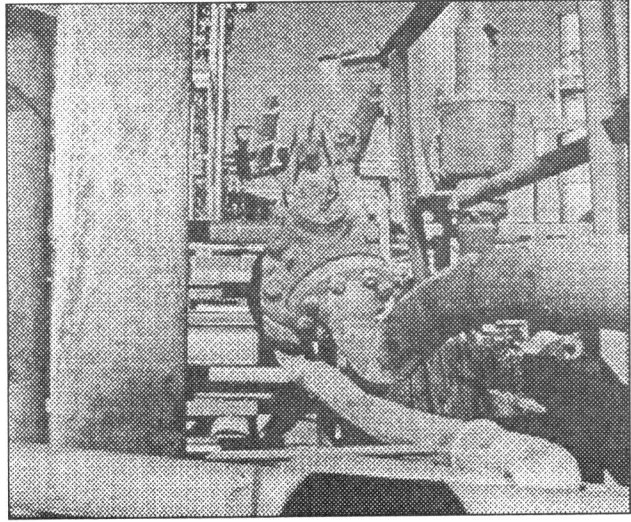

Figure 2. Photograph of failed RV-105D.

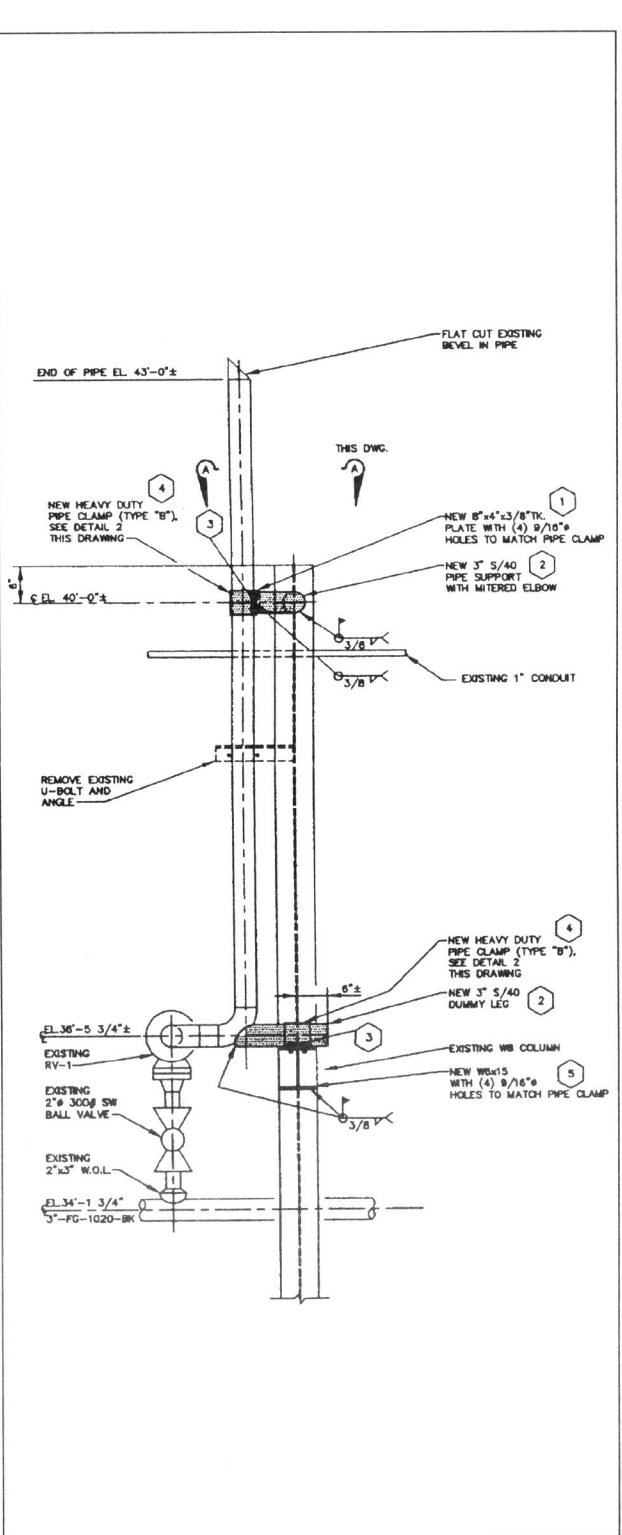

Figure 3. Sketch of appropriately supported pressure relief vent stack.

strength values for SA 516-55, the calculated minimum head thickness required for the vessel design conditions are 0.525 in. This leaves a corrosion allowance for the head of 0.012 in.

- All of the externally visible welds were sandblasted to facilitate a WFMPT. The purpose of this test is to determine whether or not cracks, which may develop as a result of heat related stresses, exist. No indications were noted.
- All of the internally visible welds were hand cleaned to facilitate a WFMPT. Three small indications were noted at three different level instrumentation nozzles. These were characterized as being original fabrication defects (porosity and lack of fusion). All three were excavated until the indications disappeared via grinding and WFMPT. None were over 1 in. in length. The final depths were measured at 3/32 in, 3/16 in, and 3/8 in. These areas were repaired using an approved welding procedure (SMAW E7018-1 electrodes). The area was preheated to 300°F using electro-resistance heaters. Welding was performed by an individual qualified to that procedure. The area was then stress relieved to satisfy service conditions (ammonia induced stress corrosion cracking) by electro-resistance heaters.
- A visual inspection of the vessel, both internal and external, revealed some minor corrosion under insulation (CUI) on the top of the drum. No heat related damage, however, was noted.
- The manway cover bolts and gasket were replaced.
- All of the small bore instrument piping immediately downstream of the vessel nozzle couplings was replaced. This was done in lieu of extensive testing to these systems.
- All of the level, pressure, and temperature instrumentation were replaced.
- As a final measure, the vessel was successfully hydrostatically tested at 72 PSIG.
- The vessel was recoated and reinsulated with urethane foam prior to returning it to service.

111F

The south side of the 111F had been exposed to heat resulting from burning synthesis gas emanating from the distorted 105D RV tail pipe. Generally, the foam glass insulation was intact; however, most of the aluminum lagging on the south face of the vessel had melted away. All of the instrumentation and associated small bore piping in the immediate area were also found to have been without insulation and exposed to intense heat.

The following measures were taken to assess the extent of damage to the 111F and assure its fitness for continue service.

- As there were no exposed sections of shell plate, no hardness measurements were recorded.
- All of the externally visible welds were sandblasted to facilitate a WFMPT. No indications were noted.
- All of the internally visible welds were hand cleaned to facilitate a WEMPT. No indications were noted.
- No significant problems were noted following an internal and external visual inspection of the vessel. No heat related damage was noted.
- The manway gasket was replaced.
- All of the small bore instrument piping immediately downstream of the vessel nozzle couplings were replaced. This was done in lieu of extensive testing to these systems.
- All of the level, pressure, and temperature instrumentation was replaced.
- The vessel was re-coated and re-insulated with urethane prior to returning it to service.

Piping

All of the lines affected by the fire were visually inspected prior to any cleanup efforts. Any lines missing insulation or with obvious coating damage were slated for additional testing. Material hardness testing was selected as the prime analysis method for evaluating line integrity. Of the lines tested, only a 6 ft section of a 10 in. ammonia line exceeded its specified hardness level. This section of line was replaced in kind with new material. The two new butt welds were radiographed to verify integrity.

A 2 in. instrument air line, a 3 in. plant air line, a 2 in. 550 psig steam condensate line, and a 3 in. 50 psig steam condensate line were replaced because the lines were badly distorted, apparently from excessive heat.

The piping to RV-105D (Figure 2), its sensing line

and its tail pipe were all badly damaged. As a result, the RV was relocated and the associated support system redesigned. These modifications were designed by Piping Analysis Inc.

(Figure 3 shows an appropriately supported pressure relief vent stack.)

All of the remaining lines were either recoated or reinstalled as required.

Structural Steel

A visual inspection of the structural steel was performed. Visual distortion was selected as the acceptance/rejection criteria. The following items were replaced based upon this criteria:
- One 8 in. pipe bent column.
- One 8 in. pipe bent horizontal beam.
- Much of the checker plate at the 110F, 111F, and 112F walkway.
- Much of the hand rails and toe rails at the 110F, 111F, and 112F walkway.

Relief Valves

The following relief valves were exposed to excessive heat and subsequently sent to their respective authorized repair centers for tear down inspections and testing:
- RV-102P;
- RV-110F;
- RV-111F; and
- RV-112F.

Prior to removal, hardness tests were performed on RV-110F, RV-111F, and RV-112F. All readings were found to be within acceptable limits.

RV-105D was replaced with a new valve from inventory.

Instrumentation

One control valve, three level transmitters, four pressure transmitters and several local pressure and temperature indicators were replaced as a result of obvious or potential fire damage. All of the instrument air tubing, regulators, conduit, and wiring in the fire affected area was replaced.

Electrical

A section of cable tray, along with several runs of conduit, sustained damage as a result of the fire.

These were replaced or repaired as required. Some field splicing was performed. Seven lighting fixtures and their associated conduit and wiring were replaced.

The inspections and repairs described above were completed in roughly 10 days. Mr. Don Hansen of Piping Analysis Inc. was retained to help evaluate the cause of the failure and evaluate other relief valve systems for potential problems.

Recalculation and Redesign of RV Piping and Restraint Structures

This RV piping and its restraining structures were not designed for the actual static + dynamic service. This is a common failure producing condition in relief valve installations. Proper design must consider all of the important forces that occur, or can occur, from the moment the RV starts to operate to the moment it closes and the entire system returns to its beginning state.

This complete cycle must operate within the appropriate ASME, AISC, or other stress/strain limiting criteria without overstress or overstrain.

The best primer on the subject of relief system design is ASME B31.1, Appendix II:

(1) How to calculate relief valve discharge pipe forces, velocities and pressures.

(2) Differences between open discharge and closed discharge systems, and important considerations for each.

(3) Vent pipe loads.

(4) Dynamic amplification of loads due to support flexibility and earthquake motions.

(5) Stress analysis of the piping and its header connections due to pressure + bending loads.

(6) General design considerations such as location, spacing, outlet pipe types, multiple installations, liquid hold-up and water seats, silencers, and supports.

Some important RV forces and restraint features are:

(1) Valve internal mechanical motion, forcing heavier air out of the exhaust pipe, initial fluid acceleration, fluid steady-state velocity reaction.

(2) Correct time-dependent force vector applied at

the discharge pipe end cut.

(3) Restraining structural and connecting piping must successfully resist all static + dynamic loadings including dynamic amplifications. RVs must be supported and restrained so that the applied loadings are grounded without causing damage.

• Long discharge pipes require heavy-duty lateral restraints to avoid shaking and columnar failure.

• The RV discharge pipe, very near the valve, is usually properly secured if "anchored," thus providing a good boundary and thrust restraint which does not pass these high loads through the RV or the RV inlet piping.

Piping reactions are usually placed in three main classes:

• Sustained reactions characterized by loadings that persist until collapse, and tend to cause continuous strain, resulting in irreversible changes and severe damage.

• Displacement reactions include thermal expansion/imposed motion, which persist to the extent of displacement. Although limited, detrimental, irreversible changes/damage can still occur.

• Dynamic loadings which are time dependent caused by forcing functions and can cause irritating noise, severe displacements, and fatigue cracking.

The mechanical connection of the earth, foundation, machinery, and piping defines the scope of structural concerns. The effect of all piping reactions is to cause internal stresses and strains.

Sustained, displacement, and dynamic vectors must be added (superimposed) to produce the total piping reaction at any particular time.

Detecting and Avoiding Excessive Piping Reactions

Successfully detecting and avoiding excessive piping reactions requires two efforts - accurately calculating the actual piping reactions and subsequently designing for an operable system, and proper construction, operation and maintenance.

The requirements for a usable analysis: Accurate calculation - the mathematical model is as close an approximation to reality as the technological method allows - and precision - the number of significant figures that result from calculation within the scope of the technological method cannot be emphasized enough. It combines competence in mechanical engineering, experience with real-world piping problems, and accuracy in both the mathematical model for computation and calculation algorithm for computer programs.

Failure-Producing Conditions

As practical as piping computer programs are, they have their limitations. Without sufficient engineering background, the user does not learn the nature of catastrophic failures, basic principles of piping behavior, or assumptions made by the computer program. This can be dangerous because it substitutes the program's judgment for that of a person. A few of the failure-producing limits from calculations include:

• Hook's Law (stress remains proportional to strain). Applies to method of elastic analysis with superposition of forces, moments, and stresses used by piping programs; actual stress is not always proportional to strain in some problems calculated with the stress theory of failure used by piping codes.

• Piping systems composed of long and slender members provide the most accurate answers. The use of close-coupled, stiff-to-flexible elements can produce undetected wrong answers.

• Expansion joints or vessel shells can be improperly described, and explicit or program-assumed stiffness characteristics (strain analysis) can be used without proper stress analysis of these elements.

• Pipe supports such as restraints, hangers, anchors are assumed to work as calculated. The typical and almost universal problem is the failure to include support stiffness and movement in the piping calculations.

Failure-Producing Conditions: Construction, Operation, Maintenance

Maintenance must examine the as-built system and compare it to as-designed requirements. A poor understanding of the nature and significance of system functions can lead to failure-producing conditions. Training programs should include input from engineering designer so the total scope is presented to those who will build, operate and maintain the real system.

All supporting elements must be checked for evidence of intended operation. This includes piping, valves, structures, foundations and anchor bolts. Substituting different materials of construction without engineering investigation is a common error. Start-up, emergency, or normal operating conditions, other than those used for design, are common causes of failure and damage.

Failing to notify a knowledgeable person about a suspicious condition is another common error. Vibrations in machinery, equipment, structures and supports can often be traced to relief valve and piping reactions. Sagging beams, broken pipe insulation, sway, noises, and leaks usually mean poor design, construction, operation or maintenance. Investigation of problems is worthwhile.

The incident, response and analysis presented here are based on circumstances that are, at least to some extent, unique. The readers may find the presentation relevant to their own situation, but must realize that each case is different.

Plugging a Bayonet/Scabbard Tube in an Ammonia Plant Waste-Heat Boiler

The primary waste-heat boilers in Kellogg ammonia plants are prone to developing scabbard tube leaks. The costly replacement of the tube bundle is the typical repair for most plants. Plugging the leaking bayonet/scabbard tube without removing the tube bundle is a much more economical alternative. The technique for finding and plugging the leaking tube is described.

Gary G. Osborne
IMC-Agrico, St. James, LA 70086

Introduction

Typical Kellogg ammonia plants built in the 1960s and 1970s have two primary waste-heat boilers (101-Cs) (Figures 1 and 2). These boilers are vertical units designed with a bayonet/scabbard tube arrangement; that is, a tube inside a tube. Figure 3 is a simplified cutaway view of a boiler showing its flow pattern. The boiler water thermally siphons down through the inner tube or bayonet to the bottom and then reverses, passing up through the annular area between the tubes. Process gas passes over the outside of the outer or scabbard tubes and provides the heat input.

This type of boiler is prone to leaks occurring in the scabbard tubes. When a leak occurs, boiler chemicals foul downstream catalyst in the high-temperature shift converter. This typically causes an unscheduled shutdown for repair to save the catalyst. When this occurred in our plant in the past, the tube bundle was pulled and replaced with a spare unit. In most cases the other boiler was also replaced because its condition was unknown. This procedure resulted in costly plant downtime of at least three to four days.

In November 1997, a leak developed in one of our boilers. The boilers had only been in service four months since replacement. The leak was thought to be due to mechanical damage during installation because of the poor condition of the metal liner inside the gas pressure shell. Overall the boiler's scabbard tubes were thought to be in good condition. Therefore, the decision was made to plug the leaking tube. This procedure had not been done in our plant before; however, other plants were known to have performed the procedure successfully. Two tubes were plugged, and the plant ran successfully.

Figure 1. 101-C in service.

Figure 2. 101-C bundle lifted.

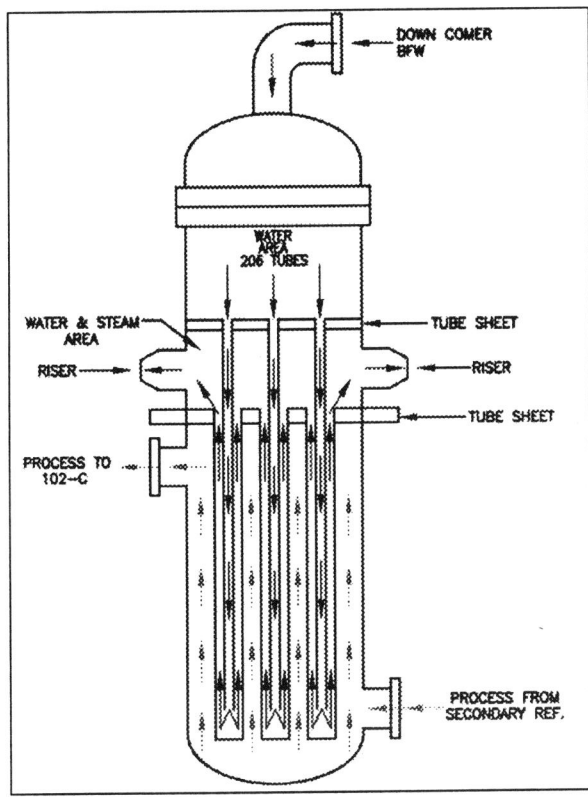

Figure 3. Simplified diagram of 101-C (primary waste-heat boiler).

A power failure occurred in January 1998 that caused our plant to trip and damaged the boilers and several other pieces of equipment. A turnaround was moved up to May, which was the earliest time all the materials could be procured. By the time the turnaround was started, tubes had been plugged on four more occasions, making a total of seven tubes plugged in each boiler. The last tube plugging was performed on April 22. Maintenance duration was 14 h, and total plant downtime was 36 h on this occasion.

The plugging method resulted in considerable cost savings. After bundle removal, the plugged tubes were found to have crystallized. They were intact and had distorted very little. No damage was found on adjacent tubes; therefore, we now have little concern that adjacent tubes will be damaged.

IMC-Agrico realized considerable savings using the plugging method. If plugging had not been used, both boilers would have been changed in November at the first leak and would have required replacement again after the power failure. These new boilers probably would have been damaged by the poor condition of the liners during installation. Conservatively, a maintenance savings of $350,000 was achieved and four days of downtime were avoided. Plugging tubes in these waste-heat boilers has proven to be a valuable tool.

Boiler Design and General Plugging Procedure

The primary waste-heat boilers in Kellogg ammonia plants incorporate a bayonet/scabbard tube design (see Figure 3). Two tubesheets exist in these boilers, one for the scabbard tubes and one for the bayonet tubes. The lower tubesheet holds 206 2-in. (5.08-cm) scabbard tubes made of 1 1/2 chrome–1/2 moly material. These tubes are seal-welded to the tubesheet. The lower end is capped. The upper bayonet tubesheet is approximately 42 in. (106.7 cm) above the scabbard tubesheet. The bayonet tubes have a diameter of 1 in. and are removable. The upper end of the tube is held in the tubesheet by a nut with a hex-shaped hole through it. The nut is tack-welded to prevent loosening during operation. The lower end of the bayonet tube is open.

To plug a tube, the top head of the boiler is first removed. If the leak is small, the boiler will still be filled with water. This water must be pumped to the level of the lower scabbard tubesheet so that each scabbard tube is isolated and holding water. The leaking tubes are found by measuring the water level in each tube (Figure 4). A tube has a leak if it has a lower water level than the other tubes. The bayonet tube is removed from the leaking scabbard. With special tooling shown in Figure 5, a plug is inserted through the bayonet tubesheet and into the leaking scabbard tube (Figure 6) and then driven into place (Figure 7). The hole in the bayonet tubesheet is then plugged with a special plug, as shown in Figure 8, and tack welded. The boiler's head is reinstalled and the procedure is complete.

Detailed Plugging Procedure

The following list is the procedure at IMC-Agrico for plugging a tube in the primary waste-heat boilers. This procedure is performed after the plant is shut down and the boilers are blinded and cool. The scabbard plugs, the bayonet plugs, the plug insertion rod, and the driving tip for the insertion rod are made before shutdown (Figures 10–13).

The plugging procedure is performed as follows:

• Hydraulically untorque the bolting on the head.

• Remove the head from the boiler.

• Pump the water remaining in the channel down to the lower tubesheet. A diaphragm pump is used with a 48-in. (122 cm) piece of 1/2-in. (1.27-cm) diameter tubing attached to the end of the suction hose. The tubing is inserted into a bayonet tube in order to pump the water down below the upper tubesheet. When the pump loses suction the water level should be below the lower tubesheet in the scabbard tube being pumped and at the lower tubesheet in the other scabbard tubes. Fill the tube used for pumping with demineralized water so that the level will be at the lower tubesheet.

• Check the water level in each scabbard/bayonet tube. The level in the tubes that are not leaking will be at the top of the lower tubesheet, approximately 42 in. (106.7 cm) below the upper tubesheet. The level in a leaking tube will be lower than this. To measure the level in a tube, a 6-ft. (2.8-m) piece of 3/8-in. (.95-cm) stainless steel tubing is inserted into a bayonet tube. The top of the stainless tubing is connected to a small, regulated air purge, and the bottom is open ended. A pressure instrument capable of measuring inches of water, such as a Crystal Model 33 Pressure Calibrator, is connected to the purge to monitor the pressure (Figure 9). Until the tubing touches water, no pressure reading will be obtained. When the open end of the tube touches water, a small pressure will develop. Measure the distance from

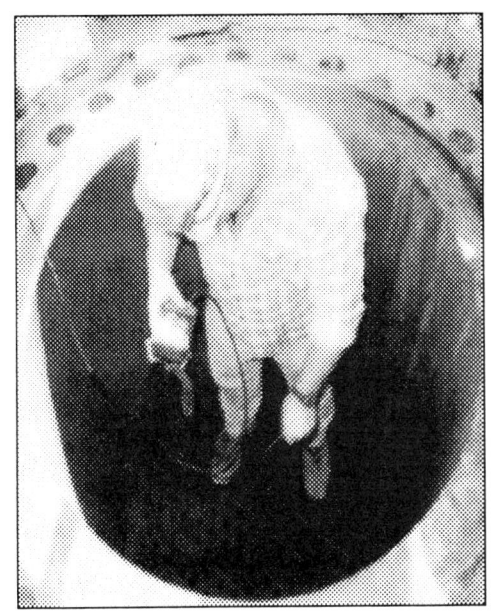

Figure 4. Water level check.

Figure 5. Special tools and plugs.

Figure 6. Plug insertion.

Figure 7. Plug being driven.

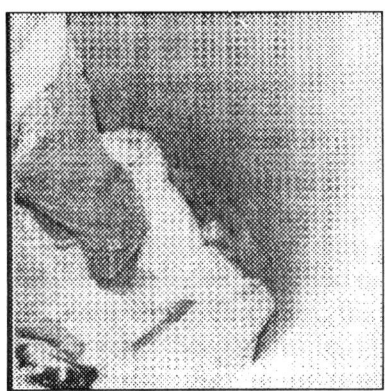

Figure 8. Bayonet plug, shown in place.

the bayonet tubesheet to the water level from the stainless tubing. Mark the bayonet tubes with a low level to indicate a leak. Refill these tubes with demineralized water and check for a dropping level to verify a leak.

• On a leaking tube, grind the tack weld from the nut securing the bayonet tube. Remove the nut with a 1-in. (2.54-cm) Allen wrench. Remove the bayonet tube and its copper gasket.

• Screw a scabbard plug (Figure 10) onto the plug insertion rod (Figure 12). Insert the plug through the bayonet tubesheet and into the top of the scabbard tube. Lightly tap the rod to secure the plug. Unscrew the insertion rod and remove it.

• Screw the plug driving tip (Figure 13) onto the insertion rod (Figure 12). Lower the rod until the tip is on top of the plug. Drive the plug securely into the scabbard tube using a maul.

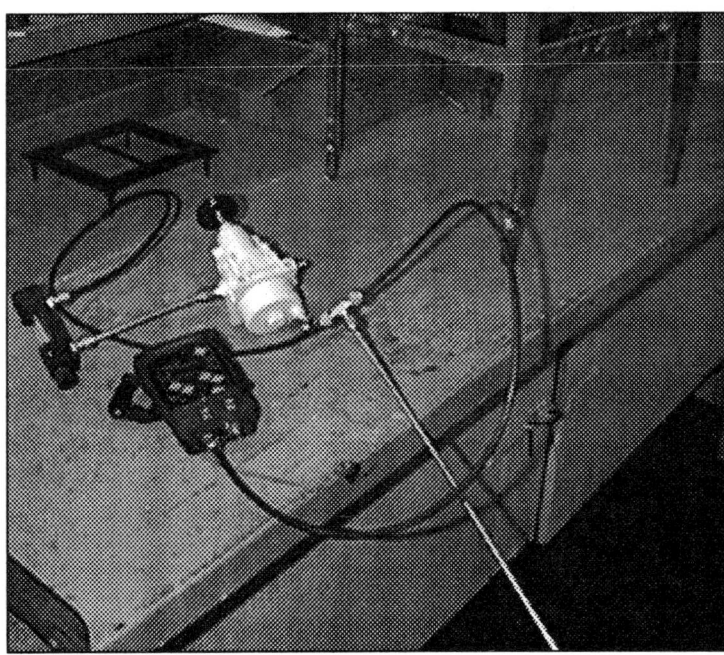

Figure 9. Tube level measuring instruments.

• Place the copper gasket removed in Step 5 onto a bayonet plug (Figure 11). Screw the plug into the hole in the bayonet tubesheet. Tighten with a pipe wrench. Tack weld the plug with E-7018 rod to prevent loosening.

• Repeat Steps 5 through 8 until all leaking tubes are plugged.

• Reinstall the head on the boiler with new gaskets. Hydrotorque the bolts.

Problems Encountered and Improvements to the Process

Three problems were encountered while plugging tubes. Two of these problems occurred during our first attempt. We have improved the process to make it faster and more reliable.

The first instrument used to measure the water level was slow and unreliable. It consisted of a resistance meter with Beldon #16 twisted pair-wire attached. The ends were secured so they would not touch. A lower resistance was measured when the ends came in contact with water that indicated the water level. Residual water on the ends caused erratic readings and low levels were indicated falsely on a few occasions. Extreme caution had to be used with this instrument. The pressure instrument currently used is fast and reliable.

The first scabbard plugs manufactured were larger in diameter than the hole through the bayonet tubesheet, which is very close in diameter to the inside diameter of the scabbard tube. An expandable reamer was used to enlarge the hole. The plug extended significantly above the scabbard tube after installation. The plug's large-end diameter was revised from the original 1.790 in. (4.55 cm) to 1.760 in. (4.47 cm). This size fits through the bayonet tubesheet and is still large enough to securely tighten in the scabbard tube.

The final and most severe problem occurred during a startup after the fourth time that tubes were plugged. One of the newly installed plugs popped out during plant startup, and the scabbard tube ruptured. The probable reason is that a tube with either no leak or an extremely small leak was plugged. Pressure would have risen inside this tube because of trapped water until the plug

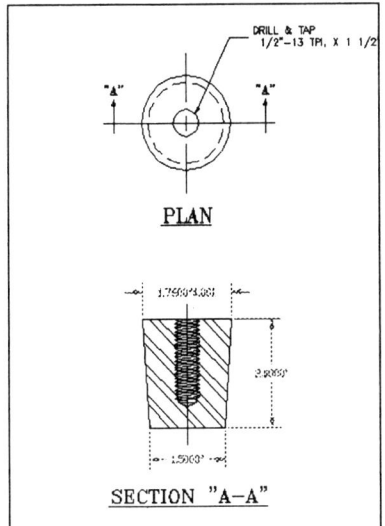

Figure 10. Scabbard plug.

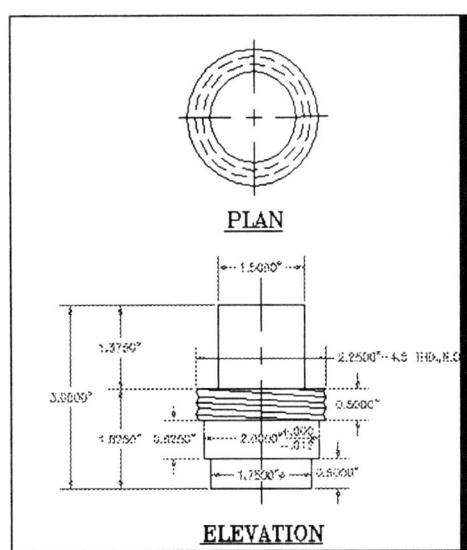

Figure 11. Bayonet plug.

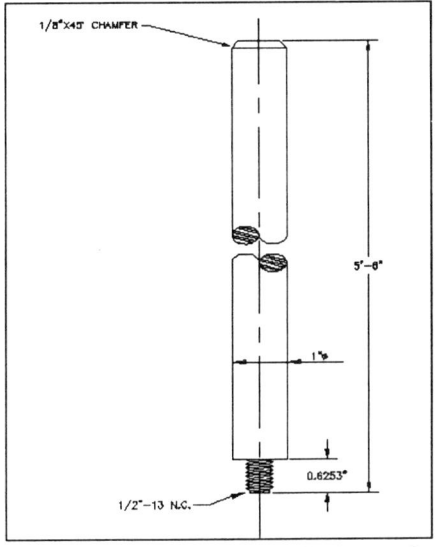

Figure 12. Plug insertion/drive rod.

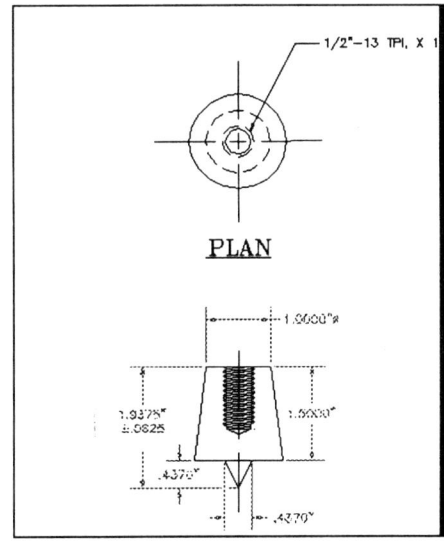

Figure 13. Plug driver tip.

popped out. Then the tube would have overheated, causing it to rupture. Extreme caution must be taken to ensure that only leaking tubes are plugged.

Conclusion

Plugging leaking tubes in the primary waste-heat boilers is a valuable procedure. This procedure can minimize unscheduled downtime, and the life of a boiler can be extended if the tubes are still believed to be in good condition. In our specific case, boiler replacement was avoided until new pressure shell liners could be procured. This procedure is not a cure-all, but it will be a valuable management tool for IMC-Agrico in the future.

DISCUSSION

I. J. Ohri, *Indian Farmers Fertiliser Cooperative Ltd.*: I would like to know what was the reason for so many failures of the tubes? Did you observe any reverse circulation in the tubes?

Brignac: We didn't have the reverse circulation. We've had that in the past. However, I could say that the initial reason for failures is that we had some poor boiler feed water quality prior to that July startup. That's about the only thing I could pin it to. They had those first few leaks. The second leaks were either nail holes or mechanical damage done, because the leaks were very small and the plugging was successful until we overheated them in January.

Ohri: You did mention the liner. Did you change the liner for the waste heat boiler?

Brignac: We changed the liner out in May. And I'm not sure if Kellogg made the recommendation, but I think that we went to Inconel 601 material for the liners. We had 310 in there before.

Ohri: I would like to share our experience in IFFCO. Our plants are quite old, say, about 20 to 25 years old. Earlier, the liner material was 310 for the bottom half of the liner and 321 for the top half and we had changed to Incoloy 800-H for total length.

Brignac: Someone had asked in a previous session about distributor nozzles in the bottom. We did not have distributor nozzles in the past, and we surveyed all the people we know. Half the people had them and half the people didn't have them. We went ahead and put them back in.

Inner Basket Failure of Ammonia Booster Reactor

A major failure involving the inner basket of the ammonia booster reactor after a scheduled turnaround caused an additional downtime of 31 days. The events leading to the leak, identification of the leak, selection of repair options and procedures, and finally the ordering of materials critical for the repair are described in this article.

Kamarudin Zakaria and Lau Nai Tuang
ABF Bintulu, Sarawak, Malaysia
Reinhard Michel
KRUPP-UHDE Fertilizer Division, 44141 Dortmund, Germany

Introduction

Bintulu Fertilizer (ABF) was established as a joint-venture project by the Association of South-East Asian Nations (ASEAN). Its primary purpose was to ensure regional self-sufficiency in the supply of fertilizers.

Commercial production began in October 1985 at a design capacity of 1,000 metric t/d of anhydrous ammonia and 1,500 metric t/d of granular urea. In 1991, its production facilities were revamped to a capacity of 1,200 metric t/d of ammonia and 1,800 metric t/d of urea. The plant was originally designed and revamped by Krupp-Uhde. ABF was accredited under the ISO 9002 quality management system in July 1993. In 1994, the first triennial turnaround was undertaken.

Further expansion was carried out in July 1997 (the seventh turnaround) to increase the plant's ammonia capacity to 1,320 metric t/d. Plant reinstrumentation to the Distributed Control System (DCS) was also carried out during this turnaround.

Description of Problem

In August 1997, while in the process of plant start-up after the turnaround, a very serious problem was encountered involving the ammonia synthesis loop (Figure 1), specifically the booster reactor, 08-R002. The inner basket of the reactor had partially failed and leaked catalyst, resulting in plugging of downstream synthesis-loop high-pressure equipment and piping.

Description of the Unit

Booster Reactor, 08-R002

The booster reactor was installed in 1991 as one of

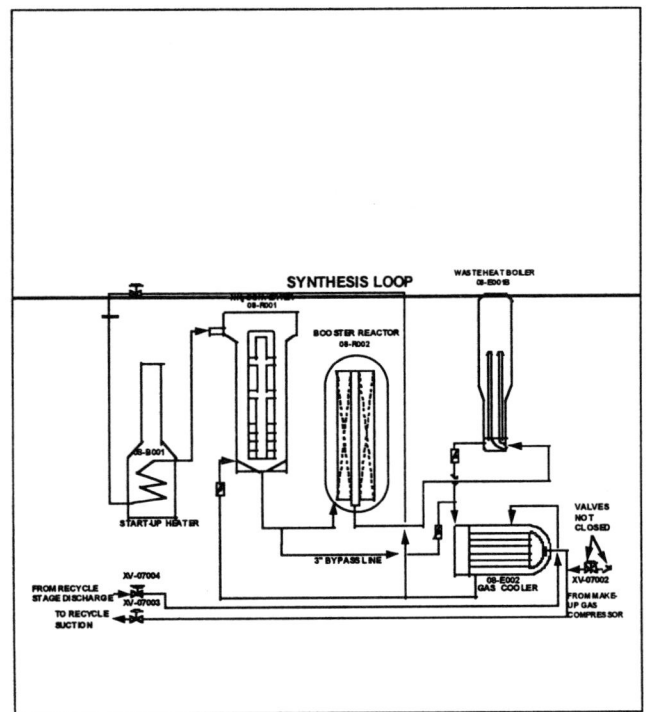

Figure 1. The ABF synthesis loop.　　　　Figure 2. The booster reactor, 08-R002.

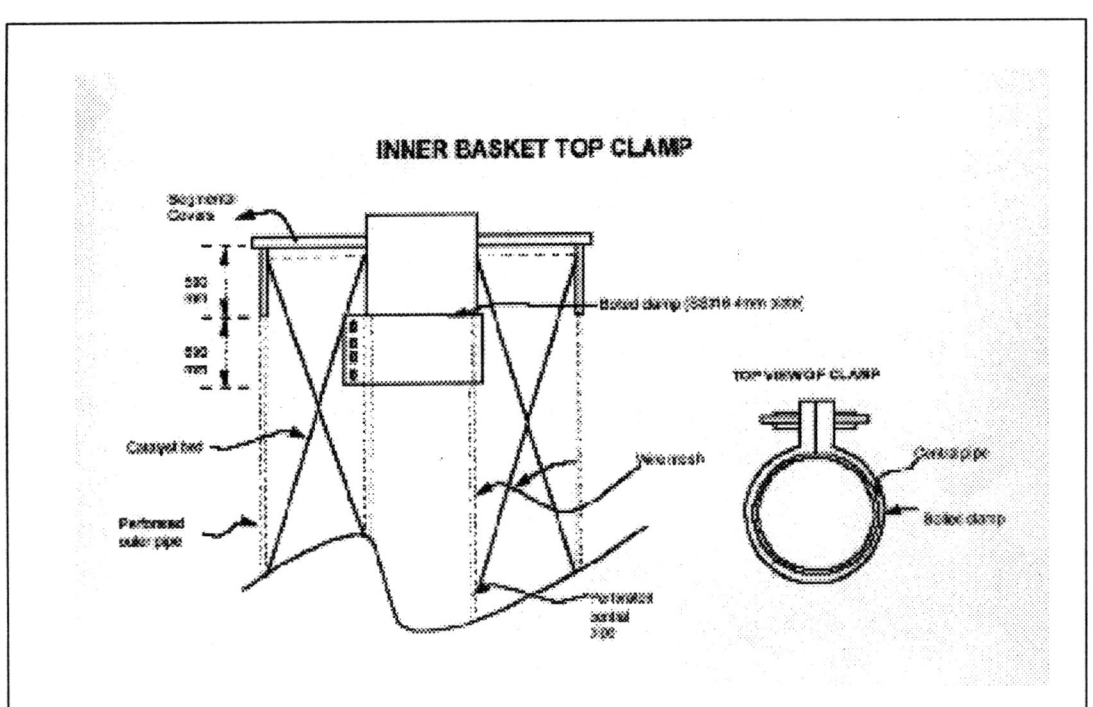

Figure 3. Clamp installation at the top-most portion of the central pipe.

many major items of equipment in the first ammonia plant revamp for a capacity increase from 1,000 metric t/d to 1,200 metric t/d. ABF's booster reactor is the first such reactor installed in the world and was followed by similar designs in BASF (Belgium), Saskferco (Canada) and QAFCO (Qatar).

Figure 2 shows a simplified layout of the reactor's internals and the gas flow direction. An annular basket in the vessel holds about 46 m³ of catalyst. Feedgas enters the annular space between the outer basket and shell, passes through the catalyst radially, and then exits through the central pipe. The outer and central pipes have 25-mm perforations to allow the gas to enter and exit. To prevent the catalyst from passing through the holes, the perforated shell and pipe are covered with two layers of wire mesh held in place at the joints by bolted strips. Altogether, there are 32 strips with 5 bolts each at the central pipe.

History of Operations

The booster reactor was opened up under nitrogen blanket as planned during the July 1997 turnaround to check for possible level drop of catalyst due to suspected compaction during service. This level drop was believed to have resulted in loss of production since it created a bypass in the vessel that allowed preferential flow of unconverted gas through the reactor. The intended remedy for this was to top up about 2 m³ of catalyst and install a 500-mm clamp at the center pipe to prevent a bypass in case the catalyst level shrank again (Figure 3).

The catalyst level had dropped by 0.68 m from the top, thus exposing the central pipe by 0.18 m. It was also found that the top-most metal strip at the exposed portion was damaged and one bolt was missing, thus creating a gap at the wire-mesh joint. This helped explain the discovery of a small portion of catalyst at one of the downstream separators. The damaged portion of the inner basket was clamped and catalyst was topped up.

Events Leading to Failure

During the plant startup after the turnaround, numerous plant trips occurred that were primarily caused by startup problems associated with the reinstrumentation project and other instrument failures. The synthesis loop was started up successfully four times before it tripped again. During these plant trips, the emergency trip valves (XV-07002,3,4) and check valve of the circulation stage were not closing. This was indicated by heavy blowing of the safety valve at the suction of the synthesis gas compressor.

During the subsequent startup, very low flow was observed through the loop. Trouble shooting was initiated and the simplest causes were initially suspected, such as wrong flow indicator, block valve lined up by mistake, and so forth. This did not yield the source of the problem.

Confirmatory Tests

Pressure survey

A pressure survey was carried out at various points in the loop. The exact location of the failure could not be determined on the basis of this survey, although the pressure profile indicated that some blockage was present between the booster reactor and the gas cooler (08-E002). It was initially suspected that the bellow/compensator at the gas cooler, 08-E002, had failed due to the backpressure created by the valves that were not closing during the previous plant trips.

The booster reactor was bypassed via a 3-in. line and flow was subsequently obtained through the loop. This confirmed the plugging of the line and equipment between the booster reactor and the gas cooler. Based on this, it was quite certain that some physical blockage, most likely by catalyst, was present downstream of the booster reactor and a major shutdown was planned. A project team was immediately set up to execute the rectification work.

Catalyst-blockage check

The drain line at the reactor's outlet was cut open and found to be full of catalyst. Equipment and pipelines downstream of the booster reactor were also opened up and catalyst was removed. This confirmed the suspicion that catalyst had leaked from the booster reactor.

Initial inspection

Since unloading the catalyst in reduced state would have been risky and expensive due to the resulting oxidation, it was necessary to first determine the level and the extent of the damage inside. It was still hoped that repair could be done from the top without removing catalyst from the vessel.

The top cover of the booster reactor was removed and its outlet pipe was disconnected for inspection. The central pipe was drained and about 2 m³ of catalyst were collected. After opening of the central pipe cover under inert, catalyst was observed leaking from the central pipe at approximately the middle of the bed height. This led to the decision to empty the catalyst out.

Problems Encountered

Spare catalyst

Expecting that the availability of spare catalyst would be critical, sourcing was put on top priority. Catalyst vendors and other ammonia plants all over the world were contacted in search for the iron catalyst. It was initially intended to reuse the catalyst and order a minimum stock. However, in consultation with the catalyst vendors, this option was not advisable because the extent of oxidation during unloading and reloading cannot be accurately predicted.

Fortunately a catalyst vendor was able to provide the full charge (47m³/143 t) of catalyst. Other problems in the catalyst transportation, such as landing rights and size of the local airport, were handled efficiently and the catalyst was delivered on-site within ten days of order. Due to the weight of the catalyst, it was shipped in two consignments using chartered Boeing 737 planes.

Equipment mobilization

The equipment was mobilized from Singapore by a handling company engaged to do the catalyst unloading and loading. The catalyst's arrival was delayed by two days because of the haze from widespread forest fire in a neighboring country that had been plaguing the region.

Removal of top cover

During the turnaround, problems occurred while removing and reinstalling the stud bolts (Figures 4 and 5) (this was the first time that the reactor was opened up since its installation in 1991).

The stud bolts were removed for ease of repairing the lip seal. Four bolts could not be fully screwed in (50% to 70% left), even with the use of a crane-assisted hammer and specially fabricated spanner. Therefore, it was decided to remove and reinstall the top cover without removing any stud bolts. The contingency plan was to replace the whole lip seal if repair was not possible.

Catalyst unloading

Dry unloading under inert was favored over wet unloading by water, as it was anticipated that drying and cleaning would be major problems later.

The problems encountered during this highly pyrophoric catalyst unloading included:

- The lack of available nitrogen affected cooling of the bed, the unloading rate, and the explosivity level inside the reactor. ABF's nitrogen plant capacity of 300 Nm³/h was insufficient. A minimum of 1000 Nm³/h was required.
- Some of the equipment broke down due to the temperature rise caused by leaking oxygen into the cycle at the suction side and the excessive temperature (~90–170°C) resulting from oxidation of the catalyst This slowed down the unloading remarkably.
- The catalyst was dangerous to the contractors working inside the vessel and was very hard towards the bottom of the bed due to compaction. A jackhammer was needed to loosen it. The wire mesh was punctured by the jackhammer and showed several holes at some portions of the bed.

Engaging a specialist to handle the unloading proved to be the right decision, as the unloading was considered a success in spite of the aforementioned difficulties.

Purchase of Other Materials

Since repair options included complete change of

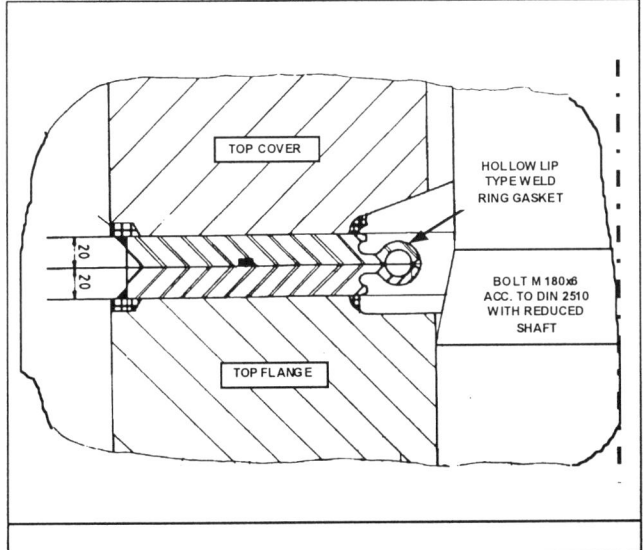

Figure 4. Top flange configuration.

Figure 5. Photo of top flange with bolts.

internal basket wire meshes, materials like wire mesh, stainless steel plates, bolts, and nuts were immediately procured or sourced from other ammonia plants. The difficulties faces in procuring the needed amount of nitrogen from the local supplier were also a constraint.

Process Cleaning

Spilled catalyst was again cleared from the central pipe, as the leak was continuous even after the plant was shut down. After vacuuming, the vessel was flushed with air under controlled conditions to oxidize any remaining catalyst dust and prepare for normal entry.

Inspection and Analysis

At several strips, one or two bolts out of five bolts per strip were missing. Two bolts failed in the thread inside the nut. A total of 28 out of 160 bolts failed. The fracture appeared similar to tensile test fracture surface. There was no indication of shear damage.

At two of the strips, two top bolts had failed, causing the strips to come off (Figure 6). This exposed the holes at the central pipe and, as a result, also exposed leakage of catalyst.

New bolts were welded in the workshop and broken by turning the nuts with a spanner for comparison. The force needed to break the new bolts was not higher than the force needed to break the actual bolts. The existing bolts were of the same strength but less ductile in comparison with the new bolts.

A nitriding test was done and it was found that only surface nitriding had occurred. Analysis showed that a very hard, embrittled, and cracked nitrided layer of 280 μm had formed on the surface of the sampled

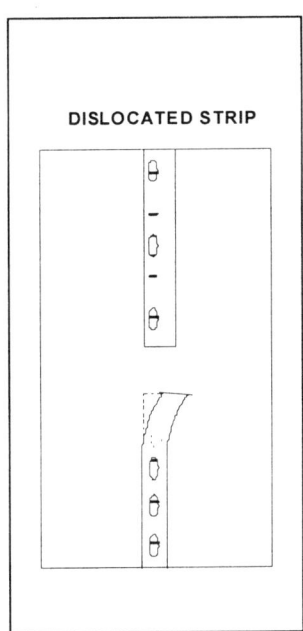

Figure 6. Dislocated strip after failure of two topmost bolts.

bolts and nuts. The embrittlement and cracking were limited to the surface layer, leaving the core intact.

During the plant trips just before the failure, backflow of the gas through the trip valves and check valve that were not closing created an undetermined pressure differential in the direction opposite to the normal flow direction for a short time period. Initial calculations indicate that the bolts may fail at a differential pressure of 13 bar. The fracture appearance of the failed bolts and extent of nitriding support this theory.

Repair

Many repair options were discussed before the catalyst was even emptied out. The options considered were dependent on the results of the damage inspection and included the following:

- Minimal catalyst removal, minor mesh repair at the top
- Partial emptying of the catalyst, minor repair of internals (catalyst reuse or new catalyst)
- Complete emptying of catalyst and minor repair/reinforcement of internals (catalyst reuse or new catalyst).
- Complete emptying of catalyst and complete change of internal basket.

The repair procedure was finalized with the help of the KRUPP-UHDE consultant after a joint inspection of the reactor internals.

- Each strip was reinforced with four additional bolts (total of nine bolts per strip).
- The strip joints were welded with close-up strips that would hold the main strips in place if the bolts failed.
- Two new circumferential clamps per strip were installed around the central pipe to hold the strips in place.
- Wire mesh damaged by jackhammer use and scaffolding activities (Figure 7) were patched with Inconel wire mesh using tack welding.
- A 1-m solid clamp was installed at the top of the central riser to prevent bypass.

Materials used for repair were as follows: strip, SS 304 (original SS 321-X6CrNiTi1810); bolts/nuts, SS 304 (original SS 321-X6CrNiTi1810); wire mesh, Inconel 600 (original); clamp, SS 304 (new) (see Figure 8).

SS304 was used because SS321 was not available.

Catalyst Loading and Box-Up

Catalyst loading was done using the Petroval Densicat Loading (dense loading) by the catalyst han-

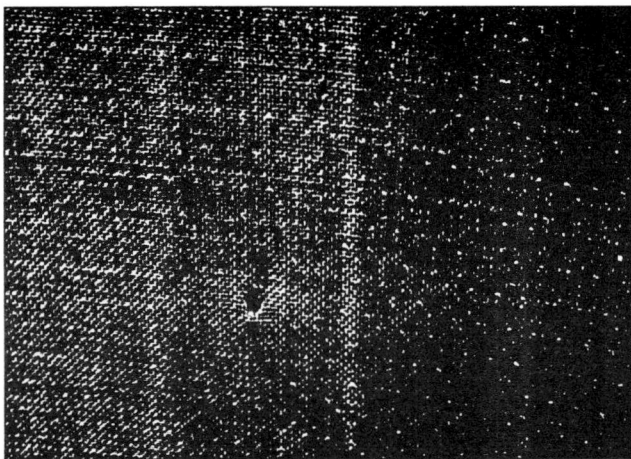

Figure 7. Appearance of wire mesh and mechanical damage from jackhammer.

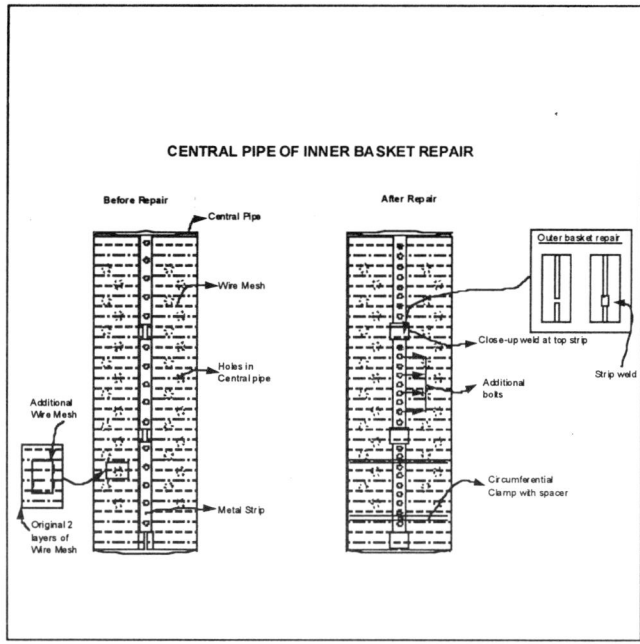

Figure 8. Repair work on basket.

Table 1. Breakdown of time of plant breakdown to recommissioning.

Confirmatory tests (pressure profile/ reactor bypassing)	2 days
Shutdown/cooling down of syn loop	2 days
Catalysts survey	2 days
Engagement of company for catalysts loading/unloading	1 days
Preparation of equipment	2 days
Mobilization of equipment, including two days delay for haze	9 days
Catalyst unloading	4 days
Repair	4 days
Catalyst dense loading	1 day
Start-up preparation and start-up	4 days
Total	31 days

dling contractor. Gouging of the lip seal was difficult because of limited clearance between the lip seal and the stud bolts. As a result of surface irregularity, the lip seal had to be cut circumferentially and rewelded under strict procedure to avoid a weld crack.

Duration

The total time taken from the point of failure to recommissioning was 31 days. The breakdown of time is shown in Table 1. All works were carried out around-the-clock with adequate staffing from contractors and ABF.

Conclusion and Future Provisions

- The numerous trips coupled with the defective valves caused the failure of the bolts and thus the catalyst leak.
- The fact that the emergency trip valves and check valve failed just after the turnaround indicates that the overhaul program for such critical equipment must be reviewed and enforced.
- This experience proved that the design booster reactor internals can be improved beyond the original design to increase their robustness.
- The possibility of installing a full bypass line around the booster reactor will be evaluated.
- The catalyst sparge policy, especially ammonia synthesis catalyst, needs to be reviewed.

Lessons Learned

- Plant startup was about 10 days behind schedule due to instrumentation problems encountered during front-end startup of the boilers and the ammonia plant. Due to the pressing need to produce ammonia on schedule to meet customer demands, the management decided to proceed with the startup while acknowledging the fact that the emergency trip valves and check valve were not closing.
- Complete reliance on vendors to do outsourced valve maintenance jobs may have been the root cause of failure of the valves just after turnaround.
- The failure of the top-most bolts, which was discovered during the turnaround, could have indicated that the bolts at other locations might have failed as well. It was predicted that, within the catalyst bed, the strips would be held in place by the catalyst if the bolts failed. The effect of backflow was overlooked.
- In future design of reactors, sufficient clearance should be provided between the stud bolts and the lip seal. This would speed up removal and reinstallation of the top cover.

Damage of Electric Motor of Benfield Solution Circulation Pump

The history of reverse rotation of the main Benfield solution circulation pump is described. Two incidents are described. During both incidents, the electric motor (6.3 Kv) was damaged beyond repair. The modification of the system after the incidents is also discussed.

Mubashar Mahmood Butt
Fauji Fertilizer Co. Ltd., Goth Machhi, Pakistan

Mogens Pedersen
Haldor Topsøe A/S, Denmark

Introduction

Fauji Fertilizer Co. Ltd. (FFC) operates two ammonia-urea plants at its Goth Machhi location. The first plant (base unit) was commissioned in 1982 with design capacities of 1,000 m tons of ammonia and 1,725 m tons of urea per day. The ammonia plant employs a late 1970s Haldor Topsøe design featuring a high steam-to-carbon ratio (3.75), a large primary reformer, a hot potassium carbonate (Benfield) system for carbon dioxide removal, and a synthesis loop with an operating pressure of 267 kg/cm^2. The base unit was successfully revamped to 122% of design capacity in 1992.

The second plant (expansion unit) was commissioned in March 1993 with design capacities of 1,100 m tons of ammonia and 1,925 m tons of urea per day. The ammonia plant is based on Haldor Topsøe's low energy process, which has the salient features of low steam-to-carbon ratio (2.5), a smaller primary reformer, a copper-based medium temperature shift catalyst replacing a high-temperature shift catalyst, a MDEA low energy process for carbon dioxide removal, a large purge gas recovery unit, and a low synthesis loop pressure of 150 kg/cm^2. Both the urea plants employ Snamprogetti urea technology.

This article describes the history of the reverse rotation of the main Benfield solution circulation pump. In a first incident which took place in July 1985, the main Benfield circulation pump tripped and the standby pump neither started on auto nor on attempts to start from the field. This led to severe damages due to the reverse rotation of the shaft. The second incident took place in January 1998, during the planned changeover of the main Benfield circulation pumps. The standby pump was started in parallel to the main Benfield circulation pump. Soon after stoppage of the main Benfield circulation pump, the standby pump also tripped. The shaft of the main Benfield circulation pump rotated in the reverse direction. In both incidents, the electric motor (6.3 Kv) was damaged beyond repair. This article also discusses the modifications of the system following the incidents.

Carbon Dioxide Removal Section

The base unit carbon dioxide removal section is shown in Figure 1, 2, and 3. The function of this section is to wash process gas, coming from the low-temperature shift converter (LTSC) by absorbing carbon dioxide in the hot potassium carbonate solution (Benfield) and stripping carbon dioxide from the same solution. Purified synthesis gas leaves from the top of the absorber for the methanator. Carbon dioxide from the regenerator is fed to the urea section.

The composition of the Benfield solution is as follows: *Potassium carbonate* = 27–29%; *Diethanol Amine* = 2–3%; *Total vanadium* = 0.8%.

Vanadium is added as a corrosion inhibitor and diethanol amine as a CO_2 absorption promoter.

The process gas enters at the bottom of the absorber (C-302) at a temperature of 110°C and 32 kg/cm² of pressure. The motorized valve (MOV-1) is installed at the inlet of the C-302 for isolation of the C-302 from the process gas side. The Benfield solution, rich in carbon dioxide (CO_2), is transferred from the high-pressure absorber (C-302) to the low-pressure regenerator (C-301) through the hydraulic turbine (XP-301). The energy recovered through this turbine supplements the main Benfield circulation pump (P-301A) drive. The prime mover is a 6.3 KV electrical motor on the same shaft. The liquid level in the C-302 is controlled by level controllers LIC-$6V_1/V_2/V_3$. The level controller LIC-6V2 regulates the flow through XP-301, and LIC-6V1/V3 are the bypasses of XP-301. In normal operation, the P-301A and XP-301 are in service to economize the energy consumption. The electric motor MP-301B, which drives the Benfield pump P-301B, is kept as standby.

The Benfield pump draws off regenerated solution at 120°C/2.5 Kg/cm² and pumps it to the absorber top at a discharge pressure of 38 Kg/cm². From pump discharge, the solution line splits into two streams. The main stream (75%) flow is controlled by the flow controller FRC-11 and is directly fed to the middle of C-302 at the high temperature of 120°C. The second stream (25%) flow is controlled by the flow controller FRC-3, and cooled to 70°C in two heat exchangers. This stream is introduced at the top

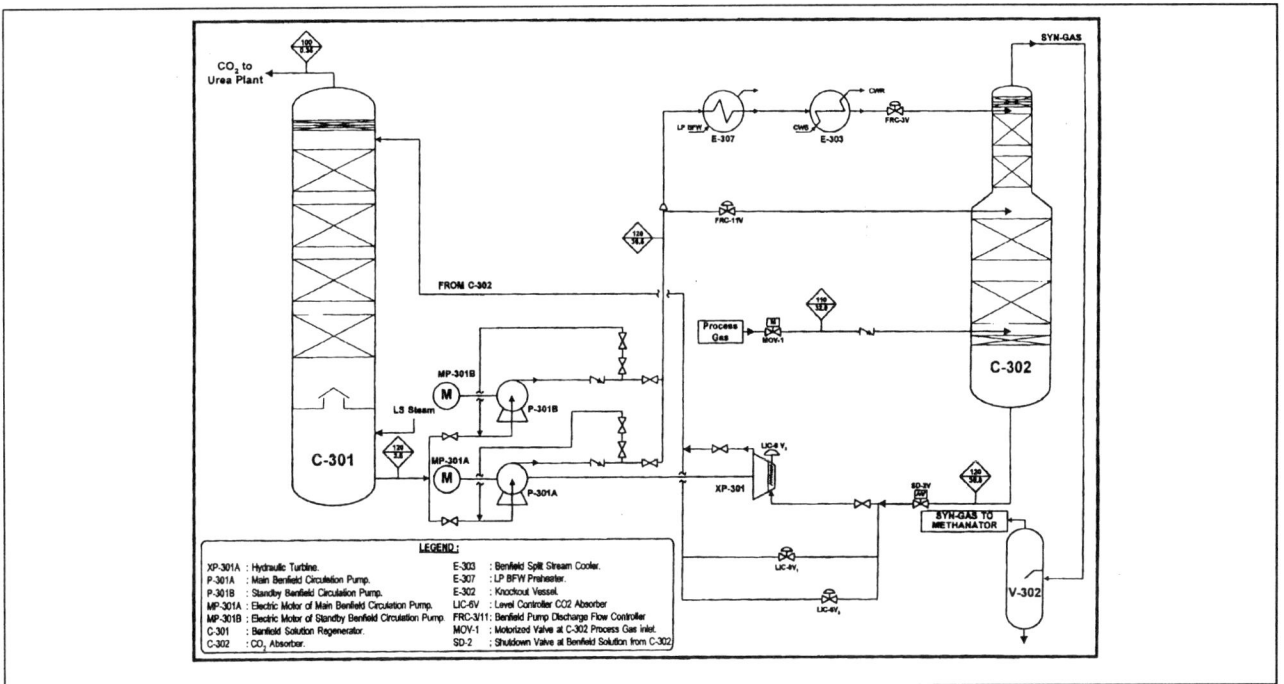

Figure 1. Carbon dioxide removal section.

of the C-302.

Reverse Rotation History of Main Benfield Circulation Pump

There have been two incidents of reverse rotation of the same Benfield pump, which was P-301A at our plant. The first incident took place in July, 1985, while the second was experienced in January, 1998. Both incidents are as described below.

First incident history

On July 8, 1985, the carbon dioxide removal section main circulation pump P-301A tripped. The standby pump P-301B did not start automatically or on attempts to start from the field. In fact, even if the standby pump had started, the sequence of events as described below would have been the same. As soon as the main pump P-301A tripped, it started to rotate in the reverse direction as its check valve (see Figure 4) did not hold the back flow of the solution. The speed of the pump in the reverse direction gradually exceeded the normal working speed of the pump. Friction due to reverse rotation at high speed caused motor bearings to fail and their high temperature ignited the lube oil flowing to these bearings. As soon as the fire started, the lube oil pump was stopped. Due to gas pressure in the absorber, the C-302 syngas started emanating from the pump P-301A seals after all the solution had escaped. After 5 min, the syngas also caught fire, which was brought under control by intensive firefighting activity and eliminating the fire source. It took 30 min to bring the situation under control. The syngas entered into the stripper through the pump's suction pipeline. After complete inspection, no damage was observed in the stripper. At the absorber inlet on the process gas side, the motorized valve MOV-1 remained stuck in a full open position.

As a result of the fire and consequential occurrences, the motor MP-301A was found damaged beyond repair, and instrument and electrical cables running above the pump were burned. No serious damage was observed to the pump P301A and the hydraulic turbine XP-301.

The following were the modifications carried out in

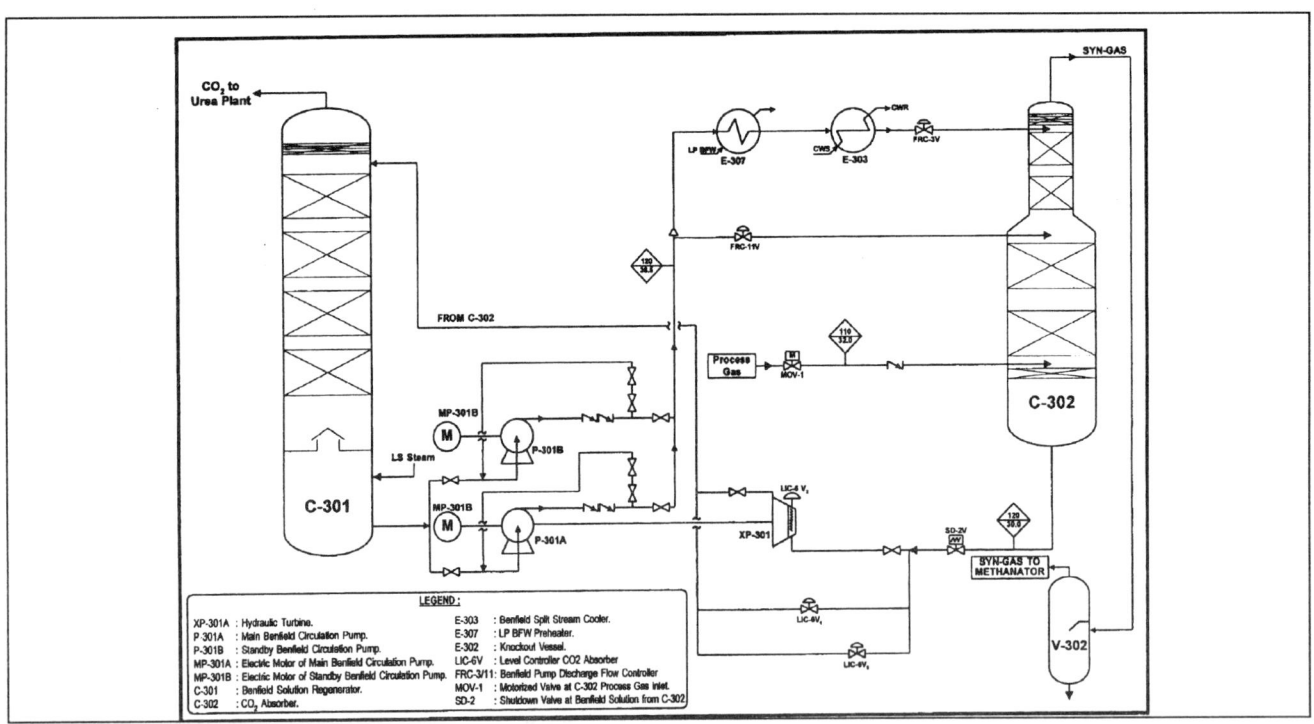

Figure 2. Carbon dioxide removal section after July 1985.

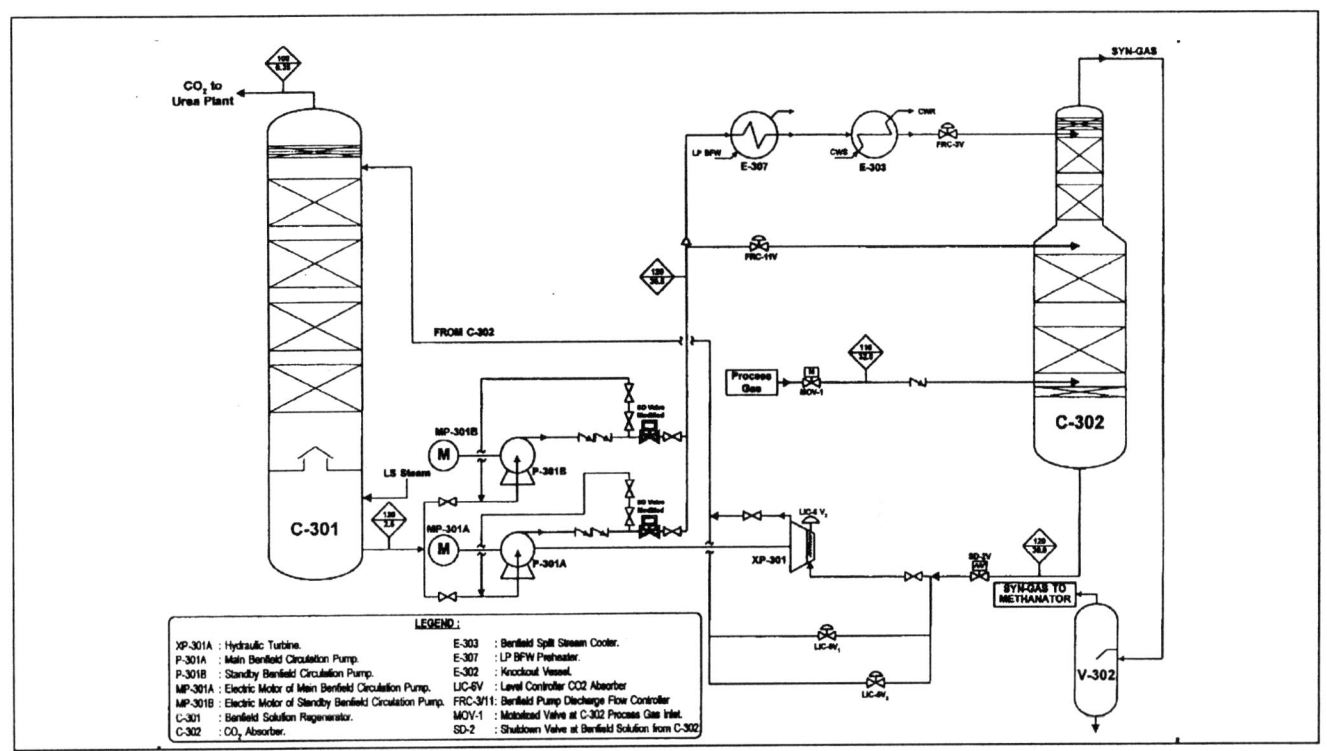

Figure 3. Carbon dioxide removal section suggesting modification after Jan 1998.

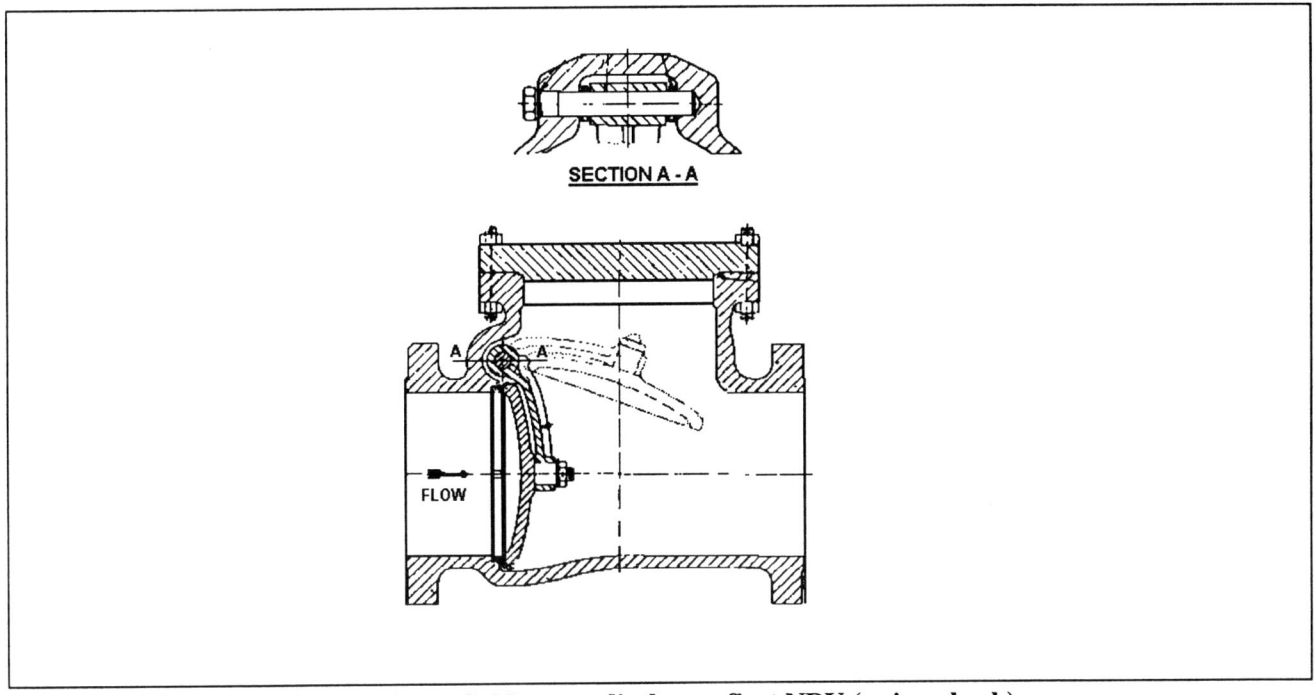

Figure 4. Benfield pump discharge first NRV (swing check).

1985:

- An anti-reverse rotating device (ARD), from Stieber GmbH (Figure 14), was installed on the new electric motor MP-301A's non-drive end.
- One additional non-return valve of a different type, the tilting disc type (Figure 5), was installed adjacent to the existing non-return valve (swing check).
- Three bolts through the arm bush of the non-return valve (swing check) were provided. The pin was extended to the body and converted into a movable lever (Figure 4a). (Figure 6 shows modification in the Benfield pump discharge first NRV.)
- The locking pin of the disc nut of the NRV (swing check) was tack welded.
- The Pointer outside the NRV (swing check) was installed indicating opening/closing of the disc.
- The operating procedure of the Benfield section, inspection frequency, and maintenance techniques were modified.
- The opening and closing limits of the motorized valve MOV-1 were redefined to ensure its operation of opening and closing on remote.
- An alarm on MOV-1 just before its full closure was provided in the control room.

Second Incident History

The plant was running normally at 129% of design capacity. Maintenance jobs were planned on the main Benfield circulation pump P-301A and on the expansion turbine XP-301. On January 11, 1998 at 9:00 P.M., the standby Benfield circulation pump P-301B was checked thoroughly just before a changeover. XP-301 was bypassed by shifting the level control from LIC-6V$_2$ to LIC-6V$_1$. After assuring that C-302 and C-301 held normal levels, XP-301A was isolated and the closing of the P-301A discharge gate valve gradually started. Observing change in the discharge flow at about an opening of 60%, its further closing was stopped as per SOP. P-301B was started in parallel to P-301A. Control of discharge flow with FRC-11 and FRC-3 on auto was attempted, but FRC-11 did not close beyond the 60% opening from the control room. The area operator immediately took the FRC-11 on a handjack, but it was hard to operate even on a handjack.

The P-301B was running smoothly and was developing normal pressure. The P-301A minimum flow line opened fully, and its discharge gate valve closed completely by the area operator. After running both pumps in parallel for about 10 min to assure that all operating parameters were at normal levels, flows, pressures and carbon dioxide slippage from the absorber, the pump P-301A was stopped using a local shutdown switch. The P-301B tripped on low discharge pressure after approximately 11 s of P-301A stoppage.

On stoppage of both pumps P-301A and P-301B, the Benfield section and downstream sections of the plant tripped. Process gas venting was shifted to the upstream of the C-302 section and the motorized valve MOV-1, at the inlet Benfield absorber, closed as per the trip logic sequence of the Benfield section. The area operator observed sparks and smoke from the outboard bearing of the electrical motor MP-301A, along with the reverse rotation of the shaft. The area operator activated the fire alarm and communicated with the control room. C-302 was depressurized from the downstream vent to avoid breakthrough of the process gas from C-302 through P-301A to the stripper C-301, and to stop reverse rotation of the shaft. MOV-1 closed to 20% of its full scale on the trip logic signal from the control room. The valve was taken on the hand jack and closed to isolate the C-302 from the process gas. The reverse rotation of the pump stopped when pressure in C-302 dropped to 16 Kg/cm^2. A major fire was avoided, because there was no breakthrough of gas from the C-302. The maximum speed in the reverse direction of the shaft was beyond its maximum range of 4,000 rpm.

Damages

In addition to the production loss of 51 h of ammonia and 40 h of urea, the incident caused the following equipment damages.

MP-301A

The electric motor MP-301A was damaged beyond repair. The electric motor shaft was seized and was removed by cutting the windings on the stator. The

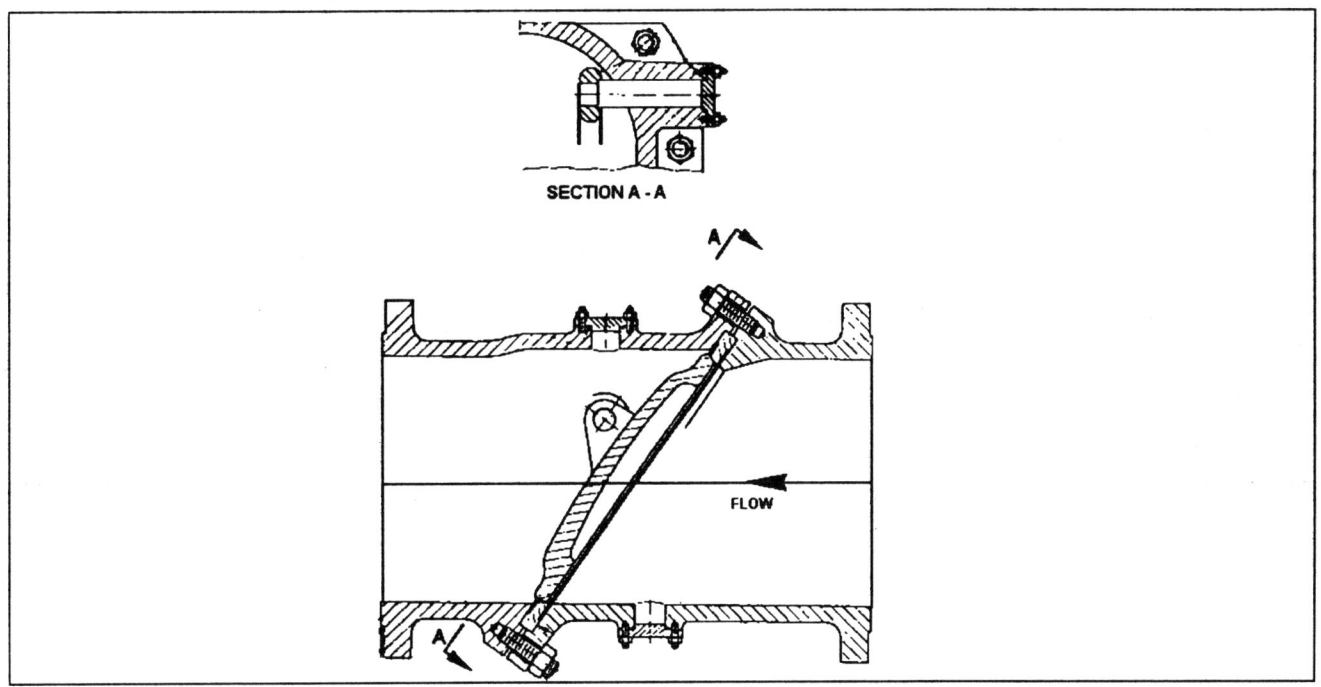

Figure 5. Benfield pump discharge second NRV (tilting disc).

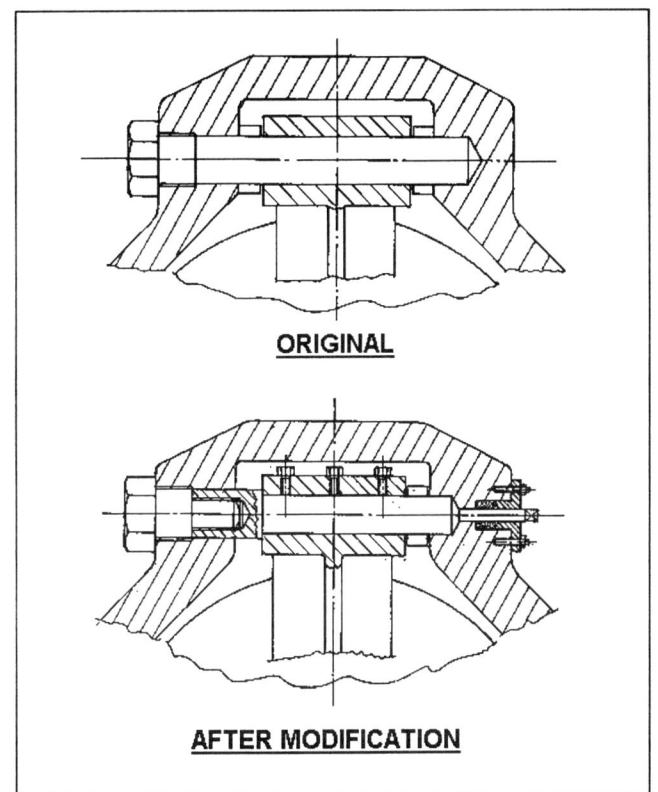

Figure 6. Modification in Benfield pump discharge first NRV (swing check).

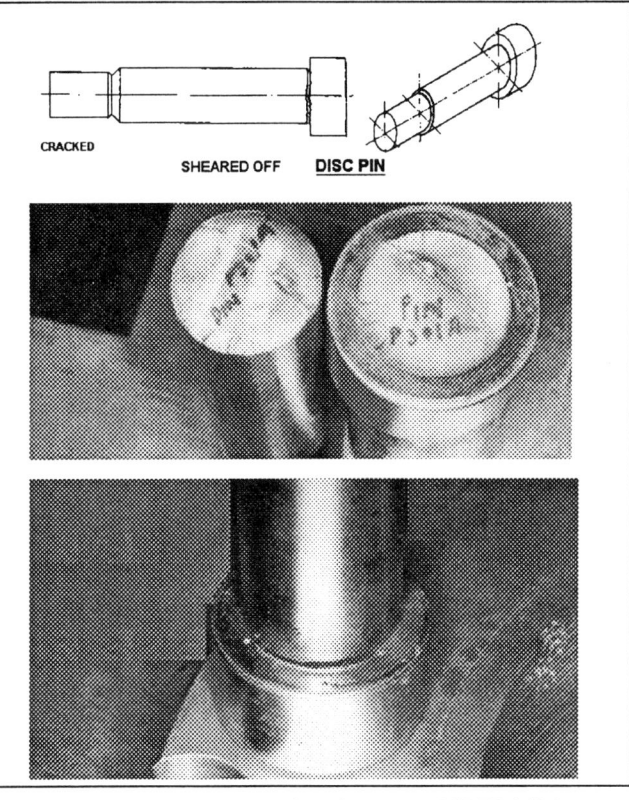

Figure 7. Damaged pin of second NRV (tilting check).

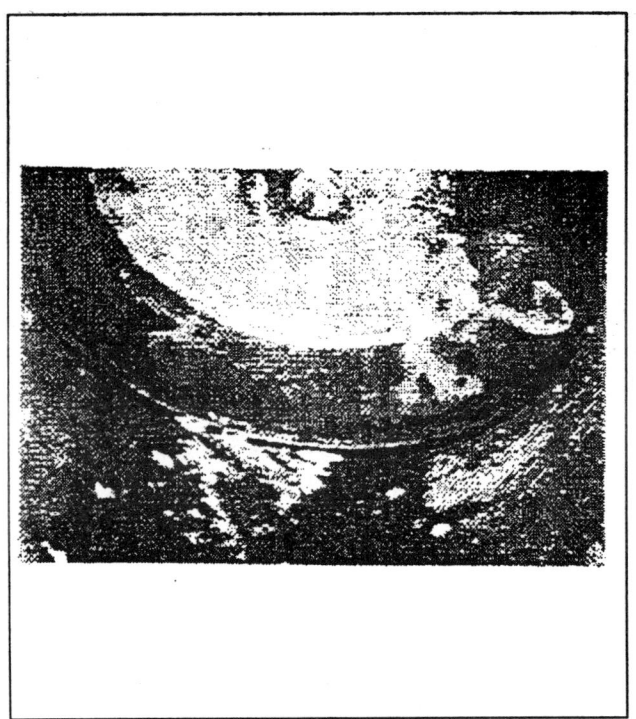

Figure 8. Clearance between disc and seat of second NRV (tilting disc type).

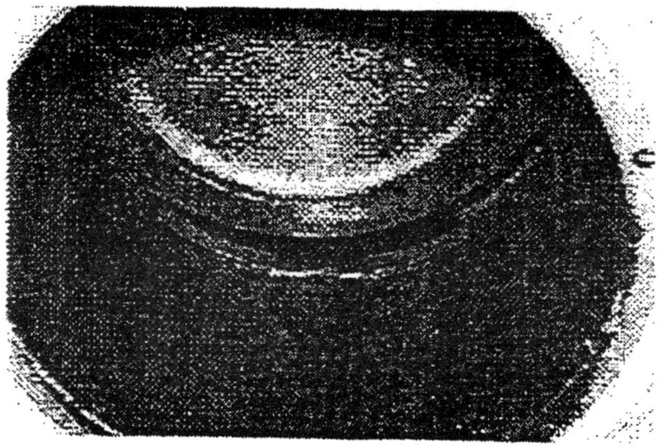

Figure 9. Clearance of discharge gate valve.

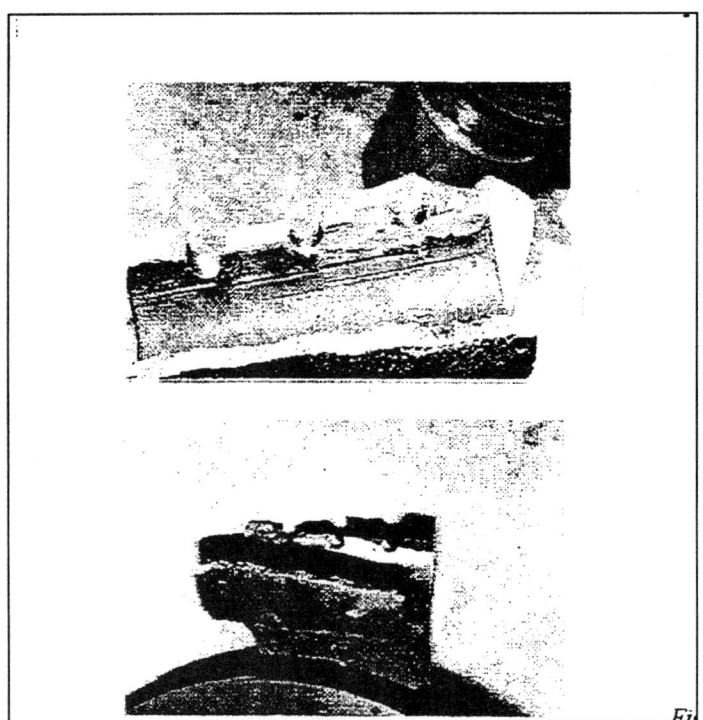

Figure 10. Damaged arm bush of first NRV (swing check check).

motor's stator windings broke and the stator slots were worn out by rubbing of the rotor at the non-drive end. At the non-drive end, the retaining ring holding the copper short circuiting ring broke into pieces. The copper short circuiting ring was badly damaged and stretched open due to excessive centrifugal force caused by overspeed. Lamination of rotor cores was twisted and damaged (Figures 11, 12, and 13). The bearings at both ends of the shaft and the oil lines at both ends of the pump were damaged.

An antireverse rotation device (ARD) mounted on the non-drive end of the shaft of MP-301A was damaged badly (Figure 14). The bolts attaching the common housing of the ARD and the non-drive end bearings to the motor frame were sheared due to excessive torque, and housing was dislodged from its mounting. Damage to the ARD caused an imbalance of the MP-301A rotor, which led to the rubbing of the rotor against the stator. The ARD failed to prevent the reverse rotation of the shaft at very high speed (more than 4,000 rpm). It is not clear from the inspection of the ARD fragments if it had cut in at all.

Benfield Pump and Hydraulic Turbine

After reverse rotation of the Benfield pump in 1985, emergency handling procedures to save the pump from process gas breakthrough were revised. This procedural awareness helped in handling the emergency. There was no fire incident at pump seals and no damage was experienced on the Benfield pump, the hydraulic turbine, or to the instrumentation.

Basic Cause of Incident

The basic cause of the incident was simultaneous failure of both the discharge non-return valves (NRV).
• First, the NRV (swing disc type) was found damaged from the three holes provided in the arm bush, as a modification in 1985 (Figure 4a). These three holes were the main stress area and cracks were initiated from their periphery, which ultimately turned into the failure of the arm bush, and it broke into two pieces (Figure 10). Two lock nuts freed from the bush were trapped in the cage of the downstream flow controller FRC-11 preventing its closure even on the hand jack.

The third lock nut was hand loose with the arm bush.
• Secondly, the NRV (tilting disc type) pin holding the disc was found sheared off (Figure 7) and the disc remained supported with one pinion leaving a clearance of one inch between the disc and the seat (Figure 8). The shearing of the pin was caused by fatigue due to high velocities in the discharge line of P-301A. Failure occurred at the point of maximum stress concentration.
• The P-301A discharge gate valve was fully tightened by the area operator. However, about a one inch opening from the bottom seat was observed during a subsequent inspection (Figure 9). Crystallization of solution in the glands increases friction of valves in Benfield service and valves become hard to operate. Additionally, the differential pressure across the valve gate makes its full closure difficult. The one inch opening of the discharge gate valve explains the tripping of the P-301B at low discharge pressure after 11 s of P-301A stoppage. The discharge flow of P-301B passed through the P-301A in the reverse direction and went to the common suction of the pumps. The ensuing low pressure caused a trip of P-301B. The back flow through the MP-301A rotated the shaft for four minutes until depressurization of the system.
• After stoppage of P-301B, flow controllers FRC-11 and FRC-3 were isolated from the area to avoid any exposure of gas to P-301A.

The installation of the second NRV at the discharge of P-301A in 1985 did not work at all. It is interesting to note that ICI, Billingham, U.K., also opted for a second NRV subsequent to reverse rotation of boiler feed water pump (Wallace et al., 1997).

Future Actions

A task force was established to investigate the incident and suggest temporary and permanent solutions. Their recommendations are as follows.

Contingency plan for nonavailability of standby pump

The base unit and the expansion unit were surveyed for the compatible electric motor for the extreme contingency. Two motors of a similar rating were identi-

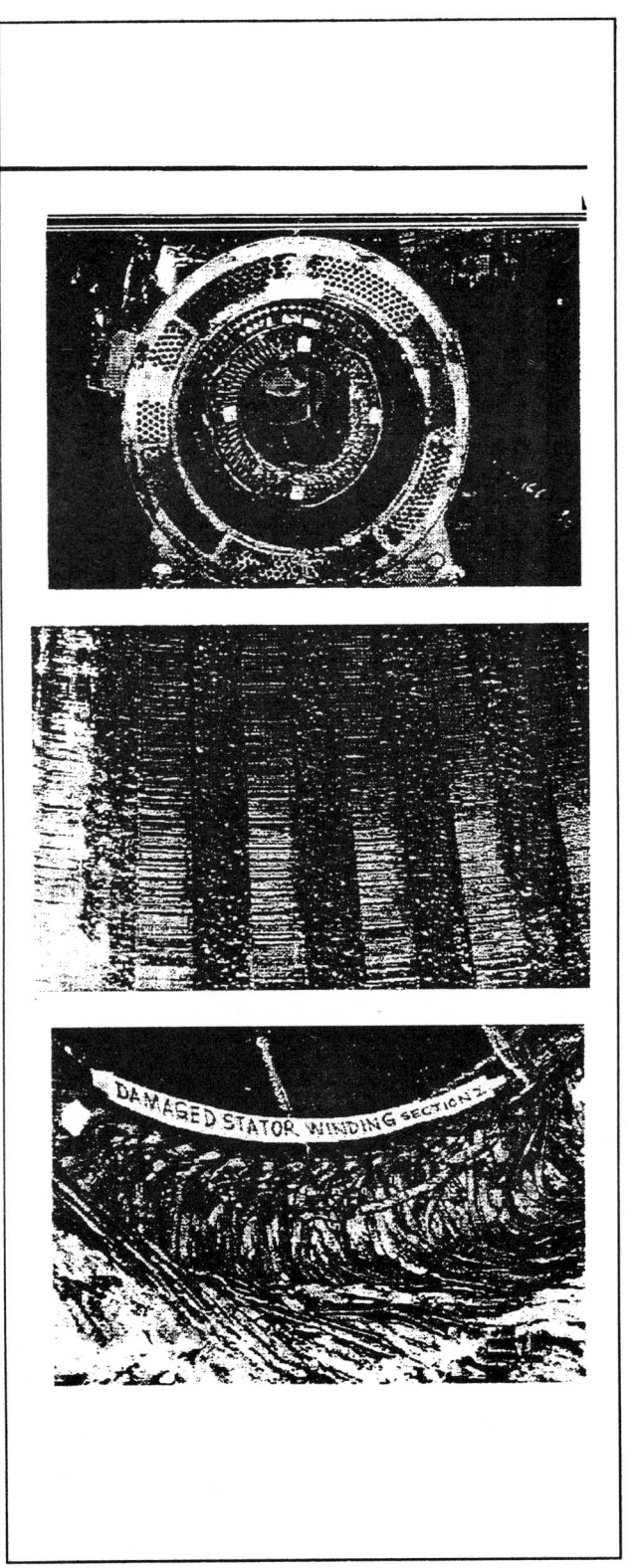

Figure 11. Damaged stator and winding of MP-301A.

Figure 12. Damaged rotor of electric motor MP-301A.

Figure 13. Damaged rotor of electric motor MP-301A.

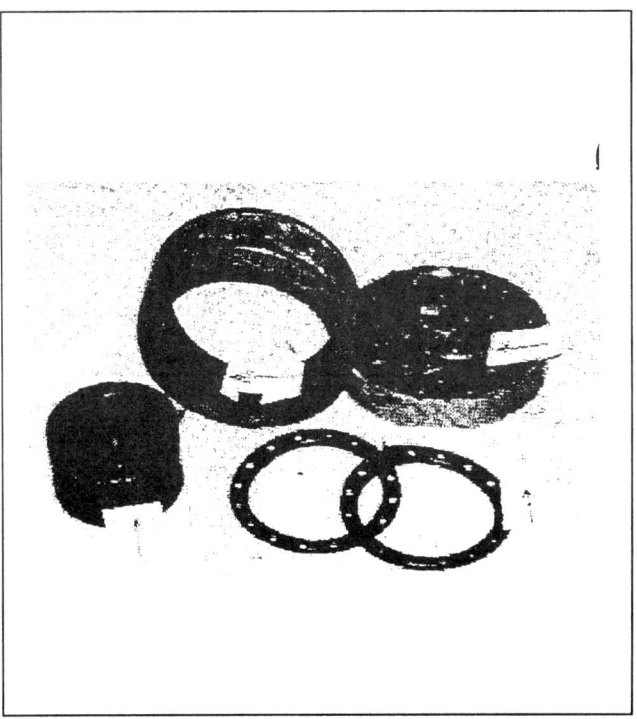

Figure 14. Damaged anti-reverse rotation device (ARD).

fied. One was of the main BFW pump at the base unit. However, its direction of rotation was opposite to that of the MP-301. The second motor is in the MDEA service at the expansion unit. The shaft elevation of this motor is shorter than that of MP-301A. A skid has been prepared to accommodate this motor. There was one more consideration to shift the MP-301B on the MP-301A pedestal so as to conserve energy by taking the XP-301 in service. This was ruled out because of undue risk of XP-301 failure, and, ultimately, ammonia and urea production loss. An emergency handling procedure for the P-301B trip was reviewed and amended.

System modification to avoid recurrence

Many brainstorming sessions were held for the revision and modification in the existing system to avoid recurrence. The outcome is as follows:

• Additional installation of a motor operated or pneumatically operated shutdown valve (ball valve type) at the discharge of the Benfield Pumps. The ball valve shall be with full bore, and straight pipe length of 1 to 1.5 m upstream and downstream shall be provided to prevent bolts, impeller fragments, and similar objects from getting stuck in the valve. The operating procedure will be modified. This valve will shut off with the trip of the Benfield pump motor. A manual opening and closing arrangement of the shutdown valves will be provided for the startup of the Benfield pump (Figure 3).

• Inspection of discharge of the NRV every six months and at any available opportunity.

• Full open and close indication on the pump's discharge gate valve.

• Benfield pump discharge line diameter is to be increased to reduce the fluid velocity. The prevailing fluid velocity is more than 5 m/s. It should be less than 3 m/s to avoid flow induced fluttering of the NRVs.

Confirmation of tight closing of flow controllers FRC-11 and FRC-3 to avoid any pass through the process gas in the case of a Benfield pump trip.

• MOV-1 at the C-302 gas inlet is to be made reliable for full closure on trip logic sequence of the Benfield section by revising its closing limits to full.

• Direction of rotation of motor is to be displayed locally.

Conclusions

• Check valves are prone to damage due to continuous movement of their flaps because of system variations. Consequently, check valves alone should not be relied upon to prevent back flow in these cases. The system should be made foolproof by additional protection like a shutdown valve, motorized valve, or any other arrangement, whichever is suitable, on the discharge of pumps.

• During normal running of the plant, control valves, where possible, should be checked for any restrictions and cleared.

• The ARD cannot be relied upon either. Nevertheless, to improve its reliability, we have increased the design torque in the replacement unit and have asked the supplier to place it on the drive end of the shaft. Also, the supplier of the ARD shall be requested to indicate the testing procedure (if possible), and the maintenance and cleaning procedure for the ARD.

• The size of piping housing and the NRV will be increased from 12 in. to 20 in. to streamline the solution flow. This will help minimize the flow induced vibrations of NRVs.

• It should be ensured at the stage of the initial design that all the major machines have the same direction of rotation. It will make it possible to swap motors in a contingency situation.

Literature Cited

Fromm, Dieter, and Wolfgang Rall, "Fire at Semi-Lean Pump by Reverse Motion in MDEA CO_2 Removal System," BASF AG, Ludwigshafen, Germany.

Wallace, D. P., P. J. Nightingale, and A. P. Walker, "Failure of Boiler Feed Water Pump Turbine Following Site Power Failure," Ammonia Plant *Safety & Related Facilities,* Volume 38, AIChE, New York (1997).

DISCUSSION

Rudy Frey, *M. W. Kellogg*: This seems to be a reoccurring problem discussed earlier in several other papers. I agree with your conclusions, and add that check valves are notoriously unreliable. You did employ redundancy in check valves, but did you consider diversity? That is providing an axial or dual flow type check valve which wouldn't be subject to the same type of failures that your swing check type check valve is subject to.

Butt: You're right, and we are looking into it. We are still looking for another type of NRV, which is safer than the swing check.

Frey: I would be glad to suggest some to you.

Khetarpal, *Khribco*: Khribco published the other paper on reverse rotation. We had faced the problem four years ago. We changed the material of construction for pump discharge isolation valves from carbon steel to stainless steel and converted it to motor operated valves. Check valves were changed to stainless steels.

Butt: We also have the valves of stainless steel 316. We also have the same set of materials; even the discharge line is of stainless steel.

Robert H. Roberts, *M.W. Kellogg*: You stated that Benfield in that section was about five meters per second?

Butt: Yes.

Roberts: Did you look at how fast they'd recommend, even with inhibitors, for Benfield? If you had the wrong material in there, that may be the cause of some of the problems with the pins and so on. That's one of the main reasons for the problems other than if the seam tracing is not good, although in the right place, or if the flush is not in the right place.

Butt: Right. The problem is because of the high velocities. The continuous fluttering of the NRV's disc can damage the pin. However, we are now modifying the system to reduce the velocities in the discharge line and to install an additional shutoff valve at the discharge of the pump, so that it can be isolated quickly. You see, for nearly 13 years between 1985 and 1998, we have revised our operational procedures. Because of operational procedure awareness, the damage was less than that in 1985.

Vanadium Recovery Solves Soot Recycle Problems

In partial oxidation (pox) operations, some 1% w/w of the hydrocarbon feedstock is converted into gasification soot. When this soot is disposed of by a total recycle, the resulting accumulation of ash components in the recycle flow is the main reason for material destruction and scaling of the pox-reactors, waste heat boilers and the related water system. A straightforward solution solves the soot disposal problem by recovering the ash forming metals as a valuable ore substitute.

W. Soyez
Hydro Agri Europe, 1200-Brussels, Belgium

Introduction

Hydro operates a combined ammonia and urea unit in Brunsbüttel, at the mouth of the River Elbe in northern Germany. The capacities are 2050 mt/d ammonia and 1600 mt/d urea. The ammonia plant is based on a partial oxidation unit with four parallel reactors and two scrubbers. The downstream gas processing uses standard technologies and is executed as a single train unit.

The partial oxidation technology has been proven to be a powerful tool for disposing the heaviest hydrocarbon feedstocks as refinery bottoms or bitumen - whatever is the cheapest. The basic principles of the partial oxidation technology are shown in Figure 1.

The feedstock, usually heavy fuel oil or visbreaker residue, is atomized and oxidized with a mixture of steam and pure oxygen. The operating pressure is about 50–70 bar (725–1,000 psig) in modern units. The reaction temperature is some 1,350°C (2,460°F). The feedstock is converted into a raw gas showing

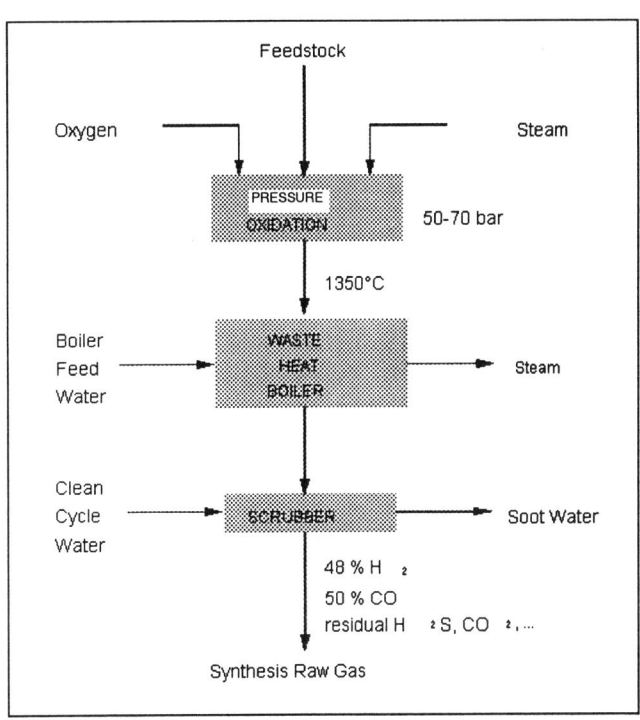

Figure 1. Partial oxidation technology.

about 50% H_2 and 50% CO. The traces of H_2S and HCN are separated first. The subsequent gas treatment stages are standard.

Technology Drawbacks

This elegant technology has two main drawbacks: soot and ash.

Soot

Conversion of feedstock to gaseous products is not complete under these operating parameters. Some 0.5 to 1% w/w of the feedstock are transformed into fine carbon particles. This soot, together with the raw gas, flows through a waste heat boiler or directly enters a quench and scrubbing stage. There the soot is precipitated and discharged as soot water.

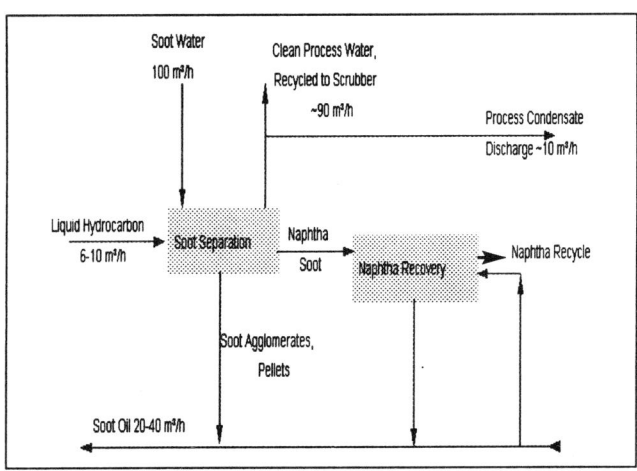

Figure 2. Soot separation stage.

The soot water, amounting to about 100 m³/h in a worldscale unit, is usually reprocessed in a soot separation stage (Figure 2).

The soot water is mixed with hydrocarbons. The hydrocarbons force the soot particles to agglomerate and to rise. The agglomerates can easily be skimmed or sieved-off. The purified water is recycled to the scrubber.

When light hydrocarbons are used for agglomeration, they have to be replaced in a 2nd stage by cheaper fuels. Whatever technology is applied, the soot ends up as a suspension in heavy fuel oil (soot oil), ready for being disposed by external firing in auxiliary boilers or by recycling the soot oil as feedstock to the gasifiers. Because soot particles drastically increase the viscosity of fuel oil as well as water, the soot concentrations have to be kept below some 3 wt. %. This explains the relatively high product flows indicated in Figure 2.

Ash

Beside soot processing, pox-operations are further complicated by the fact that traces of metals, especially vanadium, iron and nickel, are present in any crude oil. During refining, these metals accumulate in the bottom products. And the cheapest feedstocks for gasification units are in fact those residues having the highest metal concentrations.

The metals and earth alcalines entering the pox-reactor are transformed into oxides, sulfides and carbonates. Because these compounds are hardly soluble, the ash follows the soot process flow and ends up in the soot oil, too.

When the soot oil can be disposed as fuel in an external boiler, the high ash concentrations caused problems as deposits in the flue-gas duct, increase in pressure drop and downtime. These were the facts forcing Brunsbüttel to attempt 100% soot recycle operation as early as 1983.

Technology Options and Investigations

Although 100% recycle operation was mentioned as a clever solution for soot disposal in the early days of pox-technology, it took some three decades to make the process suitable for use in continuous operations.

Initially, the pox processes had no defined outlet for ashes. Thus, closed-loop operations had to lead to continuous ash accumulation. Fortunately, the metal concentration in the fresh feedstock was low and the drain of some soot/soot oil together with the process condensate discharge was enough to keep the ash levels in the soot recycle within reasonable limits (1,000–2,000 ppm V, Ni and Fe).

As a consequence of the oil crisis, the refineries had to go deeper into the barrel. Enhanced recovery of dis-

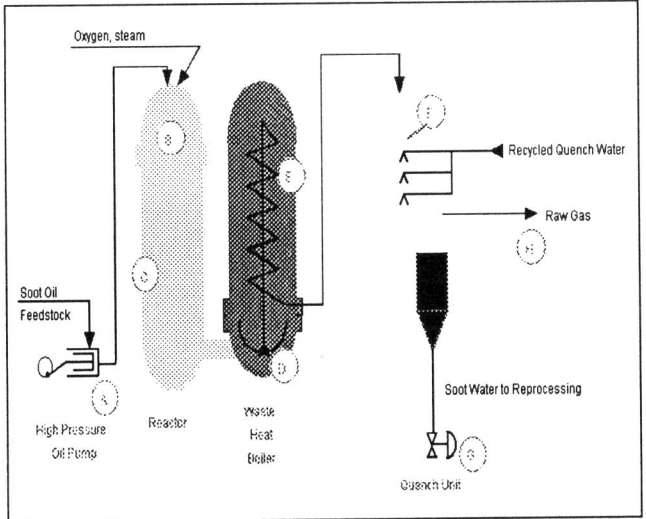

Figure 3. Locations of severe ash loads.

Table 1. Agents and Effects of Severe Ash Loads

LOCATION	AGENTS	AFFECTING
A	Al- and Si-oxides	Pumps/valves by abrasion
B	Ash	Atomizer tip, burner cooling flange
C	V-oxides	Bricklining
D	Ash, sulphides	Waste heat boiler inlets
E	Ash	Blockage of pipe diameter, causing abrasion
F	Ash	Blocking quench nozzles, pipes
G	Carbonates	Precipitation after CO_2-pressure reduction
H	Carbonyles	Subsequent gas purification

tillates caused a significant increase in the metal content in the residues. In the 1980s, the vanadium concentration in our fresh feedstock in Brunsbüttel rose to 400 ppm. To cope with the ash input, our factory had to improve the ash rejection within the cycle by enforced sedimentation and filtration stages.

Only a few years later, the metal concentration increased again to some 1,200 ppm V, Ni and Fe in sum. The existing facilities for ash rejection were incapable of controlling the ash recycle operation. The locations where the consequences of severe ash loads in pox units occur are shown in Figure 3 and listed in Table 1.

The most severe consequences are corrosion at the atomizer tip and waste heat boiler damage.

Corrosion of the atomizer tip can lead to a misalignment of the flame and in the worst case to a reactor blow-off. Corrosion of the waste heat boiler inlet nozzles causes expensive maintenance and some 10 days downtime per line and event.

The design service life of a waste heat boiler is about 40,000 h. During operation under high ash recycle conditions, the waste heat boiler service life dropped to a few weeks only. Table 2 shows the waste heat boiler mortality in Brunsbüttel during that period.

When we were faced with that situation, we evaluated possible explanations, tried to find out the corrosion mechanism and looked into new designs. But we also decided to keep one reactor line free from recycle operations.

The explanations for the increased waste heat boiler

Table 2. Record of waste heat boiler failures.

	1985	1986	1987	1988	1989	1990
A		X			X	
B					X	X
C		X			X	X X
D	X			X	X X	X

failures were overload, abrasion and ash load.

The corrosion mechanism is difficult to investigate. There is not much published and the operating conditions are hard to simulate in laboratories.

The new designs showed a double wall tube sheet, were quite sophisticated, expensive and without previous experiences at that time.

During investigations it turned out that the specific reactor line, when operated without ash recycle could stand the operating conditions without failing. This experience led to the sole explanation that only those components which have previously passed through the pox reactor can jeopardize the waste heat boiler.

Improvement Options

The consequence of this finding is to exclude the ash recycle. The soot has to be disposed in another way. Step by step we composed two technologies for soot disposal.

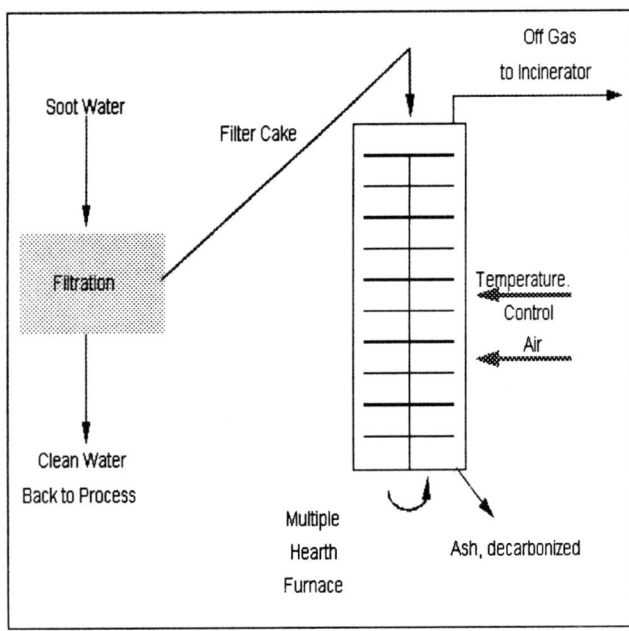

Figure 4. Process principle - Route B.

Route A: filtration followed by a multiple hearth furnace

The process principle is shown in Figure 4.
Soot is filtered from soot water, continuously or batchwise and conveyed to a multiple hearth furnace. Falling down stage by stage, the soot is first dried by the flue gases, then roasted and decarbonized under mild combustion conditions.

We had found out that soot can be incinerated and burned at quite low temperatures. This is essential to control the oxidation levels of Vanadium. Only low oxidation levels guarantee high melting oxides. If vanadium is oxidized to its highest level, low melting eutectics are formed. These eutectics attack any material and cause the disastrous V-corrosion. As long as the oxidation is well controlled and some residual carbon is left in the ash, mainly the noncorrosive oxides appear in the final product.

A process according to route A has been realized quite recently in one of the latest gasification units in Europe. During the process evaluation phase, we found that running a huge multiple hearth furnace (plus one backup unit) does not really fit into modern synthesis gas production technology. So, we developed another process for soot disposal, Route B.

Route B: filtration, drying and soot combustion

The process principle is illustrated in Figure 5.

Having already been flocculated, the soot water is filtered. In a first drainage step the clear water is withdrawn and recycled to the pox-process. The resulting soot cake, still some 90% water, is further squeezed between two belts and rollers of various diameters to get the soot cake as dry as possible. The concentration of solids in the filter cake can be increased to 20%.

The filter cake is granulated in an edge mill and falls into a fluid-bed dryer, operated under nitrogen.

After a residence time of 1 h, the soot granules are dry and ready for grinding. The size distribution after grinding is standard for coal dust burners: 90% < 90 μ. The soot dust is conveyed in an air jet to the combustor and burned. The combustor is followed by a waste heat boiler and a filter stage. The ash is quite rich in vanadium (32–45 wt. % V) and can be sold as vanadium ore substitute.

This unit was erected in 1996 and came onstream in 1997. The front end operated as designed from the very beginning. In consequence, the old soot recovery

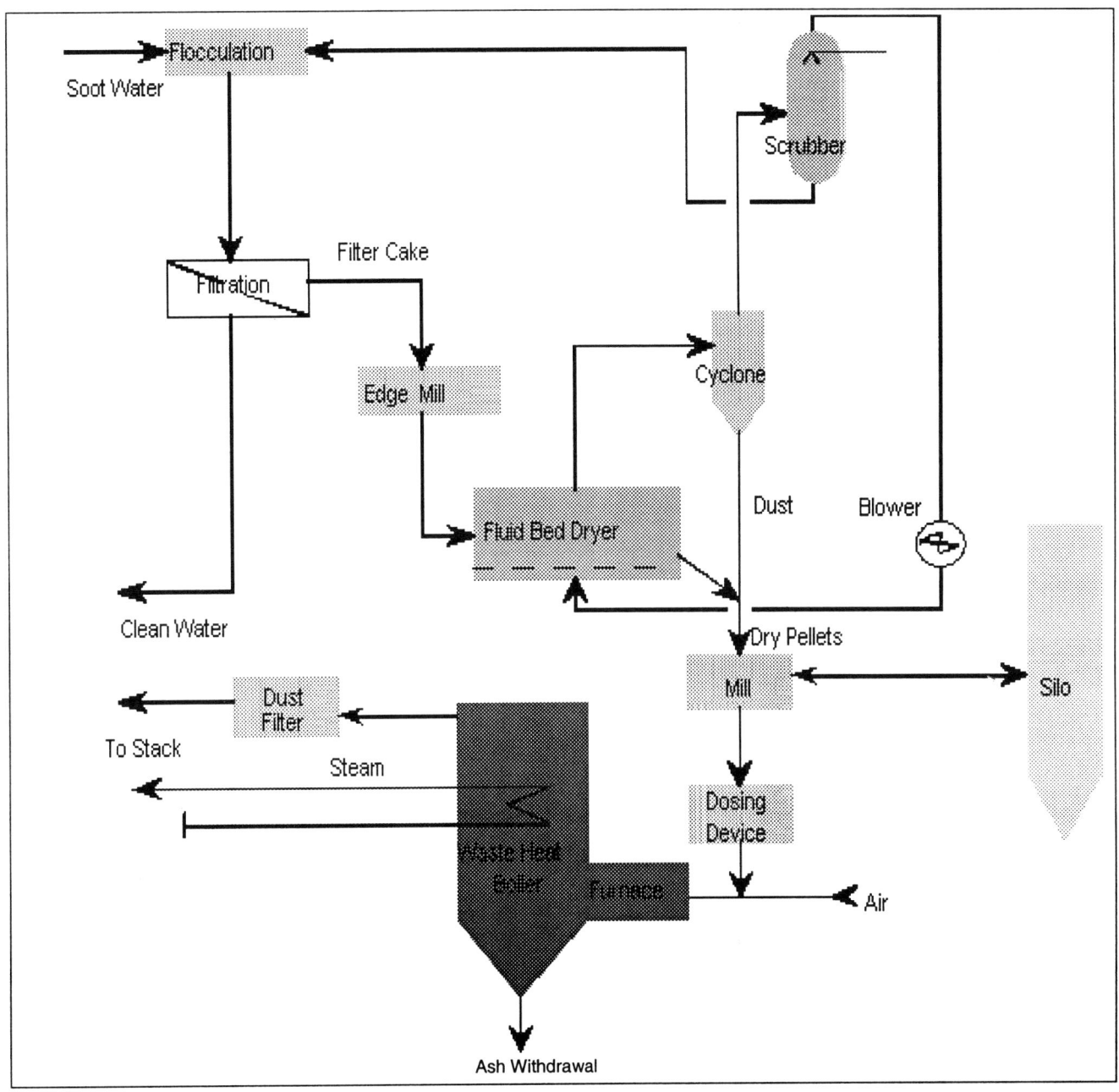

Figure 5. Process principle - Route B.

and recycle facilities were stopped and shut down, later on.

As expected, the soot combustion itself offered some challenges. The soot dust, being the fuel, consists of some 50% of ash components. To save investment costs, the burner had to be small. This requires a hot, short flame. On the contrary, the formation of higher vanadium oxides is strictly to be avoided. This calls for modest firing conditions. A technical solution for those contradicting preconditions is a circular burner (Figure 6).

The combustion air and the soot dust are injected tangentially and the flame is rolled up into a helix. To lower the flame temperature and to keep the burner lining free from ash particles, cold air is also injected.

Although the bulk of soot burned in the inner, circular flame, some soot and ash particles could reach the burner wall and agglomerated to a slag layer. This

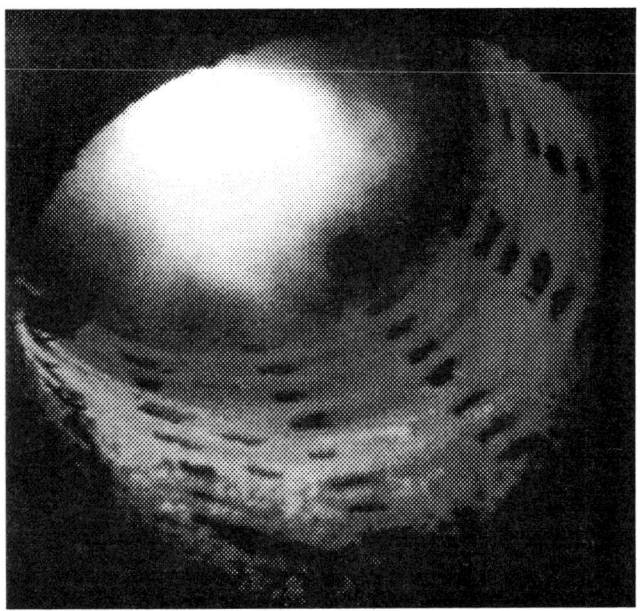

Figure 6. Circular burner.

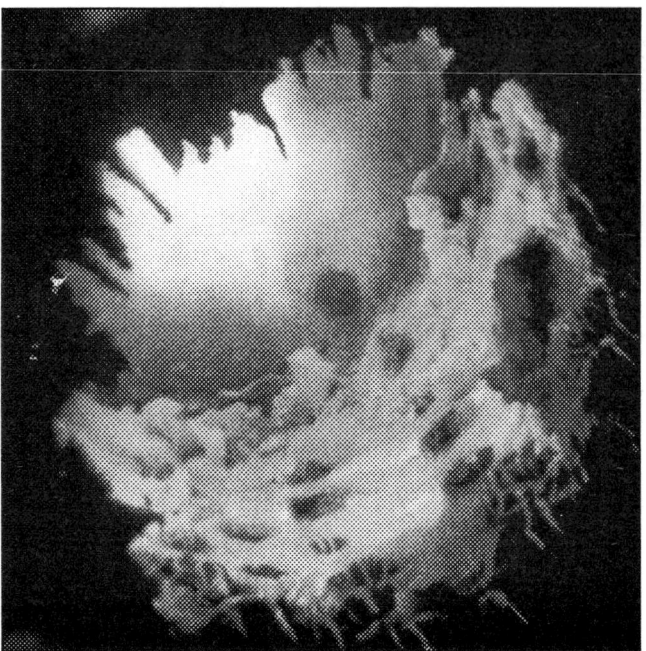

Figure 7. Circular burner showing slag formation.

high viscous layer slid down the lining (Figure 7).

Whenever cold air was injected, the viscosity increased and the slag hardened. Due to the high slag throughput, it took only a few days of operation to plug the burner. In addition, the dripping slag blocked the ash discharge. The design had only anticipated dry and fluffy products. The consequent improvements, which were required to solve the problem, lead to an overrun on project costs.

Conclusions

Operations have now stabilized and one year of experience is available. During that year we proved that the targets have justified the investment and were reached as follows

Gain in reliability

In the past, ten operating days per year were lost due to scaling and plugging of piping, valves and quench tubes. Today, the filtrate recycled to the process is nearly free from ash. No scaling can occur and long lasting pickling operations are saved.

Savings in maintenance

In earlier operations, we had to budget for one to two waste heat boiler replacements per year. Today, we are sure that a service life of more than 50,000 hours can be reached.

Savings in expensive pelletization and homogenizing oil

Pelletization and homogenization oils are more expensive feedstock compounds. A significant contribution to plant profitability can be made by saving the price differential between these fuels and high viscous residues.

Sales of ore substitute

The heavy metals vanadium and nickel were often lost or disposed of as waste. By applying our technology, these compounds are recovered in a concentration close to an intermediate for alloy production.

For future installations, a targeted soot disposal, instead of common soot recycle operations, is the preferred choice. The investment costs for the new technology are less than the total setup for soot recycle facilities.

For the existing facilities, the new process excludes

interactions between the pox-operations and the soot/ash disposal. Furthermore, it contributes to an improved process reliability and safety. Last but not least, valuable raw materials such as naphtha, pelletization and homogenization oils can be saved.

DISCUSSION

V. Jayaraman, *Southern Petrochemical Industries Corp. Ltd.*: What is the purity of Vanadium and has it got any market potential?
Soyez: There is a market potential in Europe. We sell the product at a reasonable price. The price, also quoted in newspapers, is actually close to $30 per kilogram Vanadium.
Jayaraman: What is the purity of Vanadium?
Soyez: On the market, Ferrovanadium, an alloy of iron – and Vanadium – is the standard. The concentrations are 80 wt. % Vanadium and 20% iron. Our product has about 40 wt. % Vanadium and is rated accordingly.

A Novel PSA System for Ammonia Recovery from Synthesis Gas

A modification of the ammonia recovery portion of a generic ammonia synthesis process is suggested. Removing ammonia near the synthesis temperature (and pressure) can make the operation more efficient and less costly. To accomplish this, three PSA flowsheets were devised to recover ammonia while purifying the nitrogen and hydrogen for recycle to the reactor. A set of simulations yielded results that were substantially better for one of the PSA cycles than for the other cycles.

Kent S. Knaebel
Adsorption Research, Dublin, OH 43016

Introduction

The earliest commercial ammonia synthesis process, developed by Haber and Bosch, started up in 1913. Iron was used as the catalyst at a temperature exceeding 300° and a pressure exceeding 200 atm. High temperature and pressure were needed to achieve good kinetics and also because the reaction consumed two molecules for each one it yielded. Despite that, the yield was low (about 8%) and the production rate was only 30 t/d. Today the conditions are not much different from those, although catalysts have been improved so that the yield has approximately doubled. The production rate is typically 1,500 t/d (Skolnik and Reese, 1976).

In Haber–Bosch types of processes, ammonia is recovered by condensing it (generally using refrigeration) from the dilute-reactor off-gases. After condensation, the gas that still contains 2 to 4% ammonia is reheated and fed to the reactor. The overall cost might be reduced substantially if ammonia can be recovered near reaction conditions. Likewise, in the reactor, lower conversion per pass could be tolerated if ammonia could be recovered at high purity without so many intermediate operations. In that way, it may be possible to maintain a high overall yield at a substantially lower cost. Thus, the objective is to isolate ammonia at or near reaction conditions, and to recover it as a pure product.

Methods

The major technical issue to resolve was that reactions generally occur at a much higher temperature than is normally employed for adsorption as a separation process. Therefore, the approach first involved finding a suitable adsorbent, that is, one that had acceptable capacity and kinetics to take up and release ammonia under reaction-like conditions yet was physically stable as well. Subsequently, plausible PSA cycles and associated flowsheets were created (the cycles and associated steps are explained later). Rather than build and operate these cycles, which would have required a large expense of time and cost, a process simulator was developed to predict performance on the basis of the governing material balance equations' solutions. Those simulations implied optimum operating conditions for each cycle (maximum recovery at fixed product purity). Furthermore, by comparing performance among the cycles, the simulations helped us identify the best cycle.

The remainder of this article explains the details of the aspects mentioned above. First, however, related process technology is reviewed.

Background

Sircar (1990) has suggested several PSA process concepts related to ammonia synthesis. All of these concepts dealt with preparation of the stoichiometric mixture to be fed to the converter. The process Sircar called "C" produced H_2 and N_2 in a molar ratio of 3/1. He was also concerned with clean-up of potential contaminants (CO, CH_4, A, and excess N_2) from the feed gas. The byproduct nitrogen was not retained; instead, a source of pure nitrogen was presumed. ICI and Kinetics Technology have commercialized versions of this process.

Several articles have dealt with the combined concepts of reaction with PSA by exploiting coupled catalytic and chromatographic effects. Kadlec et al. (1991) and Lu and Rodrigues (1994) have contributed to that subject and reviewed the literature. A similar report that applied to ammonia synthesis was presented by Wilson and Rinker (1982). They referred to "concentration forcing" rather than PSA, but their work could be called "partial-pressure swing adsorption coupled with reaction." Finally, Froment and Bischoff (1979) discussed modern ammonia synthesis reactors, such as those used by Haldor Topsøe, ICI, and Kellogg. The units depicted there have complex flow paths and, although no dimensions are given, the vessel walls appear to be about 1/10 of the vessel diameter. Reducing the pressure required from 150 atm to ≈ 20 atm could reduce the required thickness by a factor of 4, even if the vessel diameter is doubled.

Adsorbent Selection

The primary criterion for an adsorbent was that it offer high capacity and selectivity for ammonia compared with nitrogen and hydrogen at 300°C or higher and high partial pressures. Capacity and selectivity can be assessed via adsorption isotherms. A secondary but important criterion is the rate of uptake and release, which can be assessed via intraparticle diffusivities. A review of the literature yielded surprisingly few adsorption isotherms and no diffusion data for ammonia, despite its industrial significance as a raw material and in agricultural and refrigeration applications as well as potential health concerns associated with it. Two studies, however, provided isotherm data at elevated temperatures. One was by Shiralkar and Kulkarni (1985), who studied H-Y, Ca-Y, La-Y, and Na-Y zeolites from 60 to 210°C, although they did their experiments at pressures less than 0.5 atm. Another was by Clark et al. (1962), who studied composites of silica and alumina over a temperature range of 100 to 400°C, but at pressures less than 0.03 atm. Other studies were restricted to near-ambient temperatures. Obviously, these pressure ranges were inadequate to assess operation at synthesis reactor conditions. Thus, it was necessary to supplement the database by collecting isotherm and diffusion rate data.

The types of adsorbents considered in this study were activated carbon, silica gel, and zeolites. Subsequently, the range was narrowed to the best candidate from each of the three classes based on performance and compatibility. These adsorbents are referred to as **A**, **B**, and **C** for proprietary reasons. Accordingly, isotherms and diffusivities were measured for those three adsorbents at 30, 60, and 120°C for nitrogen and hydrogen up to 50 bar, and for ammonia up to 7 bar.

Isotherms

Isotherms are the primary means to discriminate among adsorbents. Accordingly, tests were conducted in a batchwise, volumetric apparatus up to 120°C (the upper limit of the apparatus). Deviations from ideal gas behavior were expected, so all data were corrected for compressibility effects using either the Benedict-Webb-Rubin equation (for ammonia and nitrogen) or the Peng-Robinson equation (for hydrogen). Examples of isotherms at 120°C are shown in Figure 1 for ammonia and all three adsorbents and in Figure 2 for ammonia, nitrogen, and hydrogen on adsorbent B. The isotherm data for ammonia, hydrogen, and nitrogen on the three adsorbents at 30, 60, and 120°C were fit using the Langmuir equation,

$$n_i^* = \frac{K_i C_i}{1 + L_i C_i} \quad (1)$$

In fact, the curves shown in both plots represent the fits. Those fits were very good. Furthermore, the van't Hoff equation was used to extrapolate the parameters to higher temperatures for both ammonia and nitrogen, as: $S_i = S_{i1}\exp(S_{i2}/R_T)$, where S_i represents either K_i or L_i. Isotherms at 30, 60, and 120°C and an extrapolation to 300°C for ammonia with adsorbent **B** are shown in Figure 3. Similarly, extrapolated isotherms at 300°C for ammonia and nitrogen with adsorbents **B** and **C** are shown in Figure 4. That plot shows that, among all the adsorbents examined, the most promising is **C**, since the estimated capacity for ammonia is higher than that of the nearest competitor, **B**, yet its estimated capacity for nitrogen is lower than that of **B**.

The relationship between the ammonia and nitrogen isotherms for a given adsorbent is expressed by the selectivity

$$\alpha_{NH_3-N_2}.$$

The definition is based on the Henry's law limits:

$$\alpha_{NH_3-N_2} = \frac{n^*_{NH_3}/n^*_{N_2}}{C_{NH_3}/C_{N_2}} = K_{NH_3}/K_{N_2} \quad (2)$$

where C_i is the gas-phase concentration for component **i**, and n_i is its adsorbed-phase concentration, and K_i is the Henry's law coefficient for component **i** from Eq. 1, for a specific adsorbent. Values for the three adsorbents over the temperature range measured are listed in Table 1.

Despite high selectivity and other potential advantages, adsorbent **A** was abandoned for the following reasons. Tests with ammonia at 120°C of similar materials led to their physical deterioration. Thus, the prospects for **A** at 300°C are not good. In contrast, adsorbents **B** and **C** are known to be stable under even more severe conditions. In addition, the selectivities for **A** exhibited local maxima with temperature, rather than decreasing monotonically. That is, the selectivities for ammonia vs. hydrogen and nitrogen increased (rather than decreased) from 30 to 60°C. Thus, it was decided that extrapolation of the equilibrium data was risky, and that adsorbent **A** did not warrant further attention.

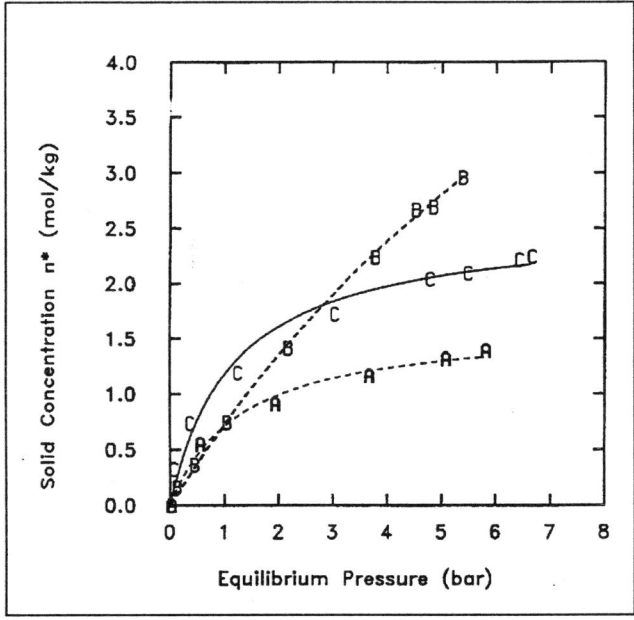

Figure 1. Ammonia isotherms of all three adsorbents at 120°C with Langmuir fits.

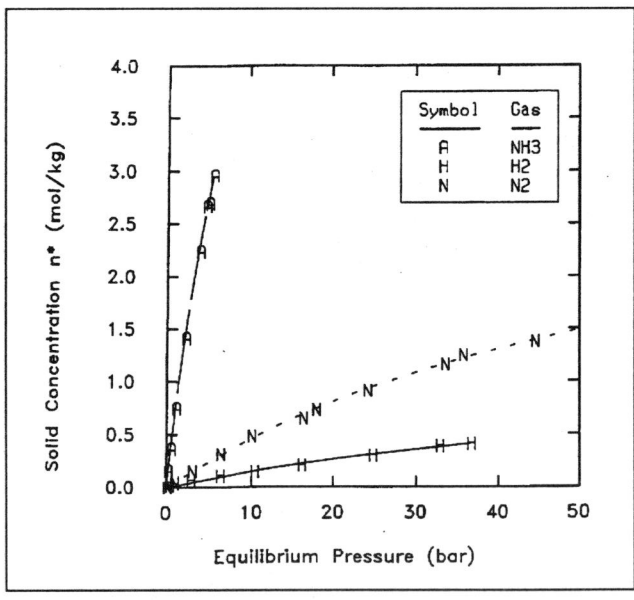

Figure 2. Ammonia, nitrogen, and hydrogen isotherms for adsorbent B at 120°C with Langmuir fits.

Table 1. Ammonia–Hydrogen and Ammonia–Nitrogen Selectivities of Adsorbent Candidates

Adsorbent	$\alpha_{NH_3-H_2}$			$\alpha_{NH_3-N_2}$		
	30°C	60°C	120°C	30°C	60°C	120°C
A	313.3	870.6	167.8	76.8	164.8	—
B	134.2	84.6	46.2	23.5	16.1	14.9
C	786.6	635.7	259.9	85.5	97.3	66.4

Table 2. Typical Diffusivities of Ammonia, Hydrogen, and Nitrogen at 120°C on Adsorbent B

	Short Time Approximation			Long Time Approximation		
	Eff. Diff. D_i^*	Pore Diff. D_I	Corr. Coeff. r	Eff. Diff. D_i^*	Pore Diff. D_I	Corr. Coeff. r
Ammonia	2.71	6.00	0.971	0.0633	0.140	−0.960
Hydrogen	8.56	2.78	1.00	73.8	240	1.00*
Nitrogen	1.93	9.24	0.92	0.584	2.80	1.00*

Note: Eff. = effective, Diff. = Diffusivity, Corr. Coeff. = correlation coefficient. The effective and pore diffusivities are expressed in $(cm^2/s) \times 10^4$. Asterisks denote data based on two data points only.

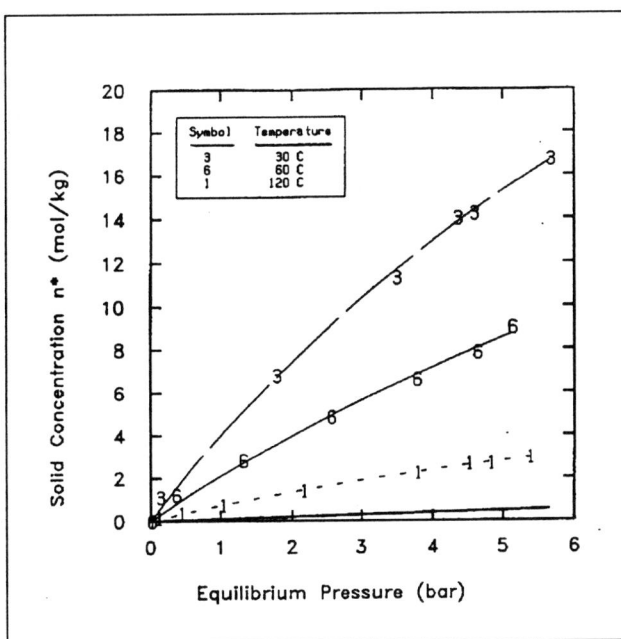

Figure 3. Ammonia isoterms for adsorbent B at 30, 60, and 120°C with Langmuir fits, and an extrapolated isotherm at 300°C.

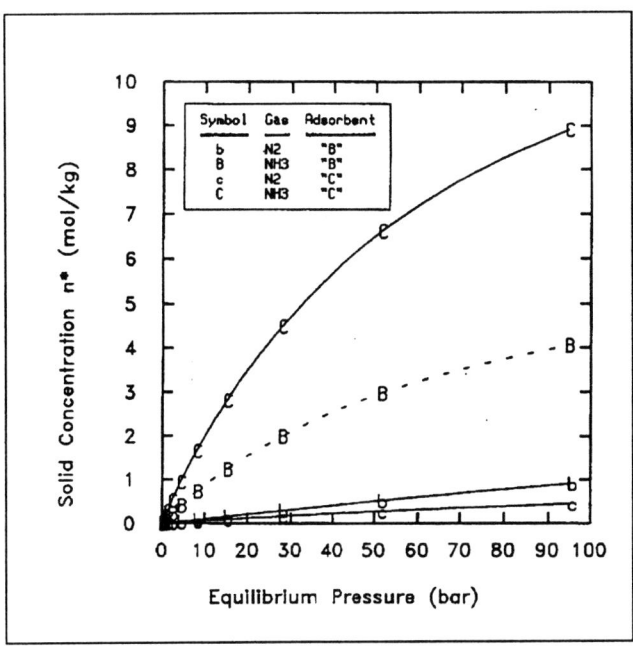

Figure 4. Extrapolated ammonia and nitrogen isotherms for adsorbents B and C at 300°C.

Diffusivities

Intraparticle diffusivities were measured using batchwise uptake and release tests in which pressure shifts were recorded with time. The values of resulting diffusivities were determined from the pressure histories by comparing the results with solutions to Fick's law, based on transient diffusion in a finite reservoir. The asymptotic solutions for both long and short times were used. The values extracted for short times were more likely to be affected by temperature variations due to the heat of adsorption. Values of both methods for all three gases at 120°C are provided in Table 2. Likewise, a plot of the transient data is shown in Figure 5 for all three gases in adsorbent **B** at 120°C. The ordinate in that plot is 1-F, where F is fractional uptake (or release). The other adsorbents exhibited similar behavior, and this was not a criterion by which a selection could be made.

It is clear from the diffusivity data that all three components diffuse rapidly during the short time (generally for fractional uptakes up to 0.8), reflecting gas-phase diffusion. The long time results show that ammonia diffuses much more slowly than nitrogen or hydrogen. Condensation of ammonia in the pores of the adsorbent during the final stages of uptake is the probable reason for this. The critical temperature of ammonia is 132.4°C. Slow diffusion is not a major concern for the proposed process, however, since it will operate at 300°C, a reduced temperature for ammonia of 1.4. In addition, these results show that short time uptake of ammonia is even faster than nitrogen, and that all three components obey Fick's law reasonably well.

Process simulation

C was selected as the basis for simulations because the differences between the diffusion rates were practically negligible and because it was clearly superior in terms of selectivity and capacity under the relevant conditions. After selection of the adsorbent and measurement of the required adsorption data, a suitable mathematical model was developed to evaluate potential PSA cycles. The simulations used ammonia and nitrogen as key components. Since hydrogen is less strongly adsorbed than nitrogen on all the adsorbents at all temperatures, lumping it together with nitrogen is conservative. Furthermore, the objective was to screen hypothetical cycles (rather than to design equipment), so the local equilibrium theory (Ruthven et al., 1994) was used because it isolates key effects without requiring empirical heat and mass transport

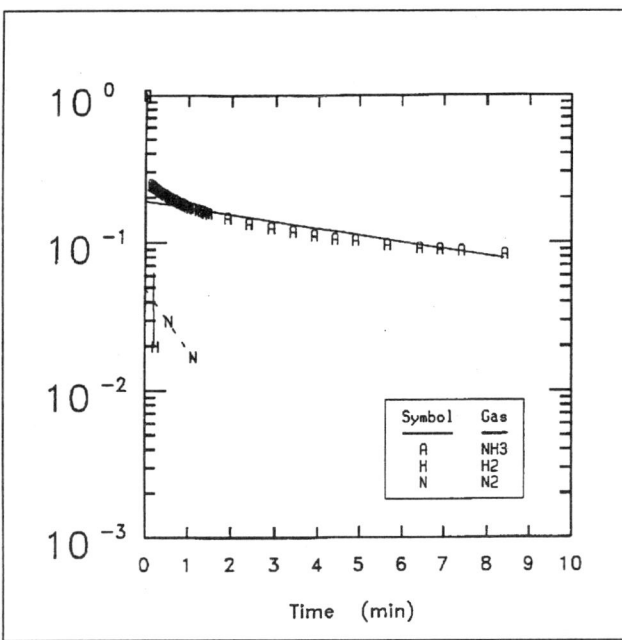

Figure 5. Fractional uptake curves for ammonia, nitrogen, and hydrogen for adsorbent B at 120°C.

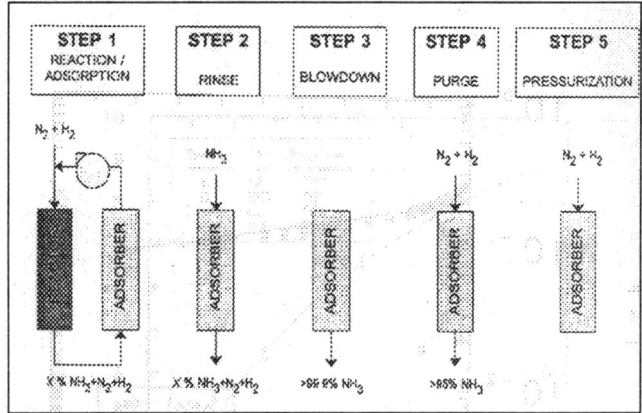

Figure 6. PSA/reactor flowsheet.
One of two to four parallel adsorbers that admit feed continuously is shown. Steps comprising each version. Cycle A = 1+3+4+5; Cycle B = 1+2+3+4+5; Cycle C = 1+2+3+5.

coefficients, which would have required even more experimental data to evaluate. When written for both components of a binary mixture, the equations that govern PSA can be solved by recasting two coupled PDEs as a pair of ordinary differential equations that apply to each step of the cycle. The resulting equations require minimal input data but yield reliable results when the inherent assumptions are valid. Details are summarized in Kayser and Knaebel (1989).

The results are presented in the following sections. No versions of this type of model account for pressure drop, mass-transfer resistances, or temperature effects attributable to the heat of adsorption or heat transfer to the surroundings. The version of the model employed here is applicable for nonlinear isotherms such as those depicted in Figure 4. It solves the partial differential, binary material balance equations that account for velocity dependence on pressure, time, axial position and composition. Although it is not completely rigorous, it accounts for the significant phenomena and allows examination of various cycle alternatives.

Proposed PSA cycles

A hypothetical PSA cycle was selected with the goal of recovering pure ammonia as the "heavy" product and recycling the unreacted nitrogen and hydrogen as the "light" product that would possibly contain some unrecovered ammonia. Once a cycle was selected, ranges of operating conditions (especially the pressures and feed compositions) were studied in an effort to identify the best possible set for the particular cycle. The operating temperature was chosen to be 300°C and the operating pressure was taken to be 100 atm. The ammonia content of the reactor effluent was varied between 10 and 30%, even though the upper limit may be beyond the ability of most reactors (shown as X in Figure 6). That range was examined to see whether the PSA separation scheme could be applied to very different reactor technologies. The range of applicability is envisioned to be set by the following two extremes. First, if relatively dilute ammonia (12 to 14%) produced by relatively simple or older ammonia synthesis systems can be recovered by PSA, those units might be made even more economical. Second, if the PSA can efficiently recover ammonia at high concentrations (17 to 20%), it could even be useful with state-of-the-art reactors, which produce relatively high yields of ammonia.

The cycles proposed here are all combinations of the steps shown in Figure 6. As shown in the left-hand side of that figure, the first step involves feeding off-gas from an ammonia synthesis reactor to an adsorber. Most or all of the ammonia is retained, and the effluent-treated gas is roughly at the stoichiometric ratio of nitrogen and hydrogen. As the adsorber reaches imminent breakthrough, valves are synchronously switched to disengage the present column and engage a parallel column. The present column then begins a sequence of steps to recover the ammonia, isolated from the reactor. In fact, three cycles comprising different sequences of steps have been devised. All share the feed step, as was mentioned previously.

The first cycle, **A**, is a conventional 4-step cycle comprised of feed, blowdown, purge, and pressurization. The next step, blowdown, yields an ammonia-rich gas as the result of desorption. The next step, purge, occurs at low pressure employing recycled light product to drive the residual ammonia from the adsorber, though a "heel" may be left in the bed. Finally, the bed is pressurized with additional recycled light product. The light product is assumed to be pure (uncontaminated by ammonia). In contrast, the ammonia cannot be recovered in a pure state, but only enriched.

The second cycle, **B**, adds a so-called rinse step following the feed step. In that step, pure ammonia is admitted to the bed and residual feed gas is recovered at high pressure and then passed to a parallel column during its feed step. In this case, the blowdown step yields pure ammonia. Under proper conditions, this yields a greater amount of pure ammonia than the feed step.

The third cycle, **C**, is the same as cycle **B**, but without the purge step. Omitting the purge step is justified by the fact that recovering a very pure light product is unnecessary, considering that current systems recycle 1 to 4% ammonia that is uncondensed. Thus, as long as the ammonia content can be kept at or below that level, purging is unnecessary. The blowdown step also yields pure ammonia.

The following symbols are used to make the following discussion concise: y_F is the feed mole fraction of ammonia, P_H and P_L are the high and low pressures, P is their ratio, (P_H/P_L), R_A and y_A are the ammonia

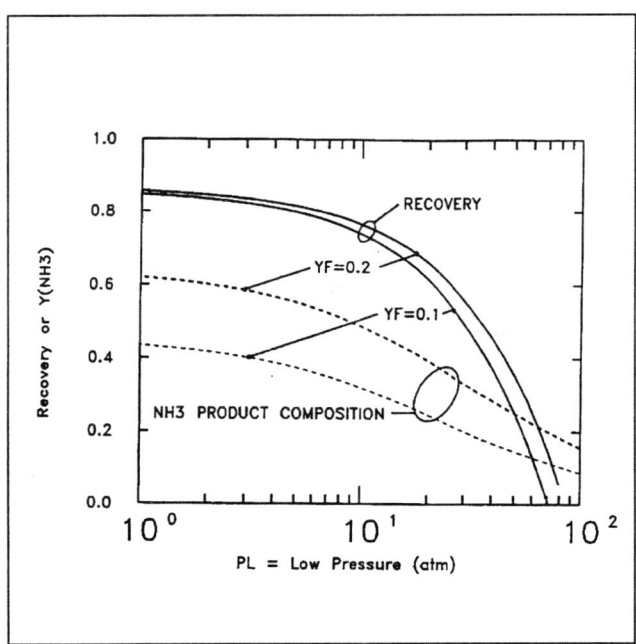

Figure 7. Cycle A.
Dependence on P_L of $N_2 + H_2$ recovery and NH_3 product composition for adsorbent C at $P_H = 100$ atm. Feed = 10% or 20% NH_3.

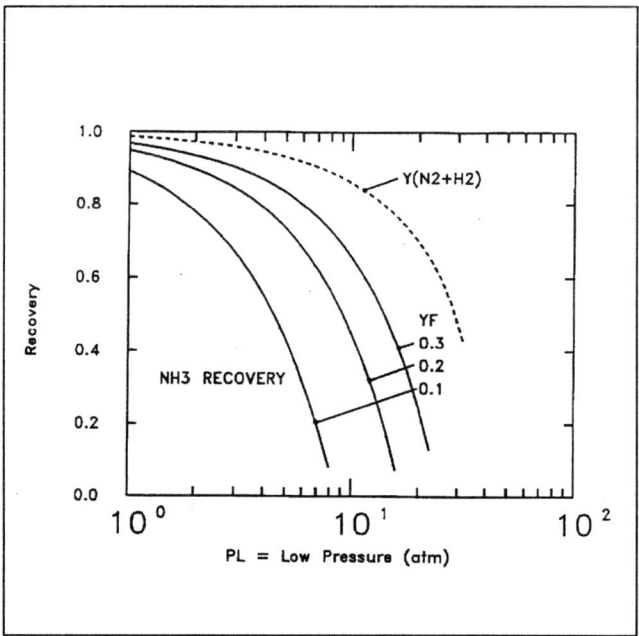

Figure 9. Cycle C.
Effect of P_L on purity of $N_2 + H_2$ (fully recovered) and recovery of 100% NH_3 with adsorbent C at $P_H = 100$ atm, for $y_F =$ 10% or 20% NH_3.

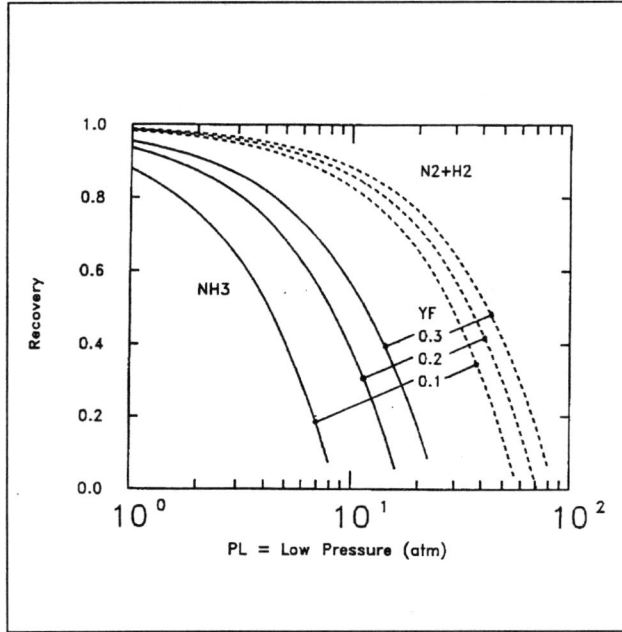

Figure 8. Cycle B.
Effect of P_L on recovery of 100% $N_2 + H_2$ and 100% NH_3 with adsorbent C at $P_H = 100$ atm, for $y_F =$ 10% or 20% NH_3.

product recovery and mole fraction, and R_B and y_B are the light product (predominately nitrogen and hydrogen) recovery and its mole fraction of residual ammonia, respectively. The recovery of a component is defined as the amount obtained as the specified product divided by the amount contained in the feed. Likewise, ammonia enrichment is defined as EA = y_A/y_F, where y_A is the average mole fraction of the ammonia that is recovered during blowdown and purge.

All of the cycles produce a mix of nitrogen and hydrogen at high pressure, but they all recover ammonia at the low pressure, P_L.

Cycle **A** (Figure 7) has the advantage of simplicity and of recovering all the ammonia, although the ammonia is impure. Its R_B increases as P_L decreases, reaching a plateau of above 80% at $P_L \approx 7$ atm, and it depends weakly on y_F. Likewise, y_A increases as P_L decreases, again reaching a plateau (that in this case depends strongly on y_F) at $P_L \approx 7$ atm.

Cycle **B** (Figure 8) has the advantage of producing two pure products, plus one byproduct stream, the effluent of the purge step, which can also be captured and recovered. R_B is significantly greater than R_A, but both increase as PL decreases, reaching a plateau as $P_L \approx 1$ to 5 atm, in which the latter depends more strongly on y_F. These results are somewhat disappointing, because most plants would prefer the ammonia product to be available at the highest pressure possible. Cycle **B** suffered in terms of performance, since ammonia was adsorbed strongly. This meant that a substantial amount of the light product was required for purging the column. Notably, the amount of purge gas required is roughly proportional to the slope of the isotherm of the heavy component, ammonia. Both the ammonia purity and the light product recovery were much higher for cycle **B** than for cycle **A**.

Cycle **C** (Figure 9) produced a significant improvement over both cycles. For all conditions, it was inherent to this cycle that: $y_A = 100\%$ and $R_B = 100\%$. In fact, the only ammonia that is actually "lost" returns with the recycle to the reactor. Consequently, the compression energy is retained as well as the material. Cycle **C** is also a relatively simple and fast cycle that recovers ammonia as a pure product. In all cases, R_A is better in cycle **C** than in cycle **B**. Again, R_A increases as P_L decreases, reaching a plateau of above 80% at $P_L \approx 1$ to 3 atm that depends on y_F. The purity of the recycle (mainly $N_2 + H_2$) stream increases as P_L decreases, reaching a plateau of above 97% at $P_L \approx 3$ atm.

Two principal trends are evident: performance improves significantly as y_F increases and PL decreases. In other words, the lower the feed content of ammonia to the PSA system for this cycle, the greater the pressure ratio must be. In addition, to obtain pure ammonia at a high pressure (so that recompression is not required) a high value of y_F is necessary. That implies that there must be high conversion in the synthesis reactor. Generally, the cycle and operating conditions must be optimized to achieve a desired combination of production efficiency and economic returns.

Conclusion

Three PSA processes are suggested that could be used to recover ammonia at or near synthesis conditions (high pressure and temperature). Performance of these cycles was predicted using a mathematical model. The model accounts for velocity dependence on pressure, time, axial position and composition, but it ignores diffusional resistance, heat transfer, and minor constituents (e.g., argon). Predictions were based on equilibrium data for relatively mild conditions: temperatures of 30, 60 and 120°C, and pressures of up to 50 atm for nitrogen and hydrogen but only up to about 7 atm for ammonia. These were then extrapolated to the relevant temperature and pressure (up to 500°C and 100 atm).

In the best of three cycles evaluated, ammonia is recovered as a pure product at low pressure, and the light product (unreacted nitrogen and hydrogen) is recycled to the reactor at high pressure, recompressed only to compensate for frictional losses in the fixed beds and fittings. The mathematical model indicates that a rinse step is essential to recover ammonia at high purity, and at moderately low pressures ($P_L < 10$ atm, when $P_H = 100$ atm). Furthermore, minimal purging leads to an improvement in recovery of both ammonia and the light product. Contrary to expectations, cooling is unnecessary between the reactor and adsorber. At $y_F = 20\%$, $P_L = 2.7$ atm, and $P_H = 100$ atm, employing no purge, the content of unrecovered ammonia (recycled with the reactants) would be in the range of just 2%. A patent has been granted (Knaebel, 1998).

Literature Cited

Clark, A., V. C. Holm, and D. M. Blackburn, *J.Catal.*, **1**, 244 (1962).

Froment, G. F., and K. B. Bischoff, Chemical Reactor Analysis and Design, Wiley, New York (1979).

Kadlec, R. H., I. D. Lee, and G. G. Vaporciyan, in A. B. Mersmann and S. Scholl, eds., *Fundamentals of Adsorption,* Engineering Foundation, New York, p. 331–344 (1991).

Kayser, J. C., and K. S. Knaebel, *Chem. Eng. Sci.*, **44**, 1 (1989).

Knaebel, K. S., "Pressure Swing Adsorption System for Ammonia Synthesis," U. S. Patent No. 5,711,926 (1998).

Lu, Z. P., and A. E. Rodrigues, "Pressure Swing Adsorption Reactors: Simulation of Three-Step One-Bed Process," *AIChE J.*, **40**, 1118 (1994).

Ruthven, D. M., S. Farooq, and K. S. Knaebel, Pressure Swing Adsorption, VCH, New York (1994).

Shiralkar, V. P., and S. B. Kulkarni, *J. Coll. Interf. Sci.*, **108**, 1 (1985).

Sircar, S, *Separation Sci. Technol.*, **25**, 1087 (1990).

Skolnik, H. and K. M. Reese, eds., *A Century of Chemistry*, American Chemical Society, Washington, DC, p. 292–293 (1976).

Wilson, H. D., and R. G. Rinker, *Chem. Eng. Sci.*, **37**, 343 (1982).

DISCUSSION

S. Madhavan, *Brown and Root:* In the ammonia plant synthesis loop, you normally have methane and argon in addition to hydrogen, nitrogen, and ammonia. Could you please comment on what would happen to methane and argon using your PSA system?

Knaebel: I could make a guess. We didn't study those in particular, but the methane and argon would be much lighter than ammonia and, therefore, would tend to be recycled with nitrogen and hydrogen. Because they would be immediately absorbed, between nitrogen and hydrogen and ammonia, they could be taken out as enriched product by timing the valves correctly. It is an intermediate cut, if you will. That would mean losing either a little bit of nitrogen, hydrogen, or ammonia in order to get that pulse of these intermediate components as they come out.

Richard Saure, *Krupp Uhde*: When you measured the equilibrium data with the ammonia with absorbents A, B, or C, did you find any hysteresis effects due to possible capillary condensation?

Knaebel: Not on the absorbents that I talked about. We did on others at lower temperatures, but not on these three.

Saure: So, they were nonporous?

Knaebel: They were porous, but they had more mesopores and macropores than micropores.

V. Jayaraman, *SPIC*: Since ammonia cooling is not required, will it eliminate the refrigeration system in ammonia production?

Knaebel: Instead of having the synthesis gas come through that loop at a mole fraction of 15% to 20%, now it can come into the condensing system basically pure, and, so, the power required would be much less. The dimensions of the refrigeration unit could be much less. If, instead, it goes to a urea plant, instead of being condensed and used as ammonia, there is no need to condense it. So, basically, it could remain at a moderate pressure gas without refrigeration.

Jayaraman: Do you expect any poisons from the absorbents?

Knaebel: We thought about those, but didn't know enough about the technology to identify what they were.

Vaidhyanathan: Would the concentration of ammonia exiting the PSA have any effect on the converter?

Knaebel: I'm sure it will. Being nearly totally ignorant about ammonia synthesis, I understood much ammonia is recycled from the mechanical refrigeration system. We should be able to do better than that to get down to 1% or 2% ammonia or lower if needed.

Liquid/Gas and Liquid/Liquid Coalescers in the Ammonia Industry

Oil contamination in liquid ammonia can cause various process, maintenance, and quality problems. The application of recent developments in high-efficiency polymer fiber coalescer technology is presented for liquid/gas and liquid/liquid separations. Commercial experience is presented along with economic considerations and process benefits.

Thomas Wines
Pall USA, Port Washington, NY
Michel Farcy
Pall Europe, 77420 Champs sur Marne, France

Introduction

Oil contamination in the synthesis and refrigeration of ammonia is a common phenomenon and can lead to adverse process effects. Oil contamination issues include off-specification product, heat exchanger fouling, catalyst bed poisoning, and large clean-up projects of storage vessels. The source of the oil is generally from the lube oil used in inlet gas, synthesis gas, recycle gas, or refrigeration compressors. A generic ammonia plant is presented in Figure 1 along with recommended locations for the installation of high-efficiency coalescers for oil removal.

In the ammonia plant, natural gas and air are first compressed to about 40 bar and then enter the primary and secondary reformers. Lube oil can enter the inlet gas streams from these gas compressors and a high-efficiency liquid/gas coalescer downstream of the inlet compressors is indicated. The gas stream leaving the reformers now contains mostly hydrogen (H_2), carbon monoxide (CO), carbon dioxide (CO_2), methane (CH_4), and nitrogen (N_2). Next the CO is changed to CO_2 in the converter unit. Here oil contamination can lead to poisoning of the converter

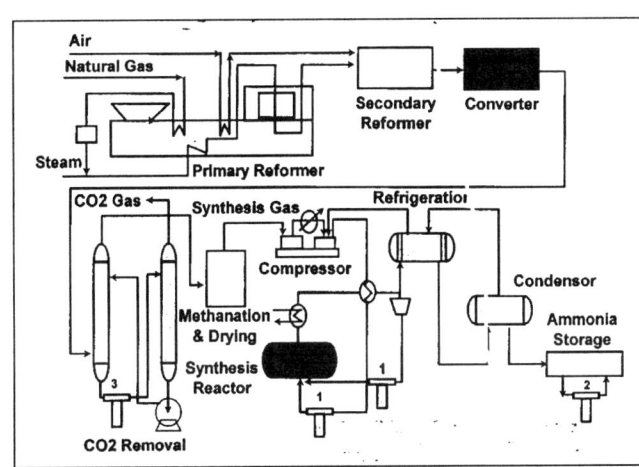

Figure 1. Ammonia plant.

catalyst. The gas stream then passes through a liquid/gas adsorber which is usually hot potassium carbonate solution or alkanolamine solution for CO_2 removal. This is followed by a final methanation and drying step to produce the synthesis gas (mostly H_2 and N_2).

The synthesis gas is then compressed to approximately 300 bar in the synthesis gas compressors. In most existing plants, the compressors are oil-lubricated reciprocating piston compressors used to achieve the high pressure required for the ammonia synthesis reaction to occur. The use of these types of compressors will generally lead to oil contamination in the final ammonia product without protection and can also poison the synthesis reactor catalyst. The use of high-efficiency liquid gas coalescers is indicated downstream of the synthesis gas compressors. A portion of the gas leaving the synthesis reactor is recycled back to the inlet of the synthesis reactor after first being recompressed by the recycle compressors. The use of high-efficiency liquid/gas coalescers is indicated downstream of the recycle gas compressor.

The ammonia gas (NH_3) product leaving the synthesis reactor then passes through a refrigeration and condenser stage that liquefies the ammonia product before reaching refrigerated storage tanks, which are maintained at about 7 bar and 0°C. The ammonia refrigeration process is presented in Figure 2. Oil contamination can lead to fouling of heat-transfer surfaces such as the condenser and vaporizer in the refrigeration process. High-efficiency liquid/liquid coalescers are indicated for the removal of the contaminant oil from the liquid ammonia stream prior to the storage tank. Thus the removal of oil contamination can be accomplished in the gas streams at high pressures using liquid/gas coalescers or in the final liquid product using liquid/liquid coalescers.

Coalescing Technology

Liquid/gas coalescer system

The separation of liquid aerosol contamination with high-efficiency liquid/gas coalescer cartridge systems has found widespread acceptance in refinery and gas plants in recent years for a number of applications (Pauley, 1988, 1991; Schlotthauer, 1991) including protection of compressors, turbo equipment, burner nozzles, amine and glycol contactors, molecular sieve beds, and hydrotreater catalyst beds. In the ammonia plant, the liquid/gas coalescers are used primarily to remove lube oil aerosols contamination downstream of compressors.

The growing trend of using liquid/gas coalescers that are more efficient has largely been the result of dissatisfaction with the traditional separation approaches of knockout vessels, centrifugal separators, mesh pads, or vane separators that have not met the requirements for aerosol reduction. The primary rationale for the use of high-efficiency coalescers is that significant aerosol contaminant that is in the submicron and low-micron size range exist in the plants (Brown, 1994). High-efficiency liquid/gas coalescers are generally constructed from glass fibers since this material allows for a fine porous structure with fiber diameters of a few microns. The small pore size is needed to achieve greater capture and separation of these fine aerosols.

Another important benefit of the liquid/gas coalescer is that this type of separation device can be operated at significantly lower flow rates than design and therefore has a high turndown ratio. This is due to the fact that the separation mechanisms are based primarily on diffusion and direct interception, unlike vane separators and mesh pads, which rely heavily on inertial separation principles. This allows the high-efficiency liquid/gas coalescer systems a greater degree of flexibility, allowing them to operate at peak performance even for high turndown ratios (reduced flow rates) that can occur during commonly encountered partial plant shutdowns and upset conditions. Generally, the high efficiency liquid/gas coa-

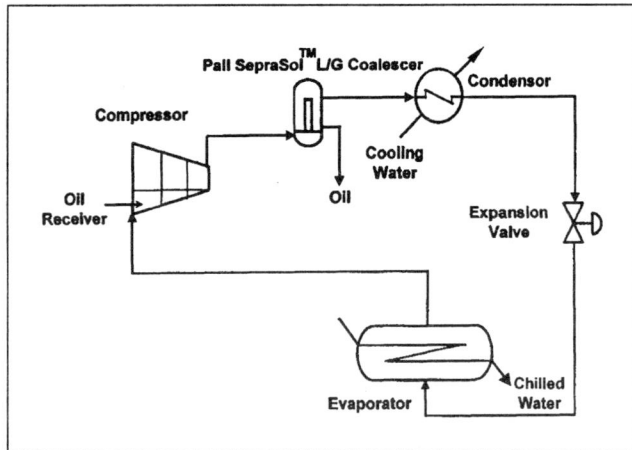

Figure 2. Ammonia refrigeration unit.

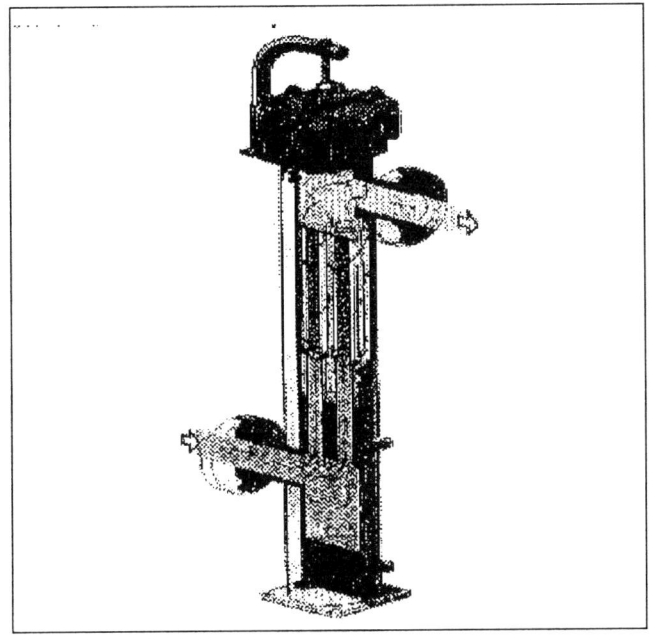

Figure 3. Pall SepraSol liquid/gas coalescer.

lescers are used for inlet aerosol concentrations of less than 1,000 ppmw (0.1%) and are placed downstream of other bulk removal separators as the final stage. Outlet concentrations for these high-efficiency liquid/gas coalescers are as low as 0.003 ppmw (Williamson, 1988; Brown, 1994).

The use of a surface treatment (Miller, 1988) on Pall high-efficiency vertical liquid/gas coalescers cartridge systems has proven to significantly enhance performance by allowing higher flow rates or smaller housing diameters compared to untreated coalescers.

Figure 3 depicts a Pall SepraSol vertical high-efficiency liquid/gas coalescer system. The inlet gas with liquid aerosol contamination first enters at the bottom of the housing into a first-stage knockout section. Any slugs or larger droplets (approximately > 300 µm) are removed here by gravitational settling. The gas then travels upward through a tube sheet and flows radially from the inside of the cartridges through the coalescer medium to the annulus. The inlet aerosol distribution is in the size range of 0.1 µm–300 µm and is transformed to enlarged coalesced droplets in the size range of 0.5–2.2 mm after passing through the coalescer medium. The advantage of flowing from the inside to the outside of the coalescer cartridge is that the gas velocity can be more easily adjusted in the annulus by selecting the optimum housing diameter to prevent reentrainment of coalesced droplets. As the gas leaves the coalescer cartridge and travels upward in the annulus it contributes to the total flow and thereby increases the annular velocity.

The coalsesced droplets immediately drain vertically downward in the coalescer medium pack. The surface treatment greatly enhances this drainage, and the coalsced droplets are shielded from the upward gas flow in the annulus in most of the coalscer cartridge length as as a direct consequence of the treatment. The coalesced droplets are first exposed to the annular gas flow when they appear on the external face of the coalescer medium pack at the bottom third of the coalescer cartridge. Once the coalesced droplets are released to the annular space they are subject to the force of the upward flowing gas. The trajectory of the coalesced droplets is modeled on a force balance between gravity settling and the drag force created by the gas flow past the droplets. This analysis leads to the calculation of a critical annular velocity for reentrainment. The coalesced drops settle into a collection sump in the bottom of the housing that can be drained manually or equipped with level control for automatic drainage.

Due to the surface treatment, the maximum annular velocity at the top of the coalescer cartridge is about three times greater than the value for an untreated coalescer. This allows for a given surface-treated liquid/gas coalescer unit to handle about three times the gas capacity of an untreated coalescer system. Given the same process gas flow rate, the surface-treated liquid/gas coalescer system can be constructed with a smaller, more economical vessel than the untreated coalescer system.

Liquid/liquid coalescer system

Currently, most ammonia producers use the storage tanks as settlers to remove contaminant oil from the final ammonia product. This method is not very efficient and occasionally the oil contamination concentration can reach levels of 10–20 ppm leaving the plant. The separation of oil from liquid ammonia using liquid/liquid coalescers is a relatively new field. The coalescer material must be able to withstand temperatures as low as –40 °C and be compatible with ammonia and oil as well as provide the required separation. Traditional liquid/liquid coalescers have used glass fiber media, which works

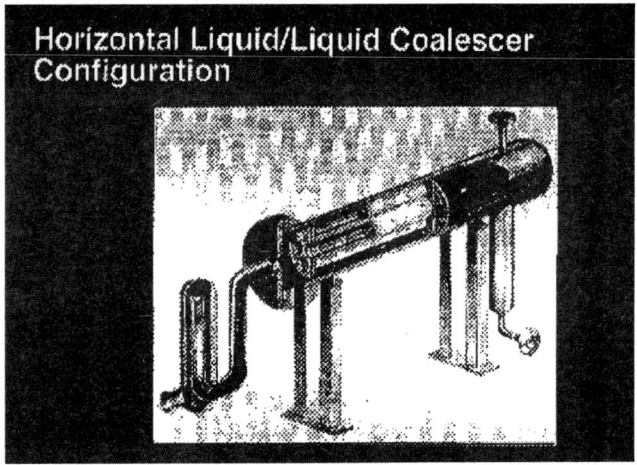

Figure 4. Pall PhaseSep liquid/liquid coalescer.

well for emulsions that have interfacial tensions greater than 20 dyne/cm and for systems at ambient temperatures. Pall has developed and constructed new coalescer media with novel formulated polymers and fluoropolymers (Brown, 1993; Wines, 1997) that are effective for emulsions having interfacial tensions as low as 0.5 dyne/cm and for low temperature service. The interfacial tension between the oil and ammonia liquid has not been characterized due to the volatility of the ammonia and difficulty in using standard apparatus for interfacial tension measurements. The Pall liquid/liquid coalescer has produced clean ammonia with oil levels down to 2–3 ppm in pilot scale tests.

A Pall PhaseSep high-efficiency liquid/liquid coalescer in the horizontal configuration is depicted in Figure 4. The system consists of a horizontal coalescer cartridge stage followed by a settling zone that relies on the difference in density for separation of the coalesced droplets. The fluid enters at the side of the housing and flows from the inside of the coalescer cartridges radially outward, causing the enlargement or coalesence of the inlet dispersion into large droplets in the outlet stream. The coalesced droplets then flow axially in the horizontal direction through a settling zone. The dispersed oil-phase coalesced droplets settle downward by gravity and are collected in a sump in the bottom of the housing. The purified liquid ammonia leaves at the top of the housing.

The liquid/liquid coalescing system operates in three stages: separation of solids, coalescence, and separation of coalesced drops.

Separation of solids

Solids can increase the stability of an emulsion and removing solids can make coalescing easier. This step can generally be achieved by a separate cartridge filter system or by a regenerable backwash filter system for high levels of solids. In addition, the filtration stage protects the coalescer and increases service life.

Coalescence

The next step in the process is the primary coalescence. In this stage, the pore dimensions begin with a very fine structure and then become more open to allow for void space for the coalescing droplets. In the primary coalescence zone, the inlet droplet dispersion containing fine droplets in the size range of 0.2 to 50 µ is transformed into a suspension of enlarged droplets in the size range of 500 to 5,000 µ.

The coalescence mechanism can be described by the following steps:
• Droplet adsorption to fiber.
• Translation of droplets to fiber intersections by bulk flow.
• Coalescence of two droplets to form one larger droplet.
• Repeated coalescence of small droplets into larger droplets at fiber intersections.
• Release of droplets from fiber intersections due to increased drag on adsorbed droplets caused by bulk flow.
• Repeat these steps with progressively larger droplet sizes and more open media porosity.

Based on this mechanism, we can predict that a number of factors will influence the coalescence performance. The specific surface properties of the coalescer fibers are critical in influencing the adsorption of droplets as well as their ultimate release after coalescing. There is a balancing act between increasing the attraction or adsorption characteristics of the fibers against the release mechanism, which strong adsorption inhibits. The fact that droplet-fiber adsorption is necessary for coalescing has been supported by a number of sources (Jeater, 1980; Basu, 1993), although it is not universally accepted. The presence of significant levels or types of surfactants to cause the disarming phenomenon has been detected by measuring the interfacial tension between the aqueous and hydrocarbon phases. When surfactants are added to the water in hydrocarbon systems, the interfacial tension is decreased. In most cases, an interfacial tension of less than 20 dyne/cm was found to cause disarming of glass fiber coalescers. When specially formulated polymeric coalescer medium was used instead of glass fiber, disarming was not observed (Brown, 1993; Wines, 1997). The coalescing performance of a polymeric medium can be greatly enhanced by modification of surface properties that cannot be accomplished by glass fiber medium.

Separation of coalesced droplets

Once the droplets have been coalesced, they are assumed to be as large as possible for the given flow conditions. During the separation stage, a settling zone is achieved that relies on the difference in densities between the coalesced droplets and the bulk fluid. The oil is then separated in a collection sump that can be manually drained on a periodic basis or equipped with an automatic level control and drain system. Estimation of the coalesced drop size and required settling zone are best determined through pilot scale tests at field conditions.

Commercial Experience of Zaklady Azotowe Pulawy, Poland

Process description

Synthesis gas (H_2, N_2) is compressed to 320 bar by multistage lubricated reciprocating compressors and fed to the ammonia synthesis reactor. Downstream of the reactor, coolers and separators containing mesh pads are in service to remove liquid aerosol contamination from the ammonia gas. Zaklady Azotowe Pulawy was interested in reducing the oil content in the ammonia product that varied from 10–20 ppm and worked closely with Pall towards a solution.

Application description and experience

A single Pall SepraSol liquid/gas coalescer system was initially installed in December 1994 on one of 16 high-pressure synthesis gas streams downstream of the compressors. The coalescer system contained 10 coalescer elements in a 580-mm- (20-in.) diameter vessel and was operated at 40,000 Nm³/h with a gas temperature of 50°C and pressure of 320 bar. The bottom stage of the coalescer housing was equipped with a mesh pad to act as a bulk removal stage for large aerosol droplets (that is, greater than 5 ppm in size). The process benefit realized was an ammonia product with no detectable oil contaminant (<1 ppm) throughout an 18-month trial. After this success, the remaining synthesis and recycle streams (16 in total) in the plant were equipped with the combined mesh pad–Pall SepraSol liquid/gas coalescer systems. This resulted in a total use of 280 Pall coalescer cartridges in sixteen vessels (10–16 coalescer elements per vessel) for process streams with flow rates ranging from 43,000–220,000 Nm³/h and oil aerosol loadings from 4–30 ppm in the inlet gas. The additional Pall coalescer units have been operation since early 1997, and have experienced service lives varying from 12 to 18 months before the change-out differential pressure of 350 mbar was reached.

Alternate solutions considered

Before adopting the Pall liquid/gas coalescers, Zaklady Azotowe Pulawy investigated a few alternate solutions. The use of an oil-free compressor was considered but had the drawbacks of high capital investment for replacing the existing compressors and the likely possibility of higher maintenance costs as compared to the existing lubricated reciprocating compressors. Adsorption columns were also considered using either activated carbon or a molecular sieve. Some of the disadvantages for this type of system included: relatively high-pressure drop across the beds, high-energy costs for regeneration, the need to monitor the bed saturation, the possibility of crushed bed particles entraining and fouling the synthesis reactor, and contamination of the final product.

Process benefits

The main process benefits are as follows:
• The reduction in oil contamination to less than 1 ppm in the final ammonia product has completely eliminated any rejected ship loads leaving the plant.
• The frequency of oil content measurement in the

final ammonia product was reduced from once a day to once a week due to the consistent low level of oil in the ammonia after installation of the Pall coalescer system. This resulted in significant safety improvements in the plant by reducing exposure to the ammonia and also reduced the manpower requirements.

• The reduction in the oil content of the final ammonia product has enabled its sale into the fine chemical markets where purity is essential. The fine chemical market will purchase ammonia at $200/metric t which is considrably more than the $120–$150/metric t the fertilizer market will pay. The total savings in this case will depend on the demand for ammonia as feedstock for the fine chemical market.

• The current lubricated reciprocating compressors can be kept in service and will not need to be replaced with oil-free compressors that are maintenance-intensive and more expensive. Less expensive lubricating oil containing higher sulfur content can also be used in the existing compressors as it will not be contaminating the synthesis reactor and final ammonia product.

BP Chemicals, North Site, Grangemouth, U.K.

Process description

An ammonia refrigeration unit consisting of a compressor, condenser, expansion valve, and evaporator is operated. In the refrigeration cycle, the ammonia is vaporized in the evaporator and heat is being absorbed from the space being cooled. The ammonia vapor is next drawn into a motor-driven compressor and elevated to high pressure. This causes the ammonia to be heated and the resulting superheated, high-pressure ammonia gas is then condensed to liquid in a water-cooled condenser. The liquid ammonia then flows through an expansion valve and the pressure and temperature are reduced to the conditions in the evaporator completing the cycle. Oil contamination was found to interfere with the refrigeration by causing fouling of heat-transfer surfaces and resulting in a loss of cooling capacity.

Application description and experience

In May 1996, a Pall SepraSol liquid/gas coalescer system containing fifteen coalescer elements was installed downstream of the compressor package to reduce the fouling of heat-transfer surfaces. The operating conditions were of system pressure of 10 bar gauge, a temperature of 86°C, and a flow rate of 1,400 actual m³/h. The system was found to work effectively by BP Chemicals as oil was being collected and drained on a daily basis from the coalescer. After the Pall coalescers were installed, the level of oil in the ammonia loop remained below the detection level of 1 ppm. Since the initial installation, the Pall coalescer elements have not been changed out and are continuing to operate effectively.

Process benefits

The main process benefits are the following:

• The heat removal capacity was improved by 10–20 % due to the reduction in fouling of the heat-transfer surfaces in the condenser and vaporizer.

• Additional costly refrigeration units (approximately $1 million) were rendered unnecessary by the increased refrigeration capacity for a relatively small investment (<$100,000).

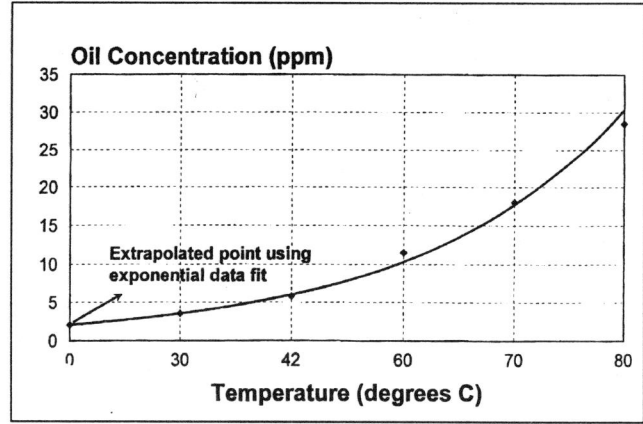

Figure 5. Oil solubilitiy in ammonia (data from Cyklis, 1971).

Hydro Chemicals France, Division of Hydro Agri France

Process description

The final liquid ammonia product is stored in cryogenic storage tanks and loaded onto trucks at a flow rate of 60 metric t/h. The process temperature is between 0 and 3°C, and the pressure is 7 bar. Hydro Chemicals France initiated a test program with Pall to determine whether high-efficiency liquid/liquid coalescers could reduce the oil content down to less than 5 ppm. The oil content of the ammonia was artificially varied to achieve a concentration from 8–15 ppm to challenge the test coalescer.

Application description and experience

In January 1998, Pall Corporation's Scientific and Laboratory Services Department conducted pilot scale tests using a test coalescer that was 10 in. in length and 2.5 in. in diameter. The test was run on site at Hydro Chemicals France at the outlet of the cryogenic storage tanks just before truck loading. The program lasted over two weeks and the average effluent concentration of oil in the final product ammonia was 3 ppm at a test flow rate of 15 L/min. A solubility curve (Cyklis, 1971) of oil in liquid ammonia at different temperatures is presented in Figure 5. The operating conditions at Hydro Chemicals France (0°C) indicates an oil solubility of 2 ppm. Therefore the Pall PhaseSep liquid/liquid coalescer was able to reduce the free oil level down to 1 ppm.

Process benefits

Hydro Chemicals France has purchased a full scale Pall PhaseSep Liquid/Liquid Coalescer to be installed on a recirculating loop on the storage tank to continuously purify the ammonia. The coalescer system has a prefiltration stage to protect the coalescers and will operate at a flow rate of 20 metric t/day using 20 coalescer elements in a 506-mm-diameter housing. This should reduce the rejected product ratio and improve the ability to sell the ammonia as fine chemical grade for higher revenues.

Literature Cited

Basu, S., "A Study on the Effect of Wetting on the Mechanism of Coalescence," *J. Coll. Interf. Sci.*, **159**, 68 (1993).

Brown, R. L., and T. H. Wines, "Improve Suspended Water Removal From Fuels," *Hydrocarbon Processing*, **72** (12), 95 (1993).

Brown, R. L., and T. H. Wines, "Recent Developments in Liquid/Gas Separation Technology," presented at the Laurence Reid Gas Conditioning Conference, Norman, OK, (1994).

Cyklis, D., and N. Gorunov, "Solubility of Lubricating Oil in Liquid Ammonia," Proc. of State Institute of Nitrogen Industry (Russia), **12**, 76 (1991).

Jeater, P., E. Rushton, and G. A. Davies, "Coalescence in Fibre Beds," *Filtration & Separation*, March/April 1980, 129.

Miller, J. D., R. R. Koslow, and K. W. Williamson, U.S. Patent 4,676,807, June 1987.

Miller, J. D, R. R. Koslow, and K. W. Williamson, U.S. Patent 4,759,782, July 1988.

Pauley, C. R., R. Hashemi., and S. Caothien, "Analysis of Foaming Mechanisms in Amine Plants," presented at the American Institute of Chemical Engineers summer meeting, Denver, CO, (1988).

Pauley, C. R., D. G. Langston, and F. Betts, "Redesigned Filters Solve Foaming, Amine Loss Problems at Louisiana Gas Plant," *Oil & Gas J.*, (1991).

Schlotthauer, M., and R. Hashemi, "Gas Conditioning: A Key to Success in Turbine Combustion Systems Using Landfill Gas Fuel," presented at the 14th Annual Landfill Gas Symposium GRCDA/SWANA, San Diego, California, March 27, 1991.

Williamson, K., S. Tousi, and R. Hashemi, "Recent Developments in Performance Rating of Gas/Liquid Coalescers," presented at the First Annual Meeting of the American Filtration Society, Ocean City, MD (1988).

Wines, T. H., and R. L. Brown, "Difficult Liquid–Liquid Separations," *Chem. Eng.*, **104** (12), 104 (1997).

DISCUSSION

John R. Dinyari, *Dyna Management*: Could you comment on the pressure drop, specifically in the first example?

Farcy: For the liquid gas coalescers, typical pressure drop is 80 millibar. We have some application which runs at 40 or 50 millibar, and recommend to change when it reaches one bar. For the liquid/liquid, when it is protected by a proper prefiltration, it is less than 100 millibar.

Gerald D. Davies: On your last application of liquid/liquid, can you handle it with any temperature? Should it be cold storage or warm storage, or can you handle both?

Farcy: We designed the liquid/liquid coalescers so it can withstand up to -33°C, which is the boiling point for ammonia. That is a low limit, -33°C, and the upper limit is 82°C.

Davies: At what temperature is the application that you're installing?

Farcy: It is 0°C, because the storage tank is at 3.2 bar gauge.

Bala Subramaniam, *PCS Nitrogen*: What will be the approximate investment, or the range of investment, for a typical case?

Farcy: For this, the cost of the coalescer all together is less than $30,000. I don't have the whole investment costs including the circulation loop, but the cost of the piece of the equipment is less than $30,000.

Ian Welch, *PCS Nitrogen Trinidad Ltd.*: Can it handle aqua solutions of ammonia, separating oil from aqua?

Farcy: We don't have any reference yet for the removal of oil from aqueous ammonia. We should perform a test to prove the capability, but we have a reference where, for the same kind of technology, we efficiently remove oil from water, and, provided that the density of the aqueous ammonia is not too close to the density of oil so as to separate in the settling zone, it should work.

Hussain A. Al-Hajari, *Saudi Arabian Fertilizer Co.*: Have you investigated whether this technology can be used, particularly the liquid/liquid coalescers, for the urea plant for lean carbonate?

Farcy: We don't have any reference yet. We have one customer who is interested in removing oil from urea, but it is still only a project.

Pan Orphanides, *Orph Anco*: Have you investigated the possibility of applying this technology, the gas liquid coalescers, to remove solution of CO_2 absorption systems from CO_2 gas prior to going to CO_2 compressor?

Farcy: This is one of the applications where we have the most references in the refineries where we have installations for amine sweetening units. Maybe, Tom, you could comment on that a little bit because you know that application better than me. However, this is definitely a typical application for what you say.

Wines: I just want to clarify that is with amine systems. With the Benfield Solutions, usually you are looking at a higher temperature than what the current technology would be capable of, however, we could be looking into some new products for that. Generally, the liquid/gas coalescers, as constructed now – and that could change, because we are constantly improving the product – are good up to about 180°F, which I think may not be quite high enough for the Benfield.

Furnace Section Optimization Using High Emissivity Ceramic Coatings

John C. Hellender
C&M Technologies of LA, Inc.

Note: The article related to this discussion was not available for publication.

DISCUSSION

Charles Ellis, *PCS Nitrogen*: In preparation work for the tubes you said it needed to be sandblasted to white metal. What preparation work do you need for the refractory?

Hellander: Right now, there's a difference between ceramic fiber and insulating brick. Let's talk about ceramic fiber first. Everybody is aware that the laws are changing for ceramic fiber. It's almost got to be classified as asbestos. What has to be done really, before we do any work in there, is actually to go in and vacuum the ceramic fiber. Several plants are already doing that in ethylene. That's a requirement now. What we're really looking at when bidding these projects is bringing in one of those large super vacs and vacuuming off the dust on the ceramic fiber. It is really just a vacuuming. There's a lot of dust on the floor and on the walls, and that is what is clogging the tubes. Discussing tubes, your tubes are centrifugally cast. Due to the thermal expansion, the creep in the operational temperature, I definitely want white metal. However, the application technique has to change because it is going to take roughly six to eight continuous coats to get two mills. No matter what anybody says, ceramics are not ductile, so I have to be careful in the application techniques to make sure everything goes well. It's a different technique because of the operational temperature of the tubes.

Ellis: What is your expected effective life of the coating for the two-mill thickness?

Hellander: On the tubes, I'd say roughly three years, and on the refractory five to seven years.

Mubashar Mahmood Butt, *Fauji Fertilizer Co. Ltd.*: What about the heat-transfer coefficient?

Hellander: Do you mean the heat-transfer coefficient of the ceramic on tubes?

Butt: Yes.

Hellander: Due to the thickness, I don't consider the heat transfer important to what I'm doing because the emissivity is absorbing heat. I've designed the system with a particle-size distribution and a density in the application to be extremely dense, and I have done some testing and it's not an insulator. It's certainly a thermal conductor, but I haven't done studies on the heat transfer of the coating. The only thing that I'm trying to accomplish is emissivity, and, certainly, at two mills, roughly on the tubes that I have done, I'm removing roughly four mills at the most of an insulating scale and applying two mills of a dense ceramic. I know that doesn't answer your question. I just don't have that data. We have done some studies in the refineries as I have developed these systems. One of the problems I've always had is with the inspection people to make sure they can ultrasonic through the coating, and that hasn't been a problem. In fact, you can detect the coating in your ultrasonic as a thin system if you have got one of the newer capabilities.

Butt: Will scale or soot adhere to the refractory?

Hellander: The units that I'm coating are natural gas, and I haven't seen a lot of soot buildup in your units. What are they firing, methane?

Butt: Natural gas. We have experienced a lot of scale and soot formation on the blocks so it will definitely reduce the transfer coefficient.

Hellander: If you are getting a soot buildup, I take it that's carbon?

Butt: Yes.

Hellander: Certainly, carbon will stick to iron or the metals, but, you know, I'm dealing strictly with dense oxides. I'm not saying they're not as nonstick as teflon, but certainly the systems are nonstick compared to metals. I haven't seen a problem with carbon buildup on the system.

P. Khetarpal, *Kribhco*: How does the coating affect the life of the tube, and, in particular, where you are sandblasting and applying the coating? What about the part of the circumference where you are not able to apply the coating?

Hellander: In refineries we're doing the work inside the furnace, which is really a pain. However, my goal certainly in your industry, due to the metallurgy, is to sandblast and to coat the tubes outside the furnace, or at least do the sandblasting outside the furnace. In reference to the life, from what I've seen, and I haven't looked at more than 20 of your furnaces, it seems to me that the metallurgical deterioration is due to temperature. You're right at the maximum operational temperature of the metal, so anything you can do to improve the radiant sections temperature uniformity from top to bottom, and improve the firing rate and temperature uniformity, will stabilize the surface temperature of the metal. Just from that alone, improved temperature uniformity will increase the tube life. My real goal in your units, where I can coat 360° of the process tubing, is to eliminate the oxidation and scale formation, improve the surface temperature profile, and help reduce creep and increase a tube life that way. I can't affect the interior tube life unless there are hot spots that are produced which deteriorate the tube. I'm not sure if that answers your question, but from a metallurgical standpoint, I'm looking at temperature stability for tube life.

Pan Orphanides, *Orph Anco*: You mentioned the case of improving the metallurgical stability of the process metal parts. Have you investigated or applied these coatings to improve specific behavior like immunity to metal dusting?

Hellander: When you say metal dusting, are you talking about deterioration due to scale?

Orphanides: Yes.

Hellander: Well, certainly when I coat a tube, and since I have more experience in refineries, as long as the coating is there, there is no scale formation. For Conoco in Billings, we did a platformer and a reboiler about 18 months ago. Typically, every three-year cycle, they come down and descale as best as they can. After 18 months, there is no visible scale formation at all, so that gives me an idea of what we're accomplishing there. For Diamond Shamrock, it's been in operation almost two years. I guess it's two years next month, and there is no scale formation after two years. These are platformers, though; they are not steam methane. There's some good and bad about that. Certainly, in your units the high firing form a better system because the ceramics function better at higher temperature. As long as the coating is there, it is bonding to the base metal, and it is certainly not going to form any scale or what you call metal dusting. I think we're talking about the same thing.

Orphanides: Not exactly.
Hellander: I'm not familiar with the term metal dusting.
Orphanides: May we discuss that outside the conference?
Hellander: OK. It's deterioration of the surface due to heat and oxidation.
Orphanides: Not specifically.
Hellander: OK.
Bali, *IFFCO*: We have a reformer with 280 tube six in. O.D. and 30 ft long. Normally, the turnaround time for a plant is around two weeks. I'd like to know if we want to insulate thermic process tubing, as well as a furnace, what is the timetable? This is because we can't afford to have more time.
Hellander: I can give you some time frames that I'm using in refineries. I can give you dimensions. The job that we finished at Citgo was 24 ft in diameter and 60 ft tall. We put two complete blast crews in there using a stable garnet, and we blasted the whole unit in three days. It took us about four days to do the whole project. Now, in your industry, it is going to be an easier blast because there is no scale. For a job I looked at for Coke Sterlington, I looked at their reformers. I would say on sandblasting those units, once it is scaffolded and all their checking and vacuuming is done, we could probably blast that unit with three crews in 2 1/2 days with a continuous cleanout. Then, coating would take another 1 1/2 days, so you are talking maybe four days.
Bali: My second question concerns that normally the life of the tube is measured by measuring the skin tube temperature by an outside hand-held instrument. When the tubes are ceramic coated, if the catalyst in the tube is blocked, or if there's a problem, and the skin temperature goes high because there is no heat transfer, will the coating make any difference?
Hellander: Well, there is one thing that it will make a difference in. I have seen several of your units; when you measure the tube temperature, I have seen several people have an emissivity rating at 1.0. We can certainly put that down to about .96, which will be accurate. That will give you more accurate tube temperature. I have seen several people taking tube temperatures using emissivity 1.0 which is incorrect. It should be .85. That will change the tube temperature right there. From a material standpoint, if the tube is running 1800, the coating will show 1800. It is not going to hide any numbers or anything.
G. R. Prescott, *G. R. Prescott & Associates*: I may have missed something. Could you tell me how you physically apply this coating?
Hellander: On your process tubing, it will be misted on, almost electrostatically. Because of the design, on the casting, it's centrifugally cast, and because of the thickness requirements on the refractory, it's spray-applied, airless.
Rizk El-Sayed, *Abu Qir Fertilizers & Chem. Ind., Egypt:* I would like to ask whether the ceramic coating can be used for old ceramic fiber or only new ceramic fibers?
Hellander: Most of the jobs that I've done are old ceramic fiber. I'd like to check the condition of them. However, if they're older, they are pre-shrunk, and they're more stable. I can still coat new fiber, but I prefer a stable fiber. The older it is, the more vacuuming you'll have to do.
El-Sayed: What is the thermic expansion of the ceramic coating?
Hellander: Concerning the refractory, I'm applying a refractory on a refractory, so it is not a variable to me. I can formulate the coating to be pretty close to thermal expansion. Of course, ceramic fiber doesn't have a thermal expansion, per se. In metals, it is a totally different thing. I can meet certain thermal expansion requirements because of the way the coating interacts with metal and forms a metal ceramic interface. You have metal, metal ceramic, a ceramic, and then a top ceramic. There is a give-and-take there. However, if your tubes are near the end of their life and you have creep occurring, the ceramic will not grow like metal. It is not a ductile material. I can meet it pretty close. There may be some microcracking with age, but there is no way I can prevent that. My standpoint is if I'm trying to maximize absorbed duty, and if you have microcracking that doesn't affect what I'm trying to accomplish.
El-Sayed: The last question: What is speed limit for the gases over the ceramic coating on fibers?
Hellander: Certainly, the ceramic coating is more stable than the ceramic fiber, because it is a thin film system. It will be almost like an egg shell. It will be more

stable than a ceramic fiber, but it is just a matter of what the velocity is. I haven't seen any problems with what I'm doing. I have seen some furnaces recently that do have erosion in crossover areas, and we're going to look at coating those. It's my contention that in the units that I've seen, instead of putting a 2,300° fiber, these should have a 2,600° fiber which is more stable and more erosion resistent in those specific areas. I think that would help too.

Secondary Reformer Burner Nozzle Design and Operating Experience

This article first describes the old design of the secondary reformer burner and problems faced in its operation. The salient design features of a new burner developed by Haldor Topsøe A/S, our experience with revamping of the old burner head to the new design, and operating experience with this new burner design are then discussed.

Sajjad Hussain and Abdul Basit
Fauji Fertilizer Company Ltd., Goth Machhi, Pakistan
Olav Holm-Christensen and Henrik Stahl
Haldor Topsøe A/S, Lyngby, Denmark

Introduction

Fauji Fertilizer Company (FFC) operates two ammonia–urea plants at Goth Machhi, Pakistan. The first plant (base unit) was commissioned in 1982 with a design capacity of 1,000 metric t of ammonia and 1,725 metric t of urea per day. The plant employs Haldor Topsøe design of the late 1970s, featuring a high steam-to-carbon ratio (3.75), large primary reformer, hot potassium carbonate (Benfield) system for carbon dioxide removal, and synthesis loop with operating pressure of 267 kg/cm²g. The plant was successfully revamped to 122% of design capacity in 1992.

The second plant (expansion unit) was commissioned in March 1993 with a design capacity of 1,100 metric t of ammonia and 1,925 metric t of urea per day. The ammonia plant is based on Haldor Topsøe low energy process. Salient features of the process are low steam-to-carbon ratio (2.5), a smaller primary reformer, copper-based medium-temperature shift catalyst instead of high-temperature shift catalyst, MDEA low energy process for carbon dioxide removal, large purge recovery unit, and synthesis loop pressure of 150 kg/cm²g.

Both urea plants use Snamprogetti urea technology. Presently FFC is producing approximately 4,200 metric t of urea per day, which is equivalent to 40% of the total urea production in Pakistan.

Secondary Reformer

The secondary reformer (Figure 1) is an integral part of the synthesis gas generation section of a conventional ammonia plant. It performs two functions within the ammonia production process. First, it provides the most suitable point for introduction of nitrogen from air for product ammonia synthesis. Second, it helps achieve a very high methane conversion, thereby reducing the methane content of synthesis gas.

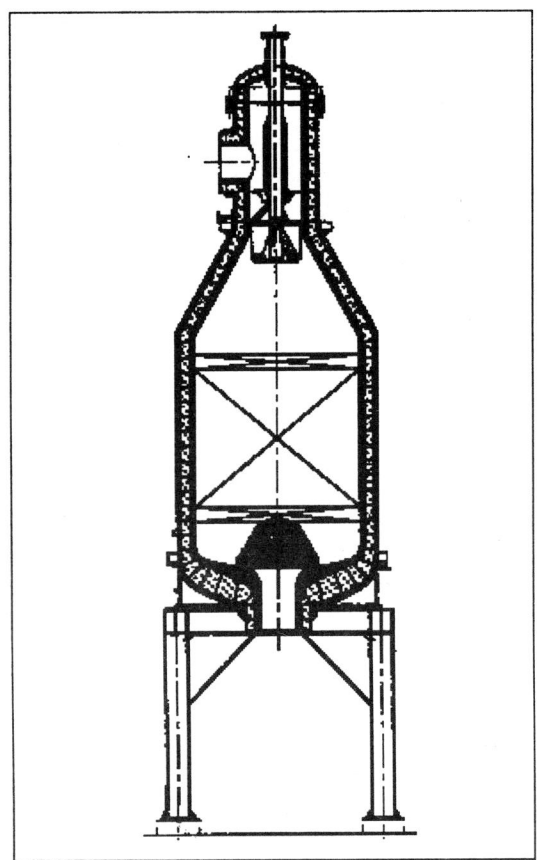

Figure 1. Topsøe secondary reformer.

Compressed air at the reformer operating pressure is injected into the secondary reformer through a burner. Process gas stream from the primary reformer also enters the secondary reformer, and mixing/combustion of gas with the air takes place in the top part of the secondary reformer.

The combustion substantially increases the temperature of mixed gas, which is then cooled as endothermic reforming reaction occurs over the catalyst bed.

The combustion process in a secondary reformer is fuel-rich and very complex. The system, therefore, requires a very rapid initial mixing for the combustion reactions, followed by further mixing of the remaining process gas into the combustion products.

Role of the Burner in Secondary Reformers

Functionally, the secondary reformer consists of two distinct regions, the combustion zone and the fixed catalyst bed. The combustion is not pre-mixed, and since the feed gas is above its automatic ignition temperature, ignition occurs instantly as the two streams mix. The performance of burners can have a significant effect on the flow pattern and the mixing within the combustion zone. Most of the poorly performing secondary reformers lack proper mixing due to inadequate burner performance.

The combustion process, which leads to the formation of a traditional diffusion flame, is controlled by the rate at which the fuel and oxidant mix. The mixing process is controlled by turbulence created by the differing velocities and densities of the two streams. Such mixing is achieved in the burner by injecting the high-pressure air into the secondary reformer at a much higher velocity than the process gas.

If the gas at the inlet of the catalyst bed is not uniformly mixed, then the average outlet composition can be far from equivalent even if each part of the gas flow is brought close to equilibrium.

Incomplete mixing of process gas and air can lead to development of hot spots in the catalyst bed or on the refractory walls. Hot spots in the catalyst bed can lead to catalyst deactivation and poor process performance, reducing the efficiency of ammonia production, while hot spots on the refractory walls are a serious safety issue that can eventually cause pressure shell rupture if they are undetected or ignored. Therefore, effective mixing through better burner design is imperative for good secondary reformer operation.

Geometry of Combustion Zone

The geometry of the combustion zone has a direct effect on the level of mixing between process gas and air. The geometry of the burner and of the vessel must ensure that the extremely high-temperature flame region is maintained at a suitable distance from the walls of the vessel. The second requirement is that a region of relatively cool process gas flows between the flame and the wall to limit the heat falling on the wall from radiation and provide the main source of cooling.

Most problems encountered within the synthesis gas industry with secondary reformers are due to problems with the combustion process that result in inadequate mixing and a variable gas composition at the catalyst bed inlet. The problem is rarely due to lack of activity in the catalyst bed, except in cases where the polant capacity has been significantly increased over the design capacity.

Construction

The secondary reformer burner is a showerhead-type sparger consisting of an elliptical head and a ring. The elliptical head is directly connected to the main (air supply) pipe, whereas the ring is connected to the main pipe through three branch pipes.

In the original design, the elliptical head and ring were furnished with holes for air supply; but in the modified design, nozzles were installed for air supply.

Experience with original burner design

We experienced burning and enlargement of air supply holes on the original burner. The enlargement of holes was irregular in shape and was severe enough in a few instances to engulf the adjacent holes. This damage was beyond the acceptable level for the burner's short operational life of four years.

The burning of hole edges showed a combination of oxidation- and carburization-type of attack. Cracks propagating from hole edges were also found on the metal surface.

The causes of burning hole edges were threefold: the poor surface finish of air holes, the presence of sharp edges and burrs on the inner edge of the holes, and the inaccurate geometrical orientation of the holes (for example, holes were not perpendicular to the metal surface).

The poor workmanship and improper geometrical orientation of holes caused generation of eddies of a very hot gas (above 1,200°C) around the hole edges. This hot gas ultimately burned the holes.

The damaged and enlarged holes affected the secondary reformer performance by disturbing the flow pattern. This caused gas impingement on the catalyst bed and resulted in increased slippage of methane in the process gas stream.

Table 1 summarizes our observations before installation of the burner with the nozzles.

Table 1. Observations Before Installation of the Burner with Nozzles

S/No	Year	Problem	Remedy
01	1985 (Turnaround)	Erosion and enlargement of holes	None
02	1986 (Turnaround)	Eroded holes deteriorated further	Welding repair
03	1987 (Turnaround)	Excessively eroded/enlarged holes	Welding repair
04	1988 (Turnaround)	Enlargement of holes	Replaced burner head
05	1989 (Turnaround)	Erosion and enlargement of holes	None
06	1991 (Turnaround)	Excessively eroded/enlarged holes	Replaced burner head
07	1993 (Shutdown)	Excessively eroded/enlarged holes	Modified burner head with nozzles installed

 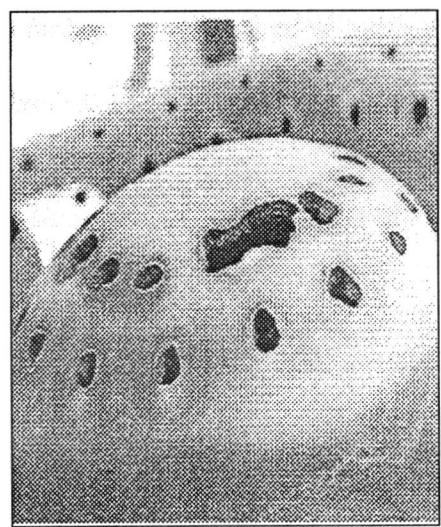

Figure 2. Burnt (on left) and enlarged (right) air supply holes.

Repairs on original burner

The burned hole edges were ground to a sound metal through exploratory grinding with a die grinder. Hole surfaces were built up with weld deposit. The subsequent drilling of holes to the required sizes was carried out by following the actual specifications and drawings. The hole backside (inner edge) was chamfered to get a streamlined flow pattern.

Modification Job

In 1991 Topsøe recommended installation of a modified burner design with nozzles to eliminate the problems associated with the existing burner at that time. Installation of nozzles on a new burner head was also recommended due to the suspected poor metallurgical health of the used burner.

Fauji had a spare used burner and were interested

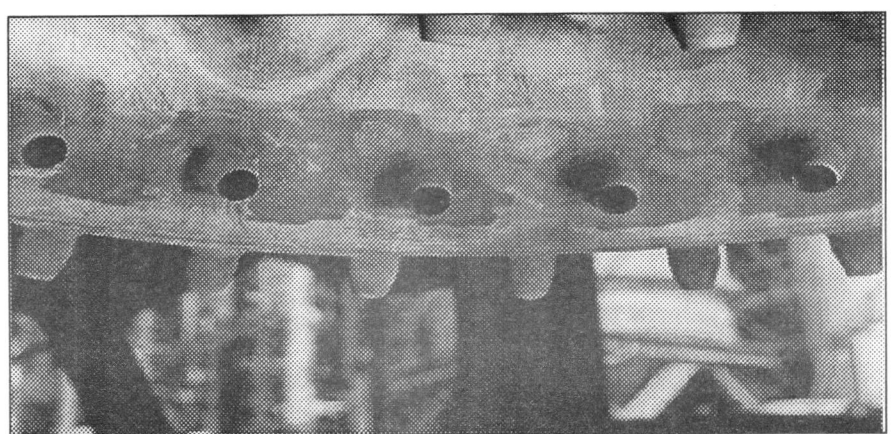

Figure 3. Plugged holes.

Figure 4a. Air supply nozzle after three years service.

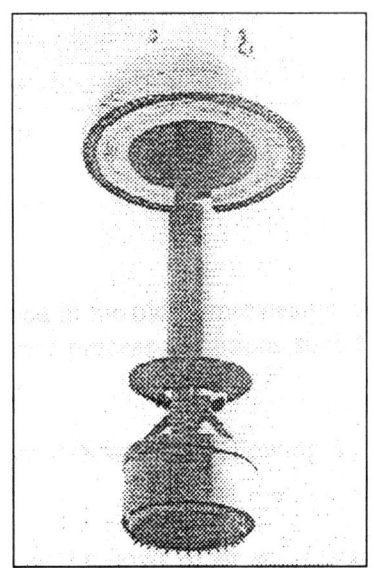

Figure 4b. Complete burner assembly.

Plugging of existing holes on burner ring

The plugging of enlarged holes (after exploratory grinding) with weld deposit was the tricky part of the job (Figure 3). The burner ring in the shape of a circular pipe was accessible only from one side for welding.

Before attempting to plug the actual burner holes, a mock-up was made from a pipe piece with drilled holes and cut into two halves. The holes on the mock-up piece were plugged with weld deposit from the front side and the weld penetration was checked from the back side of the holes. The weld penetration was satisfactory. This exercise provided the necessary confidence to undertake the plugging job on the actual burner.

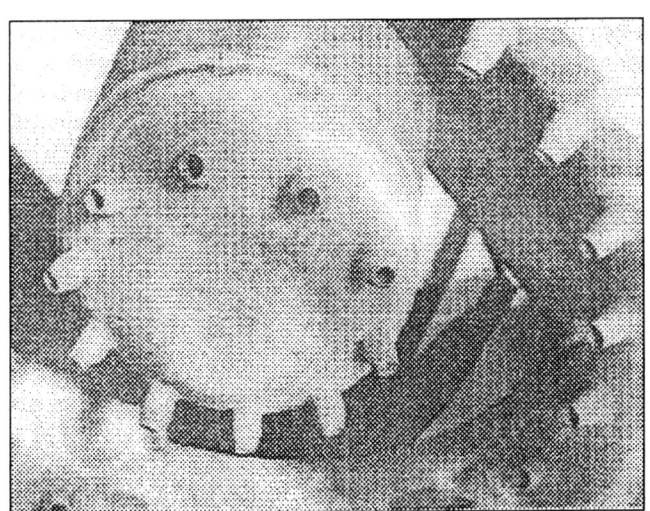

Figure 4c. Modified burner with air supply nozzles.

in modifying it for installing nozzles in order to save the cost involved in buying a new burner.

Soundness of used burner

The first step in the modification job on the used burner was to remove burnt and cracked material and make it free of defects. Enlarged and burned holes Figure 2) were ground with exploratory grinding to sound metal, and their soundness was confirmed with penetration tests.

Shroud installation

The modified burner design required installation of a shroud in the form of a cone inside the main pipe. The elliptical head (cap) was cut and a cone fabricated from a sheet was welded for the shroud installation.

The holes on the elliptical head were plugged with weld deposit.

Nozzle installation

New holes were drilled on the burner assembly before nozzles were installed. Special care was taken to drill the holes at correct angular position as provided on HTAS drawings/specifications. The accuracy achieved was ±2° of the angles as specified.

The design entailed nozzles that protruded inside the ring and elliptical head. The wall thickness available on our used burner head was 5 mm more than the designed value of the wall thickness vis-à-vis the dimension of the supplied nozzles. We ground off the outer surfaces (area where nozzles were to be installed) of elliptical head and ring to get the desired nozzle protrusion.

Conclusion on Burner Modification

The modified burner with nozzles was installed in September 1993. (See Figure 4.) Subsequent inspections have shown a burner in good working condition with no damage or erosion. An average plant load of 125% of the design capacity that might not have been possible without this modification has been maintained.

On the basis of this, the secondary reformer burner of the expansion unit was also modified in 1996.

Design of New Burner for Secondary Reformers

As was mentioned previously, damage was observed in the old burner design. The edges of most of the air supply holes were attacked by the severe process conditions to the point that parts of the pipe wall material were burned away.

A new design was developed by Haldor Topsøe A/S with the following three criteria in mind:
- The flow pattern must be controlled so that the flows of air and gas are almost unidirectional at the points where the two streams meet.
- The region of burning gas must be displaced as far as possible from all metal surfaces to avoid unintended heating of these surfaces.
- The metal parts with the shortest distance to the flames must be cooled to increase safety.

The new nozzle burner design was developed on the basis of these criteria. The nozzles, made out of temperature-resistant metal compounds and machined with extreme precision, are carefully designed on the basis of the process conditions of the individual secondary reformer. Calculations were made by Topsøe using computational fluid dynamics (CFD) to simulate the flow and temperature fields close to the air supply holes. Calculations were made for two situations characterized by air supply through both holes and nozzles. Figure 5 shows a close-up of the flow field close to the air supply ring in a plane perpendicular to the axis of this ring for situations where air is supplied through holes.

The plane cuts through the middle of an air supply hole so that the meeting point of the gas and the air can be studied with respect to flow and temperature effects. Gas can be seen flowing vertically down from the top; but when it hits the air supply ring part it undergoes a drastic change of direction so that it follows the surface of the supply ring. The flow direction of the gas is then almost perpendicular to the flow direction of the air where the two streams meet at the edge of the air supply hole.

Figure 6 shows the same situation for the case in which an air nozzle has been placed in the air supply ring. There is a clear indication—especially on the side below the nozzle—that the directions of gas and air are almost identical. This minimizes the risk of generating unintended damaging recirculation zones close to the air outlet. Recirculation zones are particularly unwanted in this area, because they will contain a mixture of gas and air in which flames will exist very close to the metal surfaces.

Figures 7a and 7b show the temperature fields for the same situations. Figure 7a shows the situation without an air nozzle. It clearly shows that high temperatures can be found close to the edge of the air supply hole. Figure 7b shows the same situation for the air support ring with air nozzle; the hot zone is

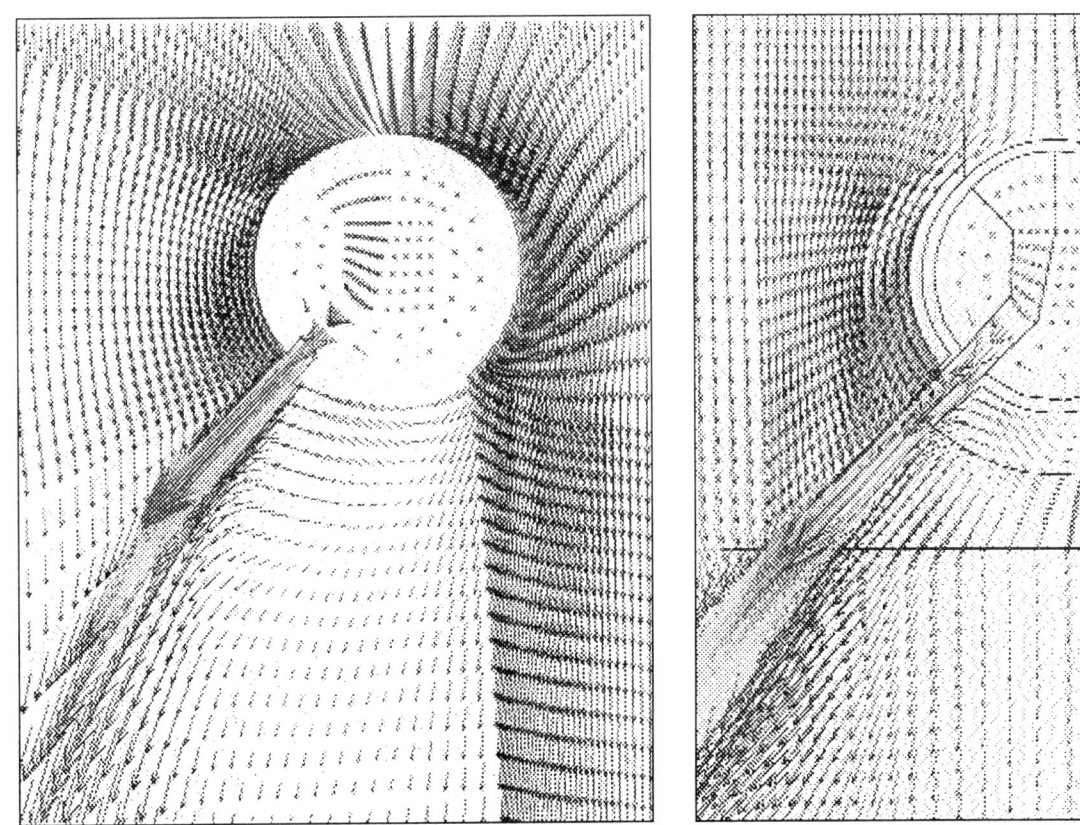

Figure 5. Flow field by CFD (air supply through holes).

Figure 6. Flow field by CFD (air supply through nozzles).

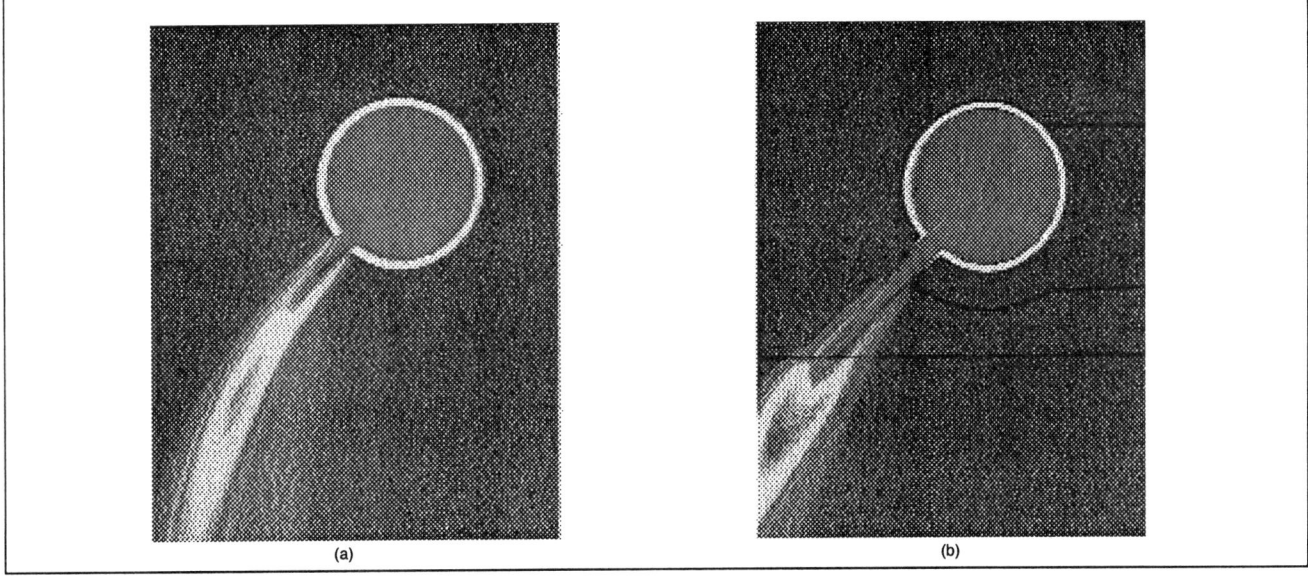

Figure 7. Temperature field. Air supply through holes is shown on left and through nozzles is shown on right.

kept away from all metal surfaces at this point. The nozzle itself has the shortest distance to the flame; but it is heavily cooled by the internal air flow, which means that it can also be kept at a low temperature. The sharpened edge of the nozzle also helps avoid recirculation zones so that little or no combustion takes place close to the nozzle edge.

Figures 7a and 7b clearly indicate why a strongly increased lifetime of the burner can be expected after the air nozzles have been added.

Apart from the flow and temperature fields close to the air supply ring, it is obviously very important to obtain an optimum distribution of temperature and composition at the inlet to the catalyst bed. Very careful studies are carried out from case to case to ensure that the right number of air nozzles is specified and that the nozzles are positioned so that the overall mixing quality is as good as possible. This ensures that the burners for secondary reformers are designed in such a way that perfect process results and long burner lifetimes can be obtained.

Conclusion

The secondary reformer of ammonia plants was improved by installing Topsøe burners with nozzles. The improvement resulted in better process behavior and longer burner lifetimes. All new burners for secondary reformers are of this design and revamp of existing burners of earlier design is possible.

DISCUSSION

Pan Orphanides, *Orph Anco*: Was the air velocity out of the nozzles in the old and the new burner configurations the same, or you have increased the air velocity? What is the range if there is an increase?

Stahl: Basically, it is in the same order and it is not increased, but when I say there is a cooling effect, it also has to do with the shape of the nozzle. The shape of the nozzle was thoroughly optimized, and the nozzle has a sharp edge of outlet. Therefore, we are sure that the air is cooling the edge sufficiently.

Jerry M. Rovner, *Kvaerner John Brown*: Could you tell us the name of the commercial software that you used for the CFD modeling? Also, on the slides for the temperature field, were you actually modeling the combustion reaction itself or were you just imposing an adiabatic temperature profile where the air and the fuel mixed?

Stahl: We are using a commercial CFD program. It is called CFX. It is the former Flow 3D, and we have added functions to make chemical reactions also. In this case the chemical reaction is very complicated, and certain shortcuts have been made. However, the combustion reaction has been taken into account when we map these temperatures. I will not say, and that is also why there were no figures on the slide. We cannot say precisely whether the temperature is 50° more or less, but we can adjust the model and refer it to experimental results of our own so that we can be quite sure that the mapping is sufficiently good for the purpose that it is used for.

Rovner: Just one last question. I'm familiar with the CFX programming, and the chemical reactions are typically modeled as a Fortran subroutine. Were you just including the heat of combustion to alter the temperature gradiant, or were you actually including the effect of the increase in volume from the combustion reaction?

Stahl: We are including the increase in volume, and we are also including the heat of reaction.

Khetarpal, *Kribhco*: Was there any effect of deformed holes of the burner on the catalyst top layer? After changing the burner, combustion has shifted slightly away from the burner. Is that going to affect the catalyst?

Sajjad: Yes, there was some effect on the top layer of the catalyst, and we skimmed and topped up the catalyst bed before burner replacement. Shifting of combustion away from the metal does not mean that it is coming very close to the catalyst, as it is optimized within the combustion zone and it is not going to affect the catalyst.

Khetarpal: Was the catalyst fused?

Hussain: Not all.

Stahl: To answer your second question, when we are talking about removing the combustion zone from the metal part, we are talking about, only, say, 10–15 millimeter or so. It is not at all which comes close to the catalyst bed.

Mukund Bhakta, *Brown and Root Engineering and Construction*: On your CFD program, could you predict what is the velocity of gas on the top of catalyst bed?

Stahl: We can predict the gas velocity at the top of the catalyst bed, and we are doing this to give an even distribution down through the catalyst. If your question is which effect it has to the reaction, yes, we can simulate chemical reactions in catalyst beds also.

Bhakta: How about disturbance of the catalyst bed due to high velocities?

Stahl: Yes. That is a difficult matter. It is possible to simulate also particle carry-away or particle movements when gas is flowing over a surface. We haven't been doing this, and I think it is doubtful if we could trust the results, but in principle yes. That means no.

Stephen Noe, *M. W. Kellogg*: It looks like you modeled the outside peripheral ring. Did you also model the center burner piece?

Stahl: Oh yes. We have been modeling the total burner in three dimensions, so everything has been included.

Noe: It seems like the centerpiece would have a greater tendency for recirculation, because you do not have a flow that can easily access the center of it. With the ring, you can have wraparound flow, of course.

Stahl: Yes.

Noe: However, did you include that?

Stahl: Yes, we did.

Ian Welch, *PCS Nitrogen*: Do you see any difference in the performance turndown rates in the plant?

Sajjad: To answer your second part of the question, after installation of the modified burner, overall process performance has increased.

Welch: Before you carry out the modification of this burner nozzle, have you noticed any hot spots on the outside of the burner nozzle or on the vessel itself?

Sajjad: There has not been any hot spot. We inspected the burner nozzle and the vessel in every turnaround and carried out necessary repairs on enlarged holes early on. Therefore, we did not have considerable damage.

Jerry Davis, *El Dorado Nitrogen Co.*: Was there a difference in pressure drop between the first nozzle on the air side and the new nozzle?

Stahl: Not as I know of. I could see no reason why it should be. However, I have no experience with it.

Dimensional Check of Catalyst Tubes to Assess Remaining Life

In this article, a procedure for measuring the remaining life of catalyst tubes based on diameter measurements, specifically ID bulging, is outlined. Our experience with this method when applied to catalyst tubes that have been in use for more than 10 years is discussed, with examples drawn from nine different plants.

T. Shibasaki, T. Mohri, and K. Takemura
Chiyoda Corporation, Tsurumiku Yokohama 230-8601, Japan

Catalyst-tube life is consumed by creep damage, which depends on stress, time, and temperature. The internal pressure on catalyst tubes and the service time are equivalent within the same plant. However, each tube has a slightly different maximum temperature, which determines the remaining life of catalyst tube. The amount of creep damage is evaluated by dimensional examination.

Figure 1 shows the change of accumulated number of catalyst tubes for each tube material. In this time, the use period of IN519 (24Cr-24Ni-Nb-0.3C iron base cast alloy) catalyst tubes is more than 12 years, which corresponds to about 100,000 h. We presented the preliminary assessment on the remaining life of IN519 catalyst tubes at the Vancouver ammonia symposium in 1994 (Shibasaki et al., 1995). After that, many IN519 catalyst tubes that were 12 years old were investigated to estimate the remaining tube life (Figure 2). Table 1 shows the design conditions of the investigated IN519 catalyst tubes.

We propose a procedure to assess the remaining life of catalyst tubes by dimensional check. This procedure consists of the following three steps.
• Measurement of hardness distribution along the tube length to find the catalyst tube's temperature profile.
• Measurement of OD and tube wall thickness at the same position to evaluate the amount of creep damage.
• Observation of microstructures from the outer surface side of the catalyst tube by the replica method to confirm the metal temperature and the creep damage.

Measurement of Hardness Distribution

Catalyst tubes are made of heat-resistant cast alloy. The hardness of heat-resistant cast alloy changes along with temperature and time. In the case of used catalyst tubes within the same furnace, the service time is the same. The hardness change in the catalyst tubes along the tube length shows the temperature profile during the operation.

Figures 3 and 4 show the results from measurement of Vickers hardness distribution for IN519 catalyst tubes that were used in a hydrogen plant for 20 years and in an ammonia plant for 19 years, respectively. The horizontal axis indicates the location of the hardness measurement (0% shows the top of catalyst tube, where it is outside the firebox, and 100% shows the bottom of catalyst tube).

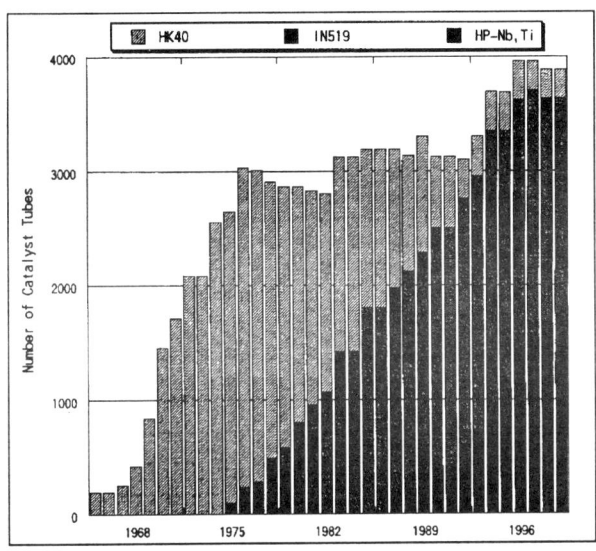

Figure 1. Change in material for catalyst tube (from 1965 to 1999).

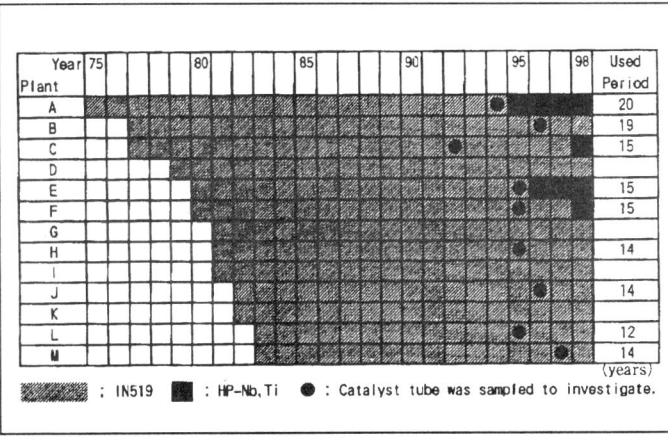

Figure 2. Used portions of investigated IN519 catalyst tubes.

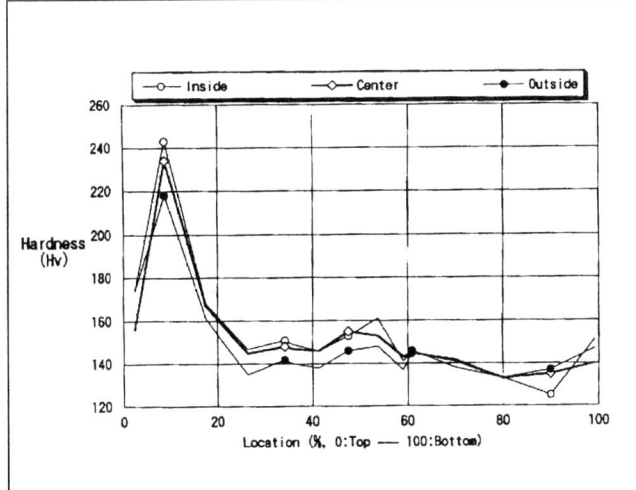

Figure 3. Hardness distribution along tube length of used IN519 catalyst tube for 20 years in plant A.

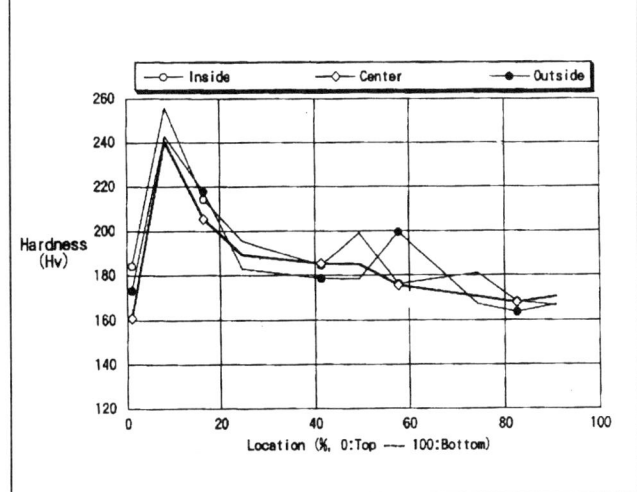

Figure 4. Hardness distribution of used IN519 catalyst tube for 19 years in plant B.

Table 1. Design Conditions of the IN519 Catalyst Tube Investigation

	Plant								
	A	B	C	E	F	H	J	L	M
Used Period (yr)	20	19	15	15	15	14	14	12	14
Design Temperature (K)	1,166	1,153	1,182	1,168	1,188	1,173	1,185	1,191	1,183
Design Pressure (MPa)	3.2	4.1	2.9	2.8	2.8	2.7	2.7	2.9	2.7
Hydrogen (H) or Ammonia (A)	H	A	H	H	H	H	H	H	H

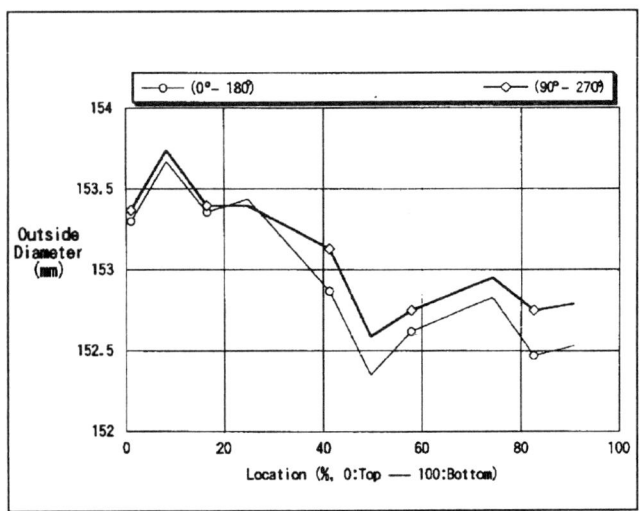

Figure 5. OD distribution of used IN519 catalyst tube for 19 years in plant B.

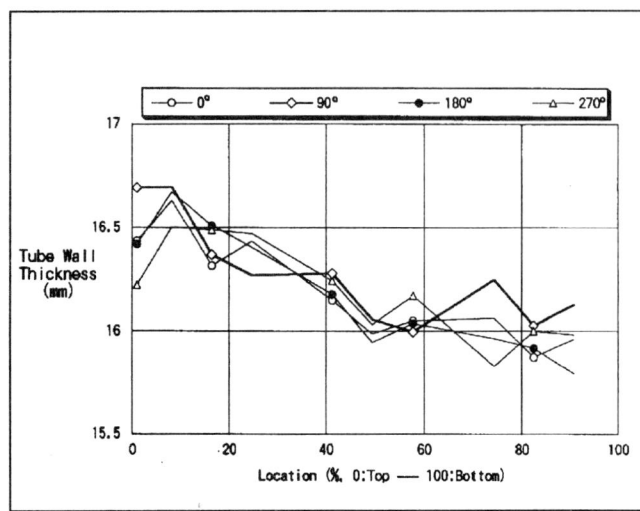

Figure 6. Tube-wall thickness distribution of used IN519 catalyst tube for 19 years in plant B.

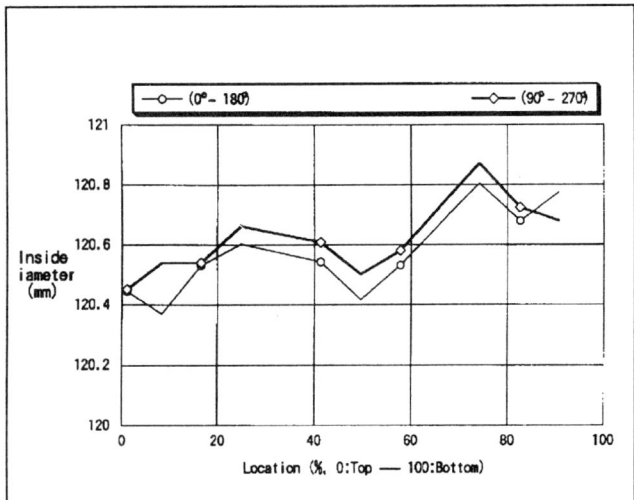

Figure 7. ID distribution of used IN519 catalyst tube for 19 years in plant B.

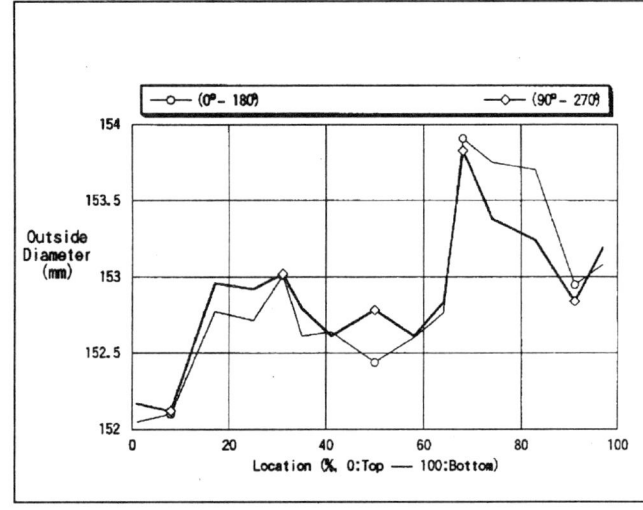

Figure 8. OD distribution of used IN519 catalyst tube for 20 years in plant A.

The temperature at the 0% portion is lower than 550°C and the hardness of its portion shows as cast condition. On the other hand, in the cell corresponding to 4% through 100% portions, the temperature of catalyst tube gradually increases in the downwards direction. Hardness changes along with the morphology of the precipitated carbides due to aging. At around the 10% portion, many fine carbides precipitate in the matrix of the catalyst tube, and the hardness indicates the maximum level. The precipitated carbides coalesce and coarsen as they progress further down the tube. The distribution density (the number of carbides in the unit area) decreases and the shape of carbides becomes massive. This mitigates the level of hardness. Therefore, the distribution of hardness corresponds to the temperature profile of catalyst tube. The softest portion corresponds to the highest tube skin temperature portion. It is important to find the highest temperature portion, since the creep damage in this portion is most severe and remaining life is shortest.

The shape of hardness distribution is almost identical in plants A and B, but the hardness levels in the

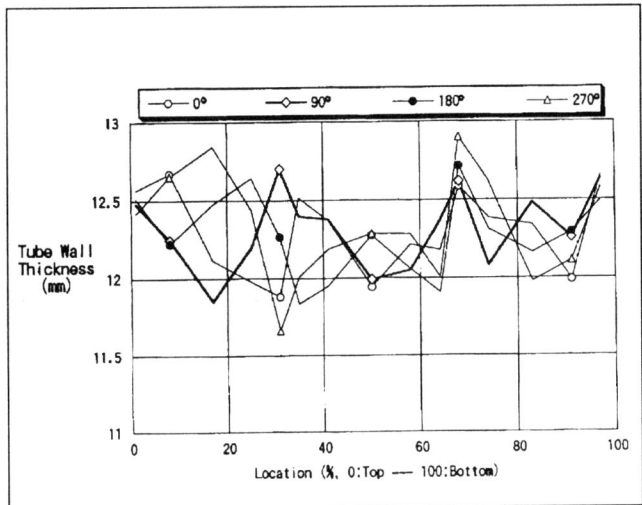

Figure 9. Tube-wall thickness distribution of used IN519 catalyst tube for 20 years in plant A.

softest portions are different. The difference in the hardness level is caused by the service temperature of the catalyst tubes. The metal temperature of the catalyst tube in plant A is higher than that in plant B. The difference of service temperature influences the behavior of creep deformation of catalyst tubes.

It is best to measure the hardness distribution for all of the catalyst tubes. If this is not practical, a minimum of two kinds of tubes should be measured. One tube should be selected randomly from the tubes used at the average service condition. The other should be selected from the tubes that have undergone something abnormal, such as severe bending or overheating.

Measurement of OD and Tube Wall Thickness

IN519 catalyst tubes bulge due to creep phenomena after long operation periods, and both the OD and the ID increase. The change of dimension corresponds to degree of creep damage, and the remaining tube life can be estimated by measuring this dimensional change.

The outside surfaces of catalyst tubes are exposed to the flue gas and attacked by oxidation, a process that ultimately reduces the OD. Therefore, the OD does not reflect bulging from creep damage. The inside surfaces are exposed to the reformed gas and are hardly oxidized or corroded compared with the outside surfaces. In fact, the original machining scratches are still present after long periods of use in many catalyst tubes. The increase in ID correlates to the creep bulging.

All of the catalyst tubes in the firebox are measured at 1 m intervals from the top to the bottom of the tube on the burner side (0–180) and the row side (90–270), that is, the side of the tube facing the next tube in the row.

The ID is calculated by subtracting both sides of tube wall thickness from the OD. Figures 5, 6, and 7 show the results of measurements of the OD, tube wall thickness, and the calculated IDs along the tube length for IN519 catalyst tubes that were used in an ammonia plant for 19 years. It is difficult to measure the pattern of dimension change caused by creep from the result of the OD measurement (Figure 5). On the other hand, it is easy to evaluate the change of dimension from ID measurement (Figure 7).

In the actual catalyst tubes, there are some manufacturing deviations in size for each unit spun-cast tube. The manufacturing deviation of the OD is larger than that of the ID because the outside surface of a unit spun-cast tube is as-cast condition and the inside surface is machined. Figures 8, 9, and 10 show the results of OD measurements, tube-wall thickness, and ID calculations for the IN519 catalyst tube after 20 years of service in plant A. The butt weld joint is located at the location 65% down the tube. A big difference in the OD was observed at the butt weld (Figure 8). The manufacturing deviation was compensated for by measuring the dimensions on both sides of the weld joint and comparing the difference of the IDs. If the difference between the IDs was 0.2 mm or more, the downstream-tube ID was corrected and then added or subtracted to the difference. A reasonable continuity in the IDs between two unit tubes is obtained as shown in Figure 10. This is another reason to measure the inside rather than the outside diameter.

The change of dimensions was calculated on the basis of the initial dimensions of the tube, rather than the dimensionS of drawing of the tube. Usually, the portion outside of the firebox retains its original dimensions. Figures 11 and 12 show the change of IDs along the tube length for the used catalyst tubes in plant A and plant B. The change of ID increases as it goes downstream.

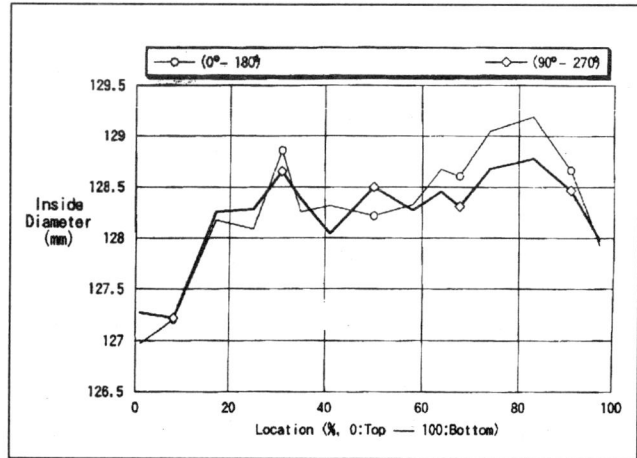

Figure 10. ID distribution of used IN519 catalyst tube for 20 years in plant A.

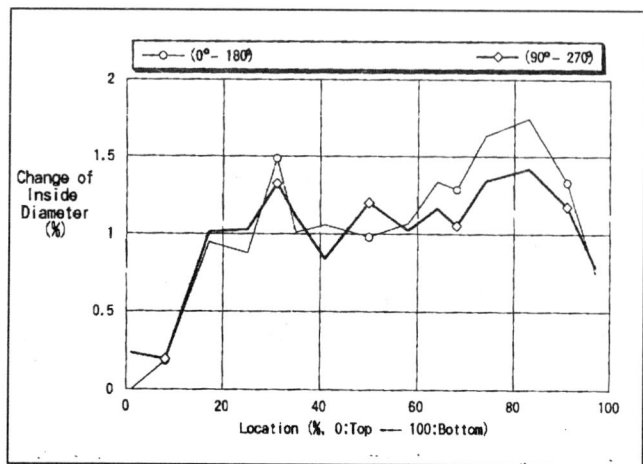

Figure 11. Change of ID distribution of used IN519 catalyst tube for 20 years in plant A.

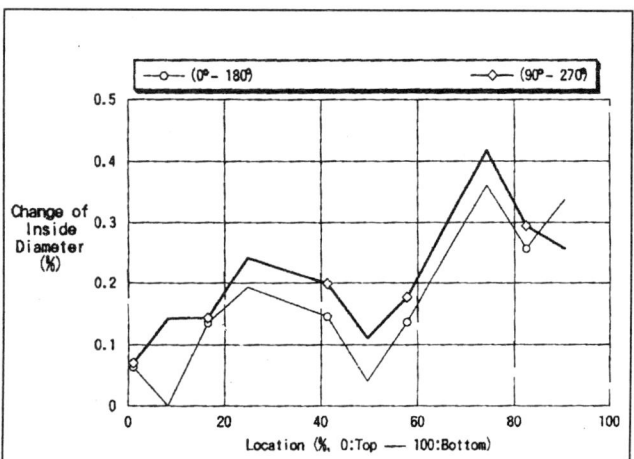

Figure 12. Change of ID distribution of used IN519 catalyst tube for 19 years in plant B.

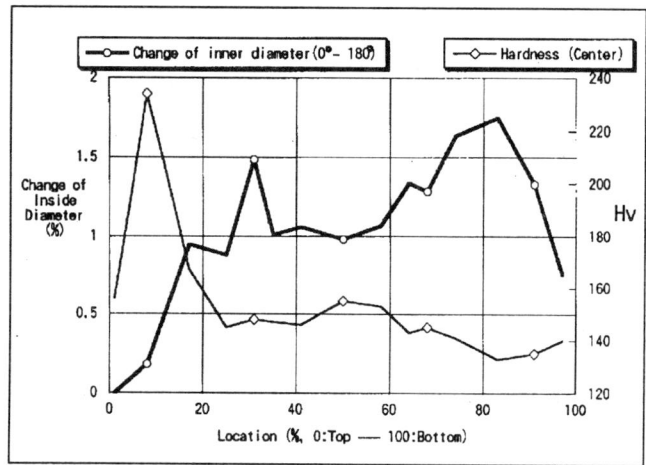

Figure 13. Relationship between ID change and hardness distribution in used IN519 catalyst tubes for 20 years in plant A.

However, if the initial dimensions before tube use and the location of the maximum tube skin temperature are known, it is not necessary to measure all parts of the catalyst tube. It is enough to check the dimensions on the maximum tube-skin temperature portion.

A method of measuring the OD and tube wall thickness for used catalyst tubes at the plant site was proposed and automatically performed with ultrasonic testing and other equipment (Shannon and Smith, 1997).

Observation of Microstructure

The microstructure of catalyst tube changes with exposed temperature and time the tube has been in use. If the database that shows the relation of the microstructures and aging conditions arranged according to an appropriate aging parameter (such as the Larson-Miller parameter) is available, it is possible to estimate the service temperature of the catalyst tube from the microstructure observation.

The creep damage of used catalyst tube material can also be confirmed through microscopic observation of the distribution of creep voids and fissures formed at the interface between matrix and primary carbides and/or sigma phases of grain boundaries.

Guide for Remaining Life Assessment

Figures 13 and 14 show the relationship between ID change and hardness distribution of used IN519 cata-

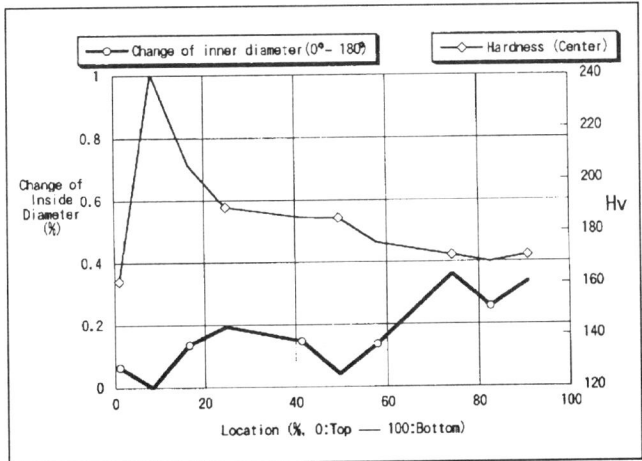

Figure 14. Relationship between change of ID and hardness distribution of used IN519 catalyst tube for 19 years in plant B.

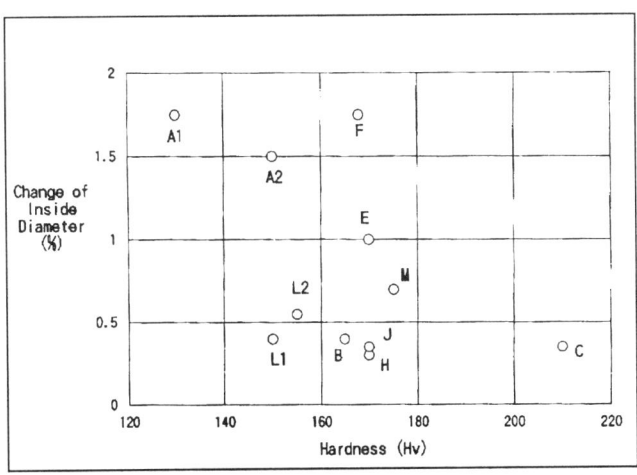

Figure 15. Relationship between hardness and change of ID of used IN519 catalyst tubes in plants A to M.

lyst tubes in plant A and plant B. The change of ID increases as it goes downstream. The portion of the tube that exhibits the maximum diameter increase corresponds to the portion of the lowest hardness. This is also the portion of the tube that has experienced the highest service temperature and the most severe creep damage. Figure 15 shows the relationship between the hardness and the change of ID at the maximum service temperature portion of used IN519 catalyst tubes in service for more than 12 years in plants A to M.

Figure 16 shows the result of relationship between the change of ID and the tube-life consumption at the maximum service temperature portion for the IN519 catalyst tube in plant A. The life of the IN519 catalyst tube in plant A was calculated by adding actual service time (20 years) to the estimated rupture time under the service condition, namely remaining life of the tube, based on the results of creep rupture tests, which were performed on the test coupons sampled from the used IN519 catalyst tube material for 20 years in plant A. The remaining life of the tube was about 2 years. The life consumption was calculated from the actual service time divided by the life of the tube, which is 22 years (20 + 2 years). Those tubes had been used for 20 years, and then were replaced with HP-Nb,Ti catalyst tubes before rupture. The 1.7% bulging of ID is considered as 95% life consumption (20 years/ 22 years).

Figure 17 shows the relationship between the change of ID and the estimated creep rupture time at 1173 K and 13.7 MPa for the test coupon sampled from the maximum service temperature portion in plants A through L. The conditions of temperature and stress for the estimation were selected for the typical service condition of those catalyst tubes. The estimated remaining life of the tube decreases as ID increases. The catalyst tube used for 19 years in plant B (noted B in Figure 17), however, had remaining life of 30,000 h, although its change of ID was less than 0.5%. On the other hand, the catalyst tubes used for 20 years in plant A (noted A1 and A2 in Figure 17) had remaining life of 15,000 and 21,000 h with ID changes of 1.7 and 1.5%, respectively. The difference of the change of ID between plants B and plant A is inferred on the basis of the service temperature difference. The service temperature of the catalyst tube in plant A is higher than that of the tube in plant B. The results of hardness distribution measurements of the tubes used in plant A and B are shown in Figures 3 and 4.

Therefore, when the remaining life of catalyst tube is evaluated by dimensional change, the effect of service temperature on creep deformation should be considered. The quantitative relationship between the creep deformation and the remaining life of the tube material is estimated from the creep data and the creep rupture data of used catalyst-tube material. To complete the quantitative information for creep deformation, many creep test data will be analyzed and the

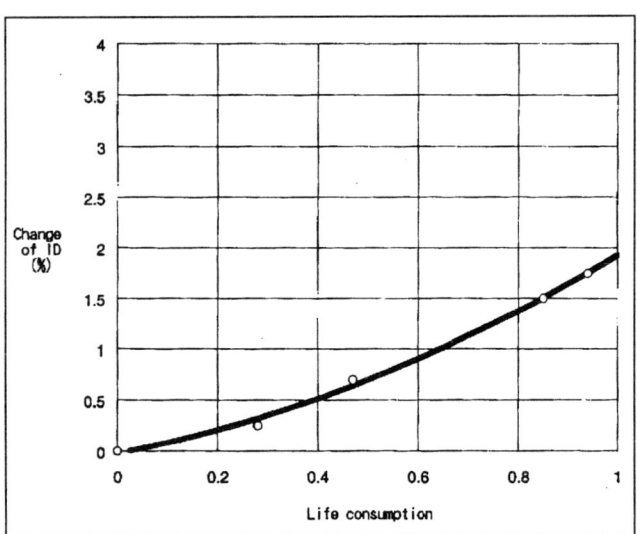

Figure 16. Relationship between change of ID and life consumption of used IN519 catalyst tube for 20 years in plant A.

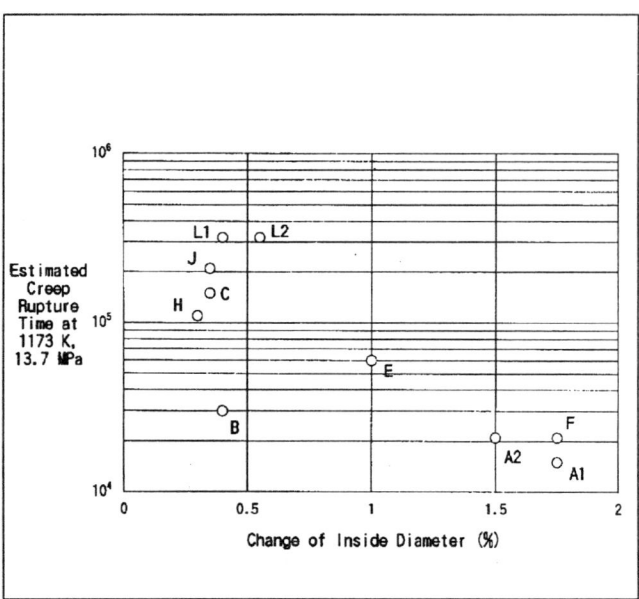

Figure 17. Relationship between change of ID and remaining life of used IN519 catalyst tubes in plants A to L.

relationship between creep strain, time, temperature, and stress will be obtained. If the application of remaining tube-life assessment were limited to IN519 catalyst tubes, the 1.7% change of ID at the highest service temperature portion might be the guideline to indicate the need for replacement within 2 years.

Conclusion

We propose the following procedure for assessing remaining tube life of catalyst tubes by dimensional check.

• A measurement of hardness distribution to find the temperature profile of catalyst tube.

• A measurement of OD and tube wall thickness at the same position to evaluate the amount of creep damage.

• An observation of microstructures from the outer surface side of the catalyst tube by replica method to confirm the metal temperature and the creep damage.

• A 1.7% change of inside diameter of IN519 catalyst tube at the highest service-temperature portion as a possible deciding factor in tube replacement.

The same procedure can be applied to other catalyst tube material, such as microalloyed HP-Nb material.

Literature Cited

Shannon, B. E., and N. Smith, "Assessing Creep Damage in Cast Furnace Tubes Using Nondestructive Examination Techniques," *Ammonia Plant Safety & Related Facilities*, Vol. 35, AIChE, New York (1998).

Shibasaki, T., T. Mohri, and K. Takemura, "Remaining Life Assessment of Nb-Containing Catalyst Tubes for Steam Reformer Furnace," *Ammonia Plant Safety & Related Facilities*, Vol. 33, AIChE, New York (1996).

DISCUSSION

Brian Shannon, *Iesco Inc.*: In any of the items that you have done your analysis on, have you seen any of the samples that had creep or dimensional damage that did not have internal microstructure deformation? If the tube had 1.7% of creep by measurement, were there any occasions where it was not damaged internally?

Shibasaki: Yes. By microscopic examination, we found creep voids and fissures in the matrix which was deformed more than 1% to 1.5%. We tried to detect by UT measures, but we could not find a change of ultrasonic properties with this method.

Robert R. Jungerhans, *Kinetics Technology International Corp.*: It appears from your graphs that the maximum skin temperature is always towards the bottom, towards the outlet part of the tube. My question is related to the heat release pattern in the furnaces you were studying, in other words, where the burners are located?

Mohri: I'm responsible for design of steam reformers for Chiyoda Corp. Your question concerns the type of reformer furnace. This type is the Haldor Topsøe radiant wall type hydrogen reformer, and we, of course, analyze the firing pattern for the reformer tube itself, and we find out some remarkable things. This reformer has six rows of side wall burners. Especially, the lower three burners were firing more than the design, so this means the firing pattern did not follow exactly the design heat input. I'm not sure of the reason why the client has been heating more than the bottom portion of the catalyst tube. This is a remarkable fact of this reformer. However, in other side wall fired reformers in Japan designed by Topsøe and Chiyoda, the heat input was more even on tube length.

Atmospheric Ammonia Storage Tank Inspection

A 36,000 tonne atmospheric ammonia storage tank was inspected. The removal and replacement of the spray on urethane foam insulation allowed for the inspection of the tank's roof plates. Small roof leaks were found that were attributed to the original tank fabrication. An out of service internal inspection was also performed. Tank decommissioning, inspections, and findings are reviewed.

Peter Jaras
Agrium Products Inc., Calgary, Alberta T2H 2PH, Canada

Introduction

Agrium Products Inc. has a 1,650 STPD (1,500 MTPD) ammonia plant and a 2,530 STPD (2,300 MTPD) urea plant located at its Carseland Nitrogen Operations (30 miles (50 km) southeast of Calgary, Alberta, Canada). The plant was constructed in 1977. Only one atmospheric ammonia storage tank (tag 2101F) was built with a holding capacity of 40,000 short ton (36,000 tonnes). This tank is 190 ft (57.912 m) in diameter and 67 ft (20.422 m) wall height with a roof dome having a 170-ft (51.816-m) radius constructed on a ring wall foundation (see Figure 1). The tank was constructed under the API 620 Appendix R guidelines and has been in service for 20 years by 1997 without an internal inspection. Acoustic emission (AE) tests were performed to determine if any severe cracking was taking place. In 1997, Agrium took the tank out of service for an internal inspection and replacement of the spray on urethane insulation.

Tank had some minor leaks in the roof plates, which were discovered by the blistering of the urethane insulation. Several tank wall areas were indicated points of concern by the past AE testing. One of the AE indications forced Agrium to reduce the overall storage capacity for the past several years. While the AE indications did not show any signs of growth, it would be prudent to go into the tank at the 20-year mark to fully investigate the integrity of the tank and verify the AE points of concern.

Planning

We were faced with several problems in taking this tank out of service to perform an internal inspection. Since the tank would have to be emptied and warmed up, it was also prudent to remove, dispose, and reinsulate the storage tank with spray on urethane foam insulation, as the original foam insulation required replacing after its 20-year life.

Some problems included:
- No nitrogen connections for purging.
- No high point vent in roof dome to allow for ammonia purge out.

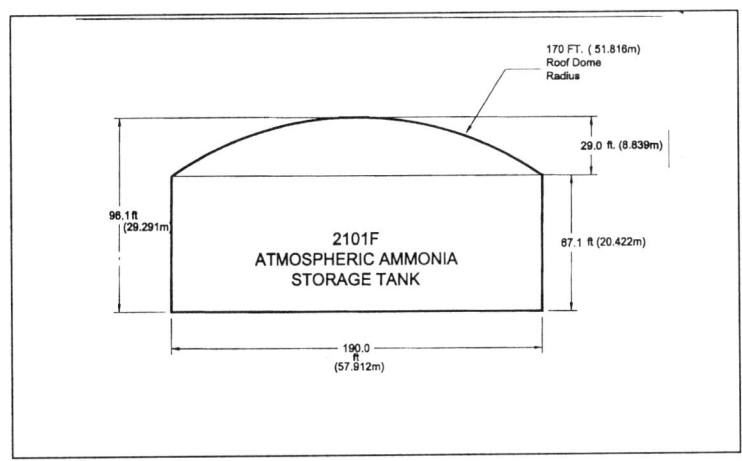

Figure 1. Tank sketch.

atmospheric storage tank. (See Figure 2.) The ammonia could then go directly to the product heater for use in the urea process, load out to ammonia trucks and rail cars, and maintain the feedstock supply to a neighboring ammonium nitrate production plant. Two 4-in. cold service piping hot taps were performed, and the appropriate valves and piping spool were installed.

The Storage tank was emptied by means of a pit pump and four 6 in. nozzles were welded to the tank wall near the bottom, equally spaced, hot tapped, and connected to temporary ring piping. The purpose of the nozzles

- There was only one 20 in. manway to gain access to tank.
- Running plants without a storage tank required piping modifications to bypass tank. Empty rail cars were required as standby storage in case the urea plant tripped.
- There was difficulty in pumping out heel of tank.

Method of Decommissioning Tank

The first step was to install a jumper line that would permit fresh cold ammonia from the on-line ammonia plant to bypass the

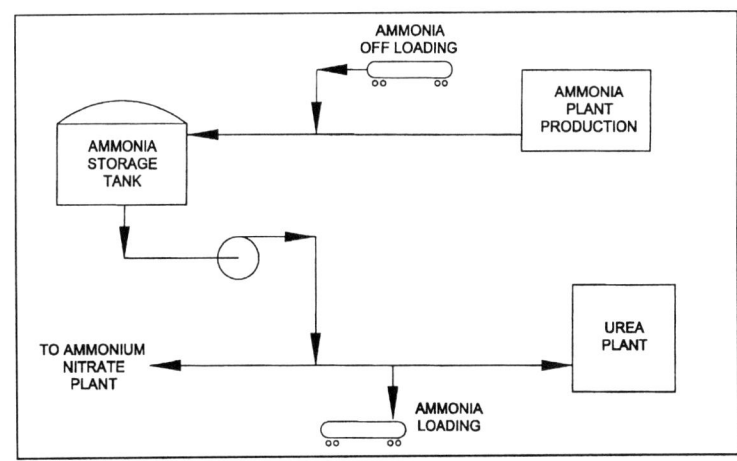

Figure 2. Ammonia flow diagram.

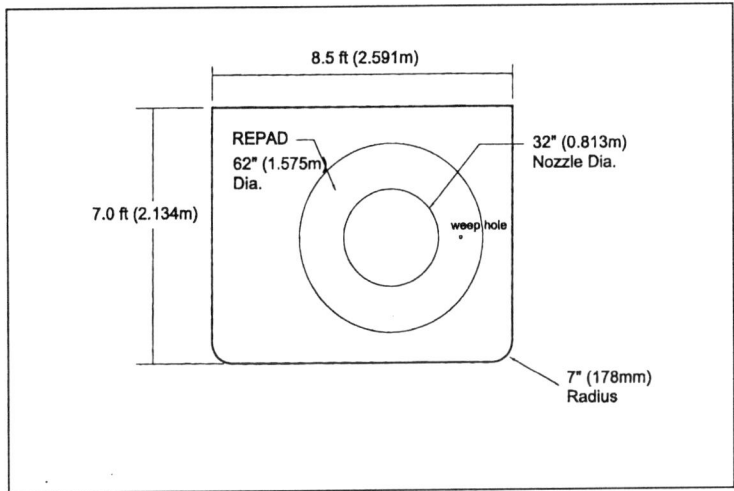

Figure 3. Manway cutout and new nozzle attachment.

was to evenly supply hot ammonia gas to the bottom of the storage tank, which would aid in the vaporization of any remaining ammonia, after the pump lost suction. The tank heaters also assisted in the final vaporization of the left over ammonia. The vaporized ammonia would be reclaimed by means of a compressor. Later, the four newly added tank nozzles would be used for nitrogen purging.

Once all of the liquid ammonia was removed, hot ammonia gas from our vapor recovery system was injected into the tank to warm it up. Once complete, a nitrogen pad was developed by injecting 4 scfm at each of the four newly installed nozzles, and maintain-

Table 1. Summary of Inspection and Insulation Project

Total Duration: 72 days: (Wednesday May 21 to Friday August 1, 1997)

Date	Activity
May 12	Main ammonia pump lost suction, switch to pit pump to pump out heel at 17 tonnes per h, 1,500 tonnes left in the tank
May 14	Installed cross-over line and tested OK.
May 21	Given the go ahead and fully committed to storage tank project
May 27	Pit pump lost suction, thought tank was empty; however, 330 tonnes still remained as could be seen by the frost line and verified by means of a boroscope camera inserted into one of the new nozzles. Require pump speed modifications and pumped out to zero back pressure at the flare knockout pot
June 8	Emptied as much of the ammonia as possible with pit pump
June 9	Started hot ammonia gassing from our vapor recovery system to vaporize remaining ammonia and warmup tank
June 13	Started building nitrogen pad at 4 scfm at each of four nozzles (laminar flow)
June 23	Bumped up N2 purge flow to 400 scfm total (100 scfm per nozzle)
June 27	Completed tank purge (1% ammonia by drager), first tank entry using fully enclosed rubber environmental suits with external air supply (Level A suit)
June 28	Large manway cutout using abrasive water jet cutting method, internal tank cleaning started
July 1	Internal tank cleaning completed
July 2	Internal tank inspection started as per API 653 on floor plates while rolling scaffold tower erection. Wet Fluorescent Magnetic Particle Inspection (WFMPI). All floor and shell weld seams and roof to shell weld were WFMPI. All floor plates and floor to shell welds were Vacuum Box Tested. Floor plates were also Magnetic Flux Floor scanned to determine underside corrosion
July 4	Rolling scaffold tower complete
July 5	Floor inspection complete, 90% of wall WFMPI complete
July 8	All internal tank inspections complete, move to external inspection
July 9	All internal scaffolding removed, started welding large manway panel back in
July 11	Door panel weld in complete
July 12	All external tank inspection complete, all internal cleaning complete
July 18	All minor internal tank weld repairs complete, final internal inspection complete and tank closed up
July 21	Recommissioning starts with 2,000 to 3,500 scfm total nitrogen purge completed in 25.5 h, used 1.5 tank volume of N_2 O_2 down to 4.6% to 4.9% by gas chromatograph
July 22	Ammonia purge started top down, -15°C at 500 scfm flow
July 29	Bumped up ammonia purge to 3,000 scfm for 18 h
July 30	Tank ammonia purge completed, pulled blinds add liquid ammonia. O_2 content down to 0.6% (90% NH^3, 9.4 % inerts) tank cold −29°C.
Aug 1	Ammonia Tank placed back into normal service

ing this laminar nitrogen flow for 10 days. Once the pad was developed, we increased the nitrogen flow to 100 scfm at each of the four newly installed nozzles, developing a plug flow, forcing the remaining ammonia gas out to the vapor recovery unit and to a high point vent. (A high point vent was created on the top of the tank by hot tapping an existing tie off post and connecting it to the tank flare stack.) The entire nitrogen purge took 14 days. When the tank was purged and the initial tank entry and inspection were completed, a second manway was cut into the tank wall, opposite to the existing 20 in. (508 mm) manway. This new manway was 8 1/2 ft (2.591 m) wide and 7 ft (2.134 m) high (see Figure 3). This large opening was used only for the duration of the internal tank inspection and made it very easy to move men, equipment, and supplies. The cutout panel was sent to a vessel shop where it had a reinforcement pad and a 32 in. (813 mm) nozzle installed. This modified cutout panel would be welded back into position upon the completion of the internal tank inspection. The panel was cut out using a high-pressure/abrasive water jet cutting method. The cut process took 8 h. The weld in of the panel took 2 1/2 days.

Roof Leak Checking

While all the preparations and modifications were done to the tank, the insulation contractor was busy removing the old spray on urethane insulation and making the roof ready for leak checking. Since a few roof leaks were apparent due to the blistering of the roof insulation, it was decided that the entire roof should be checked. A simple method was devised. The internal inspection of the tank would require it to be free of all ammonia. The decommissioning process would pump out all liquid ammonia, vaporize any remaining puddles of ammonia when the pumps lost suction, and purge all ammonia vapor from the tank before an air environment could be introduced.

Since the roof dome would be in contact with vapor ammonia from the underside for a significant period of time, the ammonia would bleed through any leak points. If an ammonia sensitive leak checking spray were applied to the outside of the dome, it would indicate all the leak locations. An ammonia sensitive leak checking paint (the product name is "Pinpoint Colormetric Developer" ADP - 219 (ammonia color change: yellow to blue) by American Gas and Chemical Co. Ltd., Northvale, NJ) was sprayed onto the outside surface of the roof dome and 21 leaks were found. Five leaks were located on roof plate lap joint seams and were a result of poor weld quality in original construction. 16 leaks were located in the center of several roof plates, typically where you would attach lifting lugs. Samples were removed and shipped to a metallurgical lab for analysis. No service related stress corrosion cracking was detected. Conclusions were that the leaks were a result of very poor quality weld repairs. During the original tank construction, lifting lugs were possibly removed by striking the lugs in such a manner that the lug to plate weld fractured and possibly tore out some of the parent plate material. These areas were then poorly weld repaired and ground smooth.

Summary of Inspection and Repairs

Table 1 provides a summary of the inspection and insulation project.

API 653 was used as a guideline for this inspection, and, in addition, nondestructive examination including WFMPI, ultrasonic thickness and shear wave, and vacuum box testing were performed. Overall, the condition of the tank was very good. Corrosion was negligible and only present in areas that had the original mill scale removed. No corrosion or pitting was noted at or above 30% of the nominal thickness. Weld repairs were only required at two locations: one topside floor plate crack and one crack located in the internal circ. weld of the original 20 in. manway.

These defects were discovered by this out-of-service tank inspection, not by the previous AE testing. No leaks were detected on the floor plate welds, and no other service related defects were found on any of the shell and floor plate welds tested.

Further NDT examination of the five areas of concern noted during the acoustic emission test did not reveal any defects deemed to require repair, however, it was decided to excavate and weld repair the one main AE area that forced Agrium to reduce filling capacity. This was done to remove the activity zone from any future AE testing. Safe fill height determinations cleared the tank of any concerns and certified the tank back to its original 36,000 tonne design fill capacity. 140 AE probe waveguides were welded to the external tank wall and then the tank was completely insulated. These waveguides will be used in the future to perform more AE testing since the next out-of-service internal tank inspection is not scheduled until the year 2017.

Conclusions

An out-of-service tank inspection was performed on Agrium's 36,000 tonne atmospheric ammonia storage tank at the 20-year mark. This allowed for the removal and replacement of the spray on urethane foam insulation, which in turn permitted the inspection of the tank's roof plates. A simple method for checking the entire external roof plates was devised. An ammonia sensitive leak checking paint was applied to the roof dome, while the ammonia tank was being decommissioned. The ammonia vapor would pass through any small roof leak area and change the leak checking paint from a yellow color to a blue color. This made it easy to locate and repair the roof leaks. A total of 21 leak areas were found; 5 were located on lap joint welds and 16 were located in the middle of several roof plates. The leaks in the middle of the roof plates were symmetrical and were probably where lifting lugs were attached to lift the roof plates into position for welding during original construction. Samples taken from these lifting lug locations ruled out stress corrosion cracking and concluded that the leaks were due to improper weld repair.

During the original tank construction, lifting lugs were possibly removed by striking the lugs in such a manner that the lug to plate weld fractured and possibly tore out some of the parent plate material. These areas were then poorly weld repaired and ground smooth.

The internal tank inspection only found two flaws that required weld repair: one topside floor plate crack and one crack located in the internal circ. weld of the original 20 in. manway. The areas of concern that were picked up by the previous acoustic emission testing showed no signs of cracking. Ultrasonic testing of these areas dismissed them as minor weld imperfections that did not warrant excavation and weld repairs. The one main area of concern was, however, excavated and found to be weld porosity. The flaw was removed by grinding and then weld repaired. AE probe waveguides were welded to the external tank wall for future AE testing. The next internal tank inspection is scheduled for the year 2017.

Acknowledgments

A special thank you to Agrium's Jerry Pyra for the excellent work both in the planning and the execution of the Ammonia Storage Tank Project. The following people also contributed important information that was used in the planning phase of the project: *Agrium Redwater*: John Mason, Reg Rudko, Darwin Serink, John Blazenko; *Agrium Fort Saskatchewan*: Don Timbres and Robert Pelletier.

DISCUSSION

Gerald A. Knazek, *La Roche Industries Inc.*: What special precautions did you take for doing the welding and cutting on the tank while it still had ammonia in it, that is, potentially ammonia vapors for the flammability explosive range?
Jaras: To reduce the risk of explosion, we hot tapped the tank nozzles so there was no flame contact with the ammonia vapor inside the tank. Before we went into the tank, we water jet cut out a large panel. The ammonia content was down to 1 ppm.

Knazek: Why did you find that the insulation had a need to be repaired? How could you tell the deterioration of it?
Jaras: We had an insulation contractor come in and do an evaluation on the existing insulation, and they did a cohesion and adhesion test. From that study, it was found that the entire insulation should be repaired or replaced. It was in service for 20 years, and that's pretty good service for urethane foam.
Ahmed Al-Badrani, *Saudi Arabian Fertilizer Co.*: I

have one comment and three questions. In the past the literature was between 6 and 10 years for inspection. Recently, it came up to 15 years. Now, you are proposing for 20 years for inspection. Isn't 20 years too long to wait for an inspection?

Jaras: When I had taken over the project, the study done before me was that the inspection would have to take place within 20 years of original construction. At the time of our study, there has been some discussion and some controversy on how often you should go into an ammonia storage tank due to oxygen-induced stress corrosion cracking, and, for that reason, Agrium waited the full 20 years and went in the tank at that time. Now, before that, at the 15-year mark, we did perform the first initial acoustic emission test, and, at that time, there was nothing very serious. There were a few points of concern that would warrant an internal inspection, and, from that, we monitored the tank every year up to the 20-year mark when we did go in and do an internal.

Al-Badrani: However, the storage tank is a very high structure. Newer storage tanks are often underground, but your tank is above the ground?

Jaras: Yes.

Al-Badrani: My first question is when you dispose of the ammonia, did the flare system take care of all ammonia?

Jaras: No. The majority of the ammonia on decommissioning was recovered by the ammonia vapor recovery compressor. However, when the inerts were too high due to nitrogen purging, we flared off the remaining volume of ammonia vapor.

Al-Badrani: Did you consider neutralizing it with acid, for example?

Jaras: No, we did not.

Al-Badrani: How much will you keep the capacity of this tank after the internal inspection? Will you keep it, as before, to full capacity or to certain level?

Jaras: Yes. During the internal inspection, we also performed a safe fill height calculation that was performed by ultrasonic crawler and that calculation cleared the tank of any flaws and brought the tank back to its original 36,000 metric ton full capacity.

V. Jayaraman, *Southern Petrochemical Industries Corp.*: The recent regulations stipulate that the ammonia tank should be double integrity, and this means that there are two tanks in service. The outside tank should be insulated, and both the tanks should have the same material. Do you have any regulation in Canada like that?

Jaras: Presently, no, we don't have a regulation for double walling the ammonia storage tank.

Jayaraman: Another regulation is that these tanks have to be inspected every six years, not 20 years. They stipulate that, every sixth year, even the double integrity tank has to be inspected for stress corrosion cracking. Do you have anything like that in your Canada rules?

Jaras: No, we don't. However, I'm interested to find out what guidelines are you referring to?

Jayaraman: Well, this requirement for testing is given in British Code of Practice for storage of ammonia, and it has been adopted in other countries, the tank should be inspected between 6 and 10 years.

Jaras: I'm not aware of that. In fact, ammonia storage tanks or cryogenic storage tanks do not have a specific code to them, and many people out there use API 653 as a guideline. And in the section 1.1, under the 653 guideline, it states that it includes cryogenic tanks. Now, I've been in contact with the API committee and have asked them if they are developing a specific code for ammonia aboveground ammonia storage tanks and they're working on it. So, to date, to the best of my knowledge, there is no specific code for ammonia storage tanks.

Jayaraman: You said that you repaired the cracks near the structural marks, removed some of the material, and, thus, created the hole. At the welded joint junctions (T-joints), did you find any stress corrosion marks?

Jaras: We did not find any stress corrosion cracking on any of the welds on the tank.

Jayaraman: Have you checked the integrity of the civil foundation?

Jaras: Yes, we did check for foundation settling by means of a laser level check and the ammonia tank settling was very minimal.

Jayaraman: Sometimes, the insulation at the bottom is not good. This will affect the civil foundation. Have you removed test samples from the civil foundation to test the integrity, the state of the concrete?

Jaras: Do you mean excavate some of the concrete

and do some sampling?

Jayaraman: I mean taking samples from the structure and then checking the integrity of the concrete whether it is porous, and whether it has got the initial strength. Was any test done like that?

Jaras: I'm not aware of taking samples from the concrete foundation. Our tank has a concrete ring wall, and we have electric floor heaters that prevent any frost from entering into the ground. That electric floor heating system is checked on-line continuously for burnouts or whatever. So, to my knowledge, during the 20 years of that ammonia storage tank, there were no foundation problems.

Jayaraman: Do you experience any problems during hot tapping? When you hot tap, when you are drilling on the roof, have you have experienced any problem, since inside the tank you have ammonia vapor?

Jaras: We follow the guidelines under API 2201 for hot tapping, and all our hot taps were performed with the required valves. We had no problems in hot tapping and removing the coupon without any excursions of ammonia vapor.

Bala Subramaniam, *PCS Nitrogen*: In answer to Mr Jayaraman's question, Simplot, Manitoba, Canada did a double wall project. Some drilling and sampling was also done at the time, but I don't have the information, although I can probably get the information to anybody who needs it.

Reducing Methanol Byproduct Formation over the LTS Converter

In this article, process parameters influencing methanol formation, potential consequences of methanol formation, and research and development efforts that have led to the development of new low-methanol LTS catalyst are identified.

Jack H. Carstensen and Birgitte S. Hammershøi
Haldor Topsøe, DK-2800 Lyngby, Denmark

Introduction

Ammonia and hydrogen plant operators are becoming increasingly concerned with methanol byproduct formation in their plants. Methanol formation takes place over the copper-based low-temperature-shift (LTS) catalyst according to the following reaction scheme:

$3H_2 + CO_2 \leftrightarrow CH_3OH + H_2O$ ($\Delta H°_{25°C} = -11.8$ Kcal × mol^{-1})

New regulations restricting the allowable levels of methanol emissions from ammonia and hydrogen plants are being implemented in many countries worldwide. In particular, plants in the U.S. are faced with severe restrictions with respect to methanol emissions.

It is possible to reduce methanol byproduct formation in some methanol plants by adjusting the LTS operating conditions. In most cases, however, the plants will be forced to implement more drastic means of reducing methanol byproduct formation and/or emissions.

Although removal of methanol from the various vent gases requires large investments and installation of additional equipment in the plant, the use of a more selective LTS catalyst may be a low-cost, highly effective alternative solution, as illustrated in this article.

A new LTS catalyst that significantly reduces methanol formation has been developed, enabling ammonia and hydrogen plants to comply with regulations without undertaking major investments in new equipment.

Why Reduce Methanol Byproduct Formation?

It is desirable to minimize methanol byproduct formation for several reasons:
• Methanol emissions are being regulated to protect the environment.

Several ammonia plants are already facing regulations that limit methanol emissions to the atmosphere.

These regulations affect plants that vent the CO_2 stream as well as plants that are not designed to recover methanol from the various vents. With growing global environmental concern, the number of ammonia plants facing restrictions of their methanol emissions is expected to increase significantly during the next few years.

• Methanol byproduct formation consumes valuable hydrogen.

In a 1,500 metric t/d ammonia plant, a traditional LTS catalyst at the start of the run may produce up to 8 t/d of methanol depending on operating conditions. Although this amount will decrease to 1–2 t/d over the lifetime of the catalyst, this byproduct formation represents a considerable loss of hydrogen. Depending on the plant design, the byproduct formation may equate to a loss of ammonia production valued at $50,000–300,000 per year (based on $150 per ton of ammonia).

• Methanol may react to form amines causing odor problems.

Part of the methanol formed in the LTS converter reacts with ammonia in the reforming section and forms minute amounts of methylamine, dimethylamine and trimethylamine. These amines are highly undesirable components because of their repulsive odor, even at the ppm level.

• Methanol may affect the quality of the process condensate.

The presence of methanol may dictate disposal of condensate, leading to significantly increased operating costs in some plants. There have also been examples of undesirable bacterial growth caused by the presence of methanol in the process condensate.

• Methanol may affect the quality of the CO_2 gas.

For CO_2 supplied to the food or pharmaceutical industries, there are no strict specifications on the level of methanol usually apply. High levels of methanol may also cause problems if the CO_2 is used for urea production.

Where Does Methanol End Up?

The methanol produced over the LTS catalyst exits the LTS reactor in the gas phase. After cooling, part of the methanol ends up in the process condensate. The remaining methanol continues with the process gas to the CO_2 removal system and leaves the plant with the CO_2 from the CO_2 stripper.

The fraction of methanol that ends up in the process condensate depends on the separator temperature. At 60°C (140°F), approximately 90% of the methanol formed ends up in the condensate, whereas at 120°C (248°F), about 40% of the methanol formed ends up in the condensate.

Methanol in the condensate is either removed in a deaerator, stripped off in a condensate stripper, or recycled directly back to the reformer via a saturator together with unheated process condensate (Figure 1). The untreated condensate may also be used as quench for temperature control upstream from the high-temperature shift (HTS) and LTS reactors.

Factors Influencing Methanol Formation

A small amount of methanol is formed in the HTS reactor, but for the most part the methanol is formed over the copper-based LTS catalyst. At typical LTS catalyst temperatures, the methanol synthesis reaction is relatively slow and far from equilibrium. Thus, the methanol formation is subject to kinetic (reaction-rate) rather than thermodynamic (equilibrium) control.

Haldor Topsøe developed a kinetic expression more than 10 years ago for the methanol formation at LTS conditions. The model has been successfully used to

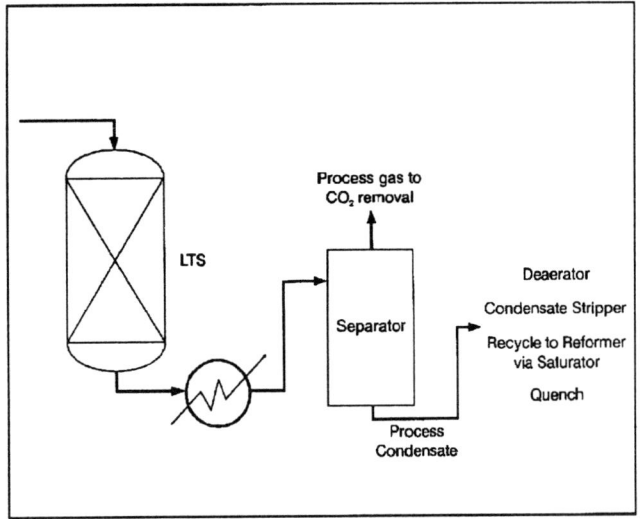

Figure 1. Distribution of methanol.

Table 1. Methanol Formation vs. S/C Ratio in an Ammonia Plant

SC Ratio (moles/mol)	S/DG Inlet LTS (moles/mol)	CO Inlet LTS (moles/mol)	Relative Methanol dry mol. %	Relative Methanol Concentration in Process Concentrate
3.6	0.45	2.4	100	100
3.2	0.38	2.9	175	200
2.8	0.32	3.5	330	450

simulate and predict methanol formation for the ammonia industry.

The various factors influencing methanol formation over a LTS catalyst are described as follows.

LTS operating temperature

The LTS operating temperature has a pronounced influence on the rate of methanol byproduct formation because the activation energy of the methanol synthesis reaction is approximately three times that of the shift reaction. The methanol reaction rate is therefore much more sensitive to variations in temperature than the shift reaction rate. Accordingly, reducing methanol byproduct formation by operating at a lower LTS temperature is possible. It has been industrially verified that a 10°C (18°F) decrease in temperature will decrease methanol formation by about 50%.

The LTS inlet temperature must be maintained at least 15–20°C (27–36°F) above the dew point of the process gas at all times to avoid pore condensation in the catalyst.

Steam-to-dry gas (S/DG) ratio

The S/DG inlet of the LTS reactor affects methanol formation because the partial pressures of the reactants vary with the steam content. Water also inhibits the methanol synthesis reaction rate.

The steam content at the LTS reactor inlet is dictated by the steam-to-carbon (S/C) ratio in the primary reformer and by additional introduction of steam/quench to the secondary reformer and the shift reactors. The steam content in the gas is normally not used to control the methanol formation.

For example, if the S/C ratio is reduced in connection with an ammonia plant, a significant increase in the methanol byproduct formation may result (Table 1). The LTS inlet temperature is assumed to be constant.

CO concentration at the LTS inlet

An increased CO concentration at the LTS inlet will increase the methanol formation. An increased CO level may be caused by a malperforming HTS catalyst or reduction of the S/C ratio in the reformer.

Operating pressure

The higher the pressure, the higher the methanol formation will be. However, pressure is not a parameter that is used to control methanol formation.

Space velocity

Methanol formation depends on the contact time because it is kinetically limited. Therefore, a higher plant load will result in a lower methanol concentration in the process condensate and CO_2 stream (although total amount of methanol formed may increase).

Catalyst volume

Reducing the LTS catalyst volume, and thereby also the contact time, may lower methanol formation. On the other hand, an insufficient catalyst volume may require the LTS inlet temperature to be increased in order to obtain satisfactory CO conversion. These two counteracting effects should be taken into account if focusing on methanol byproduct formation.

Catalyst age

Methanol formation is highest during the first few weeks of operation. As the LTS catalyst gradually

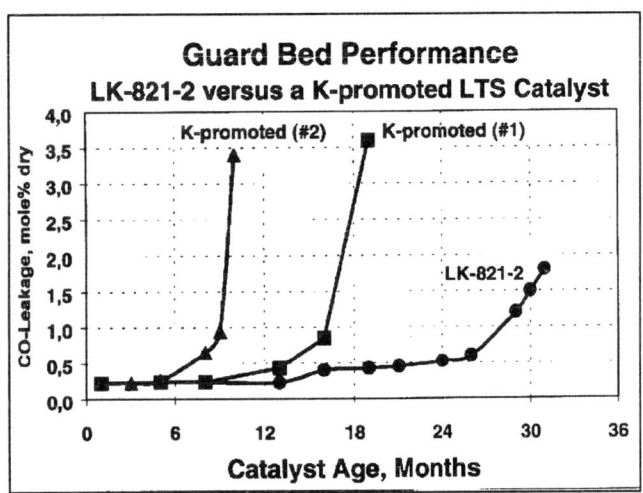

Figure 2. Industrial performance of LK-821-2 vs. potassium-promoted LTS catalyst.

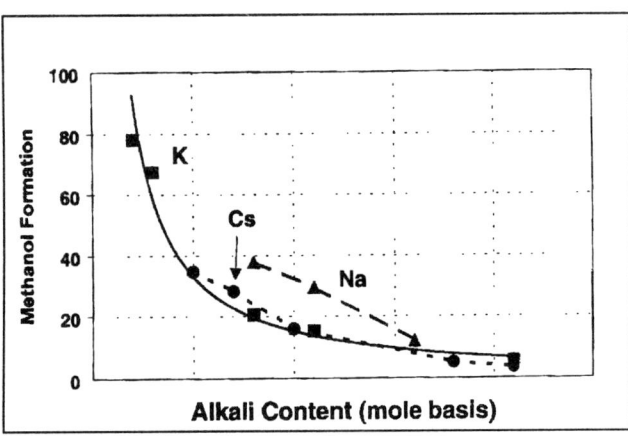

Figure 3. The effect of alkali on the methanol formation at LTS conditions.

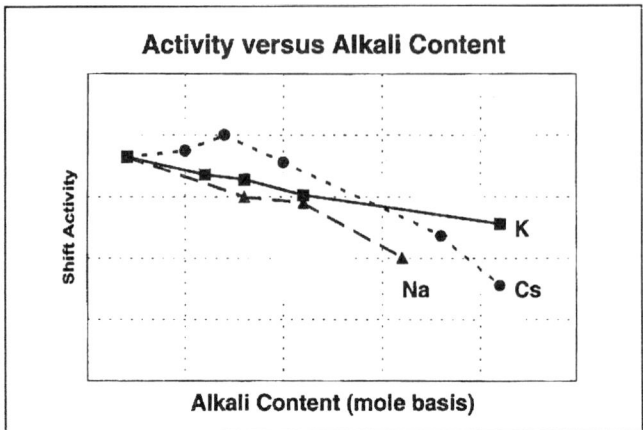

Figure 4. The effect of alkali on the shift activity at LTS conditions.

Figure 2 depicts the performance of two parallel guard beds in a 1,000 standard t/d ammonia plant in the United States. One of the guard beds was operating with two consecutive charges of potassium-promoted LTS catalyst from another catalyst supplier. The other guard bed was operating with Topsøe LTS catalyst LK-821-2.

From Figure 2, it can be seen that the LK-821-2 had a lifetime of 31 months whereas the two consecutive charges of the same type of potassium-promoted LTS catalyst in the parallel reactor had lifetimes of 19 and 10 months, respectively. Although initial performance of the potassium-promoted catalyst appeared adequate, it is apparent that the catalyst deactivated very rapidly compared to LK-821-2.

The Effect of Alkali Metals on Methanol Formation

A fundamental study aiming at elucidating the effect of alkali metals on methanol formation has been conducted in the R&D Division of Haldor Topsøe A/S. The tests were carried out by promoting Haldor Topsøe standard LTS catalyst LK-821-2 with varying concentrations of sodium (Na), potassium (K), and cesium (Cs).

The various catalysts were tested for methanol formation, shift activity and stability as well as poisoning resistance at typical LTS conditions. The physical characteristics of the catalysts were examined both before and after the tests.

The methanol formation plotted as a function of the

becomes poisoned and deactivates with time, methanol formation will decrease. The higher LTS inlet temperature required to compensate for the lower shift activity of an aged catalyst may partly counteract the effect.

Catalyst promoters

It has long been recognized that promotion of LTS catalysts with significant quantities of potassium (K) inhibits methanol formation. The problem is that in the applied quantities, potassium also significantly reduces the shift activity and catalyst stability, resulting in a negative impact on the shift conversion and catalyst lifetime.

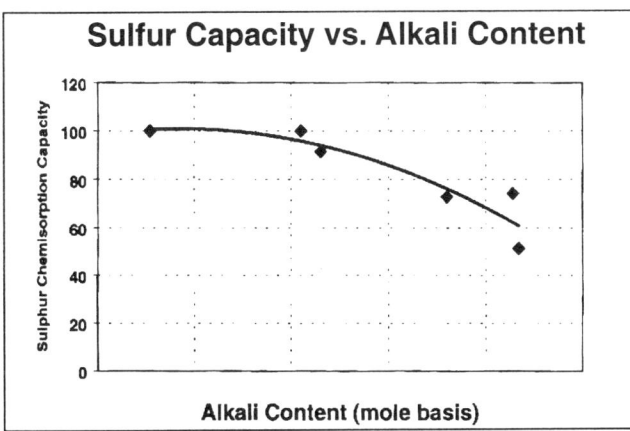

Figure 5. The effect of alkali content on the sulfur chemisorption capacity.

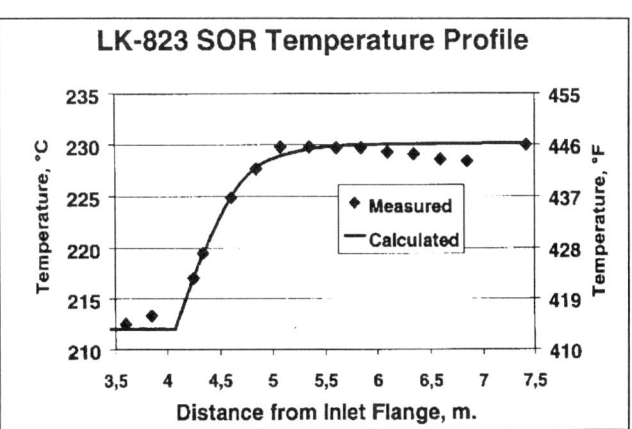

Figure 7. SOR temperature profile with LK-823 in a 1,360 metric t/d ammonia plant.

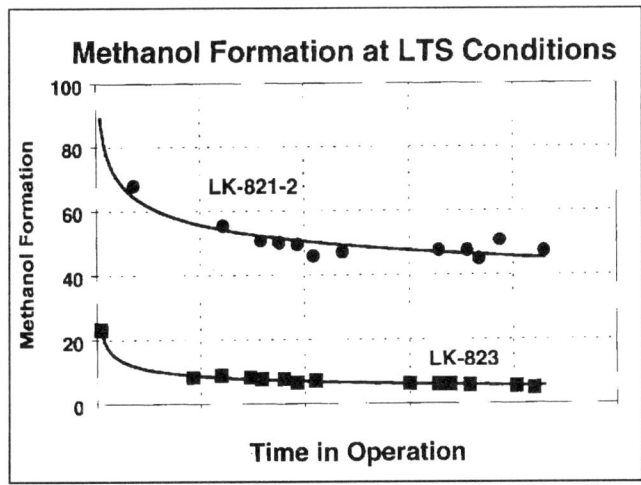

Figure 6. Methanol formation at LTS conditions over LK-823 vs. LK-821-2.

alkali content (on a mole basis) for the different alkali metals tested is illustrated in Figure 3.

Increasing concentrations of K and Cs have an increasingly and somewhat similar inhibitive effect on methanol formation (on a mole basis) whereas the effect from Na is less pronounced.

As illustrated by the previous industrial example (Figure 2), one type of potassium-promoted LTS catalyst already demonstrated inferior performance. It was therefore investigated how alkali metals in various concentrations affect the shift activity. The results of this investigation are shown in Figure 4.

It is noteworthy that Cs in small concentrations actually leads to increased shift activity, whereas high concentrations of Cs result in considerable activity loss.

Increasing concentrations of Na and K lead to decreasing shift activities, but the negative effect of K is less pronounced than for Cs in large concentrations, while the largest effect is observed for Na.

Test results also demonstrated that high concentrations of alkali resulted in increased deactivation rates.

The poisoning resistance of the catalysts was investigated in a setup in which the catalysts were exposed to sulfur in the feed gas (Figure 5). The higher concentrations of alkali had a significant deleterious effect on the sulfur chemisorption capacity, whereas the sulfur capacity was retained when the catalyst was promoted with moderate quantities of alkali.

Development of a New Low-Methanol LTS Catalyst

It is apparent that the desirable reduction in methanol byproduct formation over LTS catalysts can be achieved by promotion of the catalyst with various alkali metals. Tests have further demonstrated that both the type and quantity of alkali used influence the shift activity and poisoning resistance.

Table 2. Methanol Formation in a 40 Million scfd H$_2$ Plant Operating with LK-823

	Test Conditions		
	1	2	3
S/C Ratio	3.7	3.7	4.3
S/DG Ratio Inlet LTS	0.53	0.53	0.66
LTS Inlet Temp. (°C/°F)	206 (402)	191 (376)	190 (374)
Methanol in LTS, kg/d (lb/d)	144 (317)	71 (156)	57 (125)
Methanol in CO$_2$ Vent, kg/d (lb/d)	35 (78)	17 (38)	10 (23)
Methanol in Deaerator Vent, kg/d (lb/d)	2 (5)	2 (5)	2 (4)

Based on the experiments carried out, it was found that promotion with Cs within a definitive range of concentration is the best solution for decreasing methanol formation while retaining high shift activity and poisoning resistance of the catalyst.

Cesium has been used as a catalyst promoter for many years in the sulfuric acid industry, where Cs-containing catalyst is handled, loaded, and disposed without problems.

Manufacture of larger quantities of Cs-containing LTS catalyst provided a chance to verify whether scale-up of production changed the characteristics. Testing of the catalyst produced on an industrial scale confirmed all the results obtained in the laboratory scale, and a new catalyst was born: LK-823.

The stabilized methanol formation of LK-823 is only about one-eighth that of LK-821-2.

The fact that improvements in methanol formation have been obtained while maintaining the high activity and poisoning resistance implies that the use of LK-823 will yield the same shift conversion and catalyst lifetime as those of LK-821-2.

Many existing ammonia units are operating above 150% of the original design capacity, placing heavy demands on the physical endurance of the LTS catalyst. Accordingly, LK-823 was thoroughly tested both in the unreduced and, more importantly, in the reduced state. The tests clearly showed that promoting with cesium did not alter the excellent physical properties of the original catalyst.

Table 3. Methanol Formation with LK-823 in a 70 Million scfd H$_2$ Plant

Plant Rate, million scfd	60
S/C Ratio	4.1
S/DG Ratio Inlet LTS	0.62
LTS Inlet Temperature, °C/°F	206 (392)
Methanol Formed in the LTS kg/d (lb/d)	134 (295)
Methanol in Process Condensate wt. ppm	20
CO Leakage, mol. % dry	0.16 (Eq.)

LK-823 was introduced to the market in the summer of 1997 and was met with immediate interest from industry. The industrial experience from plants using LK-823 is summarized in the following paragraph.

LK-823: Industrial Experience

The new low-methanol LTS catalyst LK-823 is presently installed or awaiting commissioning in 10 plants (3 ammonia plants and 7 hydrogen plants). In the following paragraphs, the industrial feedback received is presented.

Plant one

The first charge of LK-823 was installed in the U.S. in a 40 million standard ft/d (scfd) hydrogen plant facing new federal regulations and even stricter local regulations limiting methanol emissions. The alternative to using LK-823 was to engineer and install methanol removal equipment at an estimated cost of close to $2 million (United States dollars).

The LK-823 was successfully reduced and commissioned in September 1997. Data from the plant has confirmed that the methanol formation over LK-823 is considerably lower than that over LK-821-2, which had been the plant's preferred choice in LTS catalyst.

Immediately after startup, the methanol emissions were found to be only one fourth of the emissions observed from a 4-year-old charge of LK-821-2.

The plant subsequently conducted a series of tests in which operating conditions (LTS temperature and S/C ratio) were varied as shown in Table 2.

The tests confirmed that with LK-823, it is possible to comply with the strict local regulations even at the start of the run. It should also be noted that methanol

Table 4. Methanol Formation with LK-823 in a 1,360 Metric t/d NH₃ Plant

Plant rate, metric t/d	1,440
S/C Ratio	3.31
LTS Inlet Temperature, °C/°F	214 (417)
CO Leakage, mol. % dry	0.24 (eq.)
Methanol Formed in the LTS kg/d (lb/d)	680 (1,500)
Methanol in Process Condensate wt. ppm	400

formation varies with changes in the LTS inlet temperature and S/DG ratio when using the low-methanol LTS catalyst.

The very high conversion (CO leakage = 0.14–0.16 mol. % dry) and the correspondingly steep temperature profile confirmed the high activity of LK-823.

After close to one year of operation, the LK-823 catalyst continues to perform excellently with low methanol byproduct formation and high activity.

We have concluded that it is possible to comply with all local and federal regulations regarding methanol emissions when operating with LK-823, thus eliminating the need to invest in expensive new equipment for methanol removal.

Plant two

This facility is a 70 million scfd hydrogen plant in the U.S. that faces the same regulations for methanol emissions as plant one.

Data from 2 months of plant operation with LK-823 (July 1998) are presented in Table 3.

This plant is designed with a high separator temperature, and as a result, only a small fraction (5%) of the methanol ends up in the process condensate.

The measured methanol formation is very close to the expected value for LK-823 using our kinetic model. All regulations have been complied with following the installation of LK-823.

Plant three

This facility is a 1,360 ammonia plant of Kellogg design located in western Europe.

The catalyst volume in the plant is comfortably large at 84.5 m³ (2,984 ft³) and has previously given an 8-year run with Topsøe catalyst.

The LTS inlet temperature in this plant is limited by the heat exchanger upstream the LTS, resulting in a LTS inlet temperature in the range of 210–215°C (410–420°F). In conjunction with the large catalyst volume and a low S/C ratio, these conditions result in the potential for considerable methanol byproduct formation.

The Selexol system used for removal of CO_2 from the process gas is operated at a very low temperature, therefore, the major part (>90%) of the methanol formed in the LTS ends up in the process condensate. Although the methanol contained in the process condensate is recycled back to the reformer, the plant personnel were interested in minimizing methanol formation and selected LK-823.

Table 4 presents operating data from a few days after startup (July 1998).

The measured methanol formation for LK-823 at the start of the run is very close to the expected/predicted figure, and close to an order of magnitude lower than the calculated methanol formation over LK-821-2.

The LTS reactor in this plant is equipped with a large number of thermocouples. This allows for an accurate assessment of the shift activity. Figure 7 illustrates the temperature profile at the start of the run.

The temperature profile confirms the high activity of LK-823 as the shift reaction has already reached equilibrium halfway down the catalyst bed.

Conclusion

Based on intensive testing, Topsøe has developed the Cs-promoted LTS catalyst LK-823 that minimizes methanol byproduct while maintaining the superior shift activity and poisoning resistance. This has provided the fertilizer and refinery industries with an effective and industrially proven technology that enables achievement of even the most stringent reductions in methanol emissions.

Literature Cited

Carstensen, J. H., J. Bøglid-Hansen, and P. S. Pederson,"Methanol By-Product Formation over HTS and LTS Catalysts," *Ammonia Plant Safety & Related Facilities*, Vol. 31, AIChE, New York, p. 113 (1991).

DISCUSSION

Al-Hajari, *Saudi Arabian Fertilizer Co.*: You showed temperatures and CO concentration at the inlet of the LTS and the pressure. Can you elaborate more on what are the minimum and maximum allowed CO concentrations at the inlet, as well as the minimum and maximum allowable inlet temperature?

Carstensen: There's no minimum or maximum. That depends on your requirements on the methanol formation. Actually, in Topsøe we have developed a kinetic expression calculating the expected methanol formation at various conditions, and given the set of operating conditions, we can calculate what is the expected methanol formation for traditional catalysts and this new catalyst.

Al-Hajari: I'm talking about the new catalyst. Is there any limit for a catalyst?

Carstensen: There's no limit as such. Of course, the methanol formation with the new catalyst would also be higher with higher pressure and higher CO concentration, but it would still be reduced by a factor of 10 compared with traditional catalyst.

Al-Hajari: What is expected methanol formation at the end of run of the new catalyst?

Carstensen: Again, you would have to look at the specific conditions in each plant. What is the steam-to-carbon ratio, the CO concentration, the pressure, and how much of the catalyst is deactivated at the end of the run? Again, we are able to make these predictions, and if you are interested, we can make these calculations for you.

Jim Huber, *Synetix*: You had stated early on in your presentation that an eight-fold reduction in methanol makes from 823 to 821-2. Was that at identical operating conditions and temperatures?

Carstensen: Yes, identical operating conditions and temperatures.

Huber: How do you explain the reduction of an eight-fold drop going to a two-fold and a four-fold drop in commercial production?

Carstensen: What do you mean two-fold and four-fold?

Huber: Well, your case studies indicated a 50% reduction over what you would expect with 821-2 in case three, and a four-fold reduction in case number 1. How come there is such a short drop from what the research data would indicate?

Carstensen: In case one I was comparing the start of run data with a new catalyst with the end of run data with the old catalyst so they are not directly comparable. The methanol formation is highest to start a run.

Huber: I understand that. So, it was the start of run conditions vs. end of run?

Carstensen: Exactly.

Huber: All other operating conditions temperature, etc. were the same?

Carstensen: Yes.

Maximizing Ammonia Production

A brief history of an ammonia plant's operation is given. The approach taken in determining and implementing the modifications for maximizing its capacity is discussed. The benefits of each of these modifications are described. The operating conditions before and after the last major revamp are also compared.

Keith Wilson
PCS Nitrogen, Augusta, GA
Mukund L. Bhakta and Michael Crowley
Brown & Root Engineering & Construction, Alhambra, CA

Introduction

A high ammonia product price is usually the main driving force behind bringing new capacity on-line. For an ammonia producer, if additional capacity becomes available, the higher price translates into higher plant profits. In such situations, the choice for the producer is between building a new plant or debottlenecking his existing one. The latter is often more economically attractive as it can usually be accomplished with less capital outlay and within a shorter time duration.

For these reasons, over its 20-year life, a number of modifications were made to PCS Nitrogen's ammonia plant at Augusta, Georgia.

This article gives a brief history of this plant's operation and discusses the approach taken in determining, and implementing, the modifications for maximizing its capacity. It also describes the benefits of each of these modifications. Finally, it compares operating conditions before and after the last major revamp project.

Plant History

PCS Nitrogen's natural gas based ammonia plant at Augusta, Georgia was engineered and constructed by C. F. Braun and Company, now a part of Brown & Root Engineering & Construction (BREC). Its design incorporates the Braun Purifier Process and the Benfield system for CO_2 removal.

Some of the key features of the Braun Purifier Proces are as follows:
• Mild reforming conditions.
• Use of 50% excess air in the secondary reformer.
• Gas turbine driven process air compressor with the gas turbine exhaust used as preheated primary reformer combustion air.
• Synthesis gas dryers and cryogenic purifier system upstream of the ammonia synthesis loop. The Purifier removes almost all impurities from the synthesis gas, and it also processes the loop purge for hydrogen recovery.
• Adiabatic axial flow ammonia synthesis converters with heat recovered externally as high pressure steam.

A simplified block flow diagram is shown in Figure 1.

PCS Nitrogen's Augusta plant was commissioned in 1978 by its original owner, Columbia Nitrogen Corporation. The plant was acquired by Arcadian Corporation in 1989 and then by PCS Nitrogen in 1997.

The original nameplate capacity of this plant was 1,500 STD (1,360 MTD). Between 1978 and 1988, some minor modifications were made to increase its capacity. These were as follows:
- Provided capability for injecting medium pressure steam into the gas turbine for obtaining additional power.
- Installed higher flow inlet guide vanes in the gas turbine to increase the inlet air flow and power.
- Removed one wheel from the process air compressor to match the gas turbine speed and increase the process air flow.

During this period, the plant's average production was 1,550 STD (1,410 MTD) during the summer months and 1,675 STD (1,520 MTD) during the winter months. This translated to an average of about 8% over the nameplate capacity. The main bottleneck limiting additional production was the process air compressor train.

In 1988, PCS commissioned BREC to investigate the feasibility of raising the plant's capacity by about 15%. The recommendations of this study are given in Table 1.

In 1990, PCS replaced the original syngas compressor turbine with a Siemens turbine to improve efficiency. The rated turbine capacity was also increased by about 6% for future expansion.

In January 1993, all the original HK-40 modified (25 Cr, 20 Ni) tubes in the radiant box of the primary reformer furnace were replaced with the thinner walled HP-40 (37 Ni + 25 Cr + niobium) tubes. This resulted in a reduced pressure drop across the primary reformer. The average plant production increased to 1,650 STD (1,500 MTD) during the summer months

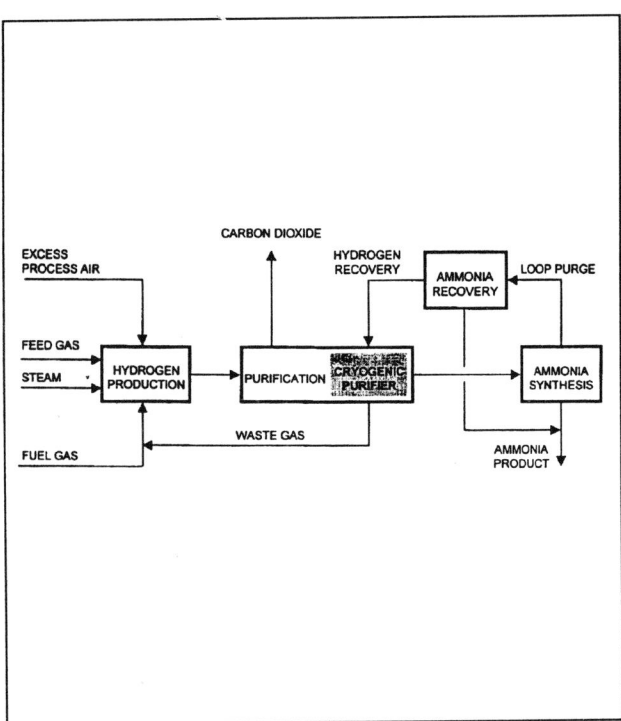

Figure 1. Overall block flow diagram of Braun Purifier Process.

Table 1. Brec Recommendations from 1988 Studies

(1) Modify process air compressor
(2) Modify Gas Turbine
(3) Increase mixed feed preheat surface in primary reformer
(4) Replace primary reformer tubes with HP-40 Niobium tubes
(5) Modify fuel oil nozzles for natural gas firing
(6) Modify CO_2 Regenerator wash section
(7) Replace impellers in lean circulation pumps
(8) Modify ammonia synthesis converters for radial flow
(9) Replace all four cores of the top coldbox exchanger

and 1,705 STD (1,550 MTD) during the winter months because of this change.

In early 1994, PCS decided to debottleneck the ammonia synthesis loop by converting the Braun axial flow converters to the radial flow design (Wilson et al. 1988). Ammonia Casale was responsible for the design of the converter internals, while BREC was responsible for the design of the waste heat boiler which was to be installed upstream of the second ammonia synthesis converter. This modification would reduce energy consumption and allow additional production in the ammonia synthesis loop. Design work for these modifications proceeded in 1994 and the implementation was scheduled for the plant's September 1995 turnaround. PCS decided to proceed with these synloop modifications even though there were no plans in place for revamping the front end of the plant.

During the 1995 period, the ammonia product price remained high and this gave the incentive to the Augusta plant management to embark on a more ambitious revamp project for debottlenecking the complete ammonia plant.

This revamp project started with PCS commissioning BREC to undertake a feasibility study in June, 1995. In August 1995, BREC issued the debottlenecking study report which concluded that 2,000 STD (1,810 MTD) production could be achieved economically. At this point, the PCS plant management decided to postpone by a year the September 1995 turnaround and the corresponding implementation of the synloop modifications. This way, all plant modifications could be made at one time. Consequently, the implementation of the total plant revamp was done during the September 1996 turnaround, about 17 months after the feasibility study release date. Previous preparatory work done on the ammonia synthesis loop by both PCS and BREC obviously helped in completing the latest revamp project within this short time period.

After the September 1996 turnaround, the plant's average production increased to about 1,900 STD (1,720 MTD) during the summer months, and about 2,050 STD (1,860 MTD) during the winter months. This averaged about an 18% increase in capacity after the revamp, or about 300 STD (270 MTD) beyond the capacity attained after changing the reformer tubes.

Revamp Project Approach and Execution

The primary goal of the 1996 revamp project was to maximize the plan's ammonia production. With the 1995 summer time capacity of 1650 STD (1500 MTD), the Augusta plant management set a target of 2000 STD (1810 MTD) for summer conditions, which was an increase of 21%. PCS also requested BREC to perform a "break-point" analysis to determine if a capacity slightly lower or higher than this target capacity was more cost effective.

The revamp project was executed in three phases.

In the first phase, the plant's operation was benchmarked and this was then used by BREC to develop a revamp scheme to fit PCS's objectives. This phase consisted of the following steps.

Data collection

BREC traveled to the plant site to gather information on the plant operation. During this visit, PCS and BREC discussed the project objectives and what was important for its success, plant operating and maintenance history, previous studies to increase production, plant modifications already planned, and PCS operators' opinion of the current bottlenecks.

PCS provided the plant operating data to be used as the benchmark for the feasibility study.

Benchmark case simulation

BREC simulated the plant operating data and evaluated the benchmark case heat, material, and steam balances. These results were then used to evaluate equipment performance under current operating conditions.

Revamp option screening

BREC first identified plant bottlenecks limiting 2,000 STD (1,810 MTD) production during summer conditions and then screened retrofit options to achieve this capacity.

From this exercise, BREC found that if the syngas compressor internals were not replaced, then the production achievable in summer would only be 1,900 STD (1,720 MTD). The cost of the compressor internals represented a step "jump" in the cost of the revamp. However, this extra cost paid out within one year at the prevailing ammonia product price.

The 1,900 STD (1,720 MTD) production rate represented the capital cost "breakpoint" for the Augusta plant.

BREC developed a revamp scheme to achieve the 2,000 STD (1,810 MTD) capacity during summer conditions. This scheme included the replacement of the syngas compressor internals.

Revamp option selection

BREC met with PCS to discuss and finalize the revamp scheme for achieving 2,000 STD (1,810 MTD) capacity during summer.

From the standpoint of availability of funds for the revamp project, PCS decided against replacing the syngas compressor internals. However, PCS requested that all new and modified equipment be sized for the 2,000 STD (1,810 MTD) capacity. PCS estimated that the yearly average capacity of the revamped plant would be about 1,950 STD (1,770 MTD) without changing the syngas compressor internals.

Revamp option simulation

Based upon the revamp scheme design basis discussed above, BREC simulated the performance of the plant. These results were then used to prepare data sheets for sizing new or modified equipment.

Capital cost estimate

BREC developed a +/- 25% capital cost estimate for the revamp project based on data prepared in the previous step.

Study report

BREC prepared the revamp study feasibility report for review by PCS.

Phase One provided the first, rough cut economics of revamping the plant for achieving the 1,900 STD (1,720 MTD) during the summer conditions, or a yearly average of about 1,950 STD (1,770 MTD).

Since this looked attractive, PCS released BREC to execute the Phase Two work in which the revamp scheme was "fine tuned" and a more accurate +/- 15% cost estimate was developed. Phase Two firmly established the economics of the revamp project.

PCS then decided to implement the project. This was done in Phase Three, which consisted of detailed engineering, procurement, and construction. BREC was responsible for detailed engineering, while PCS took responsibility for the procurement and construction activities.

Plant modifications

Table 2 lists all the recent modifications implemented at the Augusta plant to raise its production capacity. As shown in this table, most of these modifications were accomplished during the 1996 maintenance turnaround.

Some of the major points to note are as follows.

Suction chilling of air

Process air to the suction of the existing process air compressor and combustion air to the gas turbine are chilled to 69°F (20.5°C) in the revamp scheme. Chilling is provided by a lithium bromide absorption refrigeration system operating on 40 psig (2.8 kg/cm^{2}g) steam. Chilling of air to the existing process air compressor and the gas turbine contributes to about 5% more process air flow through the system.

Auxiliary process air compressor

A 6,200 bhp (4,625 Kw) motor driven process air compressor unit has been installed to provide the air required for additional capacity. This is a high speed gear-driven machine. Air from the auxiliary process air compressor train contributes to another 15% increase in the process air flow during the summer months.

No changes were made to the existing process air compressor.

Table 2. List of Plant Modifications

Item / Service	New/ Revamp	Year Implemented	Description / Comments
(1) Lithium bromide absorption refrigeration system.	New	1996	Chill air at suction of process air compressor and gas turbine.
(2) Primary reformer furnace.	Revamp	1996	(a) Increase mixed feed coil surface (b) Replace burners
(3) Auxiliary process air compressor.	New	1996	6,200 hp unit to supply additional air
(4) High temp shift effl / BFW exchanger.	New	1996	Larger unit to improve heat recovery
(5) Additional low temp shift reactor in parallel with existing one.	New	1995	This item was not essential for the expansion, but it was installed to reduce pressure drop and allow full use of catalyst.
(6) CO_2 Absorber.	Revamp	1996	Replace top 2 beds with 2" pall rings.
(7) CO_2 Regenerator.	Revamp	1996	(a) Replace top 3 beds with No. 50 IMPT packing (b) Replace bottom bed with No. 60 IMPT packing
(8) Water cooled lean solution cooler in parallel with existing unit.	New	1995	Installed for additional cooling capacity.
(9) CO_2 Removal System Solution.	New	1996	Old batch was replaced with a new charge so that UOP's ACT-1 additive can be effectively used.
(10) Process condensate pump.	New	1996	Replace existing ones for additional capacity.
(11) Additional methanator effluent cooler.	New	1996	Installed in parallel with existing unit for additionsl cooling capacity.
(12) Methanator effluent separator.	Revamp	1996	Install new vane element for higher capacity.
(13) Purifier expander.	Revamp	1996	New rotors for higher efficiency
(14) Purifier overhead condenser.	Revamp	1996	Re-pipe for better heat transfer
(15) Ammonia synthesis converters.	Revamp	1996	(a) Modify internals for radial flow design. (b) Replace catalyst with 1.5 to 3 mm

Primary reformer furnace

The main changes made to the furnace were the increase in the mixed feed preheat coil surface and the replacement of the burners in the radiant section. By providing additional surface for preheating the mixed feed to 1000°F (538°C), the increase in the radiant section load is minimized. Consequently, under the higher production case, there will be minimum change in the operating environment of the catalyst tubes.

The existing burners were replaced with the dual fuel gas, low NOx type burners. The use of separate tips for the waste gas (low pressure fuel) and natural gas (high pressure fuel) provided two advantages. First, a more stable and compact flame pattern would be obtained in the furnace and this would improve the combustion efficiency. Second, a lower back pressure on the waste gas stream would improve the Purifier coldbox performance.

No modifications were required in the existing induced draft fans.

Parallel LTS reactor

The main reason for installing this unit was to extend the time period between plant turnarounds. One side benefit was to reduce the pressure drop in the front end of the plant.

CO_2 removal system

The main changes were as follows:
- Completely replaced the existing potassium carbonate solution with a new charge.
- Used UOP's ACT-1 additive with the new charge.
- Changed the top two beds of the Absorber from 1.5 in (40 mm) to 2 in (50 mm) pall rings.
- Changed the top three beds of the Regenerator from 2 in (50 mm) pall rings to the more efficient No. 50 IMTP packing.
- Changed the bottom bed of the Regenerator from 2 in pall rings to No. 60 IMTP packing.
- Added a water-cooled lean solution cooler in parallel with the existing air-cooled unit.

The use of UOP's ACT-1 additive ensured that the increase in the solution circulation rate was lower than the capacity increase. Consequently, no changes were required in the lean solution pumps.

It should be pointed out that in early 1994, PCS had tried UOP's ACT-1 additive in the existing charge of the potassium carbonate solution. However, the results were not promising and its use was discontinued. This is the main reason why a completely new charge of potassium carbonate solution was used in the latest plant revamp.

Methanator effluent separator

This vessel was replaced with one having a higher capacity to prevent liquid carryover to the dryers.

Purifier coldbox system

The expander internals were replaced to achieve a higher efficiency operation at the 2000 STD (1810 MTD) rate. The Purifier column overhead condenser was re-piped to improve the heat transfer rate in this exchanger.

No changes were required in the coldbox exchangers.

Ammonia synthesis gas compressor

PCS decided against replacing the compressor internals to save revamp project costs. As explained earlier, PCS estimated that the revamp project was economic even at a yearly average production of 1950 STD (1770 MTD), which was expected to be achieved without changing these internals.

Ammonia synthesis converters

The internals of the two axial-flow adiabatic converters were modified for radial flow. This would result in a higher conversion per pass since a more active smaller size (1.5 to 3 mm) catalyst could now be used in these converters. The total responsibility for this conversion was given to Ammonia Casale SA. The details of the synloop debottleneck are described in a separate article (Wilson et al., 1999).

Additional synloop steam generator

The installation of this heat exchanger upstream of the second ammonia synthesis converter was essential in order to obtain the maximum advantage out of converting the ammonia synthesis converters to radial flow design. This was a forced circulation boiler, similar in design to the existing synloop steam generator.

Primary separator

The vane element in this vessel was replaced with one having a higher capacity to minimize liquid ammonia entrainment with the recycle vapor.

Additional ammonia condenser shell

This was required to obtain the additional capacity without making modifications to the ammonia refrigeration compressor. After the revamp, the plant operates on three heat exchanger shells in parallel. The new valving is such that one shell can be cleaned while the other two are in service. This arrangement is useful in tackling the cooling water side fouling problem and this helps in maintaining a high average ammonia production rate.

Plant Pressure Drop Considerations

Pressure drop through the plant is an important factor whenever a plant is revamped for capacity increase. The impact of this factor was critical in the following three sections of the plant - front end, the ammonia synthesis loop, and the steam system.

The front end of the plant covers the section from the discharge of the feedgas compressor to the suction of the syngas compressor. In this section, the primary reformer tube outlet pressure is the anchor point pressure. This was set as high as tube metallurgy allowed to minimize syngas compressor power requirements. The pressure profile for the target capacity flows was then checked to see if it fit within the mechanical limits of the existing equipment.

The ammonia synthesis loop covers the section from the syngas compressor discharge to the suction of the recycle wheel. In this section the modification of the ammonia synthesis converters to radial flow design made it possible to reduce both the synthesis loop pressure and the pressure drop.

In the plant steam system, no modifications were made to the piping. Thus, pressure drop in the high pressure steam line increased with the steam production. Also, the medium pressure steam header operating pressure had to be to be raised to allow for the higher front end pressure.

Plant Performance After Revamp

After the revamp, the plant has achieved its expected production targets. Average production is above 2,050 STD (1,860 MTD) in the winter months and above 1,900 STD (1,720 MTD) during the summer months. Currently, the major factors limiting production are as follows:
- Availability of process air.
- Availability of combustion air in the reforming furnace.
- Syngas compressor/turbine performance.

Table 3 compares operating parameters before and after the revamp.

Major factors which contributed to maximizing the plant's production are as follows.

Higher front-end pressure

Compared to the benchmark case operation, the feed gas discharge pressure was 29 psi (2.03 kg/cm^2) higher in the revamped operation. This resulted in the syngas compressor suction pressure being about 10 psi (0.7 kg/cm^2) higher at the higher production rate. PCS and BREC found that each 1 psi (0.07 kg/cm^2) increase in the syngas compressor suction pressure amounted to 3 STD (2.7 MTD) of ammonia production. Thus, in the detailed engineering phase, all efforts were made to raise the front end pressure and also minimize the front end pressure drop.

High mixed feed temperature

A higher mixed feed temperature enables higher production for the same radiant box firing rate. In the revamped operation, the temperature of the mixed feed

Table 3. Key Process Parameters

Parameter	Units	Base Case Operation	Revamp Operation
(1) Ammonia production	STD (MTD)	1,837 (1,667)	2,085 (1,891)
(2) Ambient air temperature	°F (°C)	45 (7.2)	58 (14.4)
(3) Feed gas compressor discharge	psig (kg/cm^{2}g)	522 (36.7)	551 (38.7)
(4) Steam / carbon ratio		3.2	3
(5) Radiant box inlet temp	°F (°C)	914 (490)	982 (528)
(6) Radiant box outlet temp	°F (°C)	1,229 (665)	1,207 (653)
(7) Radiant box outlet pressure	psig (kg/cm^{2}g)	445 (31.3)	456 (32.1)
(8) CH_4 (dry basis) in prim ref effluent	mol. %	29.8	33.02
(9) CH_4 (dry basis) @ sec ref effluent	mol. %	1.6	1.8
(10) Process air temp	°F (°C)	832 (444)	791 (422)
(11) Sec ref outlet temp	°F (°C)	1,550 (843)	1,560 (849)
(12) CO at LTS outlet (dry basis)	mol. %	0.27	0.34
(13) CO_2 in absorber feed (dry basis)	mol. %	16.9	16.9
(14) CO_2 at absorber ovhd	ppmv	1,100	1,200
(15) Syngas compressor suction press	psig (kg/cm^{2}g)	312 (22.0)	322 (22.7)
(16) Total front end press drop	psi (kg/cm^2)	210 (14.8)	229 (16.1)
(17) Loop pressure	psig (kg/cm^{2}g)	2,843 (2000)	2,443 (172)
(18) Final NH_3 content in effluent	mol. %	15.73	19.1

entering the radiant section was 982°F (528°C) compared to 914°F (490°C) for the benchmark operation. The mixed feed coil was modified to achieve 1,000°F (538°C) for the 2,000 STD (1,810 MTD) capacity case. Therefore, its performance for the 2,085 STD (1,890 MTD) production looked reasonable.

Higher excess air rate

In the revamped operation, the syngas exiting the methanator effluent was richer in nitrogen compared to the benchmark operation. This denoted that more excess air was being used in the revamped operation. As a result, more reforming took place in the secondary reformer and this was demonstrated by the increase in the methane conversion across this unit.

Suction chilling of air to both the existing process air compressor and the gas turbine contributed to the higher excess air rate.

In the revamped operation, the primary reformer furnace could not be fired harder due to the lack of combustion air. However, because of the availability of a higher amount of excess air, the secondary reformer took care of the shortfall in primary reforming.

Use of ACT-1 additive in the CO_2 removal system

The use of this additive in a completely new charge of potassium carbonate solution maintained the solution circulation rates within the hydraulic limits of both the CO_2 Absorber and the Regenerator.

The increase in the amount of CO_2 removed in the revamped operation was about 16% higher than the benchmark case, but the increase in the solution circulation rate was only 6%. The CO_2 leakage in the CO_2 Absorber overhead was about the same as before the revamp.

More efficient internals for the purifier expander

The replacement of the Purifier expander internals resulted in less pressure drop across this unit at the higher production rate. As mentioned earlier, minimizing the total front end pressure drop was critical in maximizing production.

Conversion of the ammonia synthesis converters to radial flow design

This resulted in increasing the ammonia content in the second converter effluent from 15.7% to 19.1% at 400 psi (28.1 kg/cm^2) lower loop pressure. The loop circulation rate was about 8% lower for the revamped operation. These process conditions showed that the ammonia synthesis loop had room for additional capacity.

For the revamped operation, the main factors, which limited production, were as follows.

Lack of combustion air in the primary reformer radiant section

The limit on the amount of fuel that could be fired in the radiant box had been reached under the revamped operation. In fact, plant operators claimed that they did not measure any oxygen in the radiant box in spite of the fact that all the gas turbine exhaust was directed to this section. Under the revamp operation, the gas turbine was running at its maximum speed. Also, during warm weather, the gas turbine operates at the maximum exhaust temperature.

The maximum temperature that could be reached at the primary reformer outlet was 1,207°F (653°C), which is 22°F (12°C) below the benchmark operation case. The corresponding methane content in the primary reformer effluent was 33%. This is about 3% higher than the benchmark case. The additional excess air available from the process air compressor train made it possible for the secondary reformer to take care of the shortfall in primary reforming.

Syngas compressor capacity

Under the revamped conditions, the first stage of the machine was operating outside its performance map and the front end pressure drop became a very critical factor in maintaining production. From this, it became very clear that a further increase in capacity will require modifications to this machine.

Future Plans

Having achieved the target capacity, PCS and BREC started to look for ways to further debottleneck the plant. A joint investigation resulted in the following conclusions:

For the plant to achieve 2,200 STD (2,000 MTD) during the summer months, the modifications required will be as follows:
- Upgrade gas turbine for higher capacity.
- Modify primary reformer to achieve 1,195°F (646°C) mixed feed.
- Temperature at the inlet to the radiant section.
- Install high efficiency packing in the CO_2 Absorber and Regenerator.
- Install new high efficiency rotors for the syngas compressor.
- Install a new secondary reformer effluent / high pressure steam superheater.
- Install a new ejector system for the CO_2 solution flash drum.

To achieve 2,300 STD (2,090 MTD) production in the summer months, the additional modifications required will be as follows:
- Installation of a prereformer.
- New internals for the syngas compressor turbine.
- New internals for the ammonia refrigeration compressor.

To Sum Up

Whenever the ammonia product price rises, every

ammonia producer considers debottlenecking his plant for maximizing profits. However, knowledge of the plant's operating history, its current bottlenecks, the approach to revamping a plant, timing, and the total project schedule are all important for the overall success of such a project.

The approach taken by PCS Nitrogen to maximize its production at its Augusta plant ensured that the revamp project was mechanically complete within 17 months of the release of feasibility study work.

The plant has achieved the expected target capacity after the revamp. PCS has also become knowledgeable of plant modifications required to further increase its capacity.

Literature Cited

Wilson, K., E. Fillipi, and J. H. Gosnell, "Upgrading a Synloop for Capacity Increase and Energy Savings," *Ammonia Plant Safety and Related Facilities*, Vol. 39, AIChE, New York (1999).

DISCUSSION

Rudolph C. Frey, *M.W. Kellogg*: I'd like to ask about the pressure increase in the front end of your plant. You indicated a 29 lb additional increase. It's always possible to increase the pressure. The question is how much you got into the design margins of the plant, the safety and reliability of the plant and the relief valve system. What sort of studies were done on this?

Wilson: Our target was to achieve 2,000 short tons per day, and we limited our operating pressures about 95% of relief valve settings. So, we increase the front end pressure to a limit where it fell within the 95% of the relief valve settings.

Frey: You reduce from 10% to 5%. Did you do anything special with regard to reducing that margin, or is that just a paperwork exercise?

Wilson: No, we limited to 95%.

Bali, *Indian Farmers Fertiliser Cooperative Ltd.*: You said normally the gas turbine exhaust is around 500 °C. In normal plants this is used for a downstream heat recovery unit. Do you have a heat recovery unit, because you said that you use most of the heat to preheat combustion air? If it is used for heating combustion air for the reformer, what is final exhaust temperature?

Wilson: We send the gas turbine exhaust directly to the reformer, so there is no heat recovery after the gas turbine. The exhaust is sent directly to the primary reformer and it contains about 16% oxygen.

Bali: What's the final exhaust temperature at the reformer?

Wilson: It's around 400° right now.

Planning and Execution of Major Revamp in Running Plant

A major revamp of a 25-year-old vintage ammonia plant was carried out to achieve an increase in capacity and substantial energy savings by incorporating recent technological innovations. The critical aspects of the challenging tasks of planning and executing the major revamp is discussed.

P. S. Neelakantan and K. V. Swaminathan
Madras Fertilizers Ltd., Chennai, India

V. K. Anil
FACT Engineering and Design Organization, India

Leif Chawes
Haldor Topsøe A/S, Denmark

Introduction

Madras Fertilizers Ltd. was operating since 1971 with one stream of an ammonia plant with a 750 MTPD capacity. The technology (by Chemico) had become outdated.

The specific energy consumption was 50% higher compared to modern plants. The majority of the plant equipment had also reached the end of its economic life.

In order to substantially reduce energy consumption, increase plant capacity, and improve reliability, a thorough study of the plant facilities was conducted in 1990 by M/s Haldor Topsøe A/S, Denmark, for the ammonia and supportive utilities plant.

The study indicated that the economic life of the plant could be extended by another 15 years by replacing certain sections of the plant, which would result in a considerable decrease in energy consumption of the ammonia plant from the present 14.2 GCal/T and simultaneously in an increased capacity of the ammonia plant by 40% to 1,050 MTPD with no additional input of raw materials.

MFL carried out the revamp at a cost of approximately $160 million. The basic engineering of the entire ammonia plant and detailed engineering of the reformer and HP-loop was provided by Haldor Topsøe, and the remaining detailed engineering was carried out by M/s FACT Engineering and Design Organisation (FEDO), India.

The total project cost for the revamp of the ammonia plant, urea plant, NPK plants, and utility plants were $160 million, of which the revamp of the ammonia plant was $103.9 million. The cost breakdown is shown in Table 1.

The entire revamp project was completed late in 1997. The plant was started up and a successful test-run was conducted in January 1998, which demonstrated a capacity of 1,062 MTPD and a specific bat-

Table 1. Project Cost Breakdown

Equipment	Million $74.6
Erection cost	Million $8.4
Financing charges	Million $19.1
Taxes and duties	Million $1.8
Total	Million $103.9

tery limit energy consumption of 7.78 Gcal/MT. Those were better than predicted (7.8 Gcal/MT).

Project Scope

The most critical aspect of this project was the need to carry out the construction alongside a running ammonia plant. The tying of the old and new sections had to be meticulously executed in the shortest possible time in order to minimize production loss. The scope included replacement, partial replacement, and retaining of old equipment in various sections of the ammonia plant as follows.

New section

• *Reforming*: New reformer system including prereformer, primary reformer, secondary reformer and RG boiler.
• *Shift section*: relocation and re-use of old HT-Shift vessel, LT-shift vessel and waste heat boilers along with new HHP BFW exchangers.
• CO_2 *removal*: new system based on MDEA process.
•*Methanation*: new methanator and exchangers.
• Compressors for process air and synthesis gas.
• 110 ata boiler.
• Process condensate boiler.

Partial replacement/additions at old locations

• *Hydro desulfurization (HDS)*: new fired primary and secondary HDS preheaters and replacement of some exchangers.
• *Synthesis*: new S50 converter downstream existing S200 converter with new synthesis loop boiler, synthesis loop exchangers, and additional electrical startup heaters.

Retained sections

• *Selectoxo*: retaining of old section.
• *Refrigeration*: retaining of old section with minor additions.
• **Continuous Polishing Unit**

Utilities

• New cooling tower cells and additional cooling water pump.
• New PSA nitrogen system.

Control system

Since the state-of-art Distributed Control Instrumentation System and a PLC-based emergency shutdown system was proposed, the necessary instrumentation changes from pneumatic to one compatible to the Distributed Control System (DCS) were to be installed for the new sections and in respect to the retained section equipment.

Implementation Methodology

Planning objectives

The new plant layout was conceived with a view to both the physical location of various sections of the old plant in operation, which was to be retained, and the partial or full replacement of other sections of the plant. The finalized layout is shown in Figure 1. Figure 2 shows the overall plot plan for new sections and, Figure 3a shows the unit plot plan of the existing area, and Figure 3b shows the plan of the new area.

While the HDS, synthesis, Selectoxo, and refrigera-

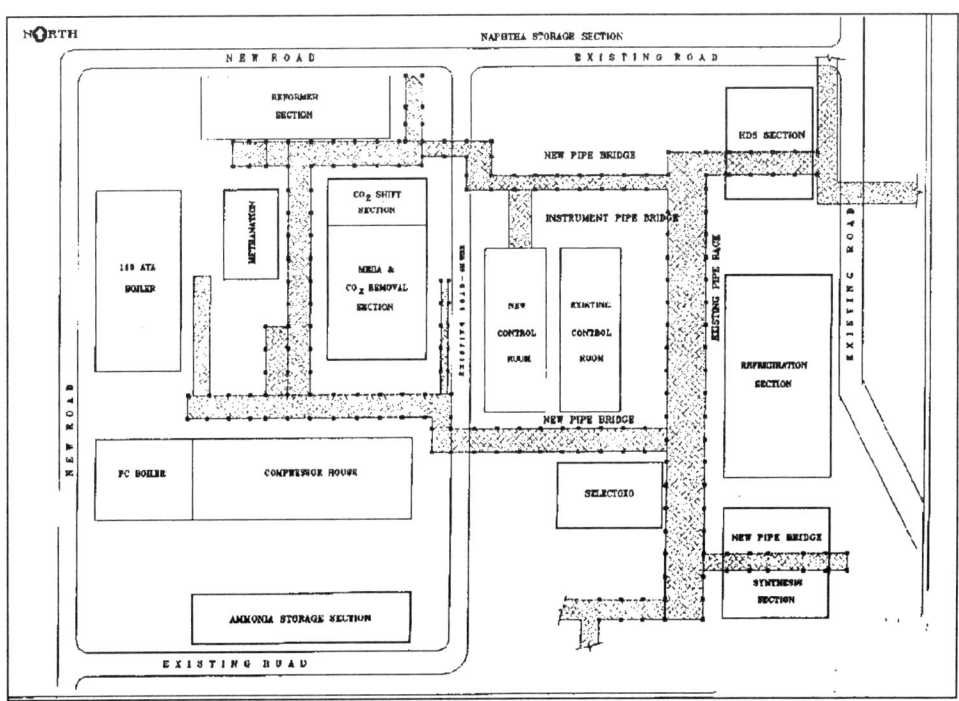

Figure 1. Overall plan layout of ammonia plant.

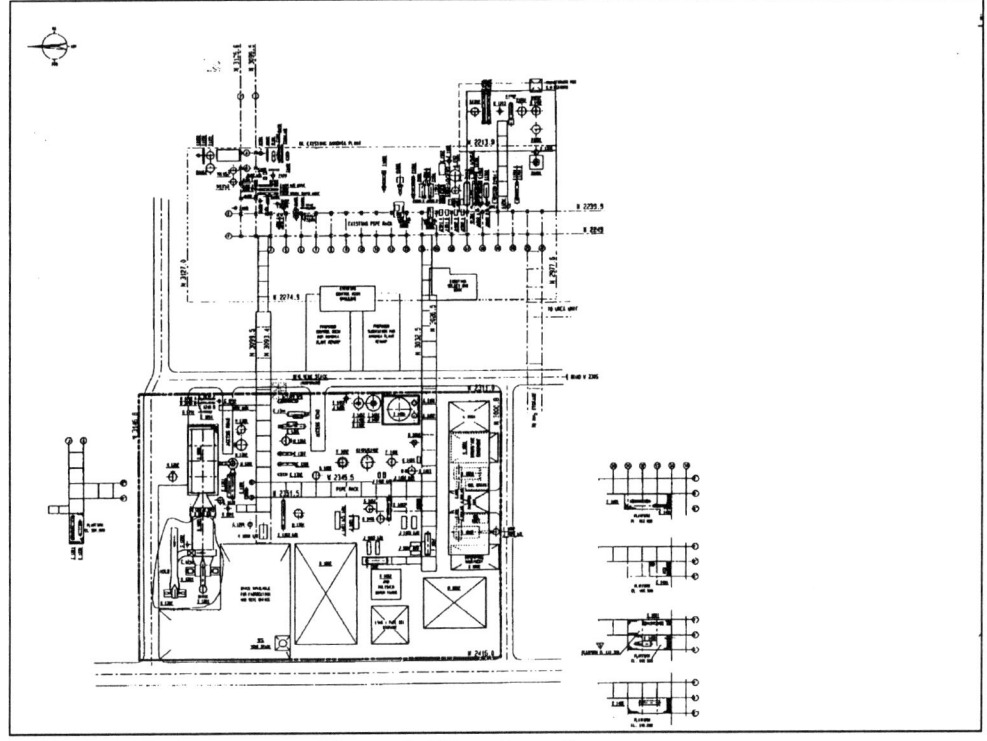

Figure 2. Plot plan of new area.

tion sections were to be retained in the old plant area, the other sections were to be built at the new area which was earmarked. Some of the old plant equipment was to be relocated to the new area, viz., CO converter, LT guard vessel, primary MP boiler, and CG WH boiler.

In order to quantify the detailed scope of work, a study was carried out, which established the following:

(1) Conformity of the existing retained piping and instrumentation with the new P&IDs developed for the revamp.

(2) Tie-in points both in respect to piping and instrumentation to connect the old retained section with the new section by providing physical tags.

(3) Finalizing locations of additional pipe bridges in the existing plant to cater for the requirements of the new plant.

(4) Feasibility of carrying out work in the existing plant areas for piling, civil foundation, mechanical equipment erection, and piping even when the plant was running.

(5) Enabling work to be carried out for the location of new equipment/pipe bridge and relocation of existing equipment.

Based on the detailed scope of work, the following main objectives were derived to plan the total execution activity:

• To complete the maximum work possible with the existing plant running.

• To ensure running of the existing plant as long as possible, keeping in line the overall project schedule to minimize production loss revenue.

• To minimize the period of final hookup shutdown of the existing plant.

• To minimize startup time.

Execution Methodology

Based on the above objectives, the jobs were sorted into three categories.

• Jobs that had to be done in the old plant in order to continue the precommissioning activities in the new plant, keeping the existing plant running. This basically included most of the utilities like steam, cooling water, and so on. A short shutdown of the plant in connection with an annual maintenance turnaround would be required to be able to complete these jobs.

• Activities that could be done while the existing plant was in operation.

• Activities that could be taken up only during the final hookup, while the existing plant was shut down. This would include most of the tie-ins on process lines and instrumentation conversion from pneumatic to electronic. Erection of relocated equipment would also fall in this category.

Determining Time and Duration for Hookup Shutdown

Having finalized the quantum of work in various categories, the timing and duration of the final hookup shutdown was firmed up. The timing for the final shutdown was to be after completion of all activities in the new sections of the plant including precommissioning activities like catalyst loading, steam blowing, water flushing, air blowing, commissioning of boiler, and solo run of turbines. These activities comprise:

• Mechanical completion of reformer and CO_2 removal sections.

• Commissioning of 110 Ata boiler.

• Charging of utility headers of steam, water, instrument air, and so on.

• Solo run of turbines especially process air compressor, MDEA pumps, synthesis gas compressor, and other small turbines.

• Catalyst loading in prereformer, primary, and secondary reformer, CO-conversion (only HT-shift), new ammonia converter and methanator.

• Refractory dry out for the reformer section.

Duration of the shutdown, based on the quantum of jobs to be done inevitably during the hookup shutdown, was planned to be 60 days. The jobs to be carried out during the shutdown were:

• Mechanical completion of CO conversion and shift sections. Even though this was located in the new area of the plant, most of the relocated equipment to be located there was equipment that required shutdown to complete this section.

• Mechanical completion of HDS, synthesis, and refrigeration section.

• Hookup of the cooling water systems between the

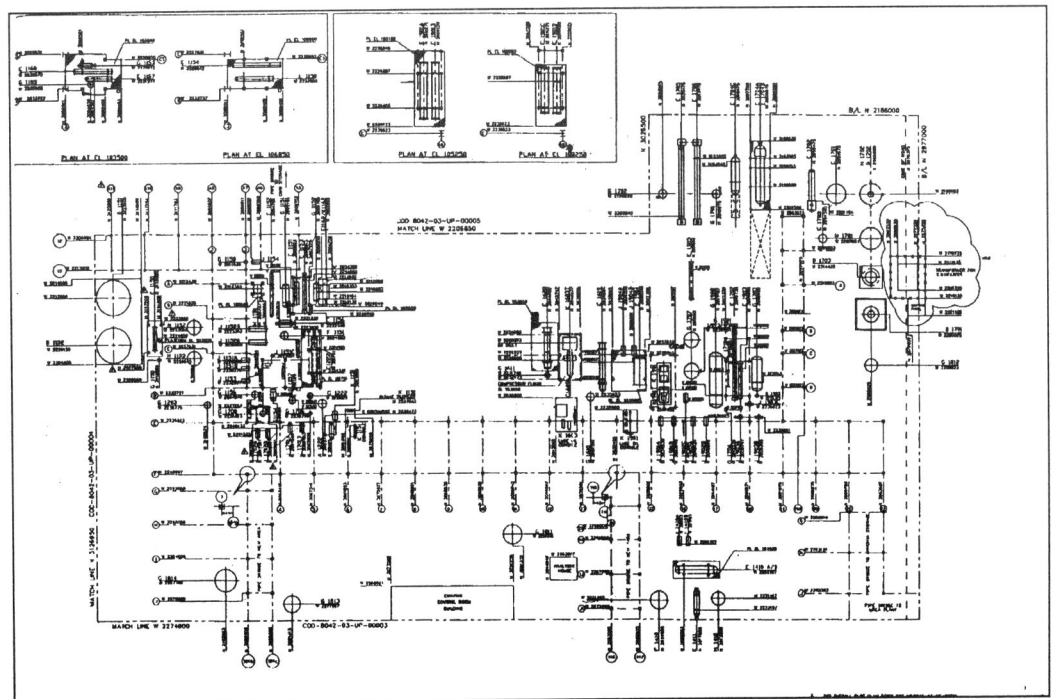

Figure 3a. Plot Plan of existing area.

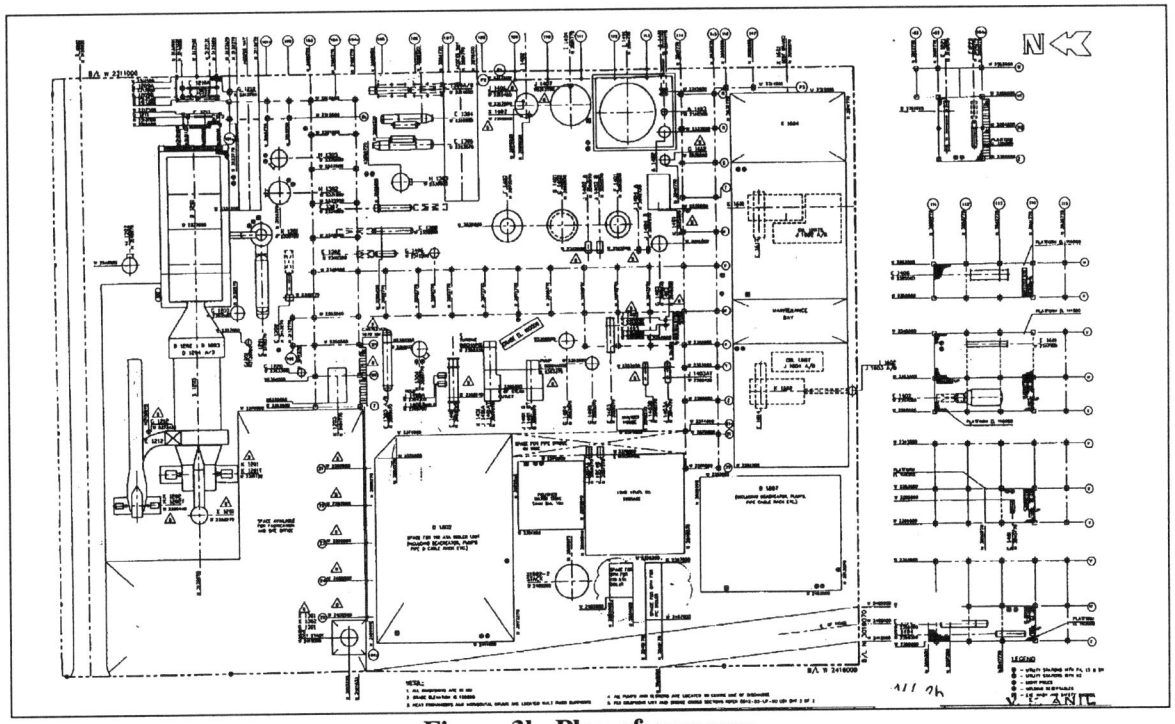

Figure 3b. Plan of new area.

old and new plant sections.
- Catalyst loading in the relocated reactors.
- Conversion of 120 closed instrument loops and 500 open loops in the existing plant from pneumatic to DCS compatible.
- Refractory dry out of HDS preheaters.
- Precommissioning of HDS, synthesis, and refrigeration sections.

The precommissioning activities were dovetailed into mechanical completion activities in order to minimize startup time. A comprehensive modular precommissioning activity chart was prepared and the time duration was assessed to be three months.

Mini-Shutdown of Old Plant for Completion of Mandatory Hookup Points

Due to restrictions in steam availability from the existing plant, steam from the new 110 Ata boiler was anticipated to be used for various precommissioning activities. Hence, commissioning of the 110 Ata boiler prior to final hookup shutdown was considered mandatory. To facilitate commissioning of this boiler and various other precommissioning jobs, piping for various utilities were required (fuel oil, DM water, instrument air, cooling water, steam, and so on), erected and interconnected with the existing facilities.

More than 20 points were identified as mandatory tie-ins for commissioning of the 110 Ata boiler, PSA nitrogen unit, and cooling water systems, which were planned for commissioning prior to stoppage of the existing plant. Accordingly, these tie-in points with necessary isolation valves were taken up and completed during the annual turnaround of the existing plant and piping continued to enable the commissioning of the above systems.

Since the complete pipe racks could not be erected to pipe racks interconnecting junction constraints, as stated above, a temporary pipeline was planned and erected on the ground. It was to run on temporary supports without affecting the running plant.

Since the existing plant RFG vent stack was located in the new plant area, it was relocated during the planned shutdown so that the work in the new section could continue without any hindrance or safety hazards.

In order to connect to the new pipes connecting the retained sections and the new sections, new pipes had to be run across the existing plant area. Hence, a third deck of the pipe racks was constructed during the above planned shutdown.

Temporary interconnection between the existing and new cooling water systems was made to facilitate flushing and passivation of the new systems using the existing cooling water pumps, even when the plant was running. The interconnection did also have the facility of using the new pump, if available, at that point of time.

Jobs Executed While Plant Continues to Operate

Piling and civil jobs alongside operating ammonia plant

Based on the soil characteristic studies carried out, all civil foundations were to be on piles. It was decided to go in for bored cast-*in situ* piles in the old plant areas to avoid interruption in the running plant and to resort to driven cast-*in situ* piles in the new plant area. The piling in the old plant area was carried out in a phased manner in order to facilitate continuous running of the existing plant since it was not possible to issue work permits in all fronts at the same time. However, in the case of a new plant area due to close proximity of various equipment and machinery nearer to the new plant site, the vibration due to the piling was felt in the running machines of the old plant. Hence, expert studies were conducted and the line of transmission of vibration was intercepted by providing trenches to dampen the effects of the piling carried out. Continuously simultaneous measurements and monitoring of vibration levels were carried out.

Because of existing underground cables and other underground pipes in the old plant area, the location of the equipment in the synthesis section was to be changed in order to achieve timely completion of pile foundation even when the plant was running. Certain jobs like the new pipe rack junctions to connect the old plant with the new pipe racks were not possible, while the plant was running. Since the piling and providing of foundations at a later date during hookup shutdown would be time consuming, interconnection of footings and provisions of large composite structures to compensate for the nonprovision of piles in these areas was resorted to, and foundations were completed well in advance.

In order to locate equipment in the existing old plant synthesis area, the existing road was moved so that the traffic and firefighting vehicle access catering for the requirements of the running plant were not affected.

New HDS preheaters in old plant area

The HDS section envisaged replacement of the primary and secondary HDS preheaters. Since these were the only two replacements in this section, they were to be located in the existing plant area itself to minimize piping, and so on. Erection of these heaters and continuing of the piping connected to the heaters meant hot work, which was restricted for safety reasons. To overcome this, an 8 Mtr high fire protection barricade was built separating the old and new preheaters to ascertain safety, and then work permits were issued and work continued (as shown in the Photos 1–6). This enabled the erection of HDS preheaters well ahead of the general plant shutdown.

Requirement of HDS flare system

The existing HDS flare system suddenly fell off during a monsoon. Since it was necessary due to concerns about pollution to flare off gases from the HDS section midway in the project, a new HDS flare system was ordered. The necessary foundation, pipe supports, piping, and so on were planned and completed in such a way that it suited commissioning prior to the main plant commissioning.

Installing S-50 converter in an operating plant

The revamp of the synthesis section included an additional ammonia converter (Topsøe S-50). The location of the S-50 was selected near to the existing converter (S-200) due to process requirements. The high-pressure piping connected to the new converter was planned to be completed prior to shutdown. However, the proximity to the running plant handling synthesis gas made it impossible to continue work. Hence, a 26 Mtr high fire protection screen was built between the two converters (as shown in Photos 1–6), and the work was completed prior to shutdown after having ascertained safe working conditions.

The S-50 catalyst, Topsøe KM1R, loading could be possible only after completion of all hot works connected to the vessel. This was not possible prior to shutdown, because the piping to the S-50 converter was to be connected to the existing S-200 converter. Hence, this hookup was a shutdown activity. In order to complete the catalyst loading prior to shutdown, the piping from the converter was completed on first priority up to the first isolation valve, and in the other line without a valve, a bladder backed up with sand was used to isolate the converter. All radiography on the high-pressure piping was completed to avoid any repair at a later stage. The catalyst was loaded, and to avoid air ingress into the converter, a continuous flow of nitrogen was maintained. The system was also continuously monitored for temperature upsets.

Reuse of redundant NH_3 converter for hydrogen storage

The reformer system included a prereformer consisting of a highly active catalyst, which had to be protected from air by nitrogen and to be provided with hydrogen in case of an unforeseen shutdown, which would otherwise impair the catalyst activity. To facilitate this, it was planned to store hydrogen and use it as a reducing agent. Since the old converter vessel (the original Chemico converter) was situated in the synthesis area of the existing plant, it was proposed to use it for that purpose. This vessel had, however, not been in service for more than ten years, and, hence, the soundness of the vessel to hold hydrogen was established by bringing in experts from the original manufacturer of the vessel. Subsequent to the final hookup shutdown, the connecting piping to the vessel was completed and hydrogen was fed to enable commissioning of the prereformer.

Photo 1. Road diversion in synthesis section.

Photo 2. Foundation view of MDEA columns.

Photo 3. Absorber (top piece) being placed alongside road before lifting to position.

Photo 4. Absorber (middle piece) being aligned with bottom piece.

Photo 5. Fire protection screen for S-50 converter.

Photo 6. Fire protection barricade for HDS preheaters.

Crash installation of water cooled ammonia condensers to revitalize old plant

During the execution of the various piping and equipment erection jobs prior to shutdown, the plant load was heavily restricted and the back end of the plant was frequently shut down due to a leakage in the water cooled ammonia condensers of the synthesis loop section. The new additions of equipment included replacement of the ammonia condensers, and, by then, they had already been received at the site. Taking advantage of this, and in order to ensure continuous running of the plant at sustained higher loads, the new condensers were planned to be erected in their proposed location and hooked up to the existing system. However, since the pipe foreseen in the post revamp was not constructed and the required pipe size for hookup was lower than the post revamp size of the pipelines, it was decided to provide a permanent pipe rack up to the extent required, and to go in for post revamp line sizes for pipelines to be laid on the pipe racks and between the exchangers. Moreover, since a cooling water connection could not be made because a shutdown of the cooling water was not foreseen, a temporary cooling water connection had to be laid on the ground.

The job involved 450 ID (inch dia.) of high-pressure piping requiring pre-weld and post-weld heat treatment, radiography and erection of 160 MT of equipment and structures. The hookup was executed as planned in 82 h, braving unprecedented rains during the hookup. This enabled continuous running of the plant at sustained higher loads for more than three months until the final hookup shutdown.

Facing onslaught of operating plant environment on new piping

Another serious problem faced during the construction stage was the heavy corrosion of pipelines and structures caused by urea dust from the urea prilling tower. The structures coated with primer started corroding immediately. The painting specification for structures had to be changed from the common enamel to a combination of epoxy and polyurethane.

The process air compressor alignment was completed and all piping was connected. At this stage, just before the commissioning of the compressor, the vendor's representative saw the condition of the pipelines and insisted on a chemical cleaning of the pipes. This would mean that all piping connected to the compressors had to be dismantled and temporary piping installed between the interstages for chemical cleaning. This would easily have set back the schedule by more than a month.

However, this was resolved by installing permanent nozzles with flanges on each of the pipelines near the compressor flanges. Temporary pipes were used to connect these nozzles, and blind plates were installed on the compressor nozzles. The chemical cleaning took place without disturbing the alignment of the piping. The nozzles were blanked with rated flanges after completion of the cleaning.

Constrained space: planning erection of tall towers

The space constraints in the plant made the erection of the MDEA columns, especially the CO_2 absorber, one of the most critical activities of the revamp. The absorber weighed 300 tons with a height of 55 m. Difficulties in transportation of the absorber in one single piece, from port to site, required fabricating and transporting it in three pieces. The next step was to explore the possibility of welding the three pieces at the site horizontally and erecting it in one single piece after the hydro test. This had to be ruled out due to the lack of clear space at which all three pieces could be placed horizontally for alignment and welding. Moreover, the absorber could be erected in one single piece only by using a derrick. The derrick structure would require piled foundations and the erection of the adjoining CO_2 stripper, and the HP/LP flash drum would have to be kept on hold. Hence, the CO_2 absorber was planned to be erected in three pieces, with the alignment and welding done vertically. Welding of the pieces vertically would be a very difficult task considering the problems that may occur in alignment and position of the weld. The stress relieving, radiography, and subsequent UT examination were planned to be done in the vertical position including the hydro test. The adequacy of the founda-

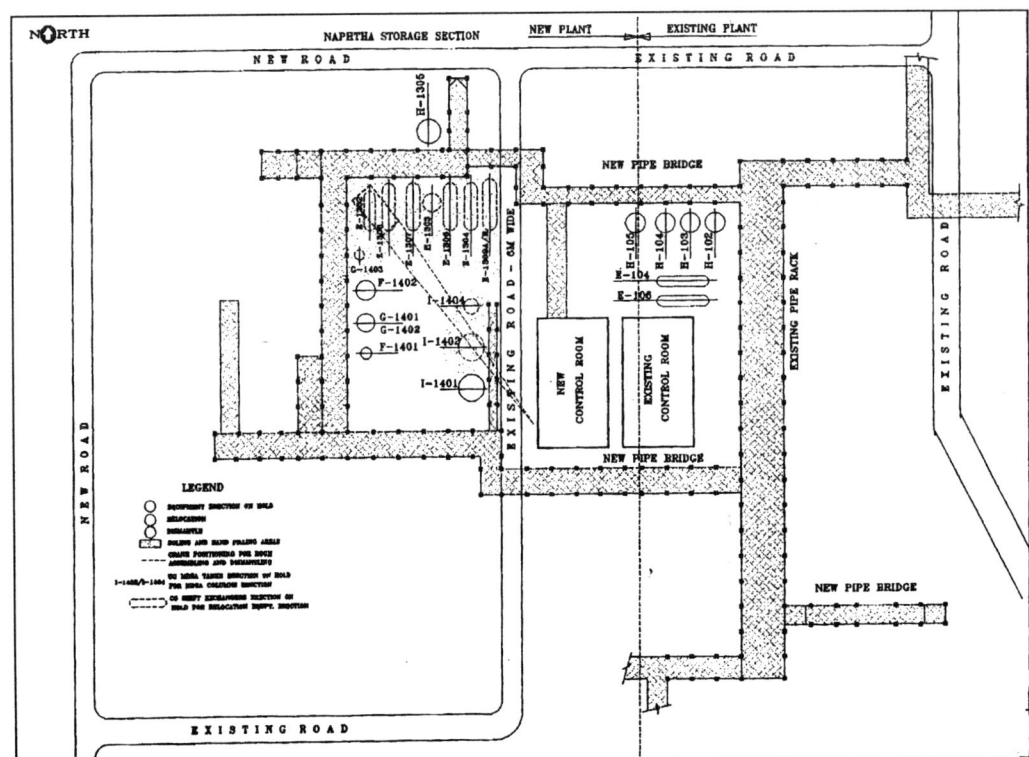

Figure 4. Space constraint in MDEA section.

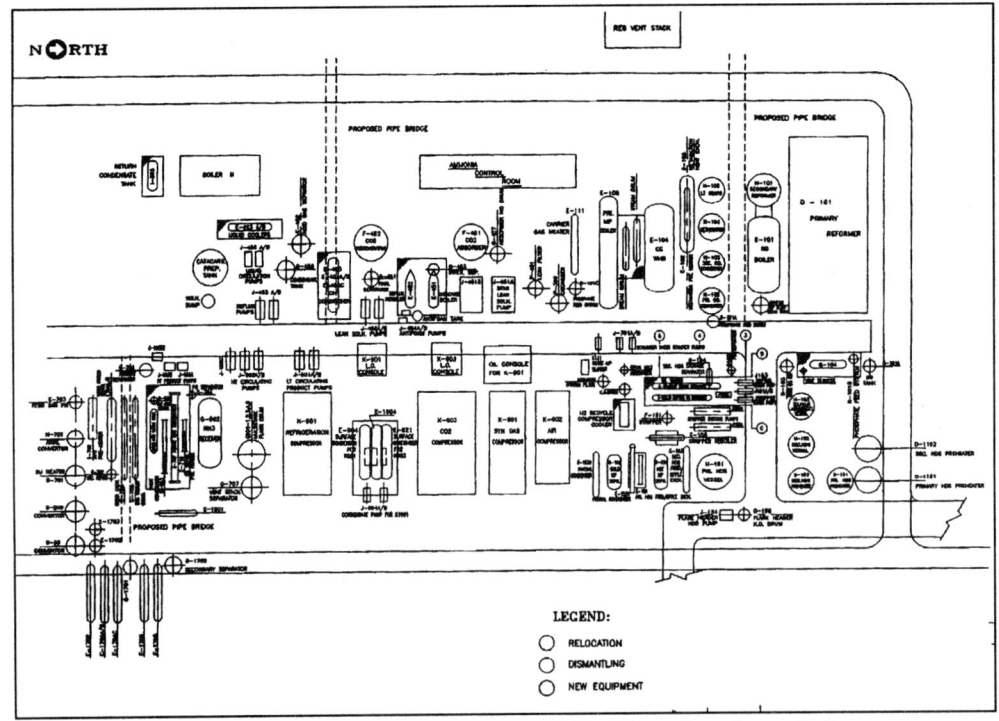

Figure 5. Existing plant plot plan showing shutdown jobs.

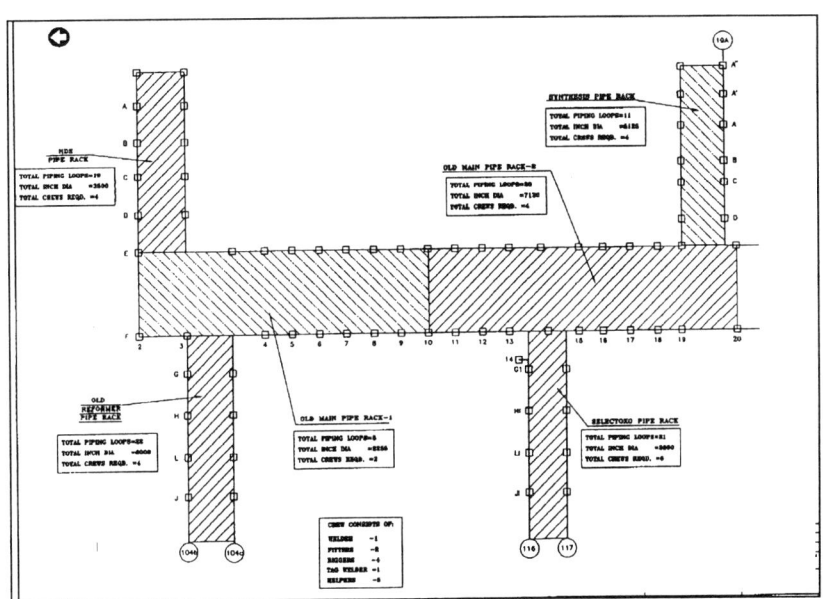

Figure 6. Piping and manpower microplanning analysis.

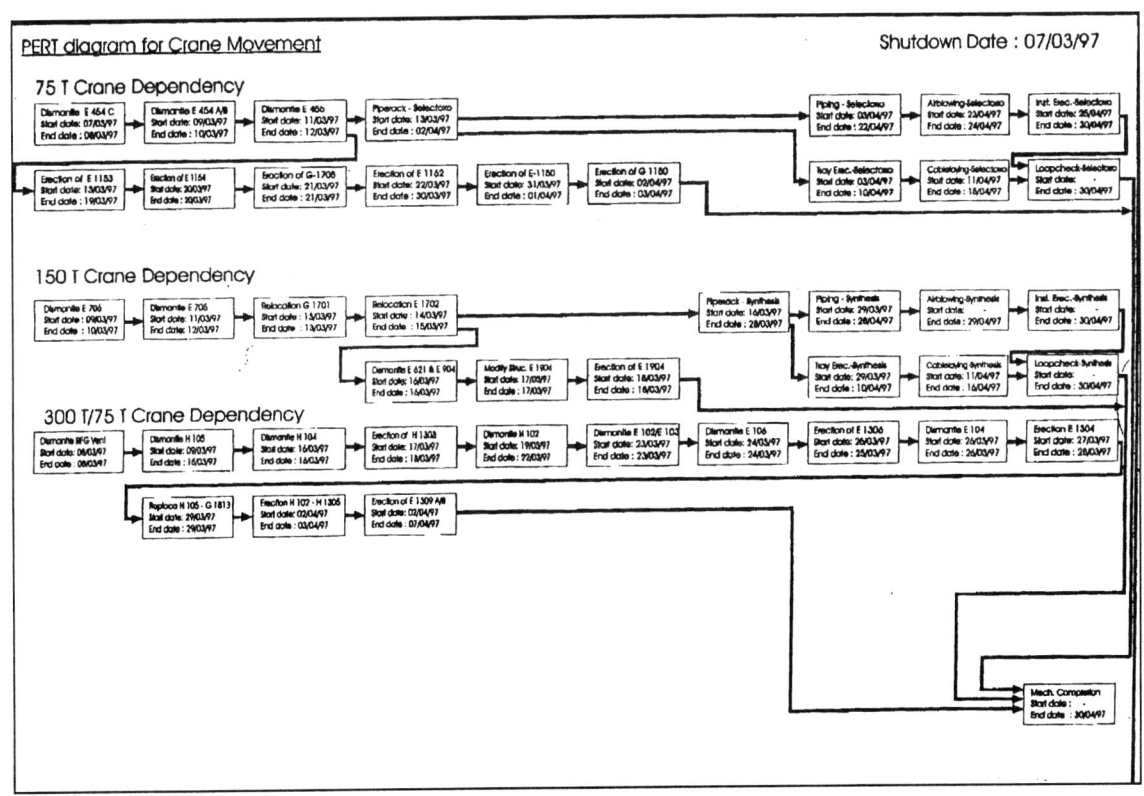

Figure 7a. Revamp of ammonia plant with crane movement plan.

tion for this load was checked and confirmed. The weight of the column with water was more than 600 tons.

A detailed schedule for the various activities and the enabling work was prepared. Both the vertical erection and the subsequent hydro test were carried out as planned without any upsets. (See attached Photos 1–6.)

Within the available space in the vicinity of the MDEA column, a crane required to lift individual pieces and a crane for tailing during lifting were to be accommodated and the lifting was to be planned. Since the available space was just adequate for assembling the boom, as indicated in Figure 4, it was planned to hold the required area from all other works during the execution of this work. (Figure 5 shows an existing plot plan, Figure 6 shows the existing area piping, and Figures 7a and 7b show the PERT diagram for crane movement.)

Nevertheless, if the crane was to move into this planned area for the erection of the MDEA column, the foundations of almost all the equipment in the shift section and methanation were to be kept under hold. Since this would mean piling and other civil foundation works after the erection had been completed and the crane was moved out of the area, it was decided to complete all piles and pilecaps of the foundations and pedestals for the equipment up to a height of 500 mm casting, leaving pockets for anchor bolts. It was also planned that the second lift of pedestals was to be cast only after the crane had been moved out of the area.

Instead of allowing for a normal curing of the concrete (21 days), special additives were planned for the second lift to speed up curing to obtain a completion within seven days.

The new MDEA tanks located below the ground in front of the absorber was an additional constraint. If they were erected, they would not allow movement of the crane when erecting the MDEA columns. The tanks were to be located below the ground on foundations with the top of the tank just above the grade level. In order to facilitate the crane movement, it was decided to backfill the space on which these tanks were to be located after casting of the foundations with sand and soiling stones. After the erection of the MDEA column, it was planned to remove sand and soiling by special earth moving equipment and erect the tanks. In order to retain the sand soiling mixture in the foundation area, a concrete retaining wall was built around the foundations.

To enable the crane movement in the above area where pedestals had been constructed up to 500 mm, the area was filled with sand and soiling up to the height of 500 mm.

Since the same crane was to erect the relocated equipment (two boilers and two reactors) in the same vicinity, it was planned to move the crane out of the area only after erection of the relocated equipment.

The necessity for holding up the erection of almost all equipment in the CO shift section had severe consequences on the availability of piping frontages. The total scope of piping in this section was around 10,000 ID out of which about 4,000 ID would have to be done in the shutdown.

Delayed equipment delivery thwarts plan for aboveground piping

Although a detailed plan for the erection of the MDEA columns was drawn up, the actual sequence of events did not occur as planned. It was foreseen that the MDEA columns should be erected and piping completed before shutting down. Delivery of the columns was, however, delayed by more than a year, and certain other exchangers, especially the stripper reboiler, methanation heat exchangers, and HHP BFW preheaters, were also considerably delayed. This was further aggravated by the delay in the receipt of piping materials, which had a crippling effect on piping and instrumentation progress. The piping progress was nearly restricted to the prefabrication work, and the erection work was restricted to laying the pipes on the pipe rack and completing the piping to the few exchangers that had reached the site on time. Since instrument works was mainly dependent of the completion of piping, progress was insignificant. The total piping scope for the revamp was around 130,000 ID and 200,000 inch meters (IM) of erection. After eight months of piping work, the progress achieved was only 50,000 ID. The section-wise status was at that stage:

(See sheets for progress of piping and equipment

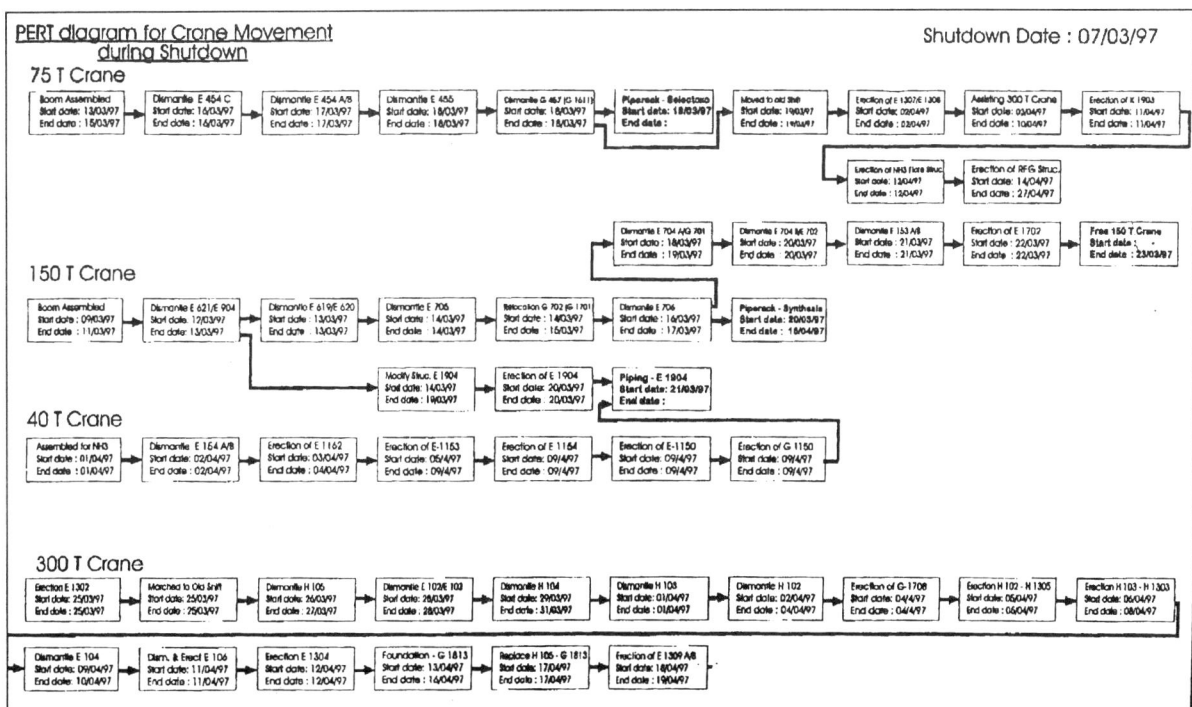

Figure 7b. Revamp of ammonia plant with crane movement plan (continued).

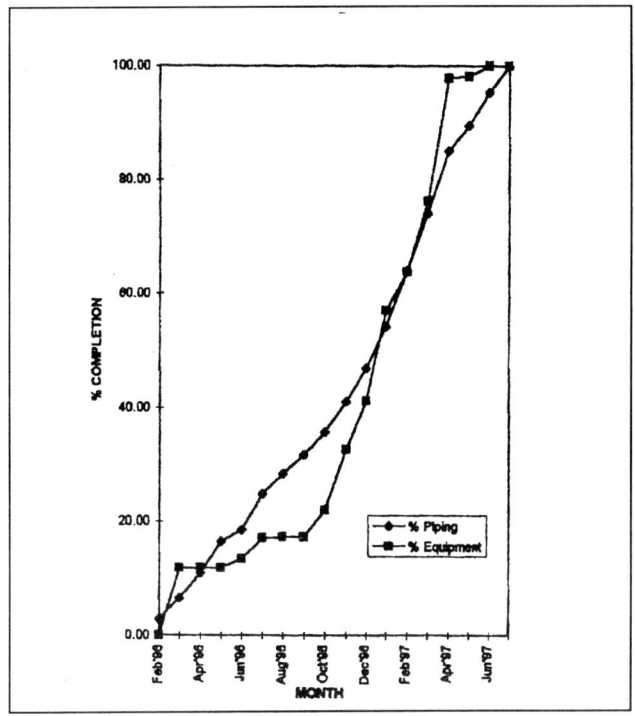

Figure 8. Progress for piping and equipment erection.

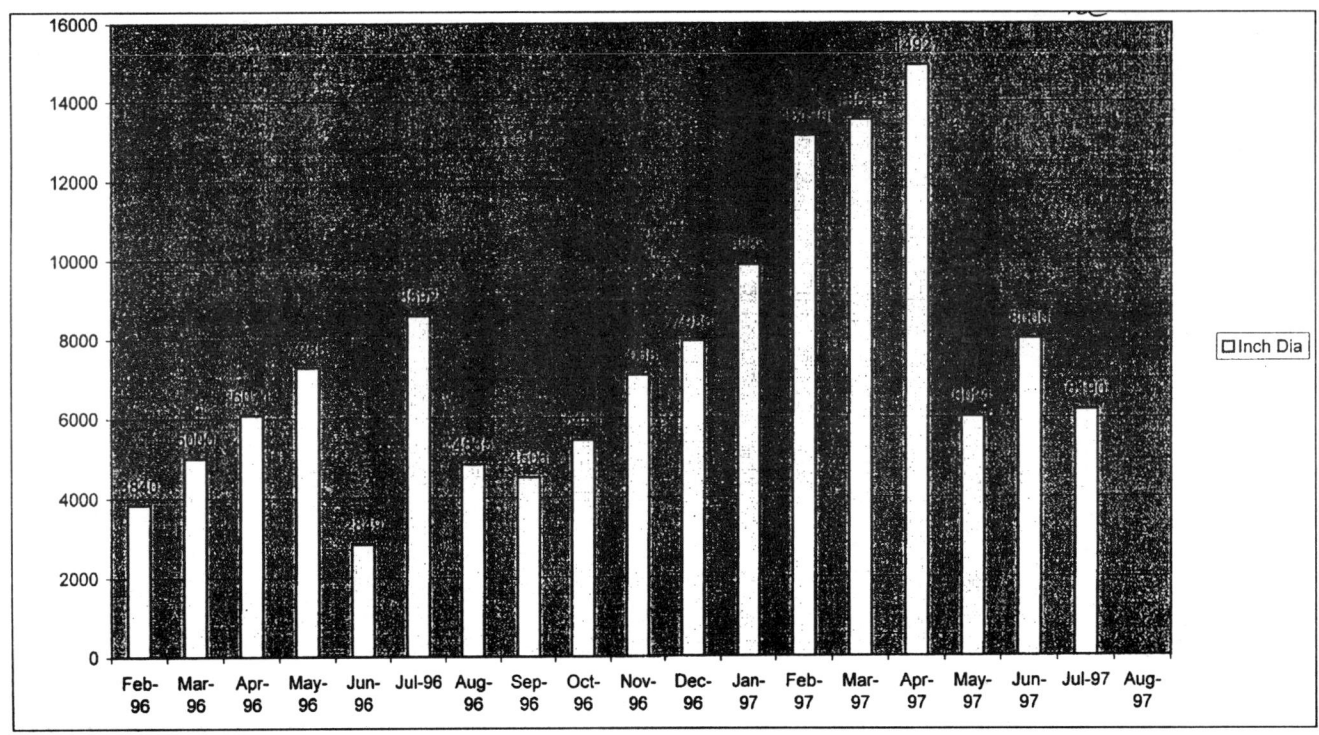

Figure 9a. Piping fabrication progress monthly.

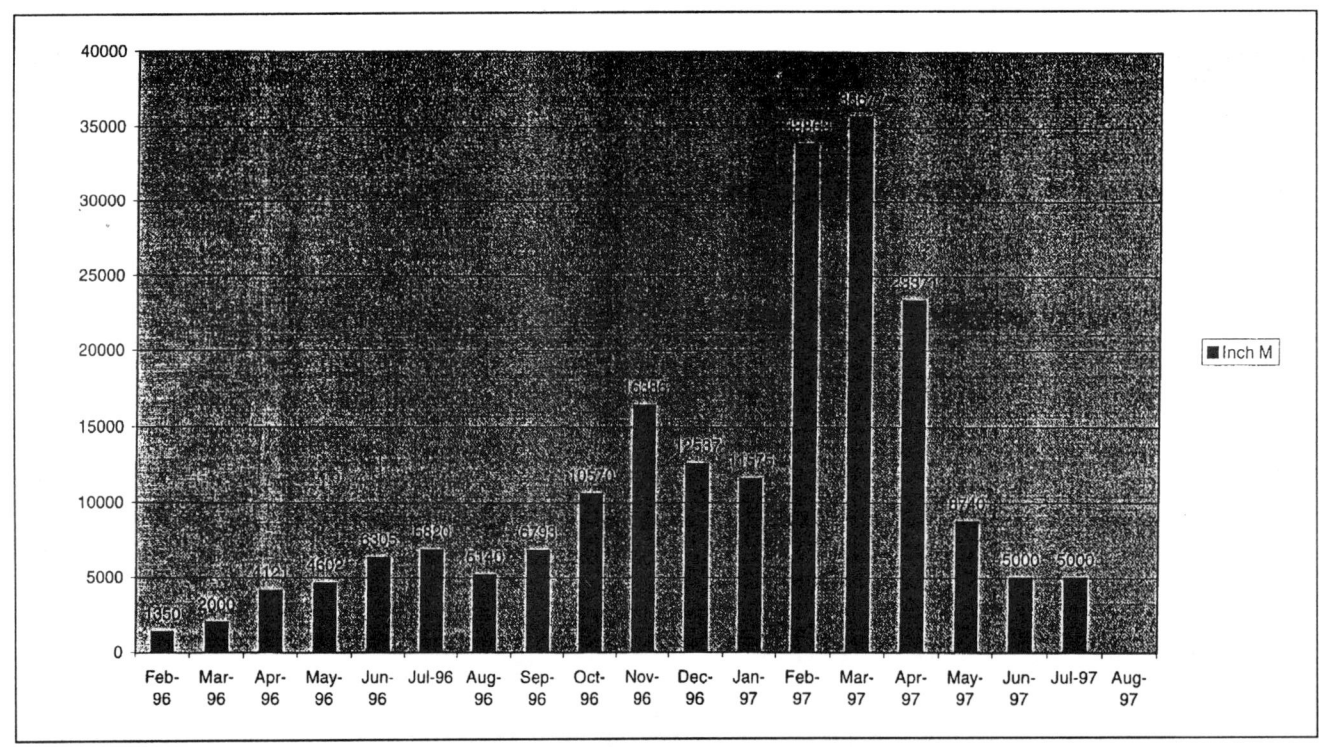

Figure 9b. Piping fabrication progress monthly (continued).

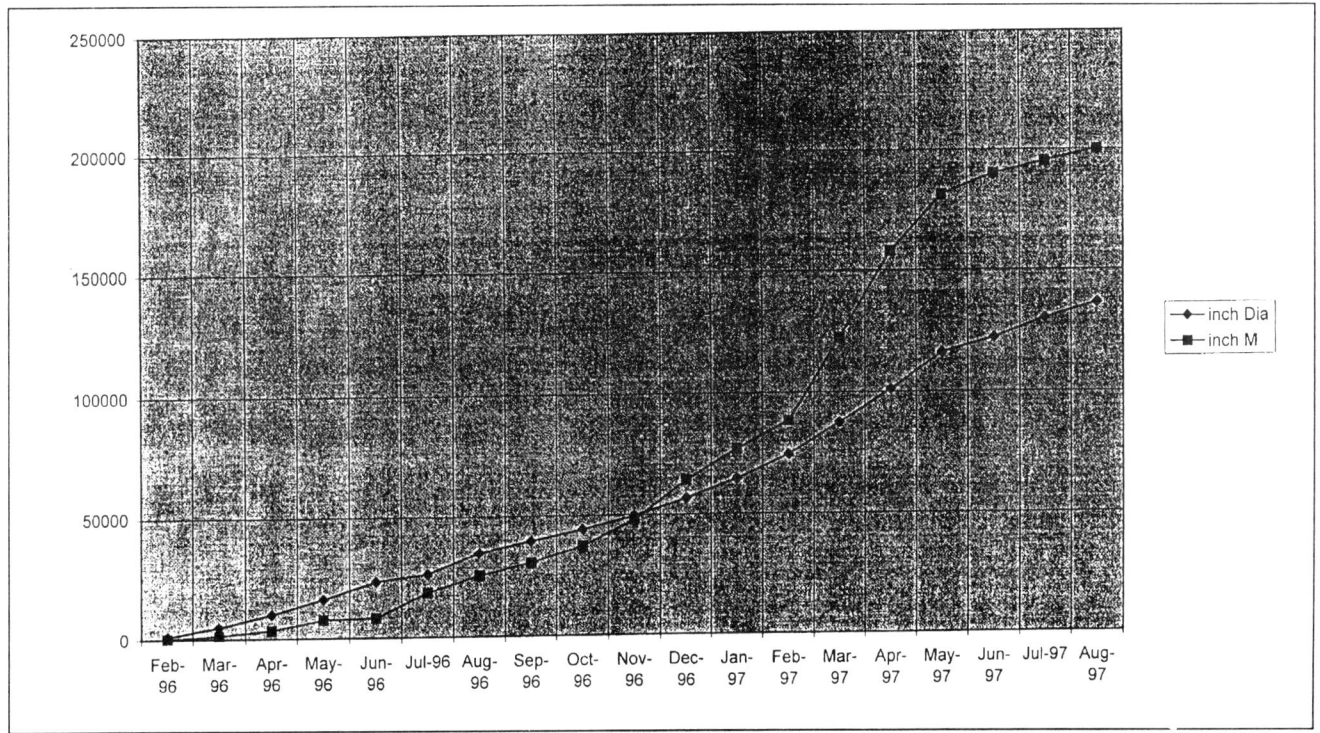

Figure 9c. Piping progress MFL.

erection in Figure 8, and piping fabrication progress, piping erection progress and piping progress in Figures 9a, 9b, and 9c.)

• *HDS Section*: None of the equipment received could be erected. Piping progress was restricted to prefabrication works.

• *Reforming Section*: The major work had been given as a turnkey package. The progress of the package was reasonably good. Piping work in this area was only hookup jobs and utility piping connected to the RG boiler and HHP BFW preheater. The BFW heater had not yet reached the site.

• *CO Shift and Methanation Section*: Out of the 13 pieces of equipment in this area, only the methanator had been received and erected. The piping progress was restricted to prefabrication. Instrumentation progress was negligible.

• *MDEA Section*: The total number of equipment in this area was 32, of which only 17 had been received and erected. Most of the major equipment had not yet reached the site, except for the CO_2 stripper, which had been erected. Almost all pumps had been received and erected on time. Piping progress was restricted to pre-fabrication and to erected equipment and pumps. Erection of the large bore piping connected to the columns had not begun.

• *Compressor Piping*: Most of the compressor piping had been completed as they were turnkey packages.

• *Synthesis Section*: Although most of the equipment had been delivered, only very limited piping could be completed due to the nonavailability of the pipe rack and the relocated equipment. Shutdown activities.

• *Refrigeration Section*: Shutdown work.

• *New Plant Pipe Racks*: The major progress in piping erection and instrument cabling works took place in this area, although restricted by the availability of imported piping materials.

The process condensate (PC) boiler, which was part of the new equipment, was foreseen to be completed along with the other plant systems, but could not be completed due to a delay in receipt of certain boiler components from the manufacturers. Since this PC boiler is dedicated for the supply of steam to the reformer by using the condensate generated in the process, this lack of steam for the reformer was to be compensated for by other means.

The existing plant has been operating with a 40 T/h capacity boiler for similar service to the existing reformer. The mechanical condition of the boiler was very poor. Replacement with a new and higher capacity boiler was thus necessary. In order to cater to the steam requirements until the time that the new PC boiler was ready, it was planned to hookup the old PC boiler to the reformer by making the necessary piping connections. Accordingly, new pipelines were laid and the boiler was connected. It supplied steam to the reformer for a period of seven months until the new PC boiler was ready and commissioned.

The delay in delivery of various equipment and the cascading effect this had on piping and instrument progress led to a review of the conditions set earlier for starting the final shutdown. The condition of the 28-year-old running plant was not encouraging. Rather than waiting for the mechanical completion and pre-commissioning of the new plant, it was felt that the possibility of executing these jobs parallel during the shutdown had to be explored. This would require considerable increase in resources from contractors as well as supervisory staff.

After a detailed review, it was decided to shut down the existing plant for the final hookup after erection of the three MDEA columns and to carry out the piping connected to the MDEA columns as part of the shutdown activity. This would increase the range of piping to be carried out during the shutdown to 9,000 ID.

As previously explained, the tallest column, the CO_2 absorber was received, assembled in vertical, and erected in position. However, the LP/HP flash drum in the MDEA section, which was planned to be received in two pieces ahead of the CO_2 absorber did not reach site on time. The LP flash drum is erected on top of the HP flash drum. While the HP flash drum was received and erected due to difficulties at the manufacturer's shop, the LP flash drum was brought to the site in rolled condition; thus, balance fabrication, hydrotest, and so on was carried out at the site. The LP flash drum was placed along the road on rollers and fabrication, stress relieving, radiography and testing was completed by resorting to continuous working and close monitoring.

Even though the above resulted in restraining traffic along the only road available, it had to be arranged in this way because of the superior force.

The timing of the shutdown plan was therefore upset, which deteriorated the condition of the existing running plant. It was decided to shut down the existing plant for a final hookup which pushed part of the erection job connected to the LP flash drum and pushed the piping located at the new area to the shutdown jobs.

Old Plant Shutdown: Execution of Equipment Relocation, Piping and Instrument Jobs

At the time the shutdown actually took place, the scope of the shutdown work had increased especially with respect to piping and instrumentation jobs. The scope of work was as given below:

(1) Fabrication completion of LP flash drum.

(2) Erection and assembly of *in situ* welding of LP flash drum.

(3) Relocation of two existing boilers and reactors to the new foundations in CO shift section.

(4) Casting of balance lift and erection of new exchangers, which were on hold for erection of MDEA columns.

(5) Erection of pipe racks in reformer, Selectoxo and synthesis section after dismantling of obstructing equipment and foundations.

(6) Completion of around 40,000 ID of piping works in the old and new sections of the plant and hydrotest of 257 piping loops.

(7) Connecting underground cooling water line from existing plant to new surface condenser.

(8) Instrument cable laying, junction box fixing, and so on for conversion of pneumatic system in old plant retained equipment to DCS compatible system.

The increase in the scope of work during the shutdown caused major logistic problems related to crane movement.

Three cranes with capacities of 300 tons, 150 tons, and 75 tons were available at the site for the relocation and erection work during shutdown. The 300 ton crane and the 75 ton crane were required for the erection of the LP flash drum and the relocation and erection of the CO shift section equipment. The 150 ton crane was utilized for the dismantling of equipment in the synthesis loop area to facilitate the erection of pipe racks after casting of foundations for the same. Since there were other small pieces of equipment to be erected, and the piping erection was to take place simultaneously, an additional 40 ton capacity crane was hired. In order to cater to the above and to streamline the crane movement in the most optimum way, a crane movement plan was worked out, as shown in Figure 7.

Since the relocation of the shift section equipment involved movement of the 300 ton capacity crane across the plant, a strengthening of the road matching the crane loading requirements was taken up prior to shutdown.

The distance to be negotiated and the space available for relocation of the CO shift section equipment was cramped and constrained, and the use of a trailer to transport the equipment was foreseen but not possible. The crane holding equipment had to be moved when this equipment was relocated. Hence, to facilitate movement of the crane, the instrument cable bridge going to the control room had to be raised to a height of 12 m.

The necessary fixtures for the transportation of the two boilers to be relocated in a single piece (including steam drums, raisers, and downcomers) were fabricated and kept ready before shutdown.

Crashing piping execution: micro plans

Because of the decision to shut down even before the completion of erecting the MDEA section and because the equipment was being kept under hold for the completion of MDEA columns, the scope of piping to be carried out for the hookup shutdown got considerably enhanced. A total scope of piping to be carried out in each section of the plant was carefully analyzed.

Since it was necessary to complete piping within the minimum possible time and the other instrumentation and precommissioning works had to go on simultaneously, the total piping, area-wise, was ascertained and necessary manpower for the piping was added to the micro planning analysis. This analysis enabled us to ascertain the extent of manpower required in each of the sections based on round-the-clock work.

Accordingly, the manpower was deployed in each section and the work was monitored on a daily basis. If any backlog was noticed, extra manpower was acquired to accomplish the work. A sample analysis is shown in Figure 6.

Use of P-Scan technique

The piping involved high-pressure piping, which had stringent quality requirements like radiography. Carrying out radiography would involve sealing off the area from human movement for long periods, resulting in a stoppage of work in and around the vicinity. To ensure continuous working and human movement in and around the affected area, an advanced NDT technique other than radiography was used. A Danish company, The FORCE Institutes, was contacted and the P-Scan technique was considered in place of radiography. This technique was adopted for at least 500 critical weld joints throughout all sections of the plant. The use of P-Scan also reduced the time used to clear the joints as it was much quicker compared to conventional radiography clearance.

Priority piping loops for precommissioning

Considering the increased scope of work for piping, the completion of various piping systems were planned based on precommissioning priorities/sequence. The piping scope was segregated into a total of 257 piping loops. Based on this, about 100 loops were identified and assigned top most priority for completion in a phased manner to suit the precommissioning requirements. This would ensure that additional time was available for completion of loops, high-pressure piping and the back-end synthesis loop section. Piping and mechanical works could be carried out in the areas as the synthesis loop even when precommissioning works were carried out in the other areas of the plant.

By this, the precommissioning activities in the front end of the plant preceded the mechanical completion of the piping in the back-end.

Table 2. SystemPrecommissioning and Commissioning Activities

System	Milestone	Relative weight	Total Weight	Date : 25-10-97 Rev. : 2 Report : PKO/RVP Completed on	% Complete
PRECOMMISSIONING ACTIVITIES					
1.Cooling Water System		4,05%			
	Precleaning of cooling water system (Part 1)	20,00%	0,81%		
	Precleaning of cooling water system (Part 2)	20,00%	0,81%		
	Discharge piping	10,00%	0,40%		
	Suction Piping	2,00%	0,08%		
	CT Screen	2,00%	0,08%		
	Power to motor	2,00%	0,08%		
	Motor Test	2,00%	0,08%		
	Coupling	2,00%	0,08%		
	Jumpovers Ready	10,00%	0,40%		
	Commissioning of cooling water pump	15,00%	0,61%		
	Flushing and cleaning of cooling water system	5,00%	0,20%		
	Jumpovers cutting /Valve fixing	5,00%	0,20%		
	Passivation of Heat Exchangers	5,00%	0,20%		
2.Instrument Air & Service Air System		0,92%			
	Leak-test of IA system	25,00%	0,23%		
	Leaktest of SA system	10,00%	0,09%		
	Airblowing of IA	25,00%	0,23%		
	Airblowing of SA system	15,00%	0,14%		
	Installation of G 1813, G 1814, A 1801	20,00%	0,18%		
	Commissioning of 18 PIC 020	5,00%	0,05%		
3.Nitrogen System		1,16%			
	Leak test completed	25,00%	0,29%		
	Airblowing of Nitrogen piping completed	25,00%	0,29%		
	Commissioning of K 1203	10,00%	0,12%		
	Commissioning of K 1801	10,00%	0,12%		
	Commissioning of X 1810	10,00%	0,12%		
	Commissioning of K 1802	10,00%	0,12%		
	Installation of G 1811	10,00%	0,12%		
4.Electrical & Emergency Systems		1,16%			
	Commissioning of 3.3 KV substation	20,00%	0,23%		
	Energizing the UPS	10,00%	0,12%		
	Energizing of MCC panels	20,00%	0,23%		
	Completion of test of electric system	50,00%	0,58%		
5.Boiler Feed Water Systems		3,47%			
	HP BFW Leak test	10,00%	0,35%		
	HP BFW Flushing	10,00%	0,35%		
	HP BFW Charging	30,00%	1,04%		
	HHP BFW Leak test	10,00%	0,35%		
	HHP BFW Flushing	10,00%	0,35%		
	HHP BFW Charging	30,00%	1,04%		
6.Steam Grids HHP, HP, MP, LP		10,40%			
	Insulation of HHP	5,00%	0,52%		
	Insulation of HP	5,00%	0,52%		
	Insulation of MP	5,00%	0,52%		
	Insulation of LP	5,00%	0,52%		
	Chemical Cleaning of HHP	5,00%	0,52%		
	Chemical Cleaning of HP	5,00%	0,52%		
	Chemical Cleaning of MP	5,00%	0,52%		
	Chemical Cleaning of LP	5,00%	0,52%		
	Steamblowing of HHP	5,00%	0,52%		
	Steamblowing of HP	5,00%	0,52%		
	Steamblowing of MP	5,00%	0,52%		
	Steamblowing of LP	5,00%	0,52%		
	Installation of PRDS (HHP/HP/MP)	7,50%	0,78%		
	Installation of PRDS (MP/LP)	7,50%	0,78%		
	Blowing I/L Reformer Header	5,00%	0,52%		
	Charging HHP grid	5,00%	0,52%		
	Charging HP grid	5,00%	0,52%		
	Charging MP grid	5,00%	0,52%		
	Charging LP grid	5,00%	0,52%		
7.Control Room & Control Systems		2,31%			
	UPS Commissioning	15,00%	0,35%		
	AC & Pressurisation Completed	20,00%	0,46%		
	Functional Test of DCS	20,00%	0,46%		

Table 2. System-wise Precommissioning Activities and Commissioning Activities (Continued)

System	PRECOMMISSIONING ACTIVITIES			Date : 25-10-97	
				Rev. : 2	
				Report : PKO/RVP	
	Milestone	Relative weight	Total Weight	Completed on	% Complete
	Cablelaying and testing in new area	20,00%	0,46%		
	Cablelaying and testing in old area	25,00%	0,58%		
8. Steam Production Systems		4,39%			
	Chemical Cleaning of 110 ATA - Loop 1	3,00%	0,13%		
	Chemical Cleaning of 110 ATA - Loop 2	2,00%	0,09%		
	Burner Testing	5,00%	0,22%		
	OPH Commissioning & Charging	5,00%	0,22%		
	Refractory Burnout	5,00%	0,22%		
	Light up & Steam blowing of Boiler	10,00%	0,44%		
	Safety Valve Floating	5,00%	0,22%		
	Steamproduction from 110 ATA boiler	5,00%	0,22%		
	Chemical Cleaning of PC Boiler	5,00%	0,22%		
	Burner Testing	5,00%	0,22%		
	Refractory Burnout	5,00%	0,22%		
	Light up & Steam blowing of Boiler	15,00%	0,66%		
	Safety Valve Floating	5,00%	0,22%		
	Steamproduction form PC-boiler	5,00%	0,22%		
	Precommissioning of E 1304	10,00%	0,44%		
	Precommissioning of E 1306	10,00%	0,44%		
9. Condensate Systems		1,50%			
	Leak test completed	25,00%	0,38%		
	Charging completed (K 1601/K 1602)	75,00%	1,13%		
10. Process Air Systems		9,25%			
	I Oil Flushing	10,00%	0,92%		
	II Oil Flushing	10,00%	0,92%		
	Final Oil Flushing	10,00%	0,92%		
	Chemical Cleaning of PAC Coolers	10,00%	0,92%		
	Insulation of Turbine	10,00%	0,92%		
	Soloruntest of the turbine for PAC	10,00%	0,92%		
	Testrun of PAC	20,00%	1,85%		
	Airblowing of Piping to Reformer	20,00%	1,85%		
11. Desulphurization & Reforming		17,34%			
	Naphtha for D 1151/D 1152	1,00%	0,17%		
	Refractory dry out of D 1151	2,00%	0,35%		
	Refractory dry out of D 1152	2,00%	0,35%		
	Catalyst unloading of H 1151	1,00%	0,17%		
	Catalyst loading of H 1151	4,00%	0,69%		
	Boxing up of H 1151	1,00%	0,17%		
	Catalyst unloading of H 1152	1,00%	0,17%		
	Catalyst loading of H 1152	4,00%	0,69%		
	Boxing up of H 1152	1,00%	0,17%		
	Catalyst unloading of H 1153	1,00%	0,17%		
	Catalyst loading of H 1153	4,00%	0,69%		
	Boxing up of H 1153	1,00%	0,17%		
	Airblowing in HDS section	5,00%	0,87%		
	Leaktest of HDS section	3,00%	0,52%		
	Power to ID/FD Fan	1,00%	0,17%		
	Precommissioning of ID-fan	4,00%	0,69%		
	Precommissioning of FD-fan	4,00%	0,69%		
	Blowing of CA System	1,00%	0,17%		
	Catalyst loading of H 1202	5,00%	0,87%		
	Boxing up of H 1202	1,00%	0,17%		
	Catalyst loading of D 1201	5,00%	0,87%		
	Boxing up of D 1201	1,00%	0,17%		
	Catalyst loading of H 1201	5,00%	0,87%		
	Boxing up of H 1201	1,00%	0,17%		
	Refractory dry out of reformer	10,00%	1,73%		
	Chemical Cleaning of RG-boiler	10,00%	1,73%		
	Airblowing of Reformer section (Process side)	5,00%	0,87%		
	Leaktest in Reformer section (Process side)	2,00%	0,35%		
	Airblowing of the fuel naphtha system	2,00%	0,35%		
	Leaktest in the fuel naphtha system	2,00%	0,35%		
	Airblowing of Start up nitrogen circuit	2,00%	0,35%		
	Leaktest in the start up nitrogen circuit	1,00%	0,17%		
	Catalyst loading of H 1222	2,00%	0,35%		
	Boxing up of H 1222	2,00%	0,35%		
	Airblowing of H2 recycle system	2,00%	0,35%		

Table 2. Precommissioning Activities and Commissioning Activities (Continued)

System	PRECOMMISSIONING ACTIVITIES Milestone	Relative weight	Total Weight	Date : 25-10-97 Rev. : 2 Report : PKO/RVP Completed on	% Complete
	Leaktest in H2 recycle system	1,00%	0,17%		
12. Shift & Methanation		6,47%			
	Catalyst loading of H 1304	10,00%	0,65%		
	Boxing up of H 1304	5,00%	0,32%		
	Catalyst unloading of H 1302	5,00%	0,32%		
	Catalyst loading of H 1302	10,00%	0,65%		
	Boxing up of H 1302	5,00%	0,32%		
	Catalyst unloading of H 1303	5,00%	0,32%		
	Catalyst loading of H 1303	10,00%	0,65%		
	Boxing up of H 1303	5,00%	0,32%		
	Catalyst unloading of H 1305	5,00%	0,32%		
	Catalyst loading of H 1305	10,00%	0,65%		
	Boxing up of H 1305	5,00%	0,32%		
	Airblowing in Shift and Methanation section	15,00%	0,97%		
	Leaktest in Shift and Methanation section	10,00%	0,65%		
13. MDEA CO2 Removal System		11,56%			
	Installation of Internals - F 1401	5,00%	0,58%		
	Installation of Internals - F 1402	5,00%	0,58%		
	Installation of Internals - G 1401	5,00%	0,58%		
	Installation of Internals - G 1402	5,00%	0,58%		
	Loading of Packing - F 1401	5,00%	0,58%		
	Loading of Packing - F 1402	5,00%	0,58%		
	Loading of Packing - G 1401	5,00%	0,58%		
	Loading of Packing - G 1402	5,00%	0,58%		
	Commissioning of J 1402 A/B	10,00%	1,16%		
	Commissioning of J 1404 A/B	2,00%	0,23%		
	Commissioning of K 1401	5,00%	0,58%		
	Commissioning of J 1401 A/B	15,00%	1,73%		
	Commissioning of J 1403 A/B	2,00%	0,23%		
	Commissioning of J 1406, X 1402	4,00%	0,46%		
	Commissioning of J 1407	2,00%	0,23%		
	MDEA section clean up	10,00%	1,16%		
	Tightness test and purge of MDEA section	10,00%	1,16%		
14. Naphtha system		1,16%			
	Commissioning of J 1110 C Pump	50,00%	0,58%		
	Flushing of Naphtha lines	50,00%	0,58%		
15. Synthesis Gas Compressor		11,56%			
	Oilflushing completed	10,00%	1,16%		
	Steamblowing completed	10,00%	1,16%		
	Chemical Cleaning of SGC Coolers	10,00%	1,16%		
	Solorun of the turbine	50,00%	5,78%		
	Testrun of synthesis gas compressor	20,00%	2,31%		
16. Ammonia Synthesis		8,09%			
	Completed I/L & O/L lines-H 1702	3,00%	0,24%		
	Catalyst loading of H 1702	30,00%	2,43%		
	Boxing up of H 1702	5,00%	0,40%		
	Commissioning of D 1703	10,00%	0,81%		
	Leaktest in ammonia synthesis	20,00%	1,62%		
	Flushing of the ammonia synthesis loop	20,00%	1,62%		
	Chemical cleaning of E 1701	10,00%	0,81%		
	Commissioning of J 1702 A/B	2,00%	0,16%		
17. Ammonia Refrigeration & Storage		4,62%			
	Commissioning of J 1903 C	10,00%	0,46%		
	Commissioning of J 1902 C	10,00%	0,46%		
	Commissioning of E 1904	30,00%	1,39%		
	Leaktest of the ammonia refrigeration system	50,00%	2,31%		
18. Effluent Treatment Liquid & Gases		0,58%			
	Precommissioning of Flare piping	20,00%	0,12%		
	Precommissioning of X 1301, G 1301	10,00%	0,06%		
	Commissioning of G 1707	10,00%	0,06%		
	Precommissioning of X 1101, G 1159	30,00%	0,17%		
	Precommissioning of G 1155, E 1160	20,00%	0,12%		
	Precommissioning of NH3 Flare	10,00%	0,06%		

Table 2. Precommissioning Activities and Commissioning Activities (Continued)

System	COMMISSIONING ACTIVITIES			Date :	25-10-97
				Rev. :	1
				Report :	PKO/RVP
	Milestone	Relative weight	Total Weight	Completed on	% Complete
1. HDS Section		10,00%			
	Tightness & Purge Test of HDS Section	10,00%	1,00%		
	Dryout of D 1151	10,00%	1,00%		
	Dryout of D 1152	10,00%	1,00%		
	Commissioning of HDS Flare	20,00%	2,00%		
	Commissioning of H2 Supply	5,00%	0,50%		
	Activation of H 1222	5,00%	0,50%		
	Reduction of H 1151	10,00%	1,00%		
	Reduction of H 1152/H 1153	20,00%	2,00%		
	Feedstock ready	10,00%	1,00%		
2. Reforming Section		20,00%			
	Boiler III Commissioned	5,00%	1,00%		
	HP Steam Available	5,00%	1,00%		
	Tightness & Purge test of Reforming Section	10,00%	2,00%		
	Fuel Naphtha System ready	10,00%	2,00%		
	Fuel Gas System ready	5,00%	1,00%		
	Combustion Air System ready	5,00%	1,00%		
	Nitrogen Circulation of Reforming Section	10,00%	2,00%		
	Dryout of WHS Section	20,00%	4,00%		
	Steam & Hydrogen to Reformer	10,00%	2,00%		
	Feedstock to Reformer	10,00%	2,00%		
	Air to Secondary Reformer	10,00%	2,00%		
3. Shift & Methanation Section		10,00%			
	Tightness & Purge test of Shift & Methanation Section	10,00%	1,00%		
	Reduction of H 1303	15,00%	1,50%		
	Reduction of H 1305	15,00%	1,50%		
	Reduction of H 1302	10,00%	1,00%		
	Taking LT Shift in line	10,00%	1,00%		
	Taking LT Guard in Line	10,00%	1,00%		
	Activation of H 1304	10,00%	1,00%		
	Commissioning of E 1304	10,00%	1,00%		
	Commissioning of E 1306	10,00%	1,00%		
4. MDEA CO2 Removal Section		15,00%			
	Leak test of MDEA System	10,00%	1,50%		
	Flushing of MDEA Section	20,00%	3,00%		
	Chemical Cleaning of MDEA Section	25,00%	3,75%		
	Preparation of MDEA Solution	5,00%	0,75%		
	Circulation of MDEA Solution	15,00%	2,25%		
	Break-in CO2 Removal	10,00%	1,50%		
	Frontend Stable	5,00%	0,75%		
	Commissioning of CO2 Blower	5,00%	0,75%		
	Commissioning of Selectoxo Unit	5,00%	0,75%		
5. Compressor Section		20,00%			
	Commissioning of PAC	25,00%	5,00%		
	PA to Instrument Air dryers commissioned	5,00%	1,00%		
	PA to Reformer Commissioned	5,00%	1,00%		
	PA to Selectoxo Commissioned	5,00%	1,00%		
	PA to IA Storage Vessels Commissioned	5,00%	1,00%		
	Commissioning of Syngas Compressor	30,00%	6,00%		
	Commissioning of MP CO2 Compressor	15,00%	3,00%		
	Commissioning of LP CO2 Compressor	10,00%	2,00%		
6. Synthesis Section		15,00%			
	Tightness & Purge test of Synthesis Section	20,00%	3,00%		
	Start of K 1601 for reduction	10,00%	1,50%		
	Heating H 1701 & Reduction of H 1702	40,00%	6,00%		
	Commissioning of Synloop Boiler	20,00%	3,00%		
	Ammonia production	10,00%	1,50%		
7. Refrigeration Section		10,00%			
	Tightness & Purge Test of Refrigeration Section	30,00%	3,00%		
	Fill Refrigeration Circuit with Ammonia	30,00%	3,00%		
	Commissioning of NH3 Vapour Blower	10,00%	1,00%		
	Commissioning of Refrigeration Compressor	30,00%	3,00%		

Instrumentation tie-ins

All the instrument work in the existing plant area was planned to be done only during the hookup shutdown. The total number of about 160 tie-in jobs for instrumentation connected to 120 closed loops and 500 open loops were to be taken during shutdown. The job involved physical verification of the condition of the old instruments including instrument air line, check of cable entry line of pressure switches, installation of new cables, junction boxes, and finally the loop checking of all systems.

Once the shutdown was taken, all instruments to be reused had to be recalibrated, and the sections of the old instrument air line which were required to be renewed were replaced. Even the reused control valves were recalibrated. The cable tray and duct for instrumentation erection started prior to shutdown. Simultaneously, all the pedestals for junction boxes were fabricated, installed, and kept ready for commissioning of the instruments.

Impulse lines, which were to be discarded, were cut down and removed. The new impulse lines were laid from the new instruments to the existing impulse piping and tubing. Minor rectification works were also required done in the existing impulse piping. New instrument air lines had to be laid from the existing tapping points to the new control valves. All instrument air lines for the reused control valve were to be changed after pneumatic testing. Once the above jobs were completed, each individual instrument loop checking for the same was taken up and completed.

Once the loop checking for individual instruments was completed, the closed loops and individual logic were also checked.

In the case of instrumentation works, the time available was even shorter, because the work on the new piping could only start after hydrotest and mechanical completion of each loop. In some cases, the instrument fixing could be done only after completion of precommissioning works like chemical cleaning, air blowing, and water flushing. This put considerable pressure on the instrumentation work, because the time available between completion of piping and starting precommissioning work was almost insignificant.

As done for the piping, the completion of instrument loops were prioritized, based on commissioning requirements. The instrument commissioning jobs and the commissioning of the system took place in tandem.

Considering the revised design conditions for the revamp plant, the existing safety valves had to be dismantled and recalibrated and reset to the new process conditions and installed.

During the shutdown, it was planned that all the existing electrical motors had to be hooked up to the MCC and the new load management system provided by HTAS.

Dove-Tailed Precommissioning/Commissioning Activities

Sequencing and scheduling

Activities were identified for precommissioning and mechanical completion plans of individual systems were sequenced in line with the requirements. A schedule was drawn for the sequential precommissioning of various systems to allow for an early commissioning. Based on the actual mechanical completion of individual systems, the precommissioning schedule also underwent changes, and the schedule was updated.

See Table 2 for precommissioning activities and commissioning activities.

The cooling water system flushing/passivation was taken up prior to shutdown using the existing cooling water pumps and hooked up during the planned shutdown. This enabled charging of the cooling water systems for various other activities like commissioning of compressor, boilers, and so on prior to shutdown.

To enable a speedy commissioning, chemical cleaning of the 110 ATA boiler was resorted to instead of conventional alkali boil out. This saved considerable time in getting the boiler ready.

All steam lines were chemically cleaned to reduce the time for steam blowing.

Steam blowing of the new steam lines was made by means of steam from the 110 Ata Boiler, as well as from boilers in the existing plant prior to the shutdown. Utility requirements for the 110 Ata boiler were made available by the tappings taken during the previous shutdown.

All the other steam nets were charged letting down steam through the pressure reducing and desuperheating station (PRDS) or from the existing plant auxiliary boiler.

All interstage lines and equipment in the process air compressor were chemically cleaned to reduce startup time, since the compressor was needed for air blowing of the process lines.

The PSA nitrogen system was commissioned well in time to cater to the requirements of nitrogen blanketing and purging of various sections.

For taking up as many jobs as possible simultaneously, along with other piping work, catalyst loading of reactors was done early by completing piping up to the isolation valves as was done for the ammonia converter.

LPG required for refractory dry out of the secondary reformer and cold collector systems could not be taken through any of the existing or new sections of the plant because of the continuing hot work being done in those areas. In order to facilitate completion of refractory dry out of the reformer system, a temporary LPG line was laid from the LPG storage facility without crossing any sections of the plant, and refractory dry out was completed well ahead of the schedule.

The MDEA column packing was received in rusted and greased condition. This required elaborate cleaning of the packing. To minimize the time required, the packing was precleaned in tanks just before being loaded into the columns.

The synthesis gas compressor was ready for commissioning well ahead of the synthesis loop section. Taking advantage of this, the compressor was started up and tested with nitrogen to look for any problems, which might arise later when it was to run with synthesis gas. The synthesis loop was leak-tested in nitrogen after water flushing using the synthesis gas compressor. This reduced the startup time of the compressor considerably, when synthesis gas was available.

New Plant Startup

A schedule for sequential startup of plant sections was drawn and updated based on completion of precommissioning. With precommissioning activities accomplished, ammonia production was reached in only 12 days after naphtha had been fed into the reformer.

Conclusion

By extensive strategic maneuvering, careful and meticulous planning, and adoption of innovative implementation methodology, the project to convert the 28-year vintage, 750 MTPD Ammonia Plant into a modern 1,050 MTPD ammonia plant was successfully executed and a smooth startup was accomplished. It is gratifying to conclude that the plant has pursued the Guarantee Test Run and proved the capacity and energy efficiency even better than guaranteed by the licensor.

DISCUSSION

Al Hajari, *Saudi Arabian Fertilizer Co.*: On the refrigeration section, how many heat exchangers were added? What about water cooled exchangers for the ammonia condensers?

Neelakantan: In the refrigeration section, we did not add anything except conversion of the old pneumatic to DCS-based instrumentation. We have a CO_2 compressor and the refrigeration compressor. We only added a common surface condenser catering to both, and, other than that, we did not make any alterations.

Al Hajari: Was there any increase in capacity of the CO_2 booster compressor?

Neelakantan: Yes, we had an additional compressor. There was no increase in the capacity of the existing compressor.

Al Hajari: Also, the paper mentioned that they have added an extra cell for the cooling towers.

Neelakantan: Exactly. We added two cells.

Al Hajari: What is cooling water supply temperature design?

Neelakantan: 34°.
Al Hajari: What is the corrosion inhibitor used?
Neelakantan: Bets Chemicals.
Richard B. Strait, *M. W. Kellogg Co.*: How long did the project take?
Neelakantan: After the start of construction, it took around 28 months. I should say that we have been running the existing plant for more than 23 months during the construction stage.
Strait: How much time was there from ammonia to ammonia, that is, the time you shut down the existing plant until you were making ammonia in the other plant?
Neelakantan: I think I took around 150 days.
Jim Grosnell, *Brown and Root Engineering and Construction*: Your paper lists the price of the revamp as $160 million. Is that number correct?
Neelakantan: Yes.
Grosnell: You must have done some studies upfront comparing the price of a new plant with the price of a revamp. I suspect for that amount of money you could have gotten a new plant with more capacity and lower energy consumption. Did you do these kind of comparisons?
Neelakantan: Yes.
Grosnell: What were your findings?
Neelakantan: If you go for a grass roots plant of this capacity, it should normally almost cost twice this amount. It was later clarified that the revamp cost was indicated for ammonia, urea and the NPK plant. For the ammonia revamp, the cost is $120 million.

Upgrading a 25-Year-Old Ammonia Plant

The revamp of a 25-year-old ammonia plant is described which resulted in a lower energy consumption and a higher production capacity. Conservation techniques have been applied to the revamp of this plant, resulting in a lower specific energy consumption. An increased ammonia production and a reduction of the environmental impact have also been obtained.

J. J. de Wit and A. Riezebos
Continental Engineering B.V., Amsterdam, The Netherlands

Introduction

The ammonia plant was originally designed and constructed by Bechtel, came on-stream in April 1971 and had a design capacity of 1360 MTPD (1500 STPD) of ammonia. The design aimed at a specific energy consumption of 34 GJ/ton of ammonia, which is equivalent to 8.1 Gcal/t or 29 MM BTU/st.

The ammonia plant uses natural gas for feedstock and fuel. The plant was designed as a stoichiometric ammonia plant, without addition of excess process air and self-supporting in steam production of high pressure steam at a 60 bar (900 psi) level and medium pressure steam at a 40 bar (600 psi) level. HP steam is applied for power steam, while the majority of the MP steam is used as process steam.

Front-end

The basic process flow diagram of the front-end is shown in Figure 1. It comprises the following sections:
- Desulfurization;
- Primary and secondary reforming;
- Waste heat recovery;
- High and low temperature shift conversion; and
- Carbon dioxide removal and methanation.

In this design the feed-gas desulfurization took place at a temperature of 390°C (734°F), while the process steam was superheated to a temperature of 645°C (1,193°F). This resulted in a mixed-feed temperature to the catalyst tubes of 575°C (1,067°F) at a steam/carbon ratio of 3.7 mol/mol. The radiant heat consumption for the steam-reforming reaction amounted to 84 MW (72 Gcal/h or 285 MM BTU/h), resulting in a temperature at the outlet of the primary reformer of 800°C (1,472°F).

The corresponding bridge-wall temperature of the flue gas was approximately 1,000°C (1,832°F). The waste heat from the flue gas was recovered by process air pre-heat, superheating of process steam and HP steam, MP steam production, natural gas pre-heat and high-pressure boiler feed water pre-heat. The flue gas

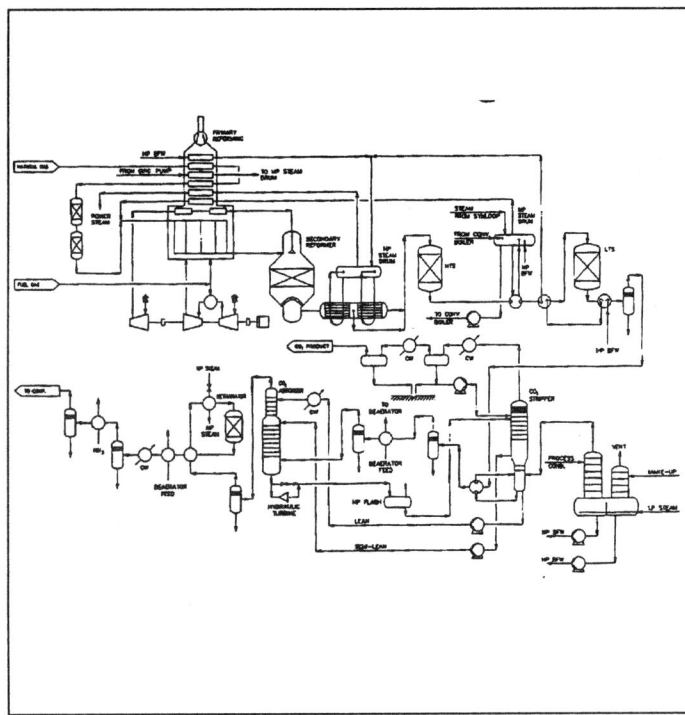

Figure 1. Front-end before revamp

leaving the convection section had a temperature of 220°C (428°F).

The process air compressor is gas-turbine driven, while the exhaust gas of this gas turbine with a temperature of 445°C (833°F) is applied as combustion air. Process air is preheated up to a temperature of 720°C (1,328°F). This gives a temperature of 985°C (1,805°F) and a methane slip of 0.3% at the outlet of the secondary reformer. Waste heat is recovered by means of HP steam production.

The high temperature shift is executed at an inlet temperature of 360°C (680°F). Waste heat recovery is obtained by MP steam production and high-pressure boiler feed water preheat.

The low-temperature shift had an inlet temperature of 200°C (392°F). Its waste heat recovery was obtained by high-pressure boiler feed water preheat, CO_2 stripper reboiler energy requirement and deaerator feed water pre-heat, resulting in a temperature of 90°C (194°F) at the inlet of the absorber.

The CO_2 removal system was a dual-stage Benfield type. The lean solution circulation pump was electricmotor driven, while the semi-lean solution circulation pumps were equipped with steam and hydraulic turbines. The CO_2 slip in the overhead of the absorber amounted to 1,000 ppmv. In addition to the already mentioned stripper energy requirement originating from LTS effluent heat, also an amount of live steam from process condensate stripper overhead was used. The specific energy consumption of this unit approached 120 MJ/kmol of removed CO_2.

The synthesis gas at the outlet of the absorber had a temperature of 70°C (158°F). It was heated up by a feed/effluent exchanger and a trim heater to the meth-anation reaction temperature of 290°C (554°F). Subsequently, the synthesis gas was cooled down by deaerator feed preheat, cooling water and chilling with ammonia before compression.

Synthesis loop

Generally, the loop pressure for a 1500 STPD plant is 200 bar (3,000 psi). This rule also applies to this case. Figure 2 shows the basic process flow diagram for this synthesis loop.

The synthesis loop contains the following elemen-

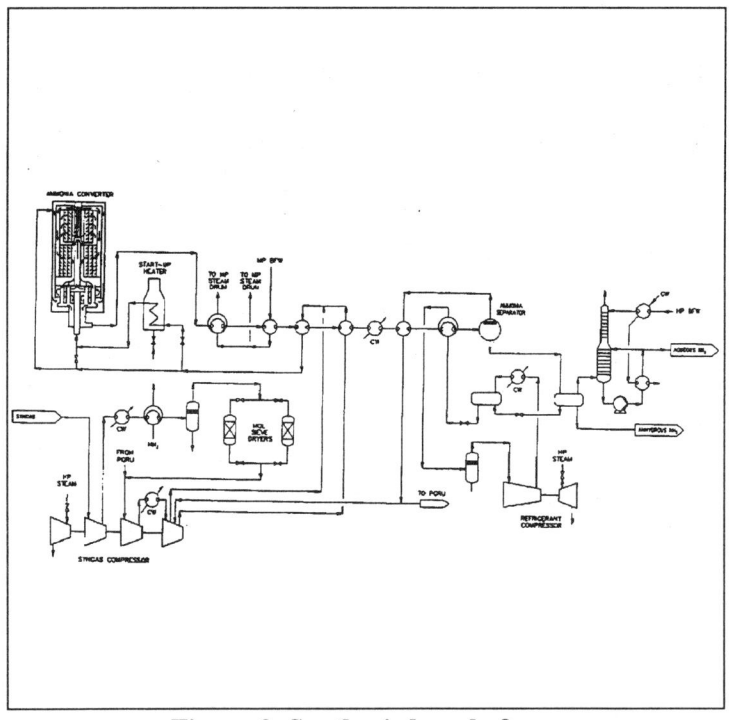

Figure 2. Synthesis loop before revamp

tary steps:
- compression, molecular sieve dryers;
- feed/effluent heat exchange;
- ammonia conversion, waste heat recovery;
- cooling and refrigeration, ammonia separation; and
- cryogenic purge gas recovery.

Synthesis gas from the methanator is compressed from 28 to 65 bar (405/945 psi). Subsequently, water is removed by molecular sieve dryers. Thereafter, the synthesis gas is further compressed in two stages to 220 bar (3,190 psi). The gas is heated in a feed/effluent exchanger and by internal heat exchange in the convertor.

Waste heat from the synthesis reaction is recovered by means of MP steam production and MP boiler feed water pre-heat. After passing the feed/effluent exchanger, the reactor effluent is cooled in a water cooler, a cold gas exchanger and an ammonia chiller to approximately 4°C (39°F). The refrigerant system consists of a single-stage chilling.

The liquefied ammonia from the separator is let down to approximately 20 bar (290 psi) where the majority of the dissolved inerts are released. The liquid product is transported to storage while the gaseous ammonia is absorbed in water.

The inerts in the synthesis loop are maintained at a standard level, while the purge gas is introduced to a cryogenic recovery unit. The recovered hydrogen is fed to the suction on the second stage of the synthesis gas compressor, while the remaining gases are used as fuel gas.

Steam system

Figure 3 shows the steam system of the ammonia plant before the revamp.

The steam system has the following characteristics:
- 79 bar (1,150 psi) steam import (minimum quantity)
- 60 bar (900 psi) and 40 bar (600 psi) steam production
- 12 bar (175 psi) and 3 bar (50 psi) steam export.

The ammonia plant is self-supporting with respect to steam production, however, a small amount is imported to keep the emergency steam-transfer line warm. The excess amount of steam is exported to other plants

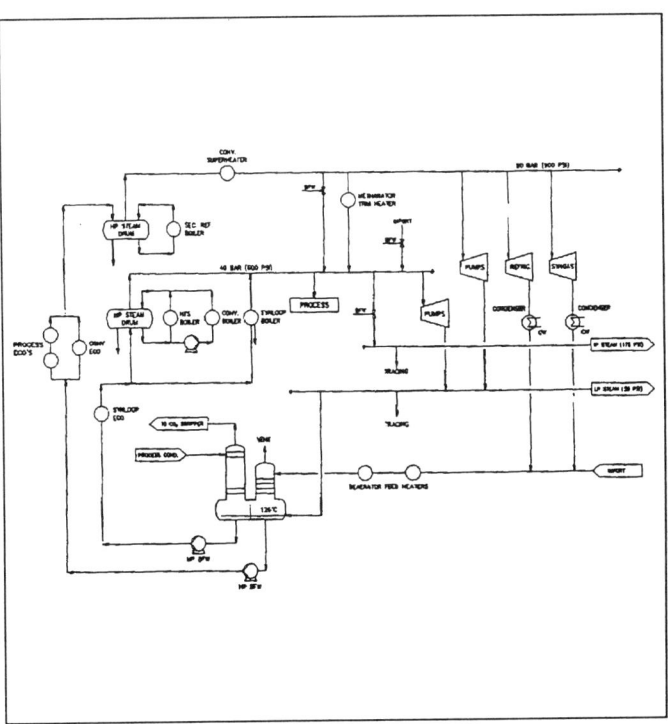

Figure 3. Steam system before revamp

on-site.

HP steam is superheated in the convection section of the primary reformer. Part of this steam is applied for the methanator trim heater, which is let down to the MP steam header. The MP steam is generated in the convection section, with HTS waste heat as well as in the synthesis loop. HP boiler feed-water preheat is performed with both process and flue-gas waste heat. MP boiler feed-water preheat is performed in the synthesis loop.

Several pumps are driven by back-pressure turbines and fed with high-pressure or medium-pressure steam. In general, these turbines have a very poor efficiency.

Starting point

In 1994 Continental Engineering was invited to study the energy conservation possibilities for this ammonia plant. The client wished to reduce the gap in energy consumption between this plant and modern plants as far as economically feasible.

Energy Conservation

General

Generally in ammonia plants, several carriers of energy can be outlined. These are natural gas, process gas, flue gas, steam and electricity.

Reduction of the overall energy consumption of the ammonia production facility is possible by performing good housekeeping with respect to the above mentioned energy carriers.

More specifically, the energy consumption of the large consumers, such as the CO_2 removal unit, steam-turbine driven compressors and pumps and process steam should be identified.

If the required amount of energy for these consumers can be reduced, and when the so-created excess amount of steam or process gas waste heat can be converted into a savings on a primary energy source, this will result in a reduction of energy consumption.

Among others, the above mentioned techniques have been applied in the existing ammonia plant, resulting in a fully integrated and flexible process design. These techniques will be outlined in the following paragraphs.

Pre-reformer installation

A pre-reformer unit performs part of the steam-reforming process while convective heat instead of radiant heat is used. Particularly, the higher hydrocarbons are converted to methane, while methane itself is converted to hydrogen and CO_2.

Therefore, the introduction of a pre-reformer unit increases the requirement for high-temperature convective heat. This heat is normally applied for steam generation or steam superheating purposes in the convection section. By excluding this steam generation and steam superheating and by using these coils for reheating the partly converted gas mixture, pre-reforming transfers excess steam production into a natural gas savings and enhances the steam-reforming capacity.

In this plant mixed feed to the pre-reformer has a temperature of 510°C (950°F). Due to the endothermic reaction, the temperature drops to 450°C (842°F). The partly converted gas is reheated to 560°C (1,040°F) by using convective heat.

For installation of a pre-reformer unit in an existing plant, it is in most cases required to modify the convection section of the primary reformer.

Convection section modification

Due to the lower steam to carbon ratio of 3.0, less process steam is required. Therefore, MP steam gener-

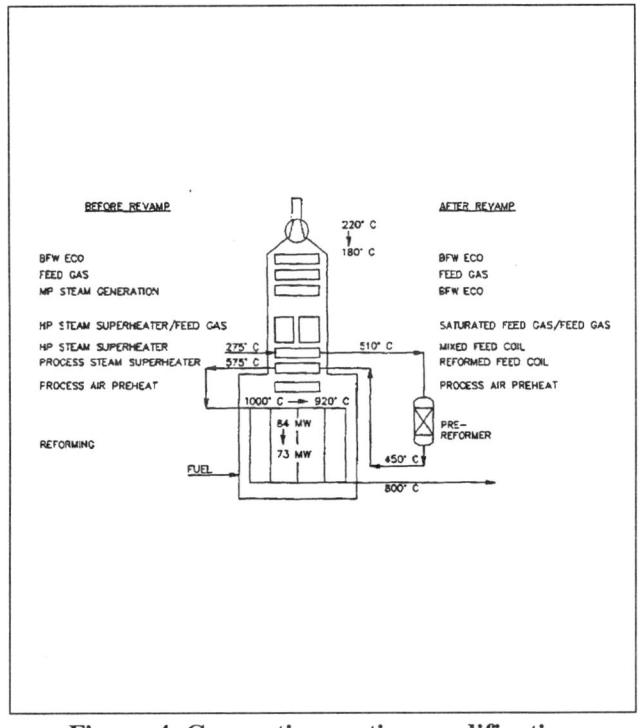

Figure 4. Convection section modification

ation in the convection section became superfluous. Figure 4 shows the convection section of the primary reformer after modification.

As a result of the lower steam to carbon ratio and the installation of the pre-reformer unit, the radiant heat requirement is reduced from 84 to 73 MW. The bridge-wall temperature is therefore reduced from 1,000°C (1,832°F) to 920°C (1,688°F).

High level heat in the convection section of the primary reformer was previously used for superheating of process steam. Now it is applied to reheat the partly reformed feed. The coil for HP steam superheating is applied for mixed feed preheat. The remaining part of the coil for the HP steam superheater is used as a saturated feed heater.

Furthermore, the coil for MP steam generation has been used to preheat HP boiler feed water. As a result from the above outlined activities, the stack temperature is reduced from 220°C (428°F) to 180°C (356°F)

Steam superheating

Before the revamp, the waste heat from the secondary reformer was recovered by producing high-pressure saturated steam. This steam was superheated in a coil of the convection section up to 430°C (806°F) prior to entering the main steam header.

As presented in Figure 4, HP steam superheating has been removed from the convection section of the primary reformer. However, part of the secondary reformer effluent, waste heat can be applied for steam superheating and excess steam production can be avoided. Figure 5 shows the high-pressure steam production and superheating after the revamp.

The HP steam superheater has been designed to superheat the total, produced amount of steam to 430°C (806°F) by means of secondary reformer effluent. For temperature control purposes, an additional bypass was introduced.

Due to the reduced steam to carbon ratio, less secondary reformer effluent waste heat is available, so less steam can be produced. Moreover, the extra amount of energy involved in superheating of steam reduces the steam production even more.

Steam balancing

Since the amount of HP steam production was reduced, several activities had to be carried out to overcome this shortage. These activities included the modification of the methanator trim heater and modification of several pump drivers. Figure 6 shows the result of these activities.

The methanator trim heater was modified from letdown to condensing and the HP boiler feed water pumps were switched to electric-motor drive. This considerably reduced the HP steam consumption.

The reduced amount of MP steam, also originating from a lower steam to carbon ratio, was compensated by the reduced amount of process steam and by modification of pump drivers. Several turbine-driven lube and seal oil pumps were switched to electric-motor drives.

This resulted in both the high-pressure and medium-

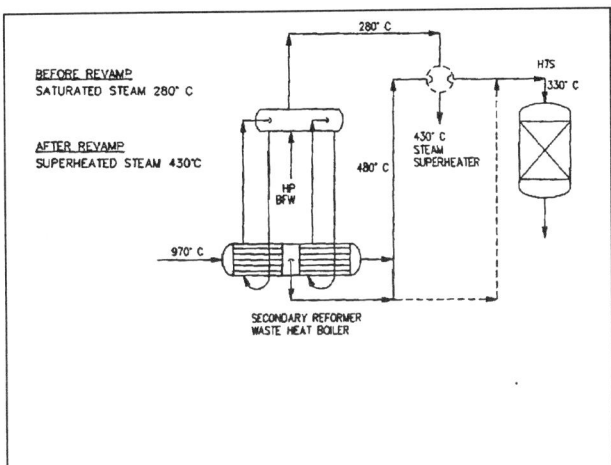

Figure 5. Steam superheating.

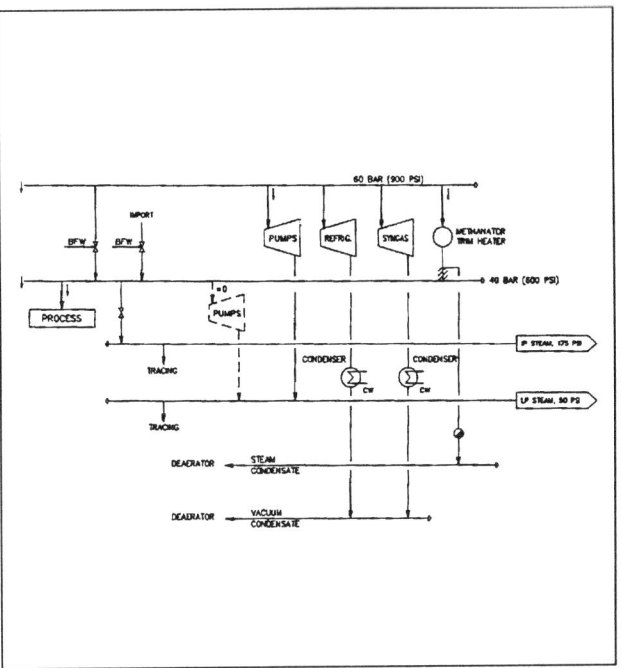

Figure 6. Steam balance.

pressure steam consumption becoming more or less in balance with the steam production.

Boiler feed water pre-heat

Before the revamp, HP boiler feed water was pre-heated with LTS and HTS effluent, in parallel with convection section waste heat.

Due to the reduced HP steam production downstream of the secondary reformer, boiler feed water preheat facilities had to be modified to recover the process waste heat. Figure 7 shows the boiler feed water pre-heat system after modification.

After the revamp, HTS and convection section waste heat is sufficient for boiler feed water pre-heat. Therefore, LTS waste heat becomes available for other purposes. The convection section MP steam boiler has been removed and the coil has been re-used as an additional HP boiler feed water economizer. All boiler feedwater heaters are operating in series.

As a result of these activities, the HP boiler feedwater temperature to the steam drum is increased from 225°C (437°F) to 260°C (500°F). Keeping in mind that the boiler feed water will start boiling at 275°C (527°F), it can be said that waste heat is recovered to its maximum extent.

CO_2 removal system

An important energy savings was reached by changing to a CO_2 removal solvent with a lower specific energy consumption per ton CO_2 removed. The original design of both the lean and semi-lean circulation pumps had some spare capacity to increase the circulation flow. This made it possible to perform a solution swap to aMDEA. The modifications which have been made are shown in Figure 8.

The solution swap to aMDEA necessitated the execution of other modifications of the CO_2 removal system such as the introduction of structured packing in the stripper, modification of the impellers of the circulation pumps, introduction of a lean/rich exchanger and a semi-lean cooler.

The overall effect was a reduction of CO_2 stripper reboiler requirement from 54 MW to 29 MW. Therefore, it was no longer required to introduce contaminated steam from the process condensate stripper, which resulted in a reduction of the environmental impact since methanol, amines and ammonia were no longer emitted.

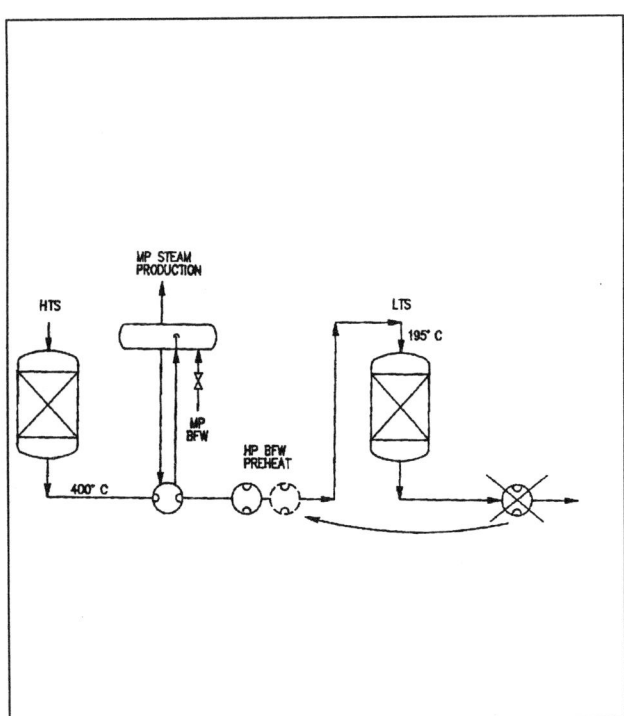

Figure 7. HP BFW preheat.

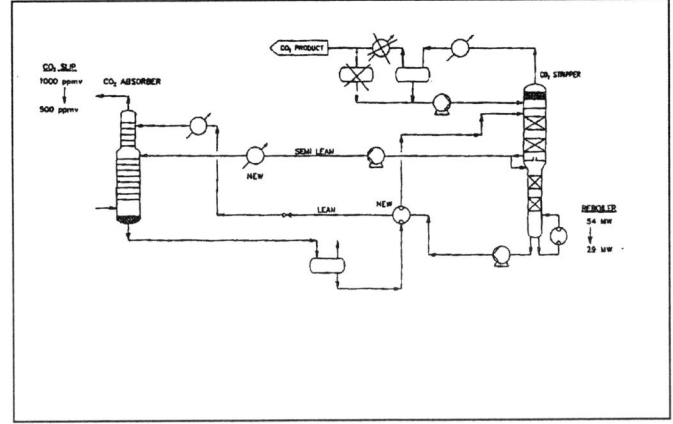

Figure 8. CO_2 removal unit.

Furthermore, due to the lower energy requirement, the temperature profile in the stripper is reduced. It became possible to modify CO_2 off-gas cooling by deleting a second overhead condenser. This caused a reduction in pressure drop across the cooling section and therefore less energy is required in the downstream plants. The CO_2 slip in the overhead from the absorber was decreased from 1,000 ppmv to 500 ppmv in the design, while in practice an even much lower CO_2 slip has been achieved. After the revamp the specific energy consumption has been reduced from 120 to 65 MJ/kmol.

Feed gas saturation

Although a lower steam to carbon ratio was applied, the reduced amount of reboiler energy requirements caused an excess LTS effluent waste heat. Moreover, for BFW preheat, only HTS effluent waste heat was applied, resulting in an even higher amount of excess LTS waste heat. This waste heat was applied for feed gas saturation. Figure 9 shows the design of the feed-gas saturation system.

When applying feed-gas saturation, process steam can be produced by using energy at a low-temperature level. This low-temperature energy is generally available in excess and can only be applied for boiler feed

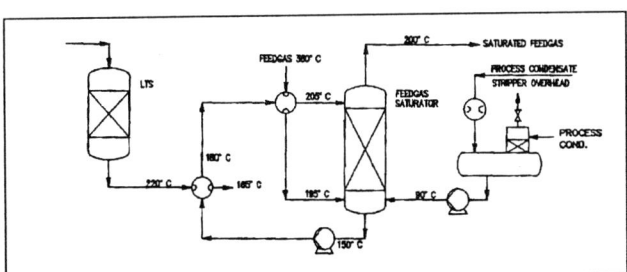

Figure 9. Feed-gas saturation system

water preheat or steam production. This design makes it possible to saturate the feed gas with 20 t/h of process steam. It can be said that by applying this technique, steam production is transferred from low pressure to medium pressure.

Besides the above mentioned energy savings, this system has another great advantage. Contaminated steam from the process condenser stripper was previously introduced to the CO_2 stripper. This has caused problems with gaseous and liquid effluent originating from this process condensate. The design of the feed-gas saturator system made it possible to apply both the contaminated steam from the process condensate stripper and some excess process condensate.

Environmental problems with the emission of methanol, amines and ammonia have been substantially reduced and liquid effluent treatment costs are reduced.

Purge gas conversion

Due to the reduced steam to carbon ratio, the amount of methane in the secondary reformer effluent was increased and also the shift conversion became less effective. Therefore, a larger amount of inerts was introduced to the synthesis loop. This resulted in an increase of the amount of purge gas from the loop and exceeded the maximum capacity of the existing cryogenic purge gas recovery unit.

Expansion of the cryogenic purge-gas recovery unit was only possible at considerable investment costs. It was therefore decided to develop a purge-gas ammonia loop. Figure 10 shows the design of this loop.

The purge-gas loop was designed to treat this purge and the purge gas from another ammonia plant at site. The introduction of this purge-gas loop, resulted in 45 MTPD of extra ammonia production and an additional high-pressure steam production of 8 t/h. Furthermore, by the extra steam production, the high-pressure boiler feed water economizers are better utilized, because of the higher mass flow.

With this purge-gas loop it is possible to transfer an excess amount of purge gas into ammonia product, and therefore improve the overall hydrogen and nitrogen efficiency.

Summary

Process design

The revamp resulted in the following process design of the front-end, which is shown in Figure 11.
The modifications in the front-end are:
• reduction of the steam to carbon ratio;

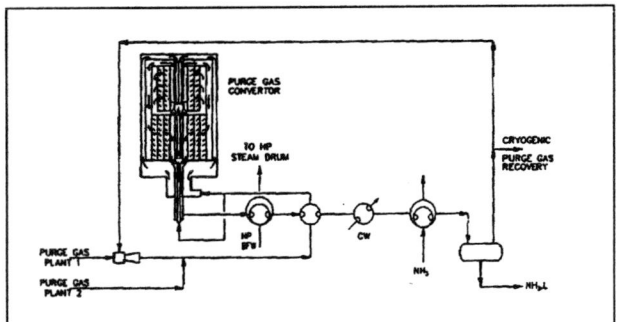

Figure 10. Purge gas conversion

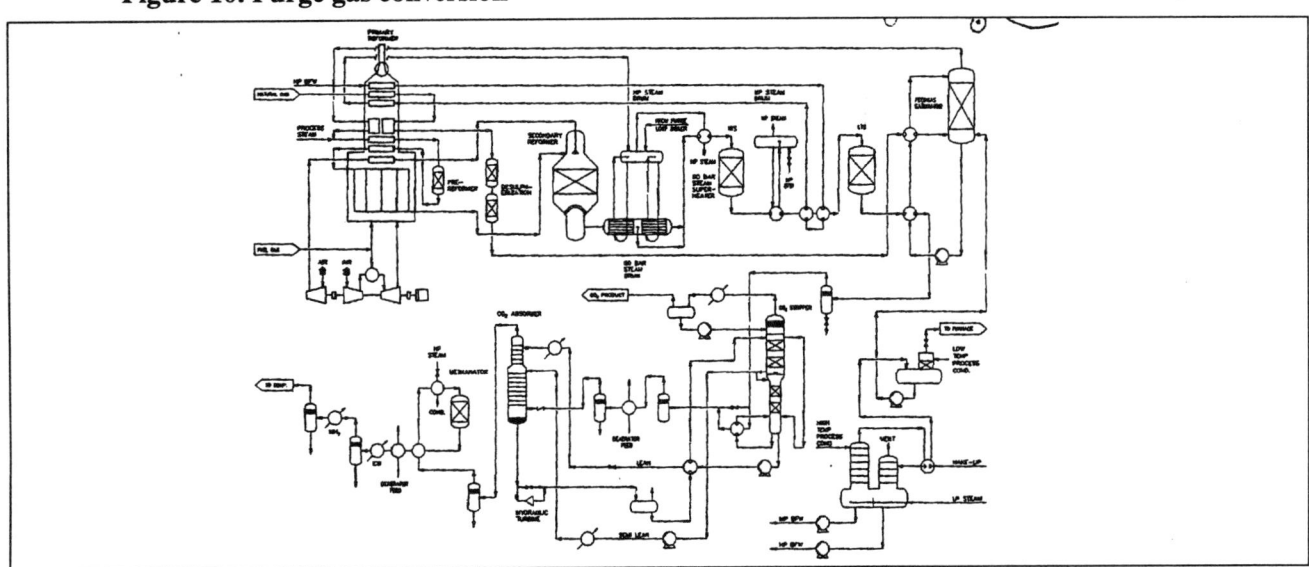

Figure 11. Front-end after revamp

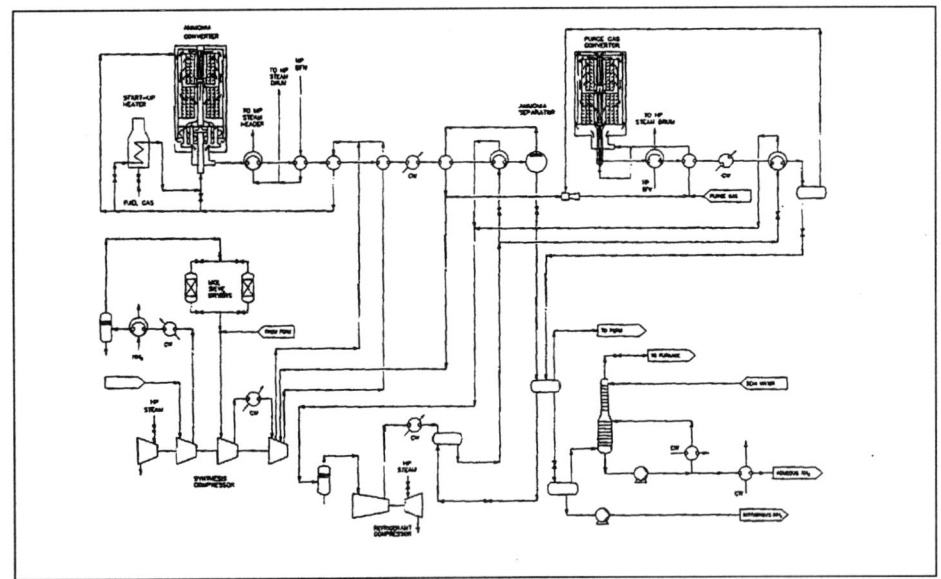

Figure 12. Synthesis loop after revamp

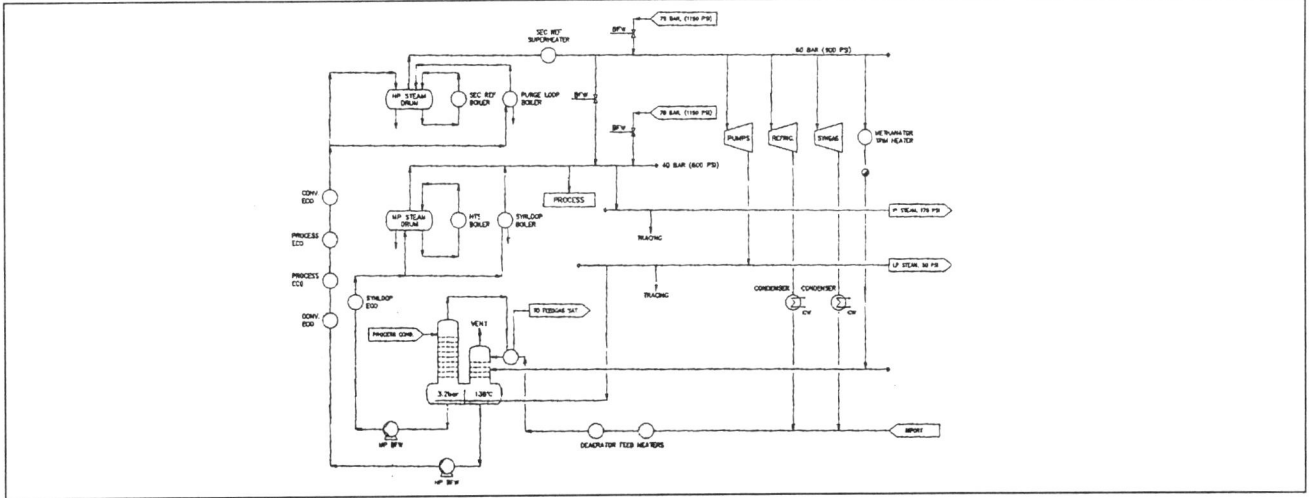

Figure 13. Steam system after revamp

- feed-gas saturation with process condensate;
- introduction of a pre-reformer unit, rearrangement of convection section coils;
- steam superheating with secondary reformer effluent;
- modification of HP boiler feed water preheat system; and
- solution swap to aMDEA, introduction semi-lean cooler and lean/rich exchanger, increased circulation flow, structured packing.

In the existing synthesis loop, no real changes have been introduced. Figure 12 shows the process flow diagram of the updated synthesis loop.

As can be concluded, an extra amount of ammonia and high-pressure steam is produced by introduction of this purge gas loop. Furthermore, an ammonia product degasification step was installed to improve the product quality.

The modifications to the steam system include:
- BFW pumps electric-motor driven;
- lube and seal oil pumps electrical driven;
- HP boiler feed-water economizer in process and convection section are operating in series;
- condensing methanator trim heater;
- removal of MP steam production in convection section;
- introduction of HP purge gas loop steam boiler; and
- Secondary reformer steam superheater.

Finally, Figure 13 shows the updated steam system in which several modifications have been made.

Both, the high-pressure and medium-pressure steam consumptions are more or less in balance with the produced amount of steam.

Revamp

The revamp of the ammonia production facility was carried out together with a regular turn-around and lasted six weeks. The plant came on-stream again in the spring of 1997.

The purge gas loop produces 15,000 tons of additional ammonia per year. The additional ammonia and steam production made the investment in this new convertor certainly worthwhile. The feed-gas saturation system causes a substantial reduction of gaseous effluent from the ammonia plant to the environment and reduces the cost for liquid effluent treatment.

The energy conservation achieved amounted to 8.6%, which reduced the original gap in energy consumption between the two ammonia plants from 11% to 2.4%. The ammonia plant now has an energy consumption figure close to that of ammonia plants built in the late 1980s and early 1990s.

Upgrading a Synloop for Capacity Increase and Energy Savings

Axial radial internals were installed in multivessel ammonia converters. One company performed the basic engineering for the loop revamping and supplied the technology for new internals for the converters. Another company performed detailed engineering for the whole project, which included other loop modifications and modifications to the front end of the ammonia plant.

Keith Wilson
PCS Nitrogen, Augusta, GA
Ermanno Filippi
Ammonia Casale, Lugano, Switzerland
Jim Gosnel
Brown & Root, Alhambra, CA

Introduction

PCS Nitrogen operates a nitrogen fertilizer complex at Augusta, GA. The ammonia unit was commissioned in 1978 with a nameplate capacity of 1,500 st/d. Designed by C. F. Braun & Co, since acquired by Brown & Root, it uses the Braun Purifier Process. The complex became part of Arcadian corporation in 1989. Arcadian was acquired by the Potash Corporation of Saskatchewan (PCS) in March 1997. PCS Nitrogen is the largest producer of nitrogen-based fertilizer and chemicals in the Western Hemisphere.

In the fall of 1996 the ammonia plant was restarted following the completion of a revamp project. A major part of the project was the installation of Ammonia Casale axial-radial internals in the multivessel ammonia converters. The plant in Augusta was the first to use the Casale concept in Brown & Root ammonia converters. The project was part of an expansion to 2,000 st/d. Details of the revamp of the balance of the plant are discussed in the article by Wilson Crowley and Bhakta. This article focuses on the debottleneck of the synthesis loop. It discusses the original design, the revamped design, and presents a comparison of the performance before and after the revamp.

Original design

Figure 1 shows the original flow scheme for the synloop. Since the loop makeup gas is cryogenically purified, it is dry and following compression is fed directly to the synthesis converters. The synthesis loop uses two adiabatic synthesis converters in series of Brown & Root design. All heat exchange is external to the converters. No quench is used.

Figure 2 shows more detail of the original flow scheme through the converters and associated exchangers, along with typical temperatures and the mode of control. The first converter effluent preheated first converter feed. After passing through the second converter, reaction heat was recovered by producing 100 bar steam in a forced-circulation boiler.

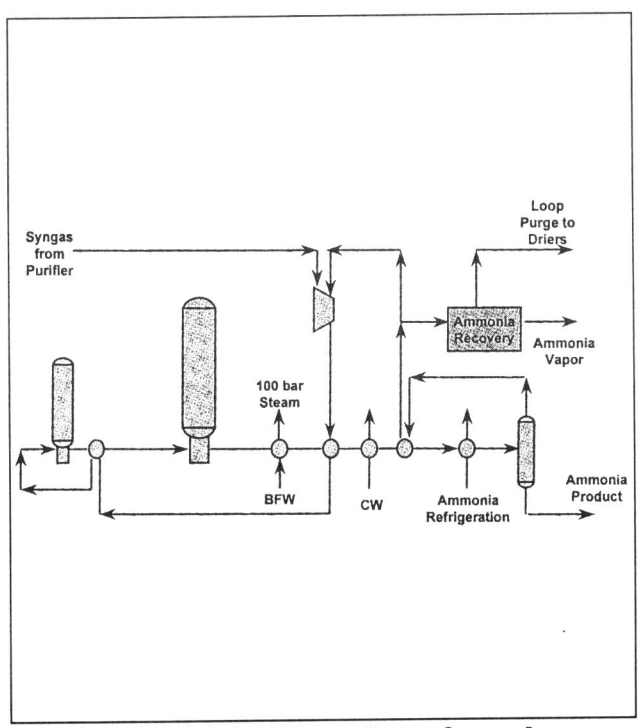

Figure 1. Original flow scheme for synloop.

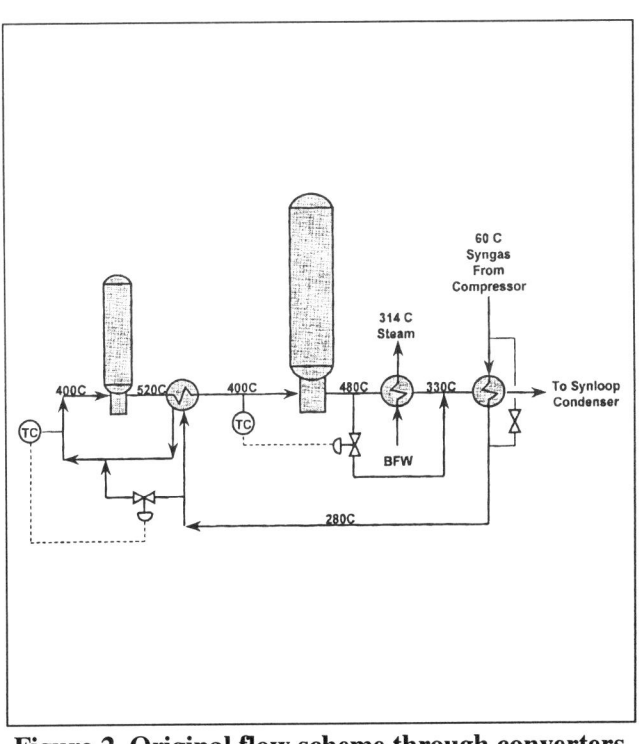

Figure 2. Original flow scheme through converters and exchangers.

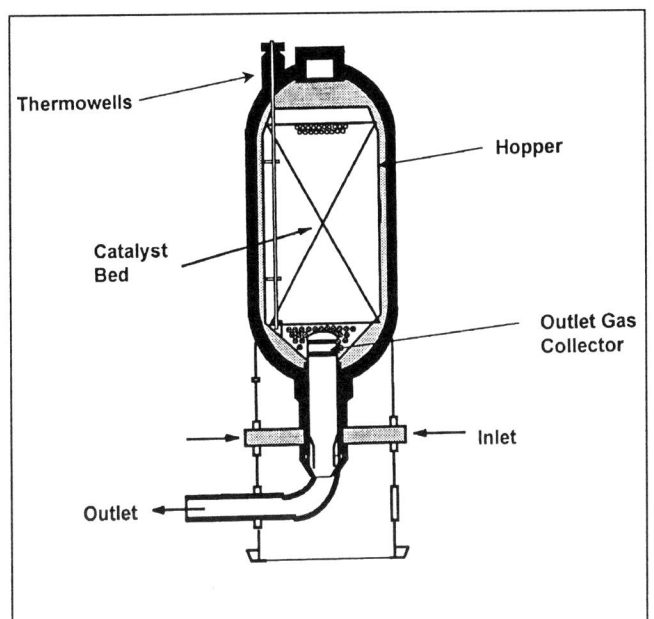

Figure 3. Original flow pattern through synthesis converters.

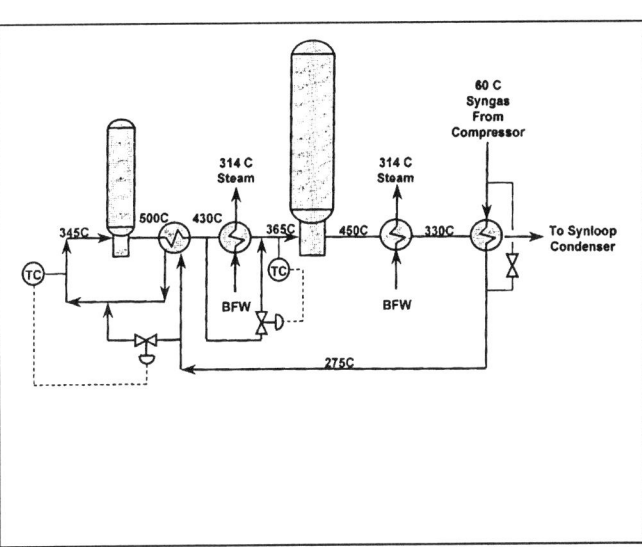

Figure 4. Revised flow scheme through converters and exchangers.

Figure 3 shows the flow pattern through the original synthesis converters. The converter internals were simple, consisting of a fixed catalyst hopper. There are no removable internals, thus only a manway on the vessel top for access. From the two inlets, gas flowed upward through the annular space between the catalyst hopper and the vessel wall. On reaching the top of the open catalyst bed, gas flowed down axially and exited through a gas collector at the bottom of the bed followed by a concentric outlet pipe. Except for the area around the gas collector, the flow was axial. The two synthesis converters were similar in design, although the second converter held a larger volume of catalyst. The catalyst size was 6–10 mm for the bulk of the beds. Around the gas collector, it was 12–20 mm.

By 1994, the original charge of synthesis converter catalyst was showing some signs of age. The charge had been in service for 15 years and some pressure drop increase was noted in the first converter, along with some deactivation at the top of the bed. The second converter temperature rise had increased, while the first converter temperature rise had decreased. This showed some shift in conversion from the first to second converters. Overall loop performance was still good. The loop could make up to 1,837 tpd, well above its 1,500 tpd design. PCS Nitrogen believed the catalyst would continue to perform well through the planned 1995 turnaround, but was concerned about possible operation through the next planned turnaround in 1998.

Scope of work

PCS Nitrogen approached both Ammonia Casale and Brown & Root about the opportunity to upgrade the synthesis converters while replacing catalyst. The vessels had never been entered before, and with catalyst life being 15–20 years, this was a rare opportunity. The goals were twofold: first, to save energy by increasing conversion, and second, to increase loop capacity to at least 2,000 tpd.

In 1994 PCS awarded contracts to, and took out licenses from both Casale and Brown & Root for their respective parts of the revamp project. The project was successfully executed by Casale, Brown & Root and PCS Nitrogen, all under PCS management. Casale provided the process design for the synloop, design and procurement of the converter internals, and supervision of the installation of the internals and catalyst loading. Brown & Root provided the design of a new high pressure steam generator, modifications to the existing interbed exchanger, and detailed design services for the loop including piping and foundations. PCS Nitrogen provided procurement, inspection, construction supervision, and installed the instrumentation.

Process Design Features

In order for PCS to obtain the expected benefits from this synthesis loop revamp, two key process design areas required special attention. These areas were the design of the new internals for the converters and the design of the converter inlet temperature control system. Each of these features and their benefits is discussed in this section of the article.

Converter internals

The proposed project called for the converters to be revamped to the Casale axial-radial pattern. There are several advantages that can be obtained from this modification such as:
• lower bed pressure drop as a result of the axial-radial flow pattern
• the use of smaller size, more active catalyst instead of the large size one, leading to an increase in conversion per pass
• better thermodynamic design when quench cooling is replaced with indirect heat exchange
• a higher installed catalyst volume due to the improved mechanical and fluid dynamic design.

All these advantages lead to a considerable energy saving and capacity increase. These advantages range from one million Btu per short ton to allowing for large increases in capacity, or some combination of both. The capacity increase results from the reduction in the circulating flow, about 30 percent, that in turn reduces the load on all loop equipment.

The most important feature of the modification to the converters was the use of the smaller, more active catalyst. Smaller sized catalyst is more active at a lower inlet temperature. With lower converter inlet temperatures an increase of the conversion per pass is possible. This results in a consequent reduction in loop circulating flow rate, which reduces the loop operating pressure and loop pressure drop. This benefit was key to meeting PCS's goals for the revamp.

Converter temperature control

In order to obtain full benefit of the low temperature activity of the smaller catalyst, a design was needed to meet the new, lower inlet converter temperatures. The existing process design could not do this. All three companies involved in the revamp recognized this need from the outset of this project. The reasons why the required low inlet temperatures could not be reached are subtle and explained in detail later in this paper. The lower operating temperature has the consequence that additional heat must be removed from the synthesis loop. This was done by generating high pressure steam in a new boiler. This addition and the better performance of the converters gave as a consequence also a considerable increase in the heat recovered, and therefore in the quantity of steam generated.

Revamped process design

As a result of the new axial-radial internals and temperature control requirement, a revised flow scheme was developed for the loop. This scheme is illustrated on Figure 4. The main differences are that inlet temperatures have been reduced by 55°C and 35°C to the two converters. Also, a new steam generator has been installed to cool the feed to the second synthesis converter.

Converter Revamp

This section of this article discusses the execution of the revamp project pertaining to the synthesis converters.

History of revamps

The revamping of the two converters has been done according to the *in situ* modification technology. This concept was introduced by Ammonia Casale in 1985 to revamp ammonia converters such as the standard Kellogg design for more capacity. Casale's first axial-radial revamp was placed on stream in 1986 for CF Industries.

Since this first installation, the Casale concept has been applied to different types of ammonia converters and also to shift and methanol reactors. It consists mainly in modifying the existing cartridge or transforming the existing axial bed into axial-radial beds. In some cases also the existing gas cooling system between beds is modified from quench to indirect heat exchange. The number of beds in the converter may also be modified.

The revamping of Braun-type converters was studied and proposed by Ammonia Casale immediately after the start up of the retrofitted Kellogg converters. It is a logical additional application of the proven Casale in situ retrofitting technology. But, its first industrial application took place only in 1996 with the PCS Nitrogen project.

Catalyst beds

For the PCS revamp, the configuration selected for the catalyst beds is typical of Casale axial-radial designs. With axial-radial distribution, most of the gas passes through the catalyst bed in a radial direction. The balance passes through a top layer of catalyst in an axial direction, thus eliminating the need for a top cover on the catalyst beds. This is illustrated in Figure 5. This feature is an essential factor for an easy and simple design of new converter internals.

To get a uniform gas flow through the bed some pressure drop must be provided across the perforated walls that distribute the incoming and receive the outgoing gas. The design provides for the correct pressure drop along the length of the catalyst bed. This is to ensure that the gas flow pattern is controlled by the design and it is not affected by possible non-uniform catalyst packing density.

A particular feature of the design is its simplicity. The catalyst-containing baskets are easily handled and have a low cost.

The axial-radial bed provides the best possible use of the catalyst volume available with the lowest possible total pressure drop. It is simple, flexible and can match most geometries and process conditions.

Mechanical aspects

The *in situ* modification of the two converters in PCS Nitrogen consisted in the removal of most of the existing internals and the installation of a new radial basket in each converter. The original internals that were removed are shown of Figure 3 and include the following:
- the vertical hopper wall containing the catalyst
- the outlet gas collector
- the thermowells
- the catalyst

The bottom pieces of the existing catalyst hopper and the exit nozzle were left in place. The new axial radial bed was then installed by introducing two cylindrical perforated walls. These walls are made in sectors to distribute the gas over the catalyst with the new flow pattern. The two walls consist of one external wall near the pressure shell, welded on the external edge of the existing hopper bottom, and one internal wall close to the converter axis. A sketch of the new radial basket is shown on Figure 6.

The new parts had to be introduced through the existing manhole, without modifying the existing high pressure shell. This meant the perforated outer wall had to be divided in to prefabricated panels to be assembled inside the converter by welding. Both perforated walls are then secured to the existing converter bottom by welding.

Of course no welding was allowed on the existing high pressure parts. The welding between new and old internals was done with a special procedure to take into consideration that the old parts were nitrided and permeated with hydrogen. The technology necessary for these mechanical aspects is the same that was developed and applied successfully for the revamping of the Kellogg-type converters.

Modifications to Synloop Heat Exchangers

This section of this article discusses the technical aspects, scope and execution of the detailed engineering part of the synthesis loop revamp project. In particular we discuss the need for the new steam generator for controlling the inlet temperature of the first converter.

Original heat exchanger system

Figure 2 depicts the original process scheme for the synthesis loop. There are two adiabatic ammonia converters separated by an interbed heat exchanger. There is also a steam generator downstream of the second ammonia converter followed by the synloop feed/effluent heat exchanger. The inlet temperature to both reactors was designed for about 400°C. Control of the inlet temperature to the first converter is through bypass of feed around the interbed exchanger. Second converter inlet temperature is controlled through bypassing its effluent around the steam generator.

Ideally the two reactors should be operated at the lowest possible inlet temperature that will still permit an outlet composition close to equilibrium, while maintaining maximum high pressure steam production. By operating this way, ammonia production and heat recovery are maximized while synthesis loop pressure is minimized. These more relaxed synthesis conditions have the disadvantage of requiring more catalyst volume as the kinetics of the synthesis reaction are slower at the lower operating temperatures. But this was not a problem for PCS as their original reactors were conservatively sized. Also, revamping the converter internals to radial flow allows the use of smaller particle size catalyst, which is more active and helps to overcome the slower kinetics at the reduced reactor inlet temperatures. To make the synloop revamp work the converter inlet temperatures needed to be lowered to about 350°C. As discussed in the paragraphs below, it was not possible to reduce the converter inlet temperatures to this level without sacrificing efficiency. A new exchanger is required.

Heat exchanger analysis

To understand why it was not possible to lower the converter inlet temperatures with the original equipment it is helpful to follow the reaction paths shown in

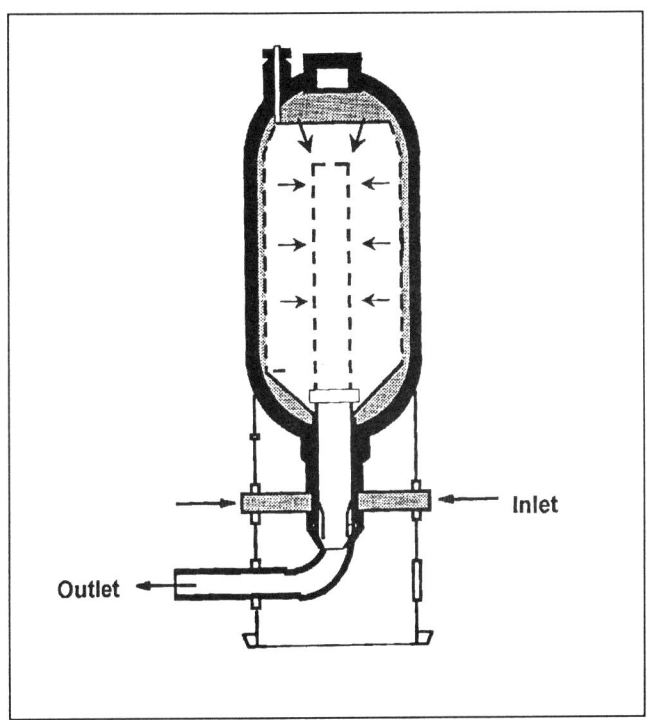

Figure 5. Revamped flow pattern through synthesis converters.

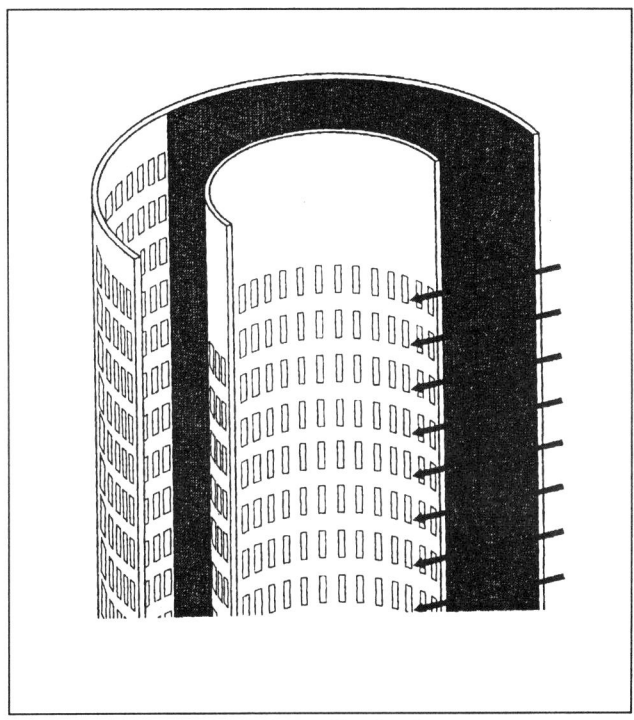

Figure 6. Radial baskets.

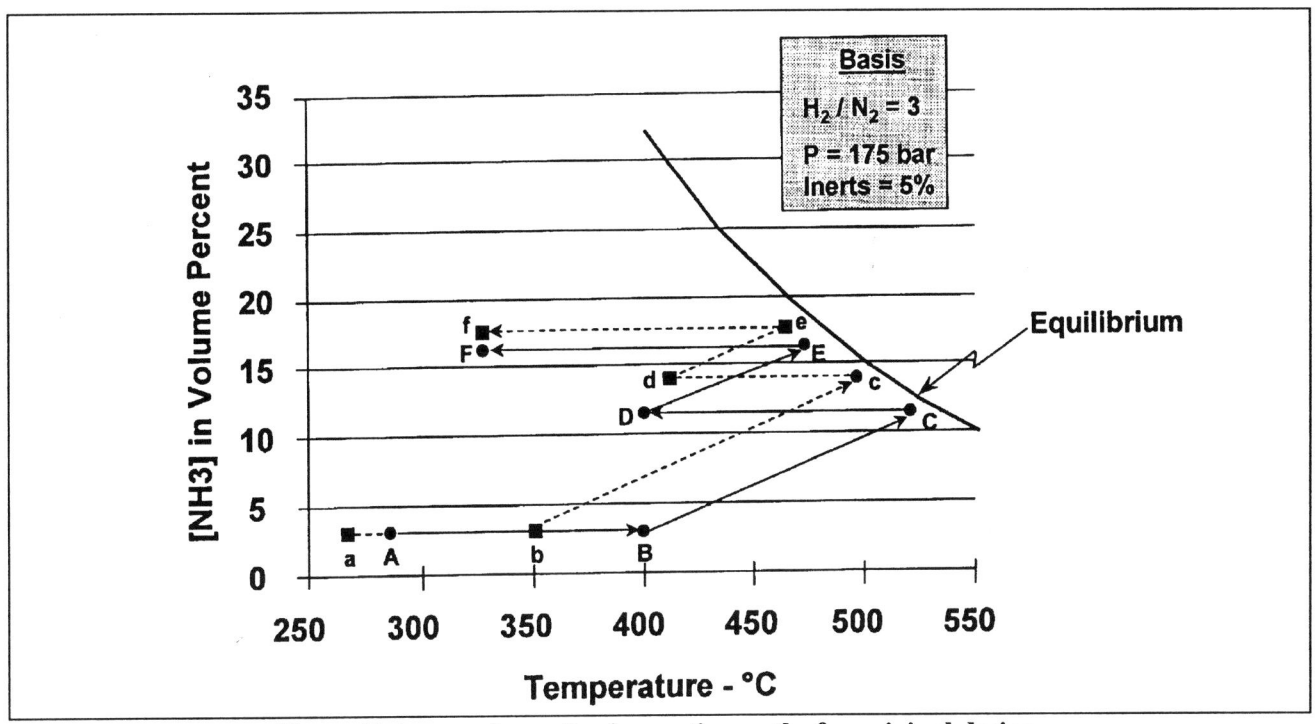

Figure 7. Ammonia synthesis reaction paths for original design.

Figure 7. This figure shows the equilibrium ammonia concentration at 175 bar-g as a function of temperature. It also shows the "reaction paths" that the synthesis gas follows as it is heated, reacted and cooled in the equipment shown in Figure 2.

Reaction paths for the original design are shown by the solid line. Feed to the interbed exchanger is heated from point 'A' at 280 C to point 'B' at 400°C. It then flows through the first converter where ammonia concentration increases from about three to twelve volume percent and the temperature rises adiabatically to about 520°C. First converter effluent is cooled by exchange with its feed in the interbed exchanger along path CD to 400°C. A key point is that the amount of this cooling is, of course, necessarily equal to the amount of heating of the feed in the interbed exchanger. In other words, the length of paths AB and CD must be equal. Thus the temperature decrease of the first converter effluent is equal to the temperature increase of the reactor feed. Both are about 120°C. The syngas then enters the second converter where ammonia concentration rises from twelve to about sixteen percent and the temperature rises adiabatically to about 480°C.

Now let's see what happens if we open the bypass around the interbed exchanger and close the bypass around the steam generator. This operation is illustrated on the dotted lines on Figure 7. Since the temperature of the saturated steam produced in the steam generator is 314°C, the lowest syngas outlet temperature we can hope for from this exchanger is 315°C. In this ideal case, the temperature of the feed to the interbed exchanger would fall to about 265°C. The temperature rise in the interbed exchanger, represented by path 'ab' will be 85°C, to give a first converter inlet temperature of 350°C. But now the cooling of first converter effluent is limited to a temperature range of 85°C, which will bring it down to only about 425°C along path 'cd'. The resulting ammonia production from the second converter is limited. The overall ammonia production rises only slightly as can be seen from the difference in final ammonia concentrations represented by points 'F' and 'f'.

We cannot efficiently reach the low inlet temperature to the second converter required to make the revamp work. This inlet temperature could be lowered at the expense of increased inlet temperature to the first converter. The lowest inlet temperatures attainable simultaneously for both converters are in the range of 380 to 390°C.

There is another way to lower the inlet temperatures further. The bypass around the synloop feed/effluent exchanger can be opened to further cool the feed to the shell side of the interbed exchanger. But opening this bypass will reduce high pressure steam generation and increase the amount of heat rejected to cooling water. As explained in a patent by Grotz, an efficient situation cannot be realized with PCS Nitrogen's original scheme. To achieve the capacity goal of the revamp project an additional heat exchanger is needed between the interbed exchanger and the second converter.

Revamped heat exchanger system

The benefit provided by the new exchanger can be understood by following the reaction path plotted on Figure 8. Feed to the first converter is heated in the interbed exchanger along path 'AB' from 290°C to 350°C. First converter effluent is cooled in the same exchanger along path 'CD' from 500°C to 440°C. The new synloop steam generator then cools this effluent further to about 370°C along path 'DE' shown by the dotted line. In this scheme the second converter can reach nineteen percent ammonia concentration as shown by point 'F'.

Engineering features

Brown & Root took Casale's process design for the synthesis loop and provided a thermal design for the new high pressure steam generator. This exchanger uses the Brown & Root forced circulation system and is illustrated on Figure 9. Boiler feed water is pumped through u-tubes where about 15 to 20 percent of it vaporizes. The hot syngas enters on the shell side and is cooled before it reaches the tube sheet. Because of the simplicity of the design a wide variety of vendors can fabricate this exchanger. Accordingly, bids were solicited from about ten vendors. A vendor was selected, the purchase order was placed and fabrication begun. Unfortunately during the fabrication the vendor

Table 1. Comparison of Synthesis Loop Parameters Before and After Revamp with Revamp Design

Parameter	Before Revamp	After Revamp	Design
Date	22 Sept. 1993	19 Feb. 1997	1994
Ammonia capacity	1482 mtd (1634 std)	1891 mtd (2085 std)	1814 mtd (2000 std)
Recycle discharge pressure	190 bar (2770 psig)	168 bar (2443 psig)	176 bar (2550 psig)
First converter inlet inerts	4.1%	4.0%	4.3%
First converter inlet ammonia	3.3%	2.8%	3.2%
First converter inlet temperature	397°C (747°F)	346°C (655°F)	353°C (668°F)
First converter temperature rise	100°C (180°F)	155°C (279°F)	155°C (278°F)
Second converter inlet ammonia	9.4%	12.6%	13.1%
Second converter inlet temp.	390°C (734°F)	364°C (687°F)	370°C (698°F)
Second converter temperature rise	93°C (167°F)	91°C (163°F)	85°C (153°F)
Second converter outlet ammonia	15.7%	19.1%	19.3%
Loop pressure drop	15.2 bar (220 psi)	11.1 bar (161 psi)	10.3 bar (149 psi)

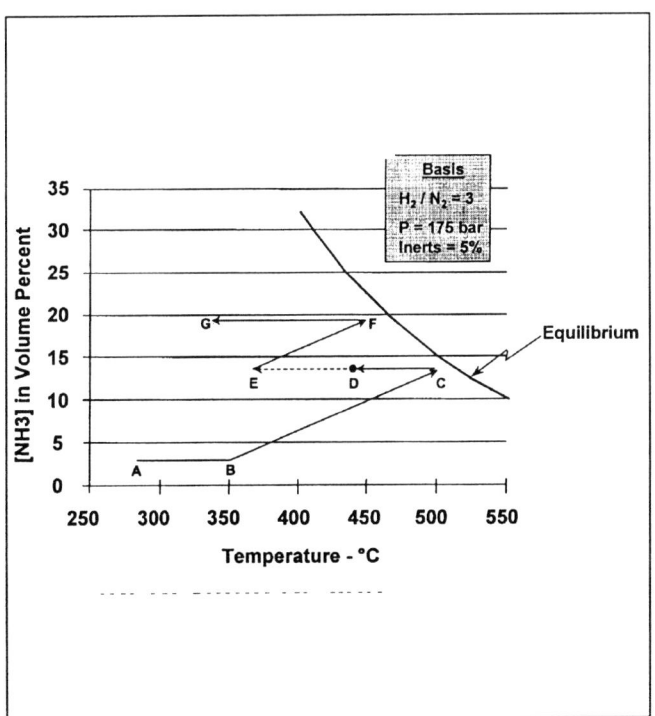

Figure 8. Ammonia synthesis reaction paths for revamp design.

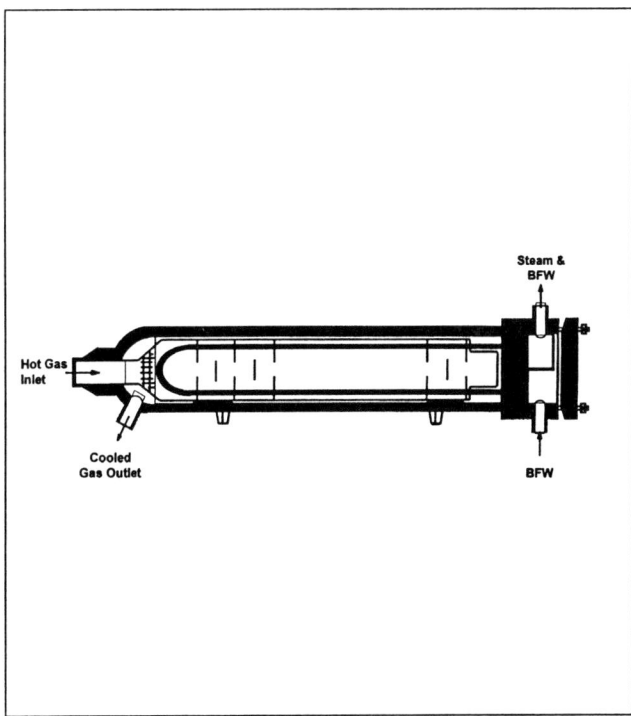

Figure 9. Synloop steam generator.

misdrilled the tube sheet and had to start over again. This caused a schedule delay in shipping the exchanger.

The reduced duty for the interbed exchanger caused a minor surprise during the engineering effort. When the final process conditions were established, we discovered that the amount of flow required through the shell side of this exchanger would have been very low. The majority of the flow had to be bypassed to control first converter inlet temperature at 350°C. As a result the temperature rise of the syngas flowing through the exchanger was so high that the outlet temperature would have exceeded the exchanger design temperature. To solve this problem the existing tube bundle was modified to reduce the heat transfer in the interbed exchanger.

Other engineering features were rather routine. The control valve in the bypass line around the original steam generator was replaced with a manual valve. It is used during startup to speed heat up of the synloop. New piping was designed for the runs between the new steam generator and the interbed exchanger and second converter. An external bypass is provided around the new steam generator to control the inlet temperature to the second converter. The existing control valve was reused in this new line. A foundation was designed for the new steam generator.

Other Synthesis Section Modifications

One of the pieces of equipment that might be expected to normally be modified in a synloop revamp is the syngas recycle compressor. The recycle compression load is provided by the last wheel on the syngas compressor high pressure case. This wheel was checked for the new flow and head conditions and it was determined that modifications were not needed. However there were two other pieces of equipment in the synthesis loop that did need to be modified. These were the following.

High pressure separator

The vanes in the high pressure ammonia separator were changed to add capacity. Although the vapor flow to the separator would be less than 10 percent above original 1500 stpd design, the lower loop pressure would result in higher velocities in the separator. Also the design liquid flow of ammonia would increase by about 33 percent over the original design rate. The vanes were replaced to add about 50 percent more area. The high pressure shell was not modified.

Refrigeration condenser

A shell was added to the ammonia refrigeration condenser. The existing condenser consisted of two shells in parallel. A third shell was added with provision to take one shell out of service at a time for cleaning. This allows for cleaning without having to wait for a maintenance turnaround. The reason we made this modification was that experience had shown that at high capacity on hot summer days, the ammonia refrigeration condenser was a bottleneck.

Performance Results

The ammonia unit was restarted after the outage in early October 1996. Pre-reduced catalyst was used in the first converter, so the catalyst reduction went quickly. As soon as the first converter catalyst was reduced, the plant was able to make original nameplate capacity of 1500 st/d. As the reduction of the second converter proceeded, the synloop settled out at a very low pressure. This was evidence of the improved conversion in the reactors.

As the front end of the plant was lined out, rates were gradually increased. The unit averaged over 2000 st/d for the month of November 1996. Cold weather allowed the plant to reach 2100 st/d for brief periods. In February 1997, a set of plant data was taken to evaluate the performance of the entire plant. The production rate was 2085 st/d. Table 1 compares before and after plant data with the 2000 st/d design.

The synloop performance is excellent. At rates above design, conversion was near design at pressures below design. The synconverters met all performance guarantees, and operating conditions suggest the converters and loop can handle additional capacity without further modification.

Literature Cited

Grotz, Bernard, Process for Synthesizing Ammonia," assigned to Santa Fe Braun Inc, Alhambra, CA, U.S. Patent Number 4,867,959.

Wilson, Keith, Michael Crowley, and Mukund Bhakta, "Maximizing Ammonia Production," *Ammonia Plant Safety & Related Facilities*, Volume 39, AIChE, New York (1999).

Experiences with Heavy Fuel-Oil Firing in a Steam Reformer

For economic reasons, the fuel used for firing a primary reformer was changed to heavy fuel oil. The paper summarizes the design modifications required for the burners, the reformer tubes, and the convection bank and reviews the problems the owner encountered during 7 years of operation.

Catela Pequeno
GDL Plant Division, Lisbon, Portugal
Manfred Severin
KRUPP UHDE Fertilizer Division

Introduction

Objective

The primary object of this article is to report on the experience gained during 7 years of operation of a top-fired primary reformer using heavy fuel oil. Experiences using this kind of fuel are very rare because very few installations worldwide are prepared to use it. In addition, some information on the background of the plant on the plant modifications carried out will be given.

More details concerning the plant modifications are available in a separate article.

Plant history

GDL was founded in 1961 as a combined ammonia and town gas producer. A second ammonia plant using the steam reforming process was constructed in 1968. This plant was operated with naphtha as feed and fuel. The plant was originally designed for the production of 550 t/d of ammonia and 45,000 nm³/d of town gas.

For economic reasons, the fuel for the reformer was changed from naphtha to fuel oil in 1987. At that time, the price of fuel oil dropped to 40% of the naphtha price (based on the lower heating value). At the same time, the plant capacity was reduced to 55%. Shortly afterwards, the production of ammonia was discontinued for economic reasons, and the plant was used to produce town gas only.

From 1987 to 1997, the plant was operated with fuel-oil firing for a total period of 51,500 h. The first 4 years of this period were used to test the plant performance with heavy fuel-oil firing (the plant was not run continuously for other reasons). Steady operation was achieved from 1991 to 1997 (Figure 1).

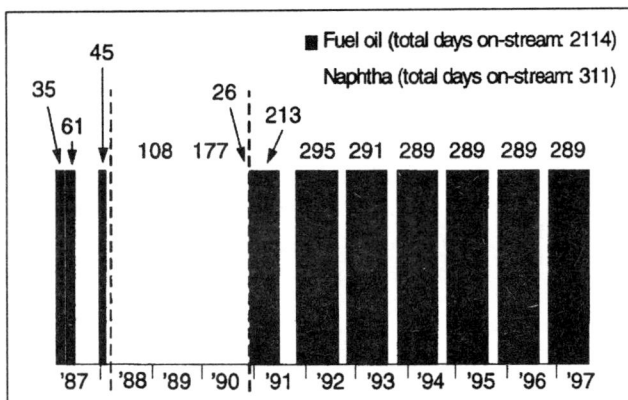

Figure 1. On-stream figures 1987–1997.

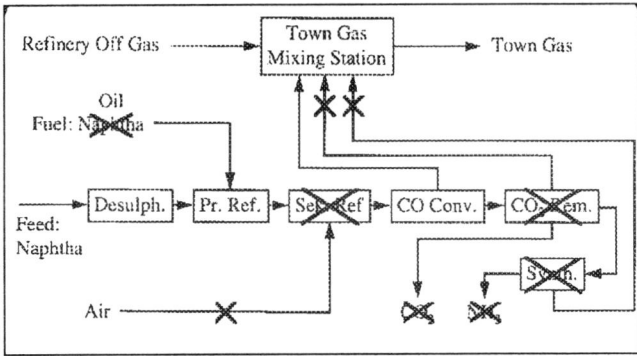

Figure 2. Simplified block scheme.

Plant description

The overall plant changes are shown in the simplified block diagram shown in Figure 2.

The conventional ammonia route was modified so that the fuel was changed from naphtha to oil, while the secondary reformer was no longer used and the CO_2 removal and ammonia synthesis sections were closed down.

The reformed gas produced was transferred to the town gas mixing station.

Oil characteristics

The typical oil characteristics are shown in Figure 3. The main impurities responsible for the expected corrosion problems were sodium, vanadium, and sulfur.

Analysis	C (mass %)	84 - 86
	H (mass %)	11 - 12
	Na + K (ppm)	59 - 108
	V (ppm)	36 - 130
	S (mass %)	1.5 - 3.5
	Ash (mass %)	0.01 - 0.03
	BS + water (vol. %)	0.05
Density	at 15 °C (kg/m³)	0.9641
Kin. viscosity	at 37.8 °C (cSt)	825
	at 50.0 °C (cSt)	341
Pour point	(°C)	6
Flash point	(°C)	97
Net calorific value	kcal/kg	9485

Figure 3. Typical fuel oil characteristics.

Plant modifications and new equipment

Changing the fuel from naphtha to oil required a number of major modifications of the existing equipment. The following modifications were carried out from August to November 1986 (Figure 4):

Desalination unit. The unit was installed to remove or reduce Na + K and other water-soluble components from the fuel oil. It was expected to reduce these elements to below 1 ppm.

To inhibit the formation of vanadium pentoxide, a dosage system for injecting a magnesium-based oil-soluble inhibitor was also provided.

Modifications of the reformer section. Most of the modifications were required in the reformer section (Figure 5).

Radiant tubes. The radiant tubes were replaced by

- Installation of an oil desalination unit
- Use of a corrosion inhibitor for the fuel oil
- Installation of bimetallic reformer tubes
- Installation of an oil circuit system
- Provision of new burner lances
- Installation of soot blowers in the convection bank
- Installation of a flame scanner and interlocks for each burner

Figure 4. Measures taken for heavy fuel oil firing.

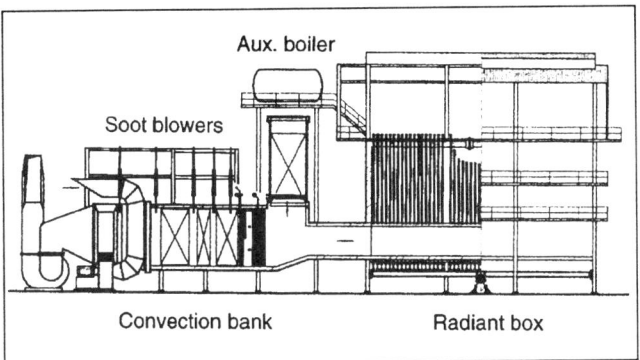

Figure 5. Steam reformer layout.

Figure 6. Refractory lined outlet manifold.

bimetallic tubes. HP-modified material was selected for the inner layer because of the requirement regarding creep resistance. The minimum required wall thickness was 7 mm for a service life of 100,000 h. To provide corrosion resistance, 50 Cr/50 Ni material was chosen for the outer layer. This resulted in a total minimum sound wall thickness of 14 mm. The corrosion layer was estimated for a minimum service life of 75,000 h based on a corrosion rate of 0.75 mm per 10,000 h.

The bimetallic tubes were very expensive, namely about US $14,000 each.

2 tubes were installed in material 45 Ni/35 Cr + Ti + Add. ("Manaurite XTM") for test reasons.

The reduced capacity of the plant required 104 reformer tubes only, instead of the 208 tubes installed originally. Therefore, only every second tube was replaced. Figure 6 shows the tube arrangement and the outlet manifold.

Burners

The burner housing and burner muffle block could be reused. Two separate interchangeable burner lances were provided, one for oil and the other one for naphtha. A pilot gas burner lance was also installed for each burner.

The fuel oil was atomized with steam while the naphtha was atomized with air.

Because of the reduced plant load, the number of burners was reduced from 84 to 42.

Each burner was equipped with an ultraviolet (UV) scanner. To avoid faulty operation of the burners during ignition and shutdown, a mechanical interlocking device forcing the proper sequence of valve actuation was installed.

The heat emission per burner was about 0.6 MW. Figure 7 shows the interlocking system for one burner.

Convection bank

Heat exchangers. Because of the reduced plant capacity, the three high-pressure (HP) steam superheaters had to be removed. Instead, only two new HP steam superheaters and one medium-pressure (MP) steam superheater (for process steam) were installed (Figure 5).

The process air preheater was used to heat up a small amount of MP steam. The BFW preheater and combustion air preheater remained unchanged.

In order to increase the wall temperature of the cold part of the combustion air preheater, two steam- and CO_2-heated air coils were installed upstream of the air preheater.

Fortunately, all heat exchangers in the convection bank were of bare tube design rather than finned and replacement was not necessary.

Figure 7. Burner interlocking device.

Figure 8. Soot blower arrangement.

Cleaning system

For the removal of soot deposits on the heating surfaces, a soot blower system was installed. The system consisted of 6 pairs of retractable rotary blowers. Each pair was installed between 2 heat exchangers. The blowers were installed in the flue-gas duct from the top or through the side walls. MP steam was used for cleaning.

The blowers were intended for intermittent operation: the operator selected the soot blowers to be used at any particular time.

The arrangement of the blowers is shown in the photograph of Figure 8.

Total investments

The total investment for the modifications was $ 9,300,000. The payout time for only the oil firing investments was calculated to be less than 2.7 years. The total payout time, including some other plant modifications, was estimated to be 2.1 years.

Emissions

At the time the project was executed, no environmental aspects had to be considered. Therefore a study of the emissions was not carried out.

Operation

During the years of operation with fuel oil, GDL experienced several problems. The most difficult problems to control were the knife-edge balance between corrosion and fouling of the radiant section tubes and the corrosion of the coils at the cold end of the flue gas duct.

A review of the problems we encountered will now be given.

Desalination unit

Process. The desalination unit used was of a conventional type, commonly found aboard vessels to clean fuel and lubricating oil (Figure 9).

In this unit, the oil was first mixed with water in a stirred vessel. The oil was sent from the stirred

Figure 9. Desalination unit.

Contaminant	Before	After
Na, ppm	22	< 1
K, ppm	3	3
V, ppm	110	88
Density (15°C), kg/m³	0.9861	0.9847

Figure 10. Fuel oil contaminants before and after desalination.

vessel to a centrifuge where it was separated from the water. The process was repeated in a second set of equipment. Sludge and metals tended to remain in the inorganic phase. The clean oil was sent to a buffer tank, from where it was pumped to the burner feed circuit.

Additives used. To make the water/oil separation easier, a deemulsifier was added to the oil upstream of the first stirred tank.

In order to control corrosion, a second additive was introduced upstream of the buffer. This additive was an organic magnesium compound that should be added in controlled quantities depending on the vanadium content of the oil. The fact that this additive was introduced on the basis of an average oil flow led to periodic overdosing and underdosing of the product. This had an adverse effect on the additive's performance.

Effluent. The operation of the desalination unit was not difficult to control, except for the fact that the density of the fuel obtainable in Portugal tends to be higher than expected. Since the operation of the centrifuges was based on the difference in density between oil and water, this tendency became a problem we had to deal with, in spite of the flexibility and adjustment capacity of the machines. The density of the oil occasionally came so close to that of the water that it was almost impossible to separate the two. In any case, the increase in the oil density reduced the capacity of the centrifuges to a considerable extent.

Sudden changes of the oil density produced problems with the effluent from the unit. Either too much oil slipped into the water outlet or too much water went to the buffer with the oil. The water was purged from this vessel periodically.

The average analysis of treated oil is shown in Figure 10.

The effluent water was loaded with sludge and organic matter. Therefore a treatment was required before disposal.

Radiant section

Reformer tubes. GDL was faced with two opposing but related problems while operating the steam reformer with fuel oil: The fouling of the radiant section tubes caused by an excess of corrosion inhibitor additives vs. the protection of these tubes against corrosion.

Initially, the inhibitor dosage rate was determined

Figure 11. Fouling of reformer tubes (1).

Figure 12. Fouling of reformer tubes (1).

by the vanadium content (Mg/V ratio of 3), as recommended by the supplier. This dosage rate was quite high, however, and the result was the deposition of a heavy layer of magnesium orthovanadate, a yellowish powder easily removable by brushing, but adversely affecting heat transfer. Within two or three weeks, the layer was so thick that the temperatures required to transfer the necessary heat to the tubes became too high. Furthermore, the powder prevented us from measuring the real tube wall temperatures. We were forced to shut down and clean the tubes.

It was decided to start the reformer using a very small quantity of additive and to gradually increase this quantity later. The intention was to determine the maximum amount of additive that could be injected without extensive formation of powder on the tube surfaces (Figure 11).

We eventually found a dosage rate that proved to be efficient in preventing extensive corrosion and that did not cause extensive fouling of the tubes (Mg/V ratio of 0.5). As indicated earlier, the dosage of the inhibitor was critical. Too much product led to fouling of the top section of the tubes inside the fire box, which was impossible to

Component	Content, %	Component	Content, %
C	0.03	Cr	20.80
S	0.34	V	4.93
Na	0.01	Fe	0.37
Mg	0.33	Mn	0.08
Ca	0.09	Zn	0.18
K	0.02	Al	0.03
Ni	26.30		

Figure 13. Composition of metallic layer on reformer tubes.

Figure 14. Samples of a reformer tube.

clean without shutting down. The absence of inhibitor, however, does not lead to self-cleaning of the excess of yellow product on the top of the tubes, but rather to rapid corrosion of the lower section, usually less evenly covered.

Operation without additive led to the immediate increase of the corrosion rate. At the end of a 10-month operating period with a low inhibitor dosage rate (Mg/V ratio of 0.07), the tubes were covered by a thin metallic layer that was hard and difficult to brush off and which formed bubbles on the surface of the tube. These bubbles ruptured when pressed, leaving a clean tube surface underneath (Figure 12).

An analysis of this metallic layer, shown in Figure 13, proved that it resulted from corrosion attack on the tubes, since it contained a high percentage of Ni and Cr.

GDL operated on a 10-month campaign schedule, followed by a maintenance period of 2 months. Each year, the tubes had to be carefully hand-brushed in order to remove the yellow powder. The use of mechanical means was not possible because of the risk of tube vibration and catalyst compacting.

With the correct level of inhibitor and careful annual cleaning, GDL was able to operate the reformer for 7 years without major problems.

A metallographic analysis was performed on one tube that was removed due to slight bending and deep localized corrosion. This tube was characterized by four different areas of external corrosion (Figure 14): uniform corrosion along the whole tube length, deep corrosion in the heat-affected zone of a weld, deep corrosion in the base material, and deep corrosion in the weld material.

The metallographic examination indicated minor uniform corrosion on sample one only. The measured wall thickness without the deposit layer was still 15.2 mm. The minimum original thickness was supposed to be 16.2 mm, including porosity.

The depth of the corrosion pitting on samples two and three was found to be 2.4 mm. This corrosion occurred because of insufficient Cr-Ni material in the outer layer. This is clearly visible in Figure 15, where the outer layer tends towards zero mm thickness on the left. The dilution of the material was obviously the result of imperfect fabrication.

The same effect was found on sample four, where Fe-rich material was discovered in the weld material. This signified that an electrode containing iron must have been used for the final covering layer.

The ideal structure of the tube wall is shown in Figure 16.

The two experimental XTM tubes installed in the reformer for testing the new alloy resisted corrosion under the severe conditions inside the fire box as efficiently as the bimetallic tubes. As opposed to the bimetallic tubes, localized corrosion was never spotted in these tubes. The XTM tubes are less expensive than the bimetallic ones.

Burners. GDL experienced no particular or unexpected problems with the installed oil burners. Adequate cleaning routines had to be implemented, which required more operator time in the penthouse.

The cold startup was performed with naphtha lances. The lances were changed over, one by one, to oil after the flue gas temperature reached 650°C. Hot startup was made directly with the oil lances installed.

In spite of the duplication of circuitry at the top of the reformer, there were no problems. The mechanical interlocking of the steam, purge and oil valves on each burner operated very well (Figure 7)

Convection section

The convection section of the furnace could be divided into two parts on the basis of the severity

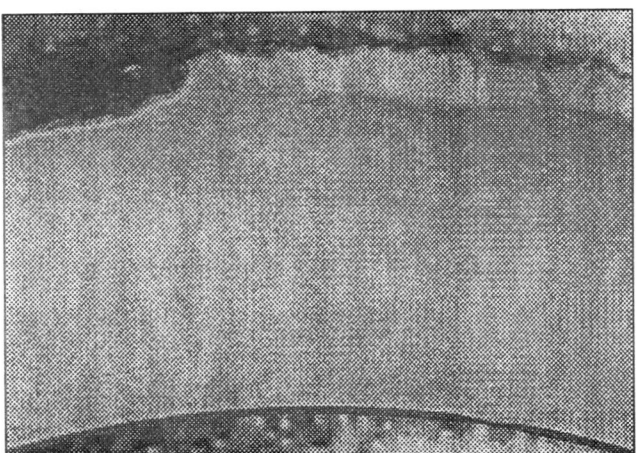

Figure 15. Reformer tube: thin 50/50 layer.

Figure 16. Reformer tube: ideal structure.

and type of problems found.

The soot blowers were of great importance for maintaining the working conditions. As can be seen from Figures 5 and 8, the blowers were evenly distributed across the length of the flue gas duct.

Soot blowers. The soot blowers were operated once per week. The cleaning effect was satisfactory for the hot section but not for the cold section.

Hot section. The hot section of the flue-gas duct can be considered to extend up to the former process air preheater. This was in much better shape than the cold end.

The deposition of magnesium orthovanadate on the coils was intense but effectively controlled by the use of steam blowers (Figure 17). Immediately after shutdown, a very thin and even yellow powder layer covered the coils.

The exception was the first two or three rows of tubes of the first steam superheater. These rows were not effectively cleaned by the soot blowers. Greater fouling could therefore be observed, causing a decrease in the heat-transfer efficiency, but this was not problematic (Figure 18).

Figure 17. Fouling of steam superheater, "hot" end.

Figure 18. Fouling of steam superheater, "cold" end.

Figure 19. Fouling of BFW preheater.

Cold section and fan. Two heat exchangers are provided in the cold section of the flue-gas duct: the BFW preheater and the combustion air preheater.

The BFW preheater was covered with a large amount of soot. This soot already contained a lot of moisture, but was still very hygroscopic. If not immediately cleaned, a sticky and strongly acid gum would result (Figure 19).

The combustion air preheater was generally in the worse condition. This heat exchanger has two passes on the air side with two soot blowers located in the middle (Figure 5). The second pass, upstream of the flue-gas duct and therefore receiving gases at a higher temperature, was heavily covered with sticky soot. The first pass, at the very end of the convection section, was usually partially blocked with debris and soot, all cemented together by the same sort of gum mentioned before (Figure 20). The tubes at the bottom of the duct were always in a poorer condition than those at the top.

It was common to have badly damaged tubes, which had to be removed and replaced (Figure 21).

The fan usually showed corrosion on the inlet vanes and outlet duct (Figure 22). An agglomeration of ash was usually found at the bottom of the fan case.

The main factor explaining the poor condition of the cold section of the flue-gas duct, each year, was the fact that the duct was too large considering the gases being produced. It must be kept in mind that the maximum plant capacity was reduced to one-half of design capacity and that, during most months of the year, the plant was operating well below that level (about 1/4 to 1/3 of the original capacity). The low velocities implied in this situation favor the deposition on the coils and the stratification of the gases in the cross-section of the duct. We were able to measure a temperature difference of the gases of 30° to 40°C, between the bottom and the top of the duct, immediately downstream of the BFW preheater.

Since the average temperature of the gases downstream of the fan was around 180°, the temperature of the gases at the bottom of the first and second pass of the air preheater was certainly far below this average. It was not surprising, though, that the lower tubes suffered most.

The failure of some tubes allowed cold air to enter the flue-gas duct directly, lowering the temperature of the gases, causing condensation of H_2SO_4, and enhancing the corrosion attack. The corrosion attack was, therefore, enhanced precisely where the situation was already worst—the lower and colder tubes at the beginning of the

Figure 20. Fouling of combustion air preheater.

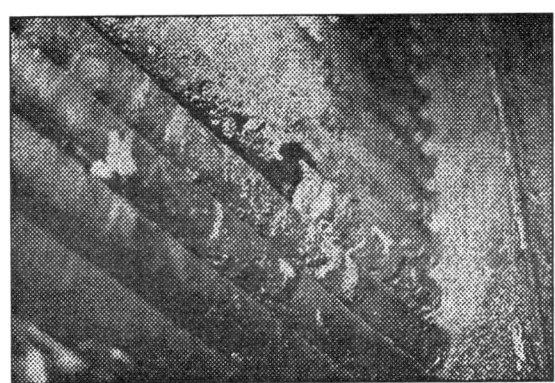

Figure 21. Damaged tubes of combustion air preheater.

Figure 22. Corrosion of fan duct.

first pass of the heat exchanger. This led to an increasing corrosion rate.

During shutdowns, these heat exchangers were subject to special cleaning measures. GDL first tried to clean the tubes by brushing, but this method was clearly inefficient, as all the soot was converted to acid gum two days after the doors of the flue duct were opened.

We then adopted the procedure of cleaning the coils chemically. This action should be taken as soon as possible after shutdown using a slightly caustic solution. This method was much more effective, but required special measures to protect the refractory lining of the duct. The effluent was full of soot and had a low pH, requiring treatment before disposal.

Emissions. Because the legislation in Portugal concerning pollution was not very strict, emissions such as NOX and CO in the flue gases were never evaluated. Occasional analysis showed a lot of excess air and, sometimes, poor burning could be observed through the peepholes in the fire box.

The worst problem, however, was the emission of soot via the stack, the effects of which were clearly visible all around the plant site. The average soot pH was 2.5, this being due to the fact that it contained a large portion of sulfur. As a consequence, metallic surfaces exposed to this soot suffered heavy corrosion.

There were several possible courses of action to overcome this problem, but, for economic reasons, the alternative of burning 1% S fuel oil was immediately discarded. There was also no space, and in any case it would have been very expensive to install a scrubber. Therefore, GDL tried to correct the pH of the soot with additives and keep the temperature of the flue gas in the stack above 180°C.

The additive used was a stabilized water suspension of magnesia. The suspension was atomized with the aid of steam directly inside the duct. The product was not easy to handle. A special pump and two injection lances had to be acquired. The lances often blocked and had to be removed for cleaning.

The results were not satisfactory. The pH of the ashes increased a little, but not sufficiently and the product finally fouled the coils downstream of the injection points.

Concerning the temperature upstream of the stack, the fact that we were operating at a very low load made the objective very difficult to achieve.

GDL could never attain positive results with regard to the soot emission. The problem was not be resolved until the plant was revamped to natural gas, in 1997.

Conclusion

The operation of the plant with heavy fuel oil firing in the primary reformer was very economical for GDL. However, the company faced several difficulties, mainly corrosion and environmental problems.

The corrosion effects on the reformer tubes and the cold part of the convection bank must be considered normal for firing this kind of fuel. The problems could be solved by cleaning. Some replacements were required in the cold part of the convection bank.

From an environmental point of view, it was not possible to find a satisfactory solution of the problem of soot emission via the stack.

Revamping Urea Plants For Large Capacity Increase

An overview is presented of the Urea Casale HEC and VRS concepts and their application for urea plant upgrading.

K. Clayton and B. Summerscales
Agrium Inc., Canada
F. Zardi
Urea Casale S. A., CH-6900, Lugano Switzerland

Introduction

Urea Casale has been active in the urea field since 1985.

From the very beginning, the activity of the company has been concentrated on revamping existing urea plants.

Currently, with a constantly increasing urea demand and economical uncertainty, there is, in fact, a high demand for plant upgrading which can give additional capacity at a lower cost through new plant construction.

Thanks to a team of very skillful people, most of them with a long experience in the urea field, Urea Casale developed several innovative and very competitive technologies to revamp urea plants to achieve:

- large capacity increases;
- energy savings;
- pollution control; and
- improvement in plant reliability.

Among these technologies we have:

- new reactor trays to reduce steam consumption and increase capacity; and
- new urea production processes (HEC, VRS) for drastic capacity increase.

The general approach of Urea Casale to urea plant revamping is to upgrade the reaction section in order to increase its efficiency, rather than just adding additional equipment, and most of the technologies have been developed to accomplish this goal.

Urea Casale revamping technologies can be applied to plants originally designed according to any kind of urea process.

The competitiveness and the success of Urea Casale revamping technologies is proven by the fact that, in the last ten years, 50 urea plants, with capacities ranging from 250 to 2,400 MTD, have been or are being revamped utilizing these technologies. Of these plants, 70% were originally designed according to stripping technologies.

Industrially Proven Technologies Available For Large Capacity Increase in Urea Plants

Casale general approach to revamping

Every revamping project has to start with the identification of client's goals and of the actual plant bottlenecks.

After this first phase, Urea Casale proposes the technical solution which it considers will reach the required goals with the best return. Urea Casale generally proposes the best combination of its technologies and of third party technologies available to it.

The general revamping philosophy as listed here below, however, is followed for every project:
- always try to upgrade the plant with new technologies;
- maximize efficiency of synthesis section;
- minimize plant shutdown;
- minimize modification to the existing plant; and
- be as simple as possible.

In case of large capacity increases, the implementation of small changes, like the introduction of high efficiency trays in the reactor, would, however, not increase the efficiency of the synthesis section enough to completely fulfil the above points and keep the investment as low as possible.

The following of the above concepts, therefore, require, in order to obtain large capacity increases, the development of new revamping approaches with an even more drastic upgrade of the synthesis section which could guarantee to minimize the addition of HP equipment, and by consequence, the investment as well.

For this purpose, the following two new technologies have been developed to obtain large capacity increases when revamping urea plants:
- the vapor recycle system (VRS) for the stripping plants revamping; and
- the high efficiency combined (HEC) for the conventional total recycle plants revamping.

Both of the above technologies will be described in the next sections.

Large capacity increase for conventional total recycle plants: Casale HEC process

Urea Casale has recently developed a new urea technology named High Efficiency Combined Urea Process. This process, based on the combination of a very efficient "oncethrough" reactor and a conventional total recycle one, presents the unique feature of having a very high average CO_2 conversion, and by consequence, a lower energy consumption than any conventional total recycle plant.

Thanks to the above features, this new concept can be very conveniently applied to the revamp of existing conventional total recycle plants for large capacity increases, achieving the following:
- capacity increase by 50% or more;
- energy consumption reduction;
- minimum investment;
- minimum modification to the existing plant; and
- minimum shutdown time.

The main concept of the HEC process is to obtain most of the urea product in a "once-through" reaction section. In the absence of recycle water, the conversion of carbamate to urea is favored and a high conversion of CO_2 to urea in single pass (75–80%) is obtained. The small amount of residual carbamate is decomposed, condensed and recycled as aqueous solution to a second reaction section (operating at lower pressure) which converts it to urea at a lower conversion efficiency (typically 55).

Now, by feeding all the fresh reactants to the high pressure reactor without any aqueous recycle, most of the product (75 to 80%) is obtained at high conversion efficiency and only a small amount (20–25%) at

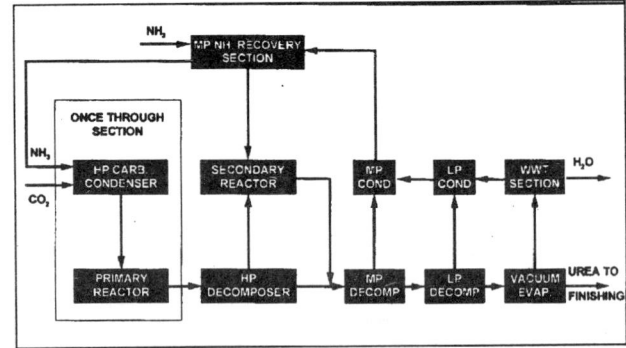

Figure 1. HEC process.

reduced efficiency. The weighted average conversion efficiency results in the 70–75% range, a value much higher than the one obtained even in modern urea plants. Consequently, the amount of steam required by the decomposition section is comparable to one of most modern plants.

Figure 1 shows the HEC process.

The "once-through" reaction section designed by CASALE for the HEC process consists of the following items in series:

- The carbamate condenser, a U-tube, kettle type heat exchanger where part of the ammonium carbamate is formed and part of the reaction heat is taken out, generating steam (with a pressure as high as 9 ata), in order to control the temperature of the primary reactor.
- The primary reactor, fitted with Casale High Efficiency Trays operating in the following condtions:

NH_3CO_2	=	3.6
H_2O/CO_2	=	0
Outlet temperature	=	195°C
Pressure	=	240 ata
CO_2 conversion	=	77%.

The second reaction section (secondary reactor) is also fitted with Casale High Efficiency Trays and operates in the following condition:

NH_3/CO_2	=	4.5
H_2O/CO_2	=	1.3
Outlet temperature	=	190°C
Pressure	=	155 ata
CO_2 conversion	=	55%.

All the CO_2 feed enters the "once-through" reaction section it reacts with NH_3 which is sent in the quantity necessary to keep the desired ratio.

About 77% of the total production is obtained in this section where steam is also generated (up to 9 ata).

The solution from the top of the primary reactor is flashed down to a pressure of 157 ata and enters the high pressure decomposer where carbamate is decomposed increasing the temperature of the solution up to 205° C by means of MP saturated steam (20 to 25 ata).

The high-pressure decomposer top vapor enters the secondary Reactor to supply part of the heat necessary to control the outlet temperature.

All unreacted CO_2 outflowing the secondary reactor and the high pressure decomposer is recycled to the secondary reactor though the MP and LP decomposition/recycling sections and an NH_3 recovery section.

The pure ammonia is obtained from NH_3 recovery section and recycled to the primary reactor.

With a two-stage (operating at 0.3 and 0.03 bar) vacuum concentration section, the urea melt which feeds the finishing system is obtained.

Features of HEC Process

Thanks to the utilization of the Casale High Efficiency reactor trays and to the fact that the main reactor is of "once-through" type, the HEC process has the following unique features:

- very high (average) CO_2 conversion, that is, ab. 72%;
- very low (average) H_2O/CO_2 ratio, that is, 0.3.

Thanks to these features, the HEC process has the following performances:

- low specific steam consumption, that is, 900 kg/MT;
- small-size of all the decomposition and recycle equipment (In particular, the size of the HP decomposer and condenser is much smaller than in the most advanced processes).

Revamping of conventional total recycle plants with HEC process

Thanks to the above features, the capacity of conventional total recycle plants can be drastically increased applying the HEC concept, and this with the addition of just a few pieces of equipment.

In order to increase the capacity of conventional total recycle plants up to 60%, Casale proposes to install its HET in the existing reactor and to apply its HEC concept as follows:

- the existing reactor (fitted with Casale HET) is used as a primary reactor; and
- a section consisting of the secondary reactor, an HP carbarnate condenser and an HP decomposer is added (see Figure 2).

The existing synthesis section will, therefore, be transformed in an HEC synthesis section.

Due to the much higher conversion obtained with

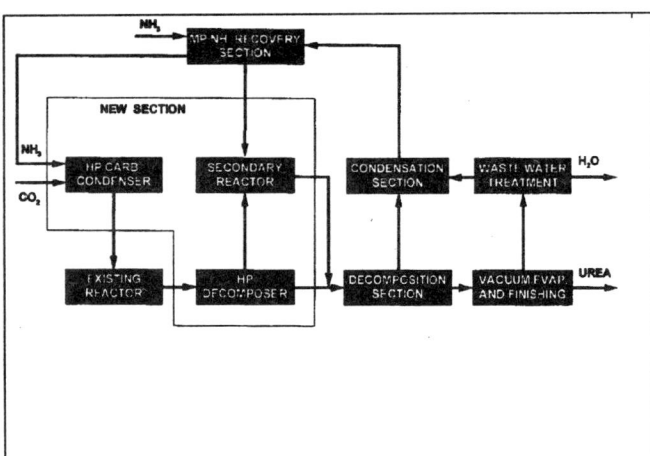

Figure 2. Conventional total recycle plant revamped with the HEC concept.

the HEC synthesis section, the amount of NH_3 and CO_2 feeding the existing sections downstream the new reaction section is, for a capacity increase up to 60%, still lower than before the revamping.

The existing back-end of the plant can take a higher production (60% or more) without being overloaded and can, therefore, be re-utilized at higher capacity with only minor modification.

With this approach, the highest possible utilization of the existing equipment is reached, keeping the investment as low as possible.

Increases of even higher than 60% can be obtained if, instead of the secondary reactor, the primary reactor is added, together with the HP condenser and decomposer. In this case the existing reactor will be used as a secondary reactor.

Large capacity increase for NH_3 and CO_2 stripping plants: Casale VRS concept

For the revamping of stripping plants, a new technology has been developed, namely the Vapor Recycle (VRS) which is described here below.

As it is well known in all "total recycle stripping" urea production plants, the residuals NH_3 and CO_2 at the H.P. loop outlet are recycled to the loop itself, after being separated from urea in the Decomposition and Finishing Sections, in the form of carbamate aqueous solution, since water is the essential carrier of such substances (in some cases, part of the NH_3 is also recycled in the form of pure liquid NH_3).

This recycled water is evidenced by the fact that the H_2O/CO_2 molar ratio at reactor inlet is higher than zero (generally in the range of 0.5 to 0.8).

The recycling of the H_2O heavily affects all the process phases, namely, the higher the amount of H_2O:

• the lower the conversion in the reactor is;
• the lower the decomposers (that is, stripper, M.P. and L.P. decomposer) performance is; and
• the higher the quantity of H_2O to be treated is.

Our new VRS concept foresees a separate circulation of recycle water and recycled NH_3 and CO_2, that is,

• The carbamate solution obtained in the downstream process sections instead of being sent the H.P. Section, is distilled in an H.P. decomposer working in parallel to the existing stripper.
• The vapors thus obtained (containing NH_3, CO_2, and a little water) are sent to the H.P. Section (H.P. Carbamate Condenser), while the distilled solution (enriched in water) is sent back to the back-end of the plant.

In this way, practically only the NH_3 and CO_2 contained in the carbamate is sent back to the synthesis while the water is almost totally sent back to the recycling and waste water treatment sections.

As a consequence, the H.P. synthesis loop will operate with very low water content with the following advantages:

• very high CO_2 conversion is obtained in the reactor (up to 70%);
• very high stripping efficiency; and
• lower amount of water to be treated in the existing decomposition, vacuum evaporation and waste water treatment sections.

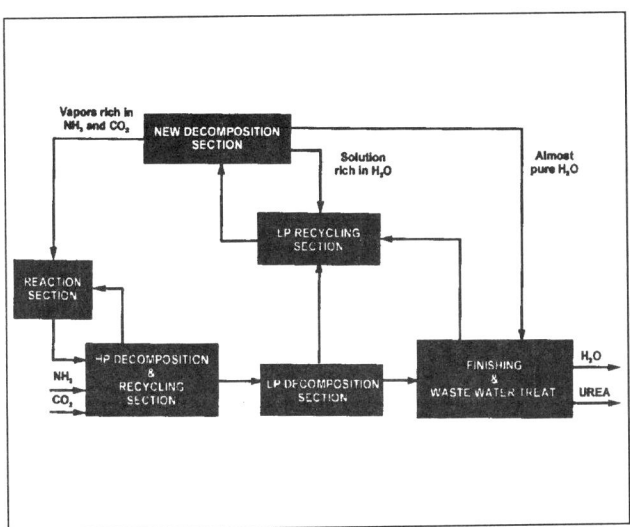

Figure 3. CO_2 stripping plant revamped according to the VRS concept.

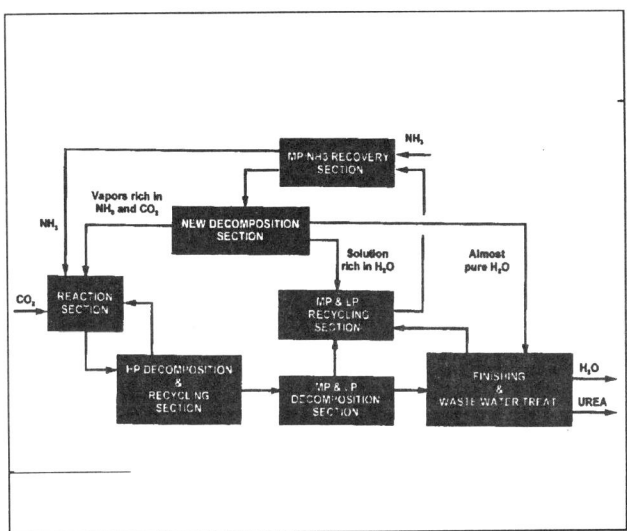

Figure 4. NH_3 stripping plant revamped according to the VRS concept.

Revamping of Stripping Plants with VRS

The existing plant is modified according to the VRS concept adding a new decomposition section parallel to the existing plant.

The HP carbamate is sent to the new section where it is decomposed.

The released vapors, rich in NH_3 and CO_2, are sent to the synthesis section, while the purified solution is sent back to the back-end of the plant.

As the existing reactor will be working with a low water content (H_2O/CO_2 molar ratio of 0.2 to 0.25), a high CO_2 conversion is obtained (65 to 70%).

Figure 3 and Figure 4 show respectively the CO_2 and NH_3 stripping plants revamped according to the VRS concept.

Due to the fact that, in the existing plant, the new conversion is much higher and the water content much lower than the ones before the modification, the existing plant can, again, be re-utilized at higher capacity with only minor modification.

With the approach, an increase in capacity up to 50 to 60% can be obtained.

One of the big advantages of the approach just described, is that the required additional section can be installed while the plant is still running, and just a few tie-ins are necessary to interconnect them with the existing plant, minimizing in this way the shut down time for the modification.

Furthermore, the solutions generated by plant upsets or shutdown can be recovered very quickly.

A Two-Case History Concerning Projects Successfully Implemented By Casale For Large Capacity Increase

From 270 MTD to 750 MTD with HEC for conventional total recycle plants

The way conventional total recycle plants are revamped using the HEC concept is illustrated using the revamping of a North American plant (Figure 5) carried out by Casale as an example.

In 1993, Casale was asked to study the revamping of a 465 MTD Toyo conventional plant in order to reach a capacity of 750 MTD. The urea was produced in two existing lines having the following capacity:
- No. 1, 195 MTD
- No. 2, 270 MTD.

Line No. 1 was a partial recycle line with a one-stage decomposition / recycling section and the NH_3 recovery section.

Line No. 2 was a total recycle line with a three-stage decomposition / recycling section and the NH_3 recovery section. A one-stage vacuum evaporation section was producing the urea solution of the desired concentration.

It was desirable to shut down line No. 1 and to have only one line for the new capacity. Due to the extremely high capacity increase required from line No. 2 (almost three times higher), a drastic increase in the efficiency of the synthesis section was necessary in order to avoid a very complicated approach with a lot of parallel equipment.

Casale, therefore, suggested to retrofit line No. 2 using a new front end, designed by Casale according to its High Efficiency Combined Process Technology (HEC), sized for 75% of the final capacity and consisting of:
- a new "once-through" reactor, working a 240 bar, 197° C
- a new HP carbamate condenser generating 6.5 bar steam upstream of the "once-through" reactor; and
- a new HP decomposer working at 157 bar fed by the once-through reactor outlet stream.

In this way, it was possible to keep the same equipment, down stream the existing reactor, and to minimize the modifications to it.

Line No. 1 was idled and some equipment used, namely the NH_3 condensation section and the machinery for NH_3 and carbamate compression. The CO_2 compression as well as the finishing section capacity were also increased.

No other new equipment was needed other than two condensers and an additional vacuum section.

This was also possible, because the stream feeding the decomposition sections then had a CO_2 conversion efficiency of almost 80%.

The revamped plant was started up in December 1995 after a modification to solve a hydraulic problem and operated at 550 MTD until June 1996 due to an unforeseen bottleneck in the existing NH_3 recovery section. In order to overcome this bottleneck, in July 1996 an idled NH_3 absorber from Line No. 1 was used and an NH_3 absorber pre-condenser and a vent scrubber were added.

From July 1996, the plant has been running with a capacity up to 800 to 810 MTD.

All guaranteed values have been met.

The expected performances of the HEC system have also been confirmed by the plant operation.

The main features obtained can be summarized as follows:
- high (average) CO_2 conversion: 70%; and
- low (average) H_2O/CO_2 molar ratio: 0.3.

A second plant revamped by Casale using the HEC technology is in operation since December 1996 in New Zealand.

This plant was originally designed to produce 480 MTD in a single line according to the Toyo conventional total recycle technology.

The revamped plant is designed to produce 750 MTD.

All guaranteed values have been met.

In addition to the just described projects, it is worth mentioning the following projects which are under implementation by Casale using the HEC for large capacity increase of conventional total recycle plants:
- Simplot (Canada) plant;

316 MTD single line Weatherly plant revamped to 625 MTD - startup by fall 1998.
- Razi Petrochemicals (Iran) plant;

500 MTD single line Vulcan plant revamped to 875 MTD - startup by 1999.
- Petrobras (Brazil) plant;

800 MTD single line Toyo plant revamped to 1,500 MTD - startup by summer of 1998.

From 1,900 MTD to 2,400 MTD with VRS for CO_2 stripping plant

Agrium's Carseland, Alberta, Canada, Nitrogen Operations were commissioned in 1977. The

Figure 5. North American plant revamped according to HEC concept.

Operation included a 1,043 mt/d Kellogg Ammonia Plant and a 1,350 mt/d Stamicarbon Urea Plant plus the design was such that was a zero discharge of effluent water to any water course. All process waters were either irrigated or evaporated on-site.

Over the years, the urea and ammonia plants had been expanded to produce 1,250 mt/d Ammonia and 1,825 mt/d Urea. It was decided to look at what the maximum that the ammonia plant could be deployed to and match that to an increase in the Urea capacity. Indications were that we could take the ammonia plant to 1,600 mt/d and the Urea to 2,350 mt/d.

Several methods of expansion were evaluated and in the Urea Plant, Carseland chose to use a novel approach referred to by Urea Casale as the "Vapor Recycle System (VRS)."

Casale proposed its VRS concept in order to fulfil all the requirements in the most economic way with this concept it was, for instance, possible to avoid any addition of reaction volume and practically no other modifications to the existing plant were needed other than the addition of a couple of vacuum condensers, some surface to the second vacuum evaporator and a few trays in the desorber.

The VRS concept was applied adding a kit to the existing plant consisting of:
- a new HP decomposer;
- a new LP decomposer; and
- a new MP separator and condenser.

The de-bottleneck of the raw material feed equipment and of the finishing section was carried out drectly by the owner. Due to maintenance reasons, the HP condenser was changed with a slightly larger one.

The designed capacity of the revamped plant is 2,400 MTD.

The project was first proposed in March of 1995, with a target startup date of June, 1996. Due to equipment delays the startup took place in October, 1996. Because of project fast tracking requirements, a team of Agrium personnel, along with the engineering companies Fluor Daniel and Urea Casale were put together. The project was designed, equipment procured and construction completed in just 14 months from actual approval to proceed.

Much of the construction occurred while the Plant was running in order to facilitate project tie-ins and completion during a three week turnaround.

This was for the most part, accomplished and with the knowledge gained by this team, improvements could be made in this area.

The results are that the plant is presently running at 1,485 mt/d on the ammonia side and 2,300 mt/d on the urea side. Areas preventing the Plants from achieving initial targets have been identified and will be rectified in due course.

The expected performances of the VRS system have been also confirmed by the plant operation. The main features obtained can be summarized as follows:

- low H_2O/CO_2 molar ratio at reactor (about 0.25);
- high CO_2 conversion even at high capacity (64%);
- high stripping efficiency; and
- high Ur concentration at stripper and LP decomposer exit.

As expected, the lower H_2O content (due to the low H_2O/CO_2 molar ratio) and the lower CO_2 content (due to the higher CO_2 conversion) allowed not only to reutilize the existing HP decomposition section without changes, but also to achieve a higher efficiency in the decomposition. This de-bottlenecked not only the LP section, but also the first vacuum evaporation stage which is now fed by a more concentrated solution.

The Urea plant use of the VRS system resulted in several operational surprises other than the tonnage gains. One was the quickness of eliminating water from the high pressure synthesis loop during startup or upset conditions.

The second is the stability of the operation at the high rates. These two items alone have the operators putting the unit on line in the startup as soon as practical.

Improvements can be made to the process and this evolution will take place as new units are installed.

Figure 6 shows a North American plant revamped according to the VRS concept.

Figure 6. North American Plant Revamped according to the VRS Concept

Conclusion

Urea Casale has developed several technologies to upgrade urea plants since the start of its activity.

Some of these technologies have proven to be real "breakthroughs" in the urea field, such as the HEC and VRS processes.

Nobody in the field would have imagined, just a few years ago, that the CO_2 conversion in urea synthesis reaction sections could be drastically increased even if at the same time the capacity is significantly increased, as Casale proved with the application of its technologies.

Thanks to its new HEC and VRS processes, Casale can offer to the Urea Industry an economical way of significantly incrementing the capacity of urea plants.

This becomes very competitive versus increasing the capacity by adding new plants.

The Casale concept, in fact, reaches the increment in capacity with an investment which is a fraction of the cost of a new plant. And, as all the new equipment can be erected with the plant running and the modifications to the existing plant are reduced to a minimum, this is obtained with a required shut down time no longer than a major maintenance shutdown.

Furthermore, the plant owner will have to operate only one plant instead of two with evident advantages to operating and maintenance costs.

These technologies, therefore, opened new horizons

in the field of urea plant modernization, making the revamp of existing stripping plants possible even when large capacity increases are required. This offers the market very competitive and flexible alternatives to the construction of new plants in today's growing demand for fertilizers, also in view of the fact that Casale technologies can be applied to almost any kind of urea process.

DISCUSSION

Dana Baham, *PCS Nitrogen*: Let me congratulate you on what I think was a very successful revamp. I have three questions, all dealing with the passivation and corrosion of the new system in the Carseland Plant. What is the metallurgy of the system? Where or how are you passivating it? Have you had the opportunity to see if there's any corrosion in the unit, and if so, what is it?

Clayton: The metallurgy of the system is similar to the metallurgy of the existing system. Actually, the high-pressure section is with 2522 stainless steel which is the same as used in the existing high-pressure section. The passivation is done with air, so it's a conventional type of passivation, and, also, with some hydrogen peroxide, which is also a passivating agent. Regarding corrosion, we have had the chance to inspect the high-pressure decomposer, and we found some higher corrosion rates than normally expected. We also found some lack of passivation agent, which was the cause for these higher corrosion rates. Otherwise, there is no other unusual corrosion in the existing part of the equipment and in the rest of the new equipment added.

Baham: Do you have some sort of analyzer on the effluent from the concentration system back to your reactor?

Clayton: The answer to your question is no. There is no analysis on the off-gas coming back into the high-pressure carbamate condenser, but we do have an N/C meter that is on-line within the plant. That gives us a very, very good control on our ratios.

Kilian, *Continental Engineering*: I have two questions for Mr Zardi and one for Mr Clayton. With your process, carbamate stripping, how much has the load on your condenser been increased? Can you give me some indication?

Zardi: Well, I think it is more or less similar to what it was. I mean that the steam consumption, which is an indication of the load of the condenser, is more or less the same as it was before. So, the load on the condenser is practically the same.

Kilian: Low-pressure steam was used for that carbamate condenser?

Zardi: Yes.

Kilian: Is the pressure lower?

Zardi: No, it is not lower.

Kilian: Is it similar?

Zardi: Yes, yes similar.

Kilian: What is the ammonia CO_2 content still present in the bottoms of the carbamate stripping that you are having installed? Can you advise me about that?

Zardi: Yes. Ammonia. It's around 10% to 15%.

Kilian: One last question for Mr Clayton. You have spoken about safety as a very important issue for this revamp. Can you advise us about the numbers of any incidents whatsoever over the period?

Clayton: The answer is I don't have them off the top of my head, but there were very few incidents. There were quite a number of first-aid incidents, but with the size of the crew, which peeked at 100 during this time, the safety was actually very good with no lost time incidents.

Maheshwari, *IFFCO, India*: I have some questions on VRS. We are operating four streams of 1,100 capacity plants. We are already operating 150% of the capacity. I want to know whether this VRS process can be considered for energy savings, savings of steam, and advantage on the moisture content of the final urea prills?

Zardi: No, I would not advise this for just energy saving purposes or moisture content. The investment would not be paid actually in the current situation of steam cost.

Maheshwari: If the capacity is increased, what payback is expected?

Zardi: Well, this depends. Of course, we should make a project. A revamping project is a case by case, so, up to now, we have seen very good returns in terms of some years. If you want to have more details, we should study your specific case and see what is the actual operating condition and what is the actual size of the plant. Then, we can make a more detailed analysis and cost benefit estimation.

Maheshwari: Say, for two streams of urea about 1,100 tons each, if you want to increase capacity by 50% by just VRS, what would the approximate cost be? Can you give some indication?

Zardi: No, unfortunately not.

Maheshwari: If I understood well, the two technologies that you have shown, HEC and VRS, are essentially dedicated with the first one for the total cycle plant and the second one for the stripping plant. However, in both you concentrated your effort to increase the conversions in order to utilize a downstream equipment and, you say, also to spare some energy. However, at least on the first one, HEC, you indicate 900 kilos per ton of urea. That is much higher than a normal stripping process and doesn't correspond to me with the high conversion you claim: 72, 75. There is something not clear to me as to which energy you would have to add the energy for the compression to the higher pressure of the once through reactor. Can you please clarify?

Zardi: Yes. First of all, regarding the compressor energy to compress to higher pressure, there is no higher pressure because we are revamping a conventional total recycle plant which has already high pressure. So, the reactor is operating between 200 and 240 bar; there is no need of additional compression to keep the same pressure. Regarding steam consumption, if you consider that we are a revamping total recycle plant, which has maybe 1,300–1,400 kg of steam consumption per ton of urea, I would say 900 is a significant reduction. Of course, it is not as low as a stripping plant because an HEC is not a real stripping plant. It's an in-between. So, this is why the steam consumption is not as low, but it still represents a significant reduction compared to what a total recycle conventional plant, which is the one we're going to revamp, had before.

Acoustic Emission Monitoring of Pressure Vessels During Proof Test

Acoustic emission is a relatively new NDT technique that offers the advantage of a global statement about the reliability of a vessel either at one precise time or continuously while in service. In this article, we summarize the technique, present the multipartner research projects (CIAPES and CRAFT), and review one of these projects in which the technique was applied to "complex vessels" through testing of a high-pressure urea reactor and a nitric acid adsorption column.

Etienne Soutif
BSL Industries, Soissons, France
Catherine Herve, Fan Zhang, and Marc Deschamps
CETIM, Senlis, France

Introduction

Pressure vessels undergo resistance tests in accordance with regulations throughout their lifetimes.

Acoustic emission (AE) provides high quality information during these tests by detecting and locating evolving defects. It is mainly for reasons of safety and accident prevention at work that industrial companies (manufacturers and users) have developed AE to both carry out resistance tests in safety and obtain detailed information on vessels.

BSL and CETIM (Technical Center of the Mechanical Industries) are involved in research projects whose aim is to develop and include AE in future European regulations.

After a brief description of the AE technique, the methodology developed for the application on pressure vessels is shown, with examples drawn from a high-pressure urea reactor and a nitric acid adsorption column.

Basic Elements of the AE Technique

Phenomenon

Acoustic emission (AE) is a naturally occurring phenomenon within materials. The term "acoustic emission" is used to define the resulting transient elastic waves when strain energy is released suddenly in a material due to the occurrence of microstructural changes within the material.

At the beginning of the 1960s, it was recognized that growing cracks and discontinuities in pressure vessels could be detected by monitoring their acoustic emission signal. Although acoustic emission is the most widely used term for this phenomenon, it has also been called *stress wave emission*, *stress waves*, *microseism activity*, and *rock noise*.

Formally defined, acoustic emission is "the class of phenomena where transient elastic waves are generated by rapid release of energy from localized sources

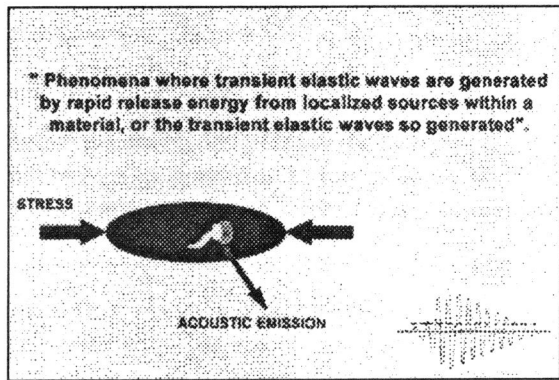

Figure 1. Classical response of materials to applied load.

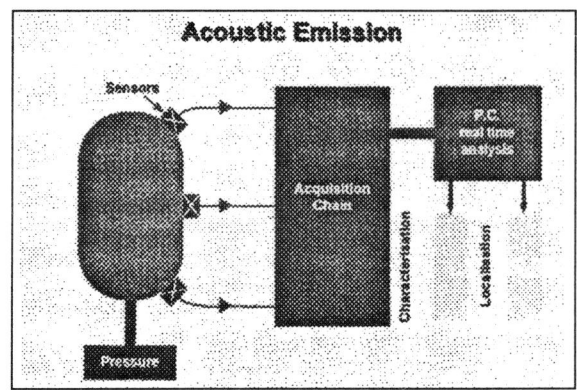

Figure 2. Acoustic emission.

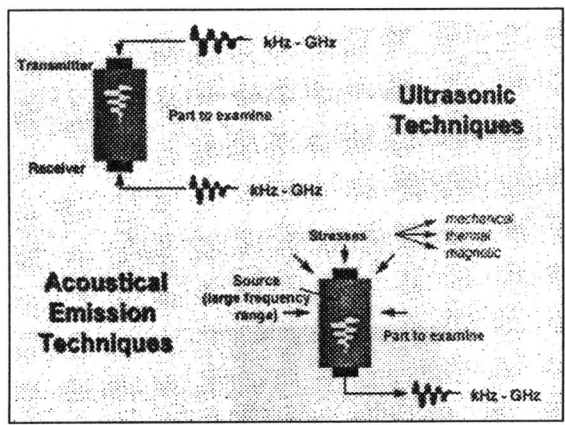

Figure 3. Acoustical emission and ultrasonic techniques.

within a material, or the transient elastic waves so generated." This is a definition that embraces both the process of wave generation and the wave itself.

Source mechanisms

Sources of acoustic emission include many different mechanisms of deformation and fracture. For metals these include crack growth, moving dislocation, slip, twinning, grain boundary sliding, and the fracture and decohesion of inclusions. These mechanisms typify the classical response of materials to applied load (Figure 1).

Other mechanisms fall within the definition and are detectable with AE equipment (Figure 2). These include leaks and cavitation; friction (as in rotary bearings); the realignment or growth of magnetic domains (Barkhausen effect); liquefaction and solidification; and solid–solid phase transformations.

These sources are sometimes called "secondary sources" or "pseudo sources" to distinguish them from classic acoustic emission, which is due to mechanical deformation of stressed materials.

A unified explanation of the sources of the acoustic emission does not exist. Neither does a complete analytical description of the stress-wave energy in the vicinity of an AE source. This is one of the major topics studied in the NDT laboratories.

Comparison with other techniques

Acoustic emission differs from other nondestructive testing methods in three significant respects.

First, the energy that is detected is released from the test object rather than being supplied by the nondestructive method as in ultrasonics (Figure 3) or radiography.

Secondly, the acoustic emission method is capable of detecting the dynamic processes associated with the degradation of structural integrity. Crack growth and plastic deformation are major sources of acoustic emission. Latent discontinuities that enlarge under load and are active sources of AE by virtue of their size, location or orientation are also the most likely to be significant in terms of structural integrity.

Usually, certain areas within a structural system will develop local instabilities long before the structure fails. These instabilities result in minute dynamic movements such as plastic deformation, slip or crack initiation, and propagation.

Although the stresses in a metal part may be well below the elastic design limit, the region near a crack tip may undergo plastic deformation as a result of high local stresses. In this situation, the propagating discontinuity acts as a source of stress waves and becomes an active AE source.

Finally, AE is nondirectional. Most AE sources appear to function as point source emitters that radiate energy in spherical wavefronts. That means that a sensor located anywhere in the vicinity of an acoustic source can detect the resulting acoustic emission.

Multipartner Research Projects

CIAPES program

Aim of the program. The program "CIAPES" (Control and Inspection of Pressure Vessels during Testing and in Service) has been created to group together industrial partners who have the same preoccupations regarding the proof test of pressure vessels.

This program has been in operation for 4 years and the partners are manufacturers of pressure vessels (BSL Industries, CETINI, Air Liquide) and users (CERN, CEA, SEP, ICI, Solvay, Rhone Poulenc, ELF ATOCHEM).

For all these industrial firms the hydraulic proof test presents many drawbacks, including the following:
- Supports and vessels need to be oversized to support the weight of water.
- Residual humidity may lead to the formation of ice, the initiation of corrosion, and hydrogen embrittlement, reducing the expected service life of the vessels.
- A long shutdown time is necessary to carry out hydraulic tests due to the change of fluid in the vessel and the required cleaning process.

To overcome these difficulties, the implementation of pneumatic tests for pressure vessels with AE surveillance is a promising method. However, a real time control of the test has to be performed since the vessel pressurized with a compressible media may explode if a dangerous fissure appears. To detect, locate and characterize defects during pneumatic tests, real time AE is used.

The aim of the program is to give industrialists the means to implement proof tests in the best conditions and with complete safety. To achieve this objective, we are developing an operational procedure for AE testing during pneumatic tests. This procedure will be validated by a surveillance system. This will enable real-time decision making concerning whether vessel pressurization should continue or be stopped when changes causing a dangerous defect are detected and located.

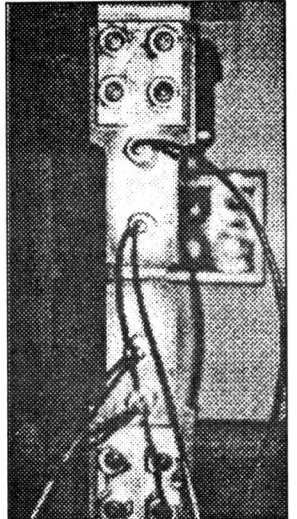

Figure 4. AE on tensile test.

Figure 5. Destructive test on small vessel.

Work program

The detection of the evolving defects will be ensured by the definition of assessment criteria. These criteria will be determined from different kinds of tests.

First, tensile tests are performed on specimens with artificial defects representative of real defects in order to compare the rates of propagation caused by various kind of flaws (Figure 4).

The next step deals with pressure tests on small vessels (Figure 5) with internal defects chosen from tensile tests to verify assessment criteria and make them more precise. After that, real structures will be controlled by either a hydraulic or pneumatic test to account for the environmental aspect of the vessels (background noise) and to validate the previous criteria.

Two kinds of materials, carbon steel and stainless steel, are studied in this program.

The data from the tests performed during the CIAPES program are collected into a database that can be used by the different partners.

For each pressure vessel tested a procedure for AE monitoring is established. This procedure describes all main actions involved in the monitoring of the structure. Particularly important parts of the procedure include the following:
- Definition of objectives.
- Normative references.
- Personnel qualification.
- Scope.
- Initial considerations and requirements.
- Preparation and execution of the test.
- Verification of the AE equipment after the test.
- Test report and documentation.

The different kinds of vessels tested include horizontal-welded vessels, vertical-welded vessels, forged bottles, columns, spheres, multilayer vessels, and reactors.

CRAFT project

A European CRAFT project, complementary to the CIAPES program described above, is being led by BSL. This project has gathered European industrial manufacturers like BARTEM (France), ARSOPI (Portugal), and CO_2-Lambach (Austria); Vallen Systeme GmbH (Germany), which is a manufacturer of AE equipment; and testing or research centers such as CETIM (France), INEGI (Portugal), and TUV (Austria). One of the objectives of this project is to develop a decision system that would allow real-time evaluation of the acoustic emission results, thus guaranteeing the safety of personnel and materials. The integration of a computer-aided decision system will place the AE method as the first nondestructive technique to be independent from the test crew and their own experience.

Testing of a "Simple Vessel": Oxygen Liquid Storage Tank

Description of the vessel: Proof test procedure

For a better understanding of the AE procedure technique, and according to the progress of the project previously described, an AE test performed on a "simple vessel" is reviewed. This vessel consists of an inner cryogenic tank for the storage of liquid oxygen (during the next stage, this vessel is introduced into an outer carbon steel jacket).

The geometry of the vessel was as follows: the body was composed of five cylindrical shell sections and two elliptical heads, the internals were six inner stiffener rings directly welded to the shell, and the supports were two permanent supports directly welded to the shell. The material was 304 L grade, volume was $300 m^3$, and the thickness was 5 mm. The dimensions of the vessel were L = 25 m × 4 m OD. The data for calculation purposes were a design temperature of 3.42 bar, a maximum operating temperature of 2 bar, and a proof-test pressure of 3.76 bar. The test was performed as an "Initial Proof Test" under air pressure in the horizontal position. The vessel had not been previously pressurized.

The sequences of pressurization are given in Figure 6.

AE testing equipment

The acquisition chain used is a VALLEN AMSY4 multichannel system for characterization and location of acoustic emission sources. Twenty-three sensors of type SE1 50-M, resonating at approximately 150 kHz,

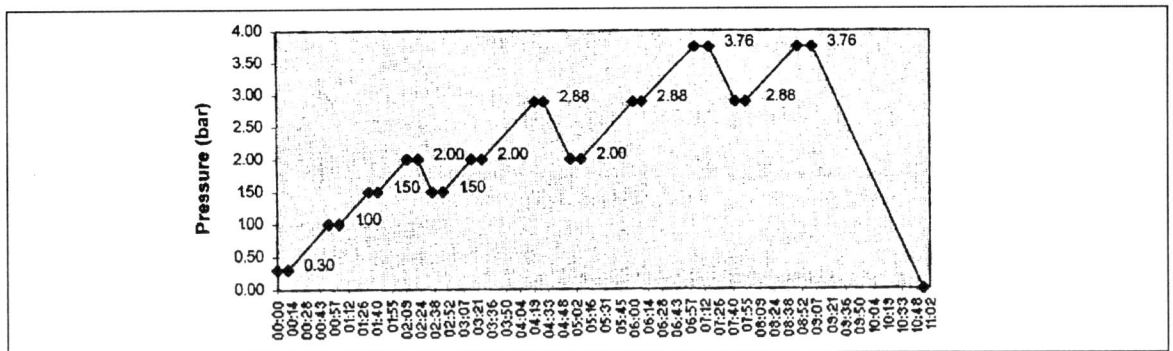

Figure 6. Pressurization cycle.

Figure 7. General view of the vessel in the BSLi workshop and location of the sensors.

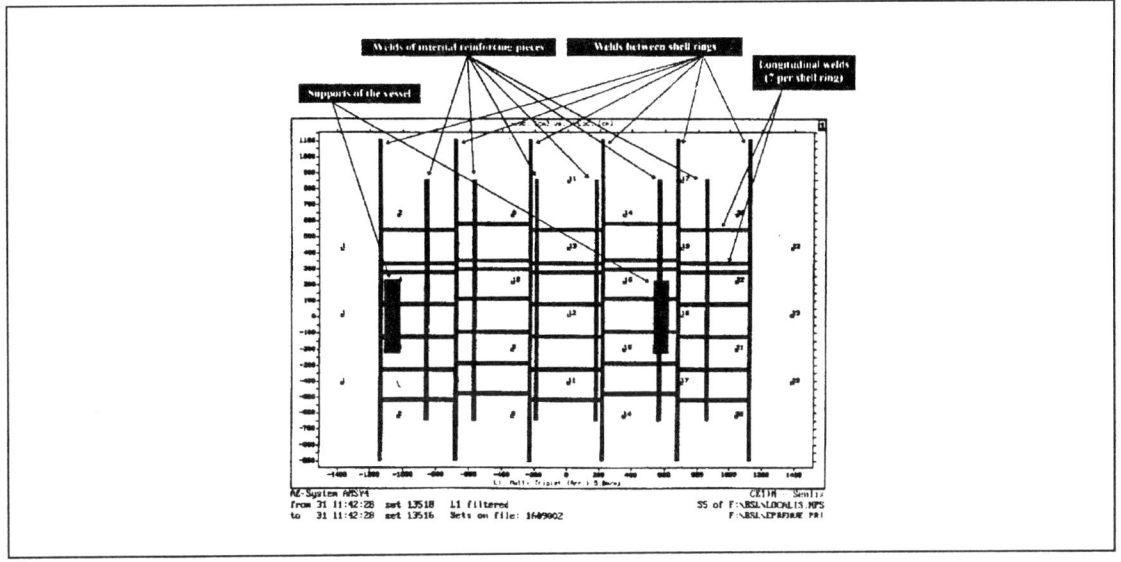

Figure 8. Location of sensors, welds, stiffener rings, and supports.

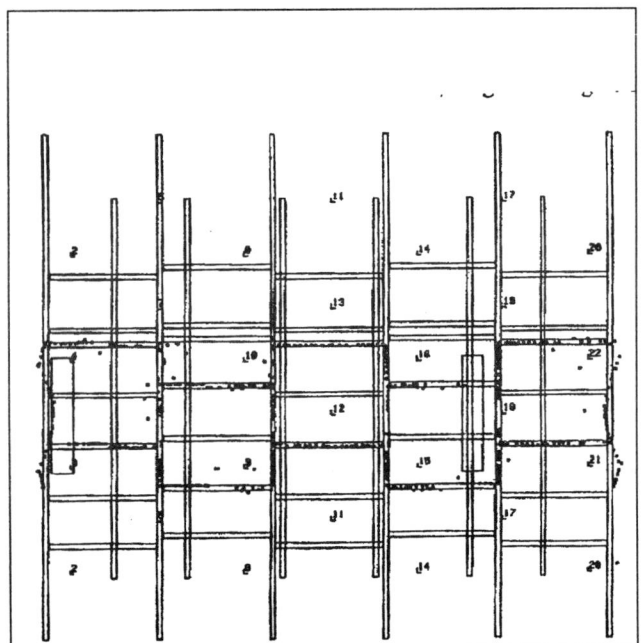

Figure 9. Preliminary location tests on welds at the bottom of the vessel.

Figure 10. Total events at the end of pressurization cycle.

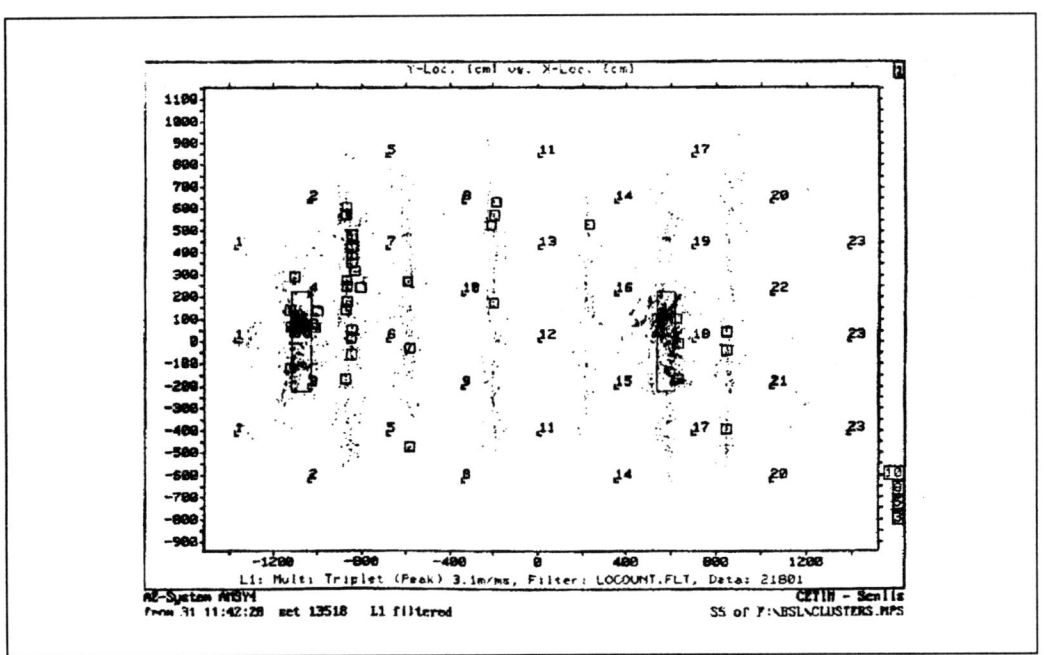

Figure 11. Location of clusters (minimum three events, over 60 dB).

were placed on the vessel according to the configuration described in Figure 7.

Figure 8 corresponds to a background view as appearing on the computer monitor. It is a flat-view representation of the vessel on which the locations of the welds, stiffener rings, and supports have been added.

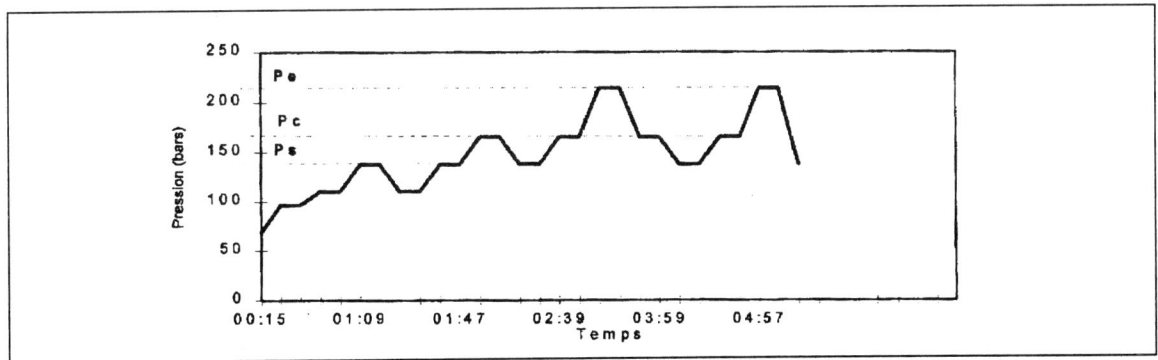

Figure 12. Pressurization cycle.

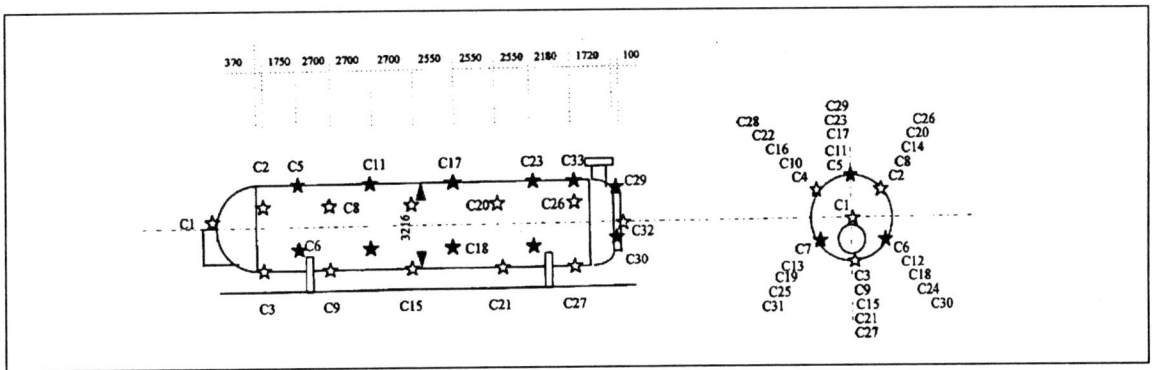

Figure 13. Location of the sensors.

Calibration

To check the sensors' sensitivity and the accuracy in the location of emission sources, some ordinary carbon pencil lead (called Hsu-Nielsen source, 2H-0.5 mm diameter) is broken directly on the vessel. Location tests are carried out in order to correlate the differences in arrival time of the signals with the true position of the source on the vessel. The graphs concerning these tests along the circular and longitudinal welds at the bottom of the vessel are given in Figure 9.

There is a very good correlation between the recorded emission signals and the expected locations. Tests are repeated at the end of the examination to check the calibration of the equipment.

At the end of the pressurization cycle, the total number of localized events is around 22,000 (see Figure 10; each black point corresponds to one event). At this stage, except for the supports location, it is very difficult to have a good interpretation of the results.

Results and comments

It is through the interpretation and the analysis of this raw data that the test diagnosis is possible. Thus, the parameters selected for *filtering* and the corresponding acceptance criteria should be considered as the key points of the technique.

The events are computerized in order to define some clusters (concentrated areas with a high AE level). Numerous parameters can be used:

• The number of events in the same area (usually a cluster is defined by a length equal to 10% of the distance between sensors).

• The amplitude of signals (in dB).

• The arrival time of the signals (during last pressure increase, for the total cycle).

• The characteristics of the signal (slope time, the number of hits over a pre-established level frequency and/or absorption length).

Figure 11 shows the clusters (squares on the graph)

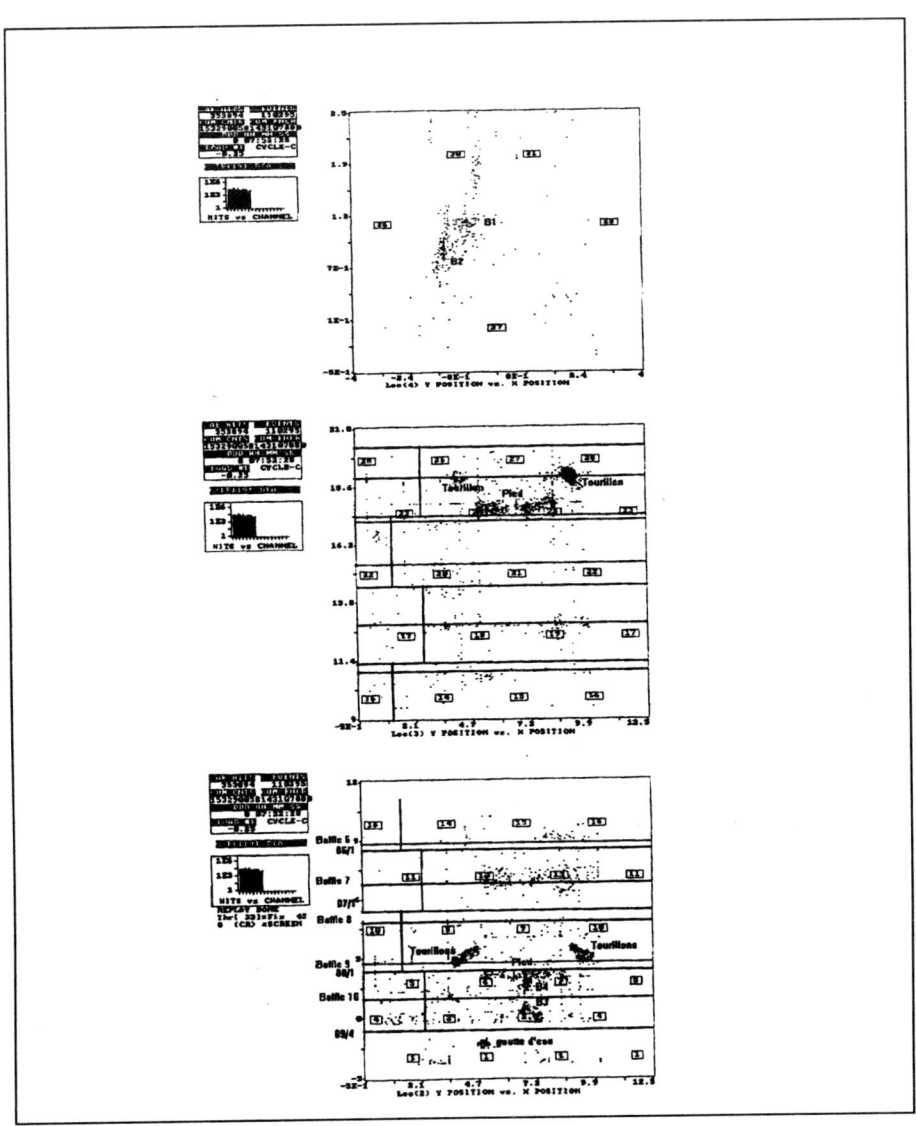

Figure 14 (a,b,c). Total events at the end of the pressurization cycle.

after filtering a minimum of three events and a 60dB amplitude. According to this filtering and in addition to the support areas, one observes some emission concentrations at the place of the stiffener rings.

After testing, the resistant pressure parts are sound. However, due to the presence of "accessories" and in spite of the relatively basic vessel geometry, a final statement is quite difficult to establish. It is expected that this phenomenon will grow in proportion to the complexity of the vessels. The two examinations described below confirm this first impression.

Testing of a High-Pressure Urea Reactor

Description of the vessel: Proof test procedure

The purpose of this test was to observe the effects due to the combination of heavy wall thickness and internals on the AE examination.

The geometry of the vessel was as follows. The body consisted of seven shell section (one longitudinal weld per section) and two hemispherical heads, with one that was directly attached to a tubesheet. The

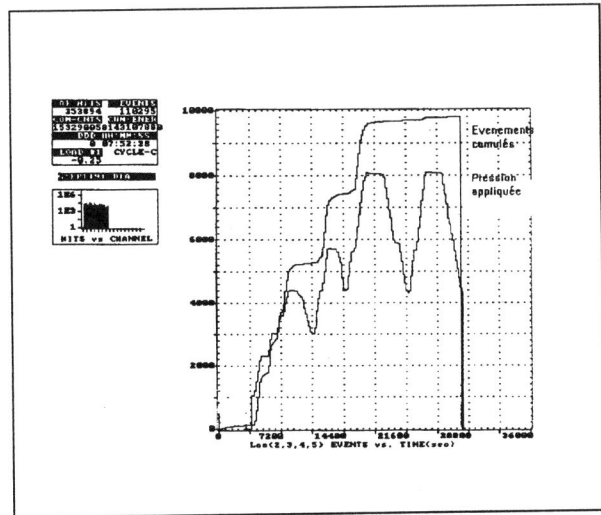

Figure 15. Signal arrival vs. pressurization cycle.

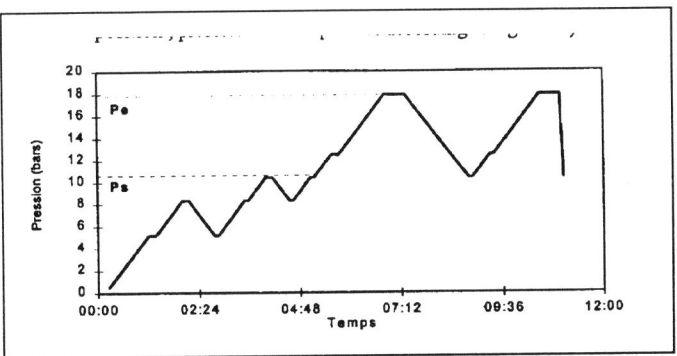

Figure 16. Pressurization cycle.

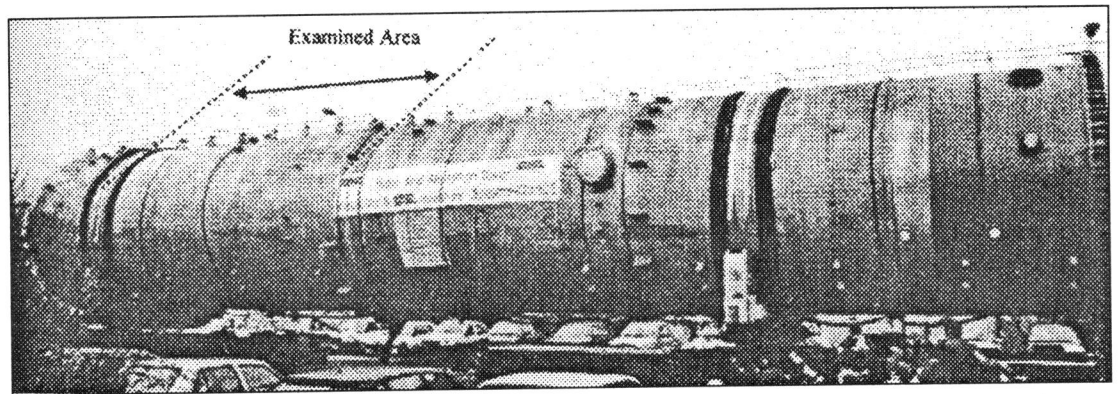

Figure 17. General view of the nitric absorption column (half section for shipping).

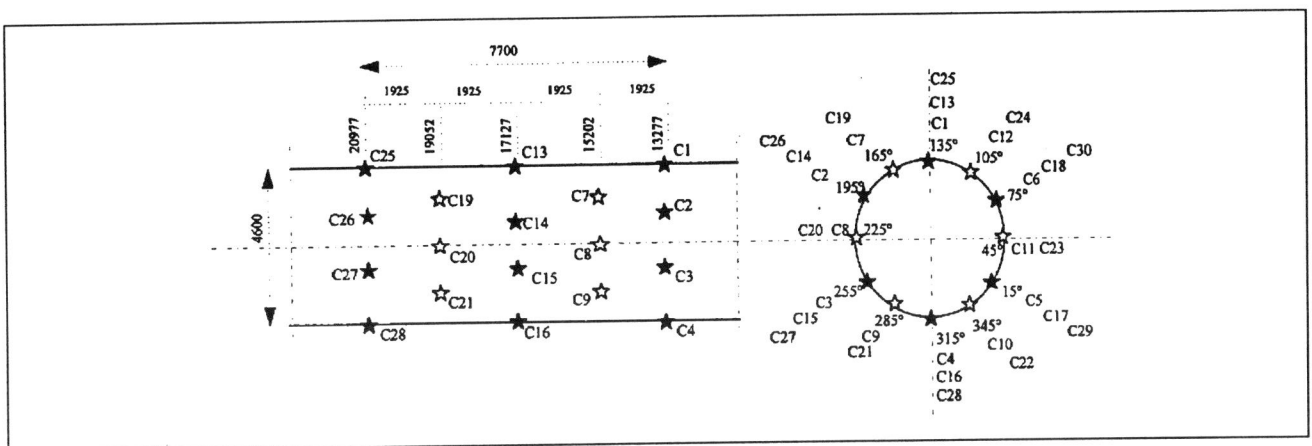

Figure 18. Length examined and location of sensors.

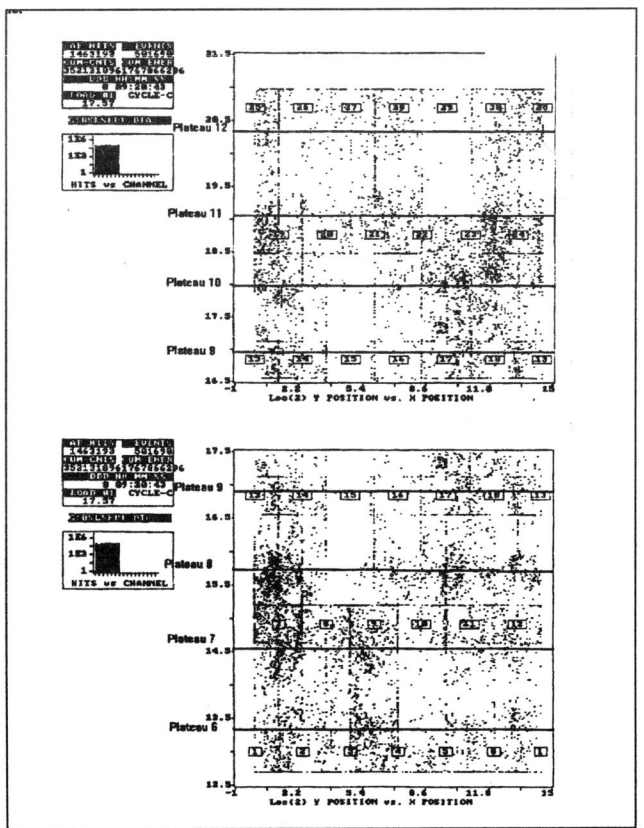

Figure 19(a,b). Total events at the end of the pressurization cycle.

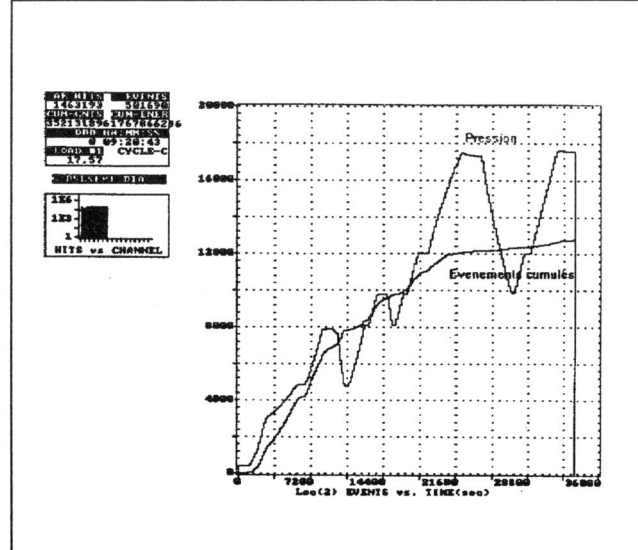

Figure 20. Signal arrival vs. pressurization cycle.

Table 1. Results of the Last AE Test

Vessel	Total of Localized Events	No. of Localized Events/m²
Cryogenic Tank	22,000	68
High-Pressure Urea Reactor	110,000	547
Nitric Acid Column	510,500	4,439

internals consisted of one U-tubes bundle and 10 baffles. Two permanent supports were directly welded to the body. The vessel was approximately 20 m × 3 m diameter and 100 mm thick. The material was low-alloy steel.

Proof test pressure was 214 bar, with the initial hydraulic test in horizontal positions and pressurization sequences according to Figure 12.

AE testing equipment

A SPARTAN acquisition chain and 33 sensors have been used. Figure 13 gives the location of the sensors.

Testing

The sensitivity of the sensors was checked using Hsu-Nielsen sources (0.5 mm-2H). The distance between two sensors (≈ 3.5 m) was acceptable in spite of the heavy wall thickness and the presence of water.

Because of the SPARTAN software structure, three grids were defined: two grids covering the shell and the hemispherical heads, and one grid covering the tubesheet.

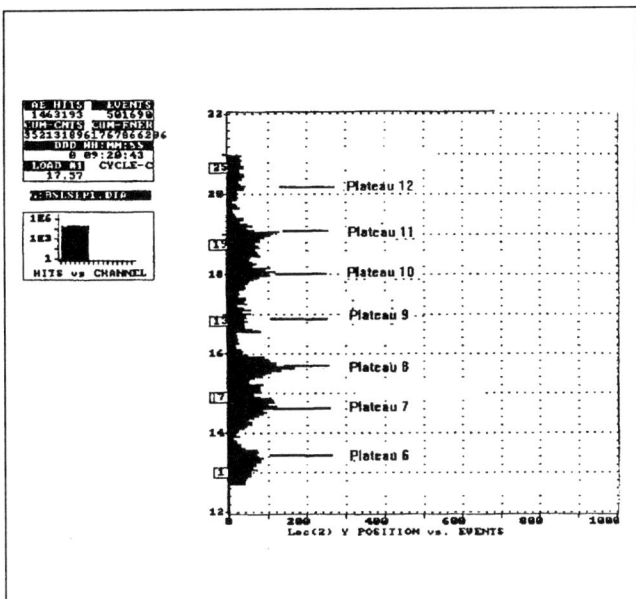

Figure 21. Signal distribution along a longitudinal axis.

Figure 14 (three grids, a,b,c) shows the results after recording the data throughout the pressurization cycle. It should be mentioned that the test was greatly disturbed when the pressure reached 200 bar because of leakage (gaskets, fittings). These leaks momentarily hid the emission sources coming from the structure.

We see that the most active emission sources are supports, lifting Lugs, nozzles B3 and B4, baffles (particularly the seventh), and the weld between tubesheet and hemispherical head (though the signals have a very low amplitude).

Filtering the amplitude of the events revealed that the clusters were plotted particularly on supports, lifting lugs, and baffles.

The level of AE during the test was very high: 110,000 localized events.

Leaks, which can be disregarded during a "classical" hydrotest, are detrimental in cases of AE monitoring and must be carefully avoided.

The supports and lifting lugs welded on the shell are the cause of friction areas, thus giving high emission, which can prevent the detection of real defects. Situating these elements as far from the welds and sensitive areas as possible is recommended.

Ultrasonic testing was used to check the emitting areas. Some specific spots were confirmed as presenting previously observed defects of a size conforming to the acceptance criteria.

Figure 15 compares the arrival of the signals with the test duration. We see that the "KAISER effect" (no AE without pressure increase) is established, so the probability of evolving defects is very low. However, during the internal inspection following the hydrotest, damage on the weld baffles was observed, which led to repair and reinforcement.

Ultimately, it is difficult to interpret the test results for two reasons: The low accuracy in the location of the events due to the heavy thickness (wave mode conversion, wave reflection in the thickness) and the inner emission sources transmitted by the water in multiple directions. To alleviate these problems, integration of a "wave mode recognition analysis" into the software with the purpose of differentiating between the signals should be one of the next developments.

Testing of a Nitric Acid Absorption Column

Because of the very complex vessel geometry, the primary purpose of the examination was to have an experimental approach that would help prepare testing parameters for the future.

Description of the vessel: Proof test procedure

The geometry of the vessel was as follows. The body comprised 18 shell sections (2 longitudinal welds per section) and two hemispherical heads. The internals were 34 perforated tray each one of which was equipped with a cooling pipe circuit connected to three nozzles. Temporary wooden supports were used. The material was 304 N grade, the volume was 966 m^3, and the thickness was 20 to 24 mm. The dimensions of the column were 62 m × 4.6 m ID.

The calculation data were design pressure of 11.7 bar and proof test pressure of 17.9 bar. The initial hydraulic test was in horizontal position, and the pressurization sequences were arranged as demonstrated in Figure 16.

AE testing equipment

Due to the large size of the vessel and the purpose of

the test, *only 3 shells were examined (those containing trays N°6 to tray N°2) for a corresponding length of 8 m* (Figure 17).

30 sensors have been used (Figure 18). The acquisition chain is a SPARTAN.

Testing

After filling the vessel with water, the sensors' sensitivity was evaluated thanks to the signals produced by a Hsu-Nielsen source (0.5mm-2H). The distance of around 2.5 m between sensors gave acceptable results concerning the accuracy in locating the emission sources.

Figure 19 shows the acquisition on the examined shell length during the whole pressurization cycle.

Results/comments

Rough observation of the graphs reveals an excessively high level of AE. This level corresponds to 510,500 localized events for the surveying of only 1/6 of the whole vessel.

Table 1 compares these results with those of the previous tests and clearly shows the high level of AE during this last test.

No clear emission concentration areas can be seen, the distribution of events being regular. Several filterings were implemented, none of them revealing particular areas.

Figure 20 shows arrivals of the events during the test cycle. The "KAISER effect" is not established since the number of events continues to grow, whereas the pressure is stable. This phenomenon would be worrisome if all the events came from the pressure resistant parts. Figure 21 clearly gives some interesting information. The graphs describe the arrival of the signals along a longitudinal shell axis (sensors 1, 7, 13, 19, 25). It is clearly shown that the arrival of the signals is concentrated where the trays are situated. So, the essential source of emission is due to the internals.

The numerous AE sources, generated by friction or stress relieving of trays and piping, make the diagnosis for the pressure resistant parts very difficult.

Following the comments of the previous tests, we are now working on the characterization of the tray signals with the purpose of neutralizing them for future AE tests. Applying pneumatic pressure to avoid the erratic transmission of the waves caused by the water is also foreseen.

To conclude, a fundamental aspect in mastering AE testing is the accumulation of data, and this is one of the major aims of the research programs. Even if the results of this test do not appear at this time to be totally convincing, the data collected will prove very useful for the progress of the technique.

Conclusion

At the beginning of the 21st century, the practice of filling vessels with water and then immediately afterwards emptying them at the time of final inspection and acceptance should be gradually replaced, at least for low and medium pressures, by pneumatic tests assisted by AE monitoring.

By increasing the safety conditions of pneumatic tests—even without insisting on the ecological benefits—the use of AE technique will give great advantages to the manufacturer as well as to the final user.

To sum up, the safety improvements are as follows:

• Reduction of supports and of vessel-wall thickness by deletion of water weight and hydrostatic pressure.

• Lower risk of initiation of corrosion.

• Deletion of drying and subsequent cleaning operations.

• Examination of the complete vessel in a very short time compared to the usual NDT techniques.

• The information collected in the AE data book enables the plant operator to perform further surveys while in service, resulting in additional safety in the plant.

However, although the AE test is now a reliable and effective technique, numerous aspects remain to be studied and improved on, in particular:

• Accuracy concerning the location of events, especially in the case of heavy wall thickness.

• Differentiation between the "true indications" (evolution defects in the structure), and "misleading indications" (friction, emission of the internals, ambient noise).

• Definition of standard acceptance criteria in order to increase the validity of the assessments.

These are but a few themes for further study and we hope that after this brief presentation you will be as eager as we are to see these problems solved.

DISCUSSION

Mark Carte, *Det Norske Veritas*: We conduct many acoustic emission tests on various pieces of equipment in the process industry. There's a growing popularity for pneumatic testing of small and large pieces of equipment. We've seen a growing increase in the desire to rerate pieces of equipment in order to increase the pressure that they normally operate at. One of the means of doing this type of testing and offering a substantial amount of savings is, again, to apply the acoustic emission. There are a couple of other very interesting aspects. Piping is applied to the piece of equipment in an adequate size and then linked to a control valve or valves. Controls to those particular valves are linked back to the operation of the monitoring of the acoustic emission works. As the pressure is applied specifically in pneumatic tests, we watch for activity that is indicative of the possibility of an injurious flaw or some type of failure of the vessel. The test can be stopped, the indications can be analyzed, and a decision can be made at that point whether further investigation should be done prior to increasing the pressure and going for the gold or whether it's erroneous in a continued pressurization. In the event these indications get extremely active, or it is felt that they are dangerous or even questionable, someone of an operations capability within the facility will activate the dump control and bleed the pressure off through these valves down to a very defined and controlled area. These applications have been done in pressure equipment. The maximum size is approximately 90 ft tangent to tangent by 32 ft in diameter. We did two of those vessels simultaneously up to right around 68 PSI. We're seeing again a tremendous growing increase in the interest in doing this on new vessels as for a proof test, for re-rating of equipment to a higher operating pressure, and also where pieces of equipment have extreme amounts of internal components such as catalysts or various types of components where the equipment needs to be tested according to PSM. However, they also do not want to remove this equipment ore hydrostatic test which is not capable because the foundation doesn't support the weight and it's a very good alternative. It's recognized throughout the industry and in most of the recommended practices in the governing bodies.

Baham, *PCS Nitrogen*: On the urea reactor that you studied, I'm assuming that it was a monolithic vessel with the stainless or 2522 liner. My question is could you determine the integrity of both layers of the vessel, the outside layer and the inner liner?

Soutif: No. In fact, we couldn't distinguish the carbon steel part of the inner overlay – because in this case it was overlaid and directly welded to the shell. If you have some acoustic emission, you cannot easily detect where it is exactly.

Baham: So, this method couldn't be used on any multilayer vessel?

Soutif: It's one point of my conclusion that for the moment, according to our practice, it is very difficult to have an idea of the location. Indeed, the thickness of the location is the event. Maybe in the future it could be possible. We hope so.

Laura Viloria, *PDVSA INTEVEP*: We have quit trying to increase the accuracy of acoustic emission inspection, but we do use it for special vessels or spheres to get the emissions and then go to ultrasonics to define what is the kind of emission we are getting. However, we don't place our trust only on acoustic emission anymore.

Soutif: Acoustic emission needs some complementing monitoring method as soon as you observe some emissions so ultrasonic testing is a very good solution.

Urea Waste Water Purification

Simultaneous thermal hydrolyzing and stripping is a technique to achieve the purity specification required for boiler feed water. This technology is only feasible where a water shortage exists. Other techniques include improved stripping with the batch/continuous stripper (BC-stripper) technology, catalytic cracking of the vapor from the vacuum evaporator or crystallizer, and enzymatic hydrolysis of urea rather than expensive thermal hydrolysis. In this article, we discuss these technologies.

J. J. P. M. Goorden and E. R. Kilian
Continental Engineering, B.V. Amsterdam, The Netherlands

Introduction

A urea plant produces approximately 0.5 kg waste water per kg of urea. 60% of the waste water is produced in the urea synthesis reaction. Environmental restrictions and possible ways to upgrade this waste stream to valuable high-pressure boiler feed water are the reasons for improving waste water purification.

Waste water generation

In all commercial urea processes, urea is formed according to the overall reaction:

$$2NH_3 + CO_2 \leftrightarrow Urea + H_2O$$

This means that, for every mole of urea, one mole of water is produced. In a production plant of 1,500 metric t/d of urea, 450 metric t/d of water is produced. Other than this water, which is formed in the reaction, water is also introduced in the process for scrubbing, ejector-steam, vacuum evaporation, washing, and other operations. An average of 0.5 t water effluent per ton of urea is released; that gives a total contaminated water flow of 750 metric t/d for a 1,500 metric t/d urea plant.

The waste water from the process always contains NH_3, CO_2, and urea. The quantity of these components can vary within broad ranges. This strongly depends on the type of plant (Stamicarbon, Snamprogetti, Casale) and on the product finishing (crystallization or evaporation). A typical waste-water composition is 4 ww% NH_3, 3 ww% CO_2, and 1 ww% urea.

Because of environmental requirements, the urea waste water has to be cleaned. To do all the cleaning through end-of-pipe treatment, like in a biological waste-water cleaning system, has not been practiced since the 1970s because of two economic reasons. The first reason is the valuable quantity of 1.15 t/h of nitrogen in the waste water. With a cost price of $200/metric t of nitrogen ($165/metric t of NH_3) this nonrecovery yields a loss of $2 million dollars per year. The second reason is that removal of nitrogen in a biological waste-water cleaning system is expensive and needs a special nitrification/denitrification treatment.

Costs of approximately $1,500/metric t of nitrogen removed are typical for these kinds of operations.

Reuse as boiler feed water

Reuse of the waste water as valuable boiler feed water gives an extra opportunity. Regaining the water is easy because of the restricted number of known components and the ability to remove them from the waste-water stream. By cleaning this waste water to lower the concentrations of NH_3 and urea to less than 1 ppm, the water can be reused in a low-pressure steam system, which mostly occurs in a urea plant.

Application of the water as high-pressure boiler feed water by further purification is a challenging possibility. It was generally accepted that for corrosion reasons the urea and NH_3 concentrations should be less than 0.1 ppm. The objective of this article is to critically review the existing and potential urea waste-water cleaning methods.

Methods

The waste-water purification method most often applied is thermal hydrolysis of urea to NH_3 and CO_2 and removing the NH_3 and CO_2 by stripping with steam. Urea and NH_3 effluent concentrations of less than 1 ppm are required in regions with a shortage of water for regaining boiler feed water. This low concentration necessitates that hydrolysis and stripping have to be partly combined; otherwise, this low concentration cannot be achieved due to the equilibrium of urea with NH_3 and CO_2.

Stamicarbon and Snamprogetti have a partly simultaneous hydrolysis and stripping method. In this method NH_3 is stripped off in a first desorber step, while the urea is hydrolyzed to NH_3 and CO_2 under stripping conditions. The NH_3 and CO_2 formed are further stripped off in a second desorber. Figure 1 shows the Stamicarbon process.

In both the Stamicarbon and Snamprogetti designs, the two desorbers can be combined into one column. These separate strippers can typically be operated with low-pressure steam.

At elevated pressures (but below 20 bar), hydrolysis and stripping can be performed by a fully combined simultaneous hydrolysis and stripping method in a specially designed column. Medium-pressure steam is minimally required. Casale and Kellogg have acquired patents in this field. Figure 3 shows the Casale design.

In the Kellogg process the fully combined simultaneous hydrolysis and stripping process is performed in combination with waste water from the ammonia plant. In this process high-pressure steam is applied. Figure 4 shows the Kellogg design.

Comparison of methods

Because low-pressure steam in a stand-alone urea process is inexpensive, partially combined thermal hydrolysis and stripping is still the cheapest method to obtain an effluent with less than 1 ppm urea and ammonia. On sites where low-pressure steam can be exported, the fully combined thermal hydrolysis and stripping is the best option with respect to investment and energy.

The perspective, requiring NH_3 and urea concentrations less than 0.1 ppm for high pressure BFW, will favor the fully combined hydrolysis and stripping method over the partly combined method. This is because the necessary temperature and pressure conditions in the hydrolyzer of the partly combined method are approaching the conditions of a fully combined thermal hydrolysis and stripping column as the concentration is lowered.

Is the conventional thermal stripping and hydrolysis still the optimal technology for 0.1 ppm effluent concentration? From an economic point of view it is not. Based on the quantity of water (0.5 metric t/metric t of urea), BFW is given at a price of $1.50/metric t of urea, which results in an additional benefit of $1.00 metric t/urea. Taking into account extra steam costs of 0.3 metric t/metric t of waste water, with a steam price of $10/metric t of stream, the extra costs add up to $1.50/metric t of urea. These investments may be attractive in regions that have high water prices.

Minimization of Waste-Water Contamination

The best method to minimize costs of urea waste-water treatment is to reduce the concentration of pol-

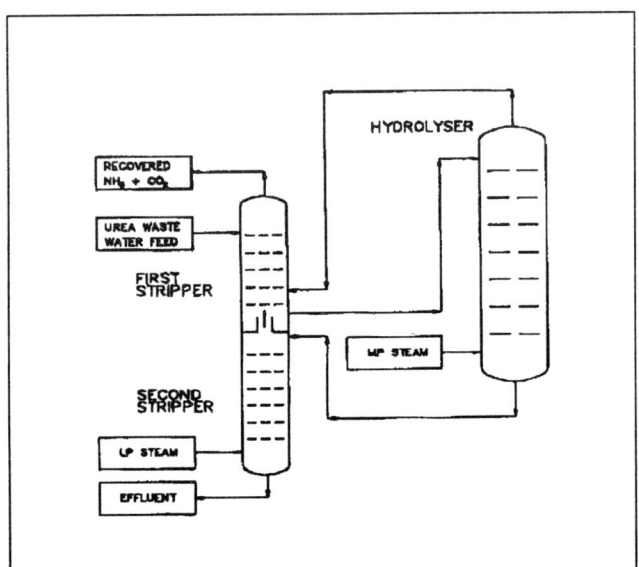

Figure 1. Partly combined hydrolysis and stripping (Stamicarbon process).

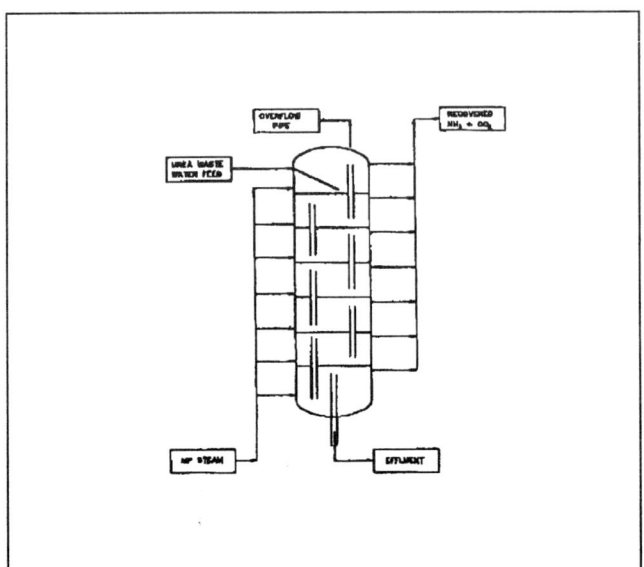

Figure 3. Fully combined hydrolysis and stripping (Casale process).

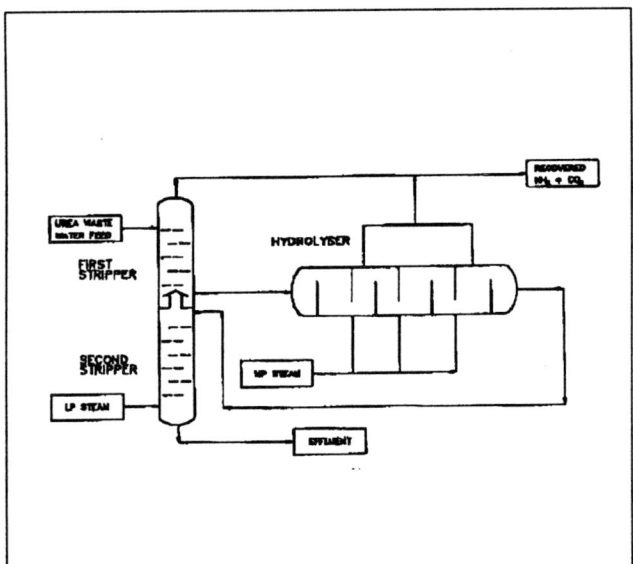

Figure 2. Partly combined hydrolysis and stripping (Snamprogetti process).

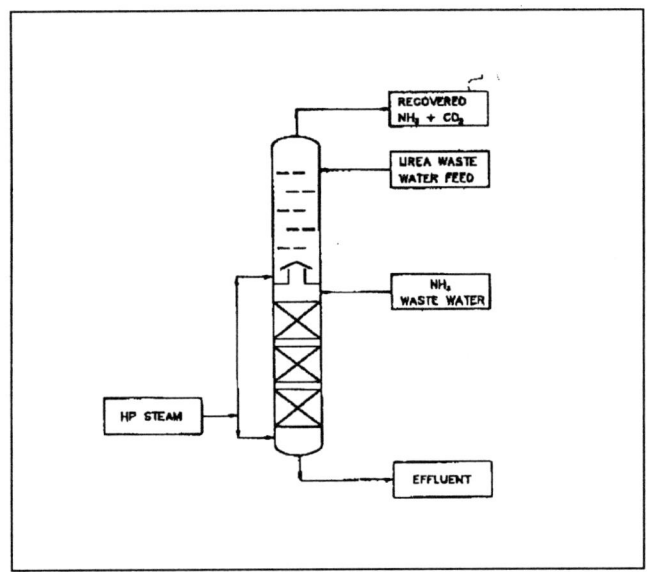

Figure 4. Fully combined hydrolysis and stripping (Kellogg process).

luting components herein. The water produced in the reaction is a fatal stream that comes from every urea plant and cannot be avoided. It is essential in this situation to minimize the concentrations of NH_3 and urea in the stream.

The fatal stream originates in the condensed overhead vapor of the evaporators and liquid effluent from the crystallizers. NH_3 and urea are present in this stream.

The secondary water stream originates from flush water from pumps, instruments and drains, and so forth. Also in this water stream the above mentioned components are present.

Avoiding urea in the waste water

Entrainment of urea is one of the main causes of the

urea component in the waste water. Various developments have improved the efficiency of separators. Separating the mist is a challenging possibility for decreasing the urea concentration in the waste water.

HNCO coming via the gas phase is the other significant urea source in the waste water. Preventing carryover of HNCO is only a partial solution due to the earlier mentioned urea entrainment.

Avoiding NH_3 in the waste water

The NH_3 mainly originates from the unreacted ammonia present in the feed to the evaporator or crystallizer. Stripping off the NH_3 can be performed under vacuum. Because there is already a stripping operation in the waste-water treatment, this only increases capital investment costs and is inefficient.

Other NH_3 sources are urea reactions like dimerization, which gives NH_3 and biuret; dissociation, which gives NH_3 and HNCO; and hydrolysis, which gives NH_3 and CO_2. These reactions can be avoided as much as possible by operating under mild conditions.

Purification Methods

NH_3 removal

NH_3 is the largest of the nitrogen components in the urea waste water. The recovery of this component by most licensors is done by steam stripping at elevated pressures. This pressure is chosen with respect to the available steam pressures or at a higher pressure in case the stripping of the waste water of the ammonia plant is combined with the urea hydrolysis.

The Stamicarbon stripping process operates at 3 to 4 bar; the Casale process operates at a pressure above 15 bar combined with hydrolysis of urea (Lagana, 1992), the Snamprogetti process operates between 1.5 to 4 bar (Granelli, 1990), and the Kellogg process is carried out at 38 tot 75 bar combined with NH_3 waste water (Czuppon, 1993). A lower stripping pressure is preferred to obtain a low NH_3 content of the effluent. Stripping to less than 1 ppm NH_3 is now a common technology. The stripping columns in today's processes contain twice the amount of trays that they did in the past, when effluent concentrations of 10 ppm were common.

Urea removal

Urea normally has to be removed to the same extent as NH_3 is removed. The urea concentration must therefore be reduced to 1 ppm. All methods are based on the overall hydrolysis equilibrium reaction:

$$H_2O + CO(NH_2)_2 \leftrightarrow 2\,NH_3 + CO_2$$

Since the hydrolysis is an equilibrium reaction, it is not possible to reach a concentration of 1 ppm urea in the effluent, without removal of NH_3 and CO_2. Hydrolysis without removal of NH_3 and CO_2 will result in an effluent concentration of 10 to 50 ppm urea. With feed concentrations of less than 1 ww % of urea and preremoval of NH_3 and CO_2, 1 ppm urea in the effluent can be achieved. This method is not practiced in modern processes because the NH_3 and CO_2 formed during hydrolysis are removed afterwards and the hydrolysis kinetics are also significantly slower with high NH_3 and CO_2 concentrations, resulting in large hydrolyzers. In the modern processes, simultaneous removal of hydrolyzed urea by stripping NH_3 and CO_2 has become common practice.

Stamicarbon and Snamprogetti have similar systems as already shown. Stamicarbon has developed a countercurrent hydrolyzer combined with strippers both upstream and downstream of the hydrolyzer. Snamprogretti has a horizontal hydrolyzer in which fresh steam is fed.

The Casale and Kellogg processes are also similar systems. Both have simultaneous hydrolysis and stripping in one column. Kellogg combines waste-water treatment of both the urea and the ammonia plants. Pressure and temperature conditions for the waste treatment of the Kellogg method are higher because of the availability of high-pressure steam from the NH_3 plant. Operating pressure in the fully combined hydrolysis and stripping lies at a minimum of 15 bar. When the pressure is less than 20 bar, the tray design has to be modified, that is, the liquid residence time on the tray has to be increased by a special downcomer design. This design is used in the Casale process.

Selection of separate stripping steps is favored by the following two factors:
- Stripping is more efficient at lower pressure.
- Low-pressure steam is present in a urea plant in excess.

The reason for prestripping the waste water before it enters the hydrolyzer is to decrease the NH_3 and CO_2 concentrations and thus enhance the urea hydrolysis.

Energy consumption and investment requirement

In this section, a 1,500 metric t/d urea production process with 1 ppm urea and 1 ppm NH_3 in the effluent is selected for the purpose of evaluation.

The energy consumption of both the partly and the fully combined processes are given in Table 1.

Combination of Hydrolysis and Stripping

Urea decrease to 0.1 ppm

Stripping of NH_3 has to be intensified to decrease the equilibrium concentration below 0.1 ppm urea. In the hydrolyzer comprising the partly combined hydrolysis and stripping, the last part of the reaction from 1 to 0.1 ppm increases the hydrolysis reaction time significantly. To avoid a significant increase of the hydrolyzer volume, higher temperatures have to be applied. These higher temperatures increase both the hydrolyzer pressure and the quantity of stripping steam. This means that conditions in the hydrolyzer become more equal to those in the fully combined stripping process.

NH_3 decrease to 0.1 ppm

Figure 5 shows a typical NH_3 effluent concentration for a 30-tray stripper. As can be seen in this figure the steam quantity in a stripper to reach this low NH_3 effluent concentration of 0.1 ppm is roughly twice the amount in the case of 1 ppm NH_3 effluent concentration.

Potential Methods to Remove Urea and NH_3

Catalytic cracking of HNCO from the evaporation

Avoiding urea entrainment from the evaporating devices and catalytic conversion of HNCO to NH_3 and CO_2 in this gas stream is a promising possibility. The gaseous HNCO can be cracked to NH_3 and CO_2 in a simple fixed bed of silica-cat, thus decreasing the urea content in the effluent. As stated earlier, the problem of entrainment should be carefully considered. To decrease the urea concentration in the waste water caused by the evaporator or crystallizer entrainment, special separation features should be applied in these designs. These features could include, among other things, a bigger knock out vessel in the off gases or a wet scrubber. Special attention should be paid to the entrainment eliminators/mist because too much pressure drop increases the energy needed to obtain vacuum conditions in the evaporator or crystallizer. The cracking method has the advantage that the purification unit only consists of a rather simple NH_3 stripper. Other small waste streams (drains, etc.), which are now small parts of the total urea waste-water stream, have to get another destination.

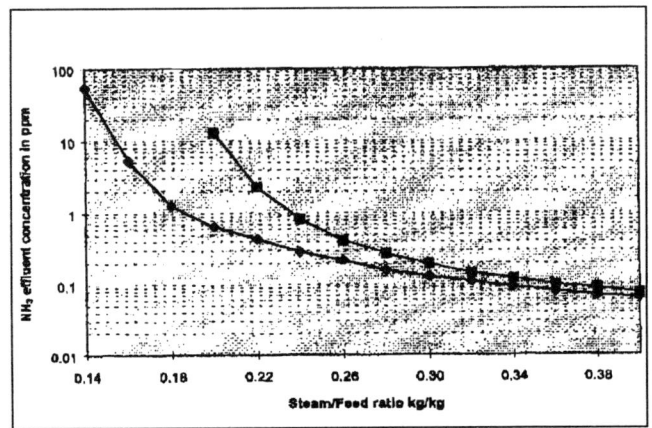

Figure 5. Comparison of stripping pressure.

Table 1. Energy Consumption and Investment Costs of the Partly and Fully Combined Processes

	Steam 3–4 bar	Steam 15 bar
Partly Combined Hydrolysis and Stripping (metric t/t of waste water)	0.3	0.15
Fully Combined Hydrolysis and Stripping (metric t/t of waste water)	—	0.4*

Required Investment Costs	
Partly combined hydrolysis/stripping	$2–4 million
Fully combined hydrolysis/stripping	$1–2 million

* Energy consumption of the Kellogg process will be slightly higher due to higher operating pressure.

Batch/continuous stripping (BC stripping)

BC stripping is an improved stripping method. The method is based on the principle that one of the countercurrent phases is flowing batch-wise through the stripping column while the other phase flows in a continuous way. The advantages in comparison with conventional stripping to get the same effluent results are that the amount of steam required is approximately 50%, the number of trays required is about half, and a combined hydrolyzer and stripper can be smaller due to a larger urea hydrolysis velocity; a consequence of better stripping.

The specification of the effluent out of the BC stripper will be, with the same quantity of steam and trays, at least a factor of ten lower than the specification out of a conventional stripper.

However, BC-stripping trays are more complicated and expensive and the concept of BC stripping is not yet commercially proven.

Enzymatic hydrolysis

With the isolated enzyme urease it is possible to hydrolyze urea at a temperature of 40° to 60°C at atmospheric pressure to NH_3 and CO_2. There is no equilibrium reaction from NH_3 and CO_2 back to urea due to the low applied temperature. The enzymatic hydrolyzer is a good alternative for the thermal hydrolyzer becoming more and more expensive in the sketched future perspective. However, very high peak concentrations of NH_3 should be prevented due to irreversible inactivation of the enzyme. This can be done by simultaneous vacuum stripping or stripping with air or inert during enzymatic hydrolysis. When dealing with high concentrations of urea it is necessary to control the NH_3 level. Some protein (an enzyme is a protein) pollution to slime building can be a disadvantage in the application of the purified waste water for boiler feed water.

Reversed osmosis

Urea can be separated from waste water by reversed osmosis. However, the separation of ammonia is less efficient. To obtain a permeate concentration of 1 ppm NH_3 in one step is not possible because of osmotic pressure of the components (back pressure from NH_3, CO_2, and urea molecules in the water). Moreover, only a restricted concentration increase of the contaminants in the retentate is possible, because the construction of reversed osmosis membranes allows pressures up to 75 bar. This means that still another "smaller," fully equipped purification installation is required for the remaining concentrate.

Ion exchange

NH_3 can be removed in an ionic exchanger. The disadvantage is that the ammonia is converted to a salt, that is, $(NH_4)_2SO_4$. This could be useful in only a few scenarios, for example, if an $(NH_4)_2SO_4$ plant is present at the site. Separate recovery or biological purification of the salt are not options.

Oxidation with hypochloride or reaction with nitrites

Urea reacts with hypochloride and nitrites according to the following two reaction equations:

$$CO(NH_2)_2 + 3NaOCl \rightarrow N_2 + CO_2 + 3NaCl + 2H_2O$$

and

$$CO(NH_2)_2 + 2HNO_2 \rightarrow 2N_2 + CO_2 + 3H_2O$$

These are purification methods for special cases and are generally not considered economically feasible treatment options for urea waste water, especially not in the case of hypochloride, since another raw material is formed and salt is an extra waste component.

Biological waste-water treatment

Urea and NH_3 are converted by living microorganisms to N_2, CO_2, and H_2O. This conversion consists of two steps. During nitrification, NH_3 is converted to NO_3^-. During denitrification, NO_3^- is converted to N_2 and H_2O. As mentioned before, this method is very expensive. Only if concentrations are equal or lower than 0.1 ww %, the method becomes feasible. Other components can also be removed in this type of purification system.

The last three methods we have discussed are not considered economically feasible at this time.

Literature Cited

Czuppon, T. A., "A Method for Treating Ammonia and Urea Condensates," European Patent Application, No. 93107098.1 (1993).

Granelli, F., "Process for Purifying the Effluent from Urea Production Plants," European Patent Application, No. 90202310.0 (1990).

Lagana, V., "Process and Equipment for the Hydrolysis of Residual Urea in the Water from Urea Synthesis Plants," European Patent Application, No. 92118068.3 (1992).

DISCUSSION

Richard B. Strait, *M.W. Kellogg*: I disagree with your conclusion that the Kellogg energy consumption is the highest. In fact, the steam used to do the hydrolysis and stripping is process steam in the ammonia plant, which is about 40 bar steam and basically free. So, I would contend that it is probably the lowest energy consumer of your options.

Killian: I wish to provide a quick response before I go into any detail. As we have heard this morning, we have talked something about revamping ammonia plants, cutting down energy costs and energy savings, and utilizing condensate from ammonia plants. We call 40 bar steam very expensive in the Netherlands, so, certainly it doesn't have a figure zero. What we know with the ammonia urea hydrolyzer from Kellogg, in fact, is that it is based on the availability of high-pressure steam in ammonia plants, and there is no high-pressure steam available in ammonia plants to save a lot more energy. I would like to have discussion with you in detail afterwards, if you don't mind.

Strait: I have two followup comments. First, the steam used is medium pressure steam, not high pressure.

Killian: However, from 60 to the higher figure is mentioned in the patent and we have utilized only available data for the records.

Strait: In practice 40 bar steam is used.

Killian: Maybe you can help me with some more information to update the article.

Strait: I would be happy to. The other item is that, in fact, the process steam will be used in the ammonia plant in any event. The only thing that the tower does is saturate the steam and the stripping is performed at very low cost. The total cost is not the cost of a pound or ton of 40 bar steam. It's very much less.

Maheshwari, *IFFCO, India*: In the Kellogg combined hydrolytic process you are showing two waste streams, one entering at the top containing urea, and the other at the middle containing ammonia. What are these two separate waste water streams?

Killian: We strictly have reviewed only with respect to the urea plant so we have made some assumption ourselves. That's the reason these are very rough figures. However, we have made some homework to split the ammonia and urea plants.

Maheshwari: Are these separate streams coming from the ammonia plant and the urea plant?

Killian: Yes.

S. Balu, *PCS Nitrogen*: During the investigation, I had while getting proposals from Kellogg or Casale or others, no one was prepared to guarantee one ppm ammonia and urea in the final process condensate after the hydrolysis. I'm surprised to see a figure of one ppm or even point one ppm ammonia and urea.

Kilian: Continental Engineering is an independent company, and we are not bound by any other firm. We are sure at least for presentations that we held here, for one company that we have seen on paper, and utilized for a client in Kuwait, that there is a guaranteed figure of one ppm ammonia and one ppm urea.

Tom Czuppon, *M.W. Kellogg*: I'd like to answer the gentleman who asked a question about the combined streams or combined treatment of the streams. Indeed, the Kellogg condensate stripper handles process condensate from the urea plant in the top portion of the vessel and ammonia plant condensate in the bottom portion. A single tower serves to hydrolyze urea and strip the two condensates. To further comment on the energy consumption, the medium pressure steam is essentially free because it is process steam. There is a small penalty to preheat the condensates and to desuperheat the process steam.

Risk Management Plans for Existing Control Rooms

With the advent of API RP 752 "Management of Hazards Associated with Location of Process Plant Buildings" in the U.S., and new Chemical Industry Guidance in the U.K., there is a need to reexamine the vulnerability of control rooms and other occupied buildings on our existing plants. The way one part of ICI Chemicals & Polymers at Teesside, England, has approached this problem at its sites in the northeast of England is described with specific application to ammonia and similar plants.

R. A. McConnell
ICI Chemicals and Polymers, Wilton, Middlesbrough, England

Introduction

The vulnerability of control rooms was brought into sharp focus in England by the Flixborough disaster on June 1, 1974 (U.K. Department of Employment, 1975) when 28 people were killed in the plant control room in the Nypro Chemical Works explosion. The Flixborough accident caused ICI to reexamine its control rooms at its Teesside, England sites and make some improvements to reduce the vulnerability of their occupants in the event of serious plant explosions. These improvements included reducing the size of windows, protecting glass by covering with a plastic film, and replacing heavy light fittings. A new ICI Design Guide was produced that specified how the risk to new plant control rooms should be assessed, and a suitable design process established to protect the occupants of vulnerable buildings. The U.K. Chemical Industries Association (CIA) published a design guide in 1980 (U.K. Chemical Industries Association, 1980), which specified the degree of protection required for new buildings in vulnerable areas. This did not apply retrospectively to existing controls rooms at that time.

Despite these developments, many older buildings continued to be used in locations where modern standards would only allow blast-resistant buildings to be occupied. Arguments put forward at the time (Kletz, 1981) were that the large sums of money required to strengthen existing buildings would probably be better spent on ways of reducing the probability of an explosion. Memories of Flixborough faded during the 1980s, and the urgency to update standards of older buildings became less of a priority.

The Piper Alpha disaster in the North Sea, which occurred on July 6, 1988 (Cullen, 1990), caused 167 fatalities, most of whom were in the accommodation module on the oil rig. This event had substantial impact on the layout and protection required for offshore oil platforms in the North Sea, including retrospective improvements to existing installations to protect the people working and living on them.

Then, on September 21, 1992 a jet flame flash fire occurred at the Hickson and Welch chemical factory in

Castleford, England. The force of the flame demolished a lightly-constructed control building, killing four people inside it, before striking an office building 55 m away, killing one person inside the building. The effect of this incident was to reawaken the chemical industry and the regulators in the United Kingdom to the risks of locating occupied buildings near to hazardous plants. One of the lessons from the United Kingdom Health and Safety Executive Inquiry report (HSE, 1994) was:

"The design and location of control and other buildings near the chemical plant, which process significant quantities of flammable and/or toxic substances, should be based on the assessment of the potential for fire and explosion and/or toxic releases at these plants.

Companies should assess the suitability of existing control buildings and if they are found to be vulnerable, reasonably practicable mitigating action should be taken."

The U.K. CIA Guide produced in the late 1970s needed revision, so the CIA set up a working party of industry members to revise and rewrite U.K. Industry Guidance. This revision has now been issued (U.K. Chemical Industries Association, 1998).

Earlier developments in the United States led to the publication of API RP 752 (American Petroleum Institute, 1995) in January 1995. This contains extensive guidance on managing the risks of existing occupied buildings. The common element in these and other similar publications is the need to reassess existing occupied buildings by carrying out a risk assessment.

In addition a CCPS Book (CCPS, 1996) was published that gives comprehensive guidance on all aspects of evaluating plant buildings. This article describes how one part of ICI Chemicals & Polymers in the northeast of England has approached this issue.

Guidance from API RP 752 and the New U.K. CIA Guide

API RP 752 provides guidance for identifying hazards that may affect process plant buildings, and for managing the risks related to those hazards. It is not intended for use in designing and locating safe refuge from the effects of fire, explosion, and toxic release. It provides a framework that can be used to address facility sitting within the process hazard analysis requirements of OSHA 29 CFR 1910-119 as applied to buildings.

Section 2.3.1 describes a three-stage analysis process for identifying hazards and managing risk to building occupants from explosion. These stages are:
• *Stage 1 Building and Hazard Identification*: Provides an initial identification of the buildings to be selected for further investigation due to their proximity to processes which have the potential for explosion, and their occupancy level.
• *Stage 2 Building Evaluation*: Describes three approaches that can be used to evaluate the potential hazards to buildings.
• *Stage 3 Risk Management:* Outlines the use of qualitative and quantitative risk assessment tools to perform a more complex evaluation for buildings, coupled with proposals for reducing and controlling risk where required.

Figure 3.1 in the API report gives the stages and alternatives for evaluation within this three-stage framework. The guidance continues by describing the evaluation processes, risk criteria, and risk-reduction measures. It also covers fire and toxic release threats to occupied buiidings.

This approach was helpful when we were assessing the risks to ICI's vulnerable occupied buildings.

The new U.K. CIA "Guidance For The Location And Design Of Occupied Buildings On Chemical Manufacturing Sites" was published in February 1998. It is intended to provide advice on the location and design requirements for both new and existing occupied buildings. ICI has contributed to the development of this guidance within the U.K. CIA.

It recommends that companies should have a policy statement to commit the company to the following:
• The company should consider the vulnerability of people in buildings to the hazards that can affect them.
• The company should set criteria against which acceptability of risk can be judged.
• The company should carry out a risk assessment when a building is erected (permanent or temporary) or when a site change is made that increases the risk to people in a building.

The guidance contains flow charts for considering

new and existing buildings, risk criteria and tolerability, methods for assessing vulnerability of occupants, and worked examples. We are now assimilating this guidance into the approach used for assessing occupied buildings.

Risk Assessment Process

A strategy for existing occupied buildings on our sites was devised in line with API RP 752. This is shown as Figure 1.

The first stage of the risk assessment was to identify all those occupied buildings within the hazard zone near hazardous plants. This requires definition of where occupied buildings and hazardous plants are, and how far the hazardous zone extends in hazardous plants.

Guidance on assessing the occupancy of buildings is discussed in API RP 752, and it was decided that the definition of an occupied building should be as simple as possible, so any building in which people are present for more than 2 h in any 24-h period is defined as an occupied building.

For this part of the risk assessment process, it was decided that hazardous plants were defined as those which can threaten buildings due to explosions. Toxic release events and protection for buildings have already been well defined for our plants, and suitable toxic shelters have been provided. For most plants serious fire events have a lower hazard distance than explosion events. However, one or two cases emerged during the study.

Evaluation of the explosion hazard zone around the plants has been a more difficult task, and two different methods have been used to help define the hazard zone: the TNT-equivalent method and the TNO multi-energy method. Details of these methods are described in the CCPS Guidelines for Evaluating Vapour Cloud Explosions (CCPS, 1994).

They allow the modeling of possible gas releases followed by delayed ignition to be analyzed to give an estimate of explosion overpressure blast wave die-off with distance from the explosion. Having studied the effects of overpressure on buildings we decided to define the limit of the hazard zone as the point where the overpressure declines to 1 psi (69 mbar). This judgement of 1 psi (69 mbar) was made after considering various references, including Table 3.3 in API RP 752, and the CCPS Guidance (CCPS, 1994, 1996).

Having identified the list of occupied buildings within the explosion hazard zone, the next stage was to calculate the actual overpressure to which the building could be subjected and then to assess the structural vulnerability of the building.

The TNT methodology was generally used to assess the possible overpressure for specific buildings, although in later cases the TNO multi-energy method was used. The building assessment was carried out by an architectural & civil engineering engineer within ICI.

The next stage was to assess what measures could be taken to reduce the risk, such as moving people out who do not need to be in the building or reducing the risk from the plant itself. In some cases it has been possible to move substantial numbers of people out of vulnerable buildings and close them down.

Finally, the results of the overpressure study and the building survey were considered.

If there were very few sources of explosion, the

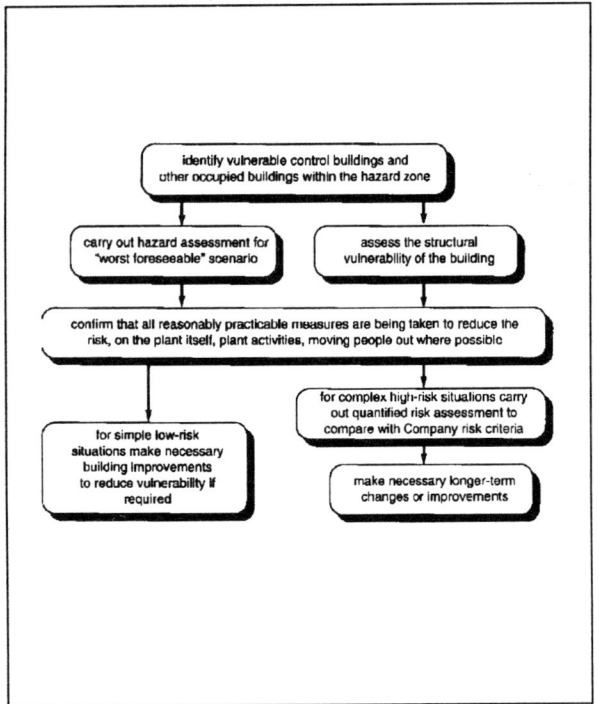

Figure1. Strategy for occupied buildings inside the explosion hazardous zone.

overpressure at the building was low [less than 2 psi (138 mbar)], and the number of people at risk was low, the potential for simple modifications to reduce the vulnerability of the building was examined. These included bricking windows or doors in, minor strengthening to the roof structure, and providing a double entry door with a lobby.

If a significant number of explosion hazard events occurred, and the building was subjected to higher overpressures from some of these events, a quantified risk assessment was carried out. This required the following procedure:
- Identify the possible release scenarios that could occur.
- Assess the approximate frequency of each event.
- Calculate the overpressure at the occupied building for each scenario.
- Define the survivability of the building such that it was assumed that the event would cause serious injury to persons in the building above a certain threshold pressure. An estimate of the survivability of each building was obtained by consultation with the architectural/civil engineering engineers, and varied from 1 to 3 psi (69-207 mbar) depending on the type of frame and strength of the building.
- Add up the frequency of all the events causing overpressure above the threshold. This required an assessment of each event to decide its approximate frequency using generic data, the probability of an explosion, and then the use of the TNO multienergy method to assess the overpressure at the building depending on its distance from the explosion.
- If the total frequency exceeds 35×10^{-6} per year limit, the existing situation is not tolerable and improvements need to be made. For individuals who spend 36 hours per week at work, the Individual Risk is approximately 7×10^{-6} per year which is below the level suggested in the new CIA Guidance on Risk Tolerability (U.K. Chemical Industries Associates 1998).
- Using the table of scenarios it was simple to raise the threshold overpressure to see what overpressure the building must survive to give a frequency below 35×10^{-6} risk limit.
- This then defines the degree and type of strengthening required to improve the building.

The 35×10^{-6} hazard frequency limit derives from internal guidance that the risk from a specific process hazard should be as far below an FAR of 0.4 (per 10^8 hours) as is reasonably practicable (Kletz, 1978). The result of carrying out this more complex assessment has resulted in considerable expenditure (>$100,000) for three existing control rooms to improve their structural survivability.

The results of applying this approach are described below.

Application to an Ammonia Plant

Vapor cloud explosions have occurred on ammonia plants due to sudden release of hydrogen-rich synthesis gas followed by a delay in igniting the released jet of gas. However, an analysis of incidents by Hawksley (1986) indicates that unconfined hydrogen vapor clouds have unusual characteristics:
- Large clouds of gas are unlikely to form due to the rapid diffusivity in the atmosphere. As a result there is not usually a large quantity of gas in the flammable region of 4 to 74% v/v.
- There appear to be no recorded major damage explosions resulting from unconfined or semiconfined releases of hydrogen-rich gas on ammonia plants. Damage effects from observed explosions indicate a low yield energy. However, confined explosions such as in enclosed compressor houses have caused very serious damage.
- Hydrogen in small clouds of a few kg can cause aerial explosions that rattle windows but cause little damage. This is different from typical hydrocarbon explosions.

Hawksley summarizes the features of the explosions reviewed as follows:
- No explosion appeared to cause damage indicative of high overpressures being generated.
- The explosion yield (based on TNT) on average was about 1%.
- Quantities of the order of tens of kilograms (or less for 100% hydrogen) can cause explosions sufficient to cause window damage over a wide area.
- The incidents involving gases in which more than 50% is hydrogen are characterized by rapid ignition (10 s on average) that limits the quantity of gas

exploding.

In many respects this analysis suggests that the TNT-equivalent method may be more appropriate for estimating hydrogen-rich gas explosions rather than the TNO multi-energy method. High energy releases are unlikely to fill large structures in an ammonia plant before ignition occurs. However, application of the TNT method requires judgement to take into account the factors observed in actual explosion events.

For the worst case scenario for an ammonia plant, it could be assumed that all the synthesis loop is released at once and contributes to an explosion containing the synthesis loop inventory of, say, 13.5 metric t (14.85 t) of gas. Taking a typical overall heat of combustion for the gas of 23 MJ/Kg (9,888 BTU/lb), the TNT equivalent for 1% yield (as suggested by Hawksley) is 675 kg (1,488 lb), giving a distance to 1 psi (69 mbar) of about 154 m (505 ft).

The method used for our assessments was considerably more sophisticated than this but gave a similar answer.

The 1980 U.K. CIA Guide suggested that to identify a "worst foreseeable" scenario, a possible weak branch or similar weak point should be identified within the synthesis loop. The location of a vulnerable branch might be identified as 4-in. (10.16-cm) diameter branches on the air-cooled condensers upon exit from the synthesis gas compressor.

The estimate of initial release rate is about 180 kg (397 lb) per second. Assuming this high velocity jet forms a flammable cloud, the maximum size of flammable cloud forms after 8 s., and contains approximately 517 k (1,140 lb) of flammable gas between the flammable limits. This is derived using equations derived by Marshall (Marshall, 1977). Assuming 10% of the heat of combustion translates into blast energy (which is typical for energetic gases such as hydrogen having calculated the amount in the flammable cloud), the TNT equivalent is 694 kg (1,530 lbs), giving 155 m (508.5 ft) to the 1 psi (69 mbar) explosion hazard zone, similar to that calculated above.

Application of the TNO method is simpler in that it does not require the judgement of what the release rate could be, but does require an estimate of the volume of semiconfined structure within the plant which could be filled with flammable hydrogen rich gas. It is then assumed that this gas is at stoichiometric composition and ignites to cause the explosion. A typical scenario might be that the 150 Nm^3 (5.565 mscf) of gas in the synthesis loop at 200 barg (2,900 psig) expands to fill 2,000 m^3 (70,630 ft^3) of structure in the plant, in which case the distance to the 1 psi (69 mbar) explosion hazard zone is 170 m (558 ft) for the highest strength (7–10) explosion. However, the physics of a high-energy gas jet release, combined with subjective assessment techniques, make this method difficult to apply until more guidance on its application is available.

Other scenarios for causing an explosion include a release of methane-rich feed gas into the plant and a release from the front-end shift, CO_2 removal, or methanation sections. In each case, the potential explosion effects are less severe that the synthesis loop gas explosion. Methane has a significantly lower flame speed, which may allow use of strength 5 explosions within the TNO method, giving significantly lower explosion overpressures.

Having examined the possible scenarios, an assessment was made on the effects on the plant control room, which was located approximately 75 m (246 ft) away from the expected explosion location. The worst case overpressure is between 2 and 3 psi (138 and 207 mbar) at this point, and the building was originally built to withstand this. There are no windows in the control room, and the support structure and wall thickness is substantial. No further work was required for this building.

Application to Other Plants and Buildings

A total of 12 vulnerable buildings and control rooms was identified during this study and different levels of assessment were carried out for each. Specific actions resulting from this were:

A 2-story maintenance building with offices and a conference room was located centrally in a petrochemical plant. Quantified risk assessment clearly showed that the risk to people in this building was too high, so the building has now been closed and the people moved out.

The control room for an LPG terminal, located

40–60 m (130–200 ft) from compressors and storage units, was located in a small building that was in poor condition and unlikely to survive more than 1 psi (69 mbar) overpressure. A new project to install an additional processing unit was proposed. A quantified risk assessment for 30 scenarios for the existing units, combined with an additional 4 scenarios from the proposed new unit, showed that the risk to people in the building was above the guidance on risk limits. The building needed to be strengthened to withstand 1.75 psi (121 mbar) to reduce the risk below the limit, so expenditure to improve the building was sanctioned as part of the new project.

A petrochemical unit constructed in the 1960s has its control unit located 30–40 m (100–130 ft) from the edge of the processing units containing boiling hydrocarbons. A systematic quantified risk assessment using the approach described above and calculating semi-confined volumes using the TNO multienergy method indicated that the risk to people in the control building was above the limit. Extensive strengthening work, including installing a steel support frame outside the building, removing windows, adding a porch and lobby, and providing internal supports at the wall to roof joints has been carried out to reduce the vulnerability of the building. These modifications reduced the risk below the limit.

For some of the other buildings, minor changes such as removing windows in buildings and adding double doors with lobbies have been carried out. In a few cases no changes were required.

During this process, it was most important that open and direct discussions were held with the people concerned. In many cases there was some surprise that the risks that had existed for the past 15 or 20 years had now become unacceptable, particularly when there was an apparent loss of amenity such as blocking out windows facing the plant. In general, the argument that greater awareness of these hazards, the very low residual risks involved, and the improvement in safety resulting from the changes, was accepted and agreed to by plant staff.

Conclusions

Recent developments have increased the awareness of the potential risk to occupants of vulnerable buildings located close to hazardous chemical plants. Due to continued operation of older plants, there are still vulnerable occupied buildings in use today. Publications such as API RP 752 and the recent 1998 U.K. CIA Guidance have provided a framework for managing these risks by carrying out a systematic review. This article has described the way ICI Chemicals & Polymers in the northeast of England is managing and reducing these risks.

Literature Cited

American Petroleum Institute, API Recommended Practice 752, "Management of Hazards Associated with Location of Process Plant Buildings" (Jan. 25 1995).

Center for Chemical Process Safety, "Evaluating Process Plant Buildings for External Explosions and Fires," AIChE, CCPS, New York (1996).

Center for Chemical Process Safety, "Guidelines for Evaluating the Characteristics of Vapor Cloud Explosions, Flash Fires and BLEVEs," AIChE, CCPS, New York (1994).

Cullen, Hon Lord, "The Public Inquiry into The Piper Alpha Disaster," Department of Energy, London, HMSO, ISBN 0 10 113102 (Nov. 1990).

Hawksley, J. L., "Unconfined Vapour Cloud Explosions Involving Hydrogen-Rich Gases Estimating the Blast Effects," Institute of Chemical Engineers Loss Prevention Bulletin 068 (April 1986).

HSE Books, "The Fire at Hickson & Welch Ltd. A Report of the Investigation by The Health and Safety Executive Into the Fatal Fire On 21 September 1992," ISBN 0 7176 0702 X (1994).

Kletz, T.A., "How Far Should We Go In Bringing Old Plants Up To Modern Standards?," *Loss Prevention*, AIChE (1981).

Kletz, T. A. "Practical Applications of Hazard Analysis," *Chem. Eng. Prog.* (Sept 1978).

Marshall J. G., "The Size of Flammable Clouds Arising from Continuous Releases into the Atmosphere," IChemE Series No 49, pp. 99-109 (1977).

U.K. Chemical Industries Association, "An Approach

to the Categorization of Process Plant Hazard and Control Building Design," (1980),

U.K. Chemical Industries Association, "Guidance for the Location and Design of Occupied Buildings on Chemical Manufacturing Sites," Kings Building, Smith Square, London SW1 P 3JJ ISBN 1 85897 077 6 RC21   Reference CIA/CISHEC/9802/CP/500/2M (Feb. 1998).

U.K. Department of Employment, The Flixborough Disaster Report of the Court of Inquiry,"HMSO, ISBN 011 361075 0. (1975).

DISCUSSION

R. Squire, *Dupont*: We did a quantitative risk assessment approximately 10 years ago and decided to review our ammonia distribution and handling system using inherently safer processes. We concluded that we needed automatic shutdowns on several process lines. We thought we had adequate temporary safe havens, but they turned out not to be. Depending on the response time, if you can respond in 3 min to leaks, great. If you have a 15 or 20 min response like most, you can be in trouble if you don't have automatic shutdowns.

McConnell: Yes. I certainly go along with that.

Risk Management Plan Modeling for Ammonia Retailing

The modeling of off-site consequence of ammonia releases in support of The Fertilizer Institute (TFI) guidance is discussed. Worse-case and alternate ammonia release scenarios were modeled for both anhydrous ammonia and aqueous ammonia. Results for storage for 10,000–170,000 lb of anhydrous ammonia and 10,000–200,000 lb of aqueous ammonia (30%) are presented.

Gale F. Hoffnagle and Steven E. Zell
TRC Environmental Corporation, Windsor, CT 06095

Introduction

The Degadis model was selected for two primary reasons: it compares favorably and conservatively to the only full scale field tests of ammonia releases (Desert Tortoise), and because it is not a proprietary model and thus available to all parties to use or replicate the results. Because the accidental releases being modeled are of short duration, that is, 2 and 10 min., Degadis was run in the transient mode rather than steady-state mode. The steady-state mode assumes that the release occurs continuously at a constant emission rate. The transient mode allows for short-term releases to be followed as a puff downwind or for variable release rate problems such as evaporation from a pool to be modeled. The input variables are as shown in Table 1.

Degadis has been run with three different surface roughness coefficients: (1) rural (3 cm); (2) agricultural rural (20 cm); (3) urban (100 cm). For anhydrous ammonia, the quantity of ammonia flashed as vapor was calculated and the remainder was assumed to be entrained as aerosol, and the density of the cloud was adjusted accordingly. For aqueous ammonia (30% ammonia), it was assumed that a pool having a depth of one cm was formed, and the ammonia evaporation rate was calculated as a function of time. The alternate-case anhydrous ammonia released from a broken pipe assumes the flow out of the pipe can be calculated using the Bernoulli equation. This method is conservative since it ignores the choking action of two-phase flow in the pipe.

The EPA-specified "toxic endpoint" for ammonia is derived from the ERPG-2 value of 200 ppm which is a

Table 1. Input Variables.

	Worst Case	Alternate Case
Windspeed m/s	1.5	3.0
Stability class	F	D
Temperature, °C	25	25
Relative humidity, %	50	50
Release time, min	10	2

Table 2. Modeled Release Scenarios

Figure	Substance	Release Time	Release Amount
1	Anhydrous Ammonia	10 min	10,000–180,00 lb
2	Anhydrous Ammonia	2 min	3,000–46,000 lb
3	Aqueous Ammonia (30%)	Evaporation	10,000–200,000 lb
4	Aqueous Ammonia (30%)	Evaporation	1,000–10,000 lb

one-h average value. We evaluated the distance to the endpoint as follows:

- The 200 ppm 1 h average is equivalent to a one-half h average of 283 ppm and a 5 min average of 690 ppm.
- A post-processor was coded to use the Degadis output to calculate 1 h, 1/2 h and 5 min averages.

The maximum of the three distances was used to select the result.

The other model results were obtained from EPA documents for comparison.

Table 2 provides a summary of the scenarios modeled with Degadis. Figures 1–4 provide the Degadis results and Figure 5 shows the comparison to other models. Figure 6 shows the Degadis predicted concentration for each time step for a 100,000 lb release.

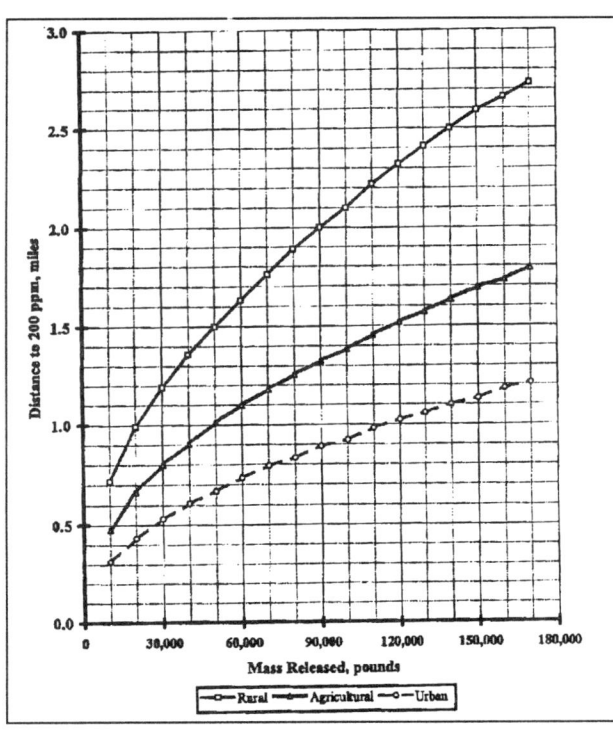

Figure 1. Worst-case ammonia releases.
Distance to 200 ppm as a function of mass released (60-min averaging time).

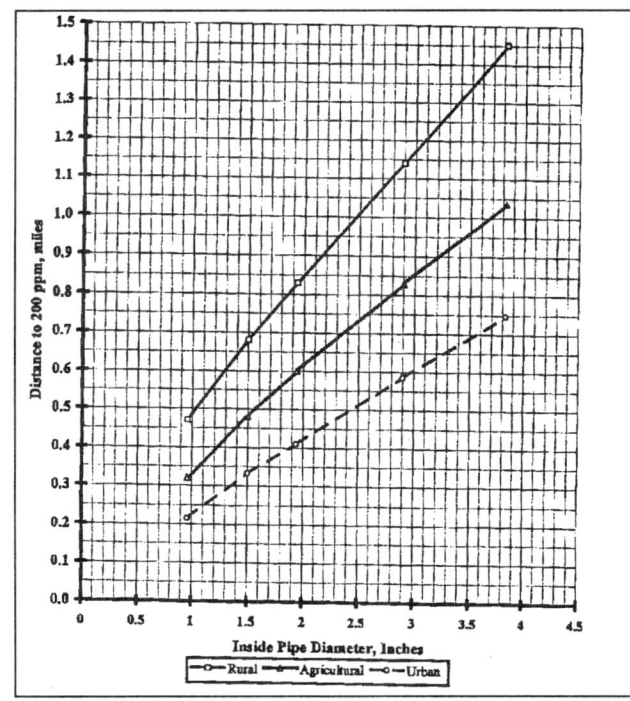

Figure 2. Alternate case anhydrous ammonia releases.
Distance to 200 ppm (60-min average) as a function of pipe diameter (3.0 m/s wind speed, D stability).

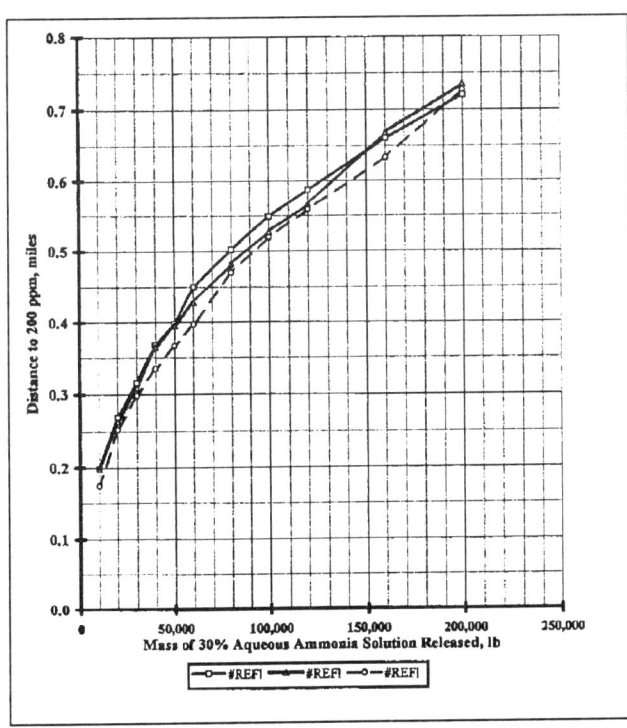

Figure 3. Worst-case 30% aqueous ammonia releases.

Distance to 200 ppm (60-min average) as a function of mass of solution released (1.5 m/s wind, F stability, 77°F at 50% humidity).

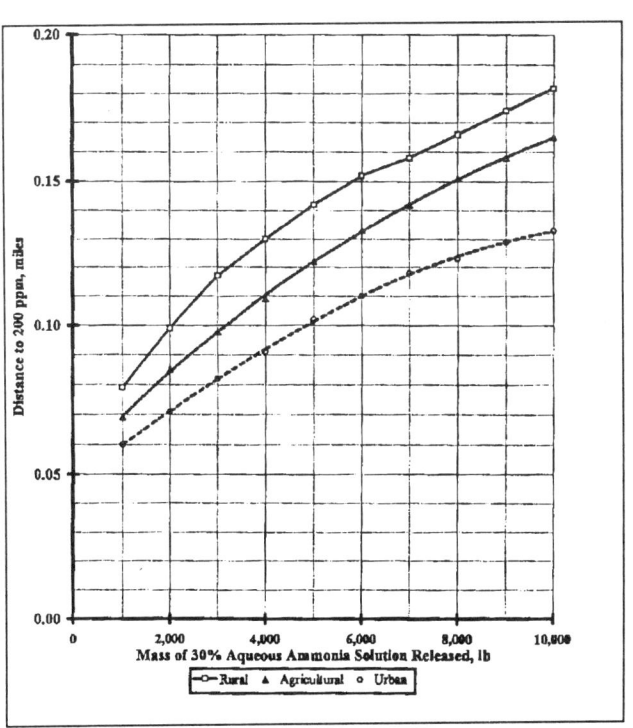

Figure 4. Alternate case 30% aqueous ammonia releases.

Distance to 200 ppm (60-min Average) as a function of mass released 3.0 m/s wind, D stability, 77°F at 50% humidity.

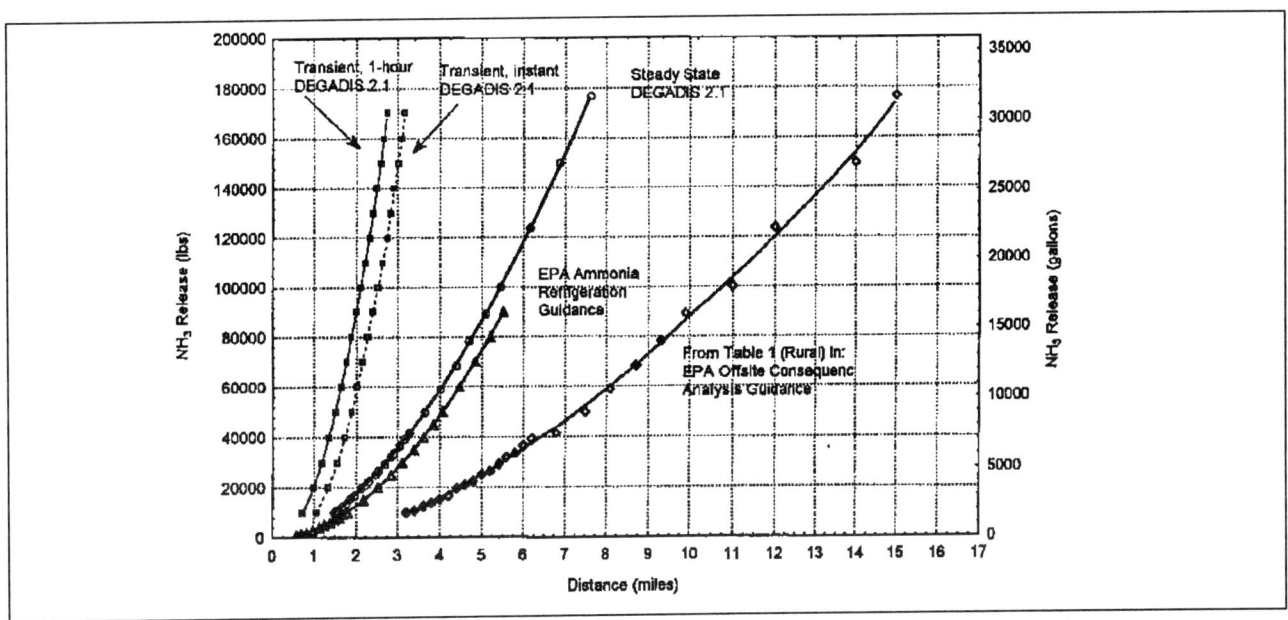

Figure 5. "Distance to endpoint" for worst-case anhydrous ammonia releases in rural terrain settings.

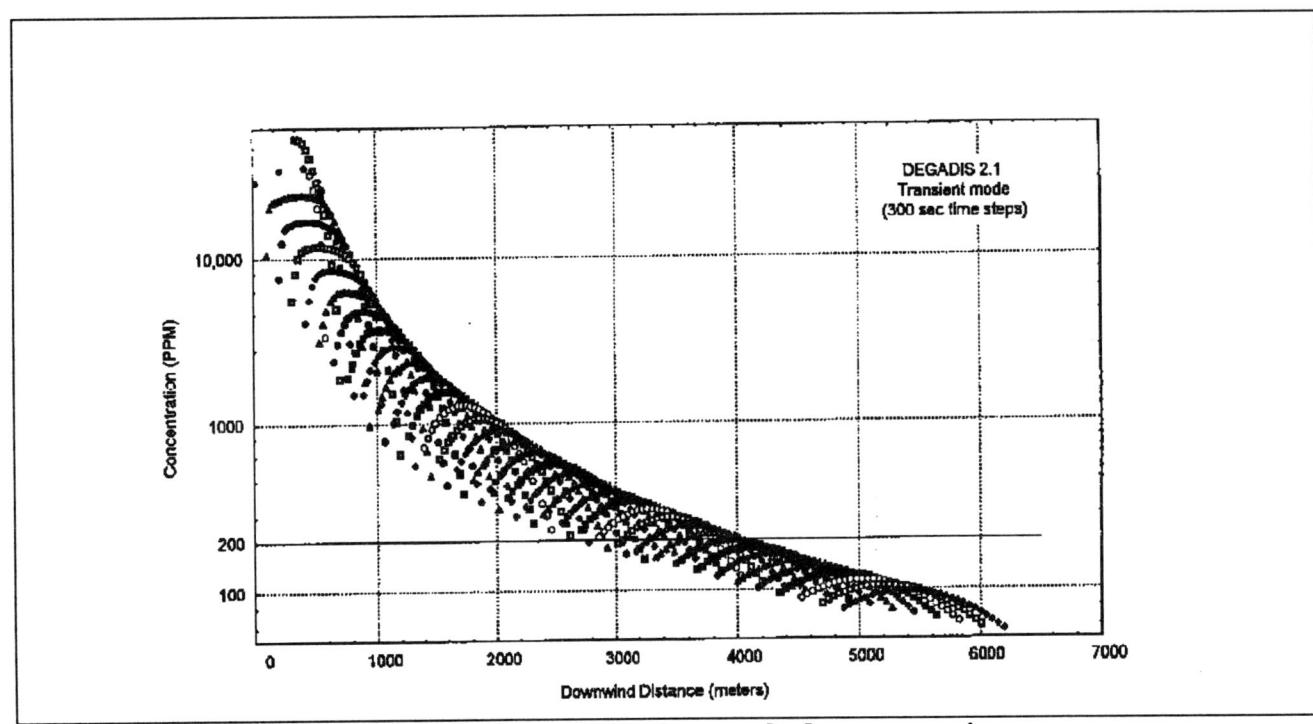

Figure 6. TFI Worst case release: anhydrous ammonia.
(100,000 lbs in 10 min.)

Results

The Degadis results, in the transient mode, give much shorter distances than the Degadis steady-state results. This is to be expected since the release times are short compared to steady-state and the plume is in reality a puff moving downwind. The diffusion in the x (downwind) direction is very important for these short-term releases. Both EPA Guidance documents overstate the distance to the toxic endpoint because of the use of steady-state and/or other conservative assumptions.

A comparison of the exposure based on various averaging times yields the longest distance for the 1 h average. This is also primarily due to the spreading of the puff in the x (downwind) direction.

Conclusions

(1) Degadis can be used for the determination of off-site consequences of ammonia.

(2) Degadis, in the transient mode, yields distance which are more accurate for short-term releases.

(3) The Degadis post-processor shows that 1 h averaging time gives the longest distances for offsite consequence analysis.

Literature Cited

40 CFR Port 51, Appendix W, Appendix B.

Lawrence Livermore National Lab., "Desert Tortoise Series Data Report, Report UCID-20562 (Dec. 1985).

DISCUSSION

R. Squire, *Dupont*: Interesting information. I guess what always concerns me as I struggle with this regulation is what do you report to the government and what do you try to do for the local LAPC. The distances you report may be "correct" but misleading if you have cryogenic ammonia, which will flash and rise over the top of things. Warm ammonia will autorefrigerate, and the cloud will hang on the ground for as long as it takes to get the energy to vaporize the ammonia aerosol, for it to warm and then rise So, there are considerably different cases for cryogenic or warm ammonia.

Gale F. Hoffnagle: Yes. When you get to larger tanks or bullets, for instance, that we have in manufacturing facilities, you can get substantial differences. Large refrigerated tanks, for instance, up to 30,000 t or so in a diked area run two to three to four miles depending upon the meteorology. Although they look to the public and they look to everyone else like they're a gigantic problem, they are not really a gigantic problem if they are diked properly. But, your question was, yes, we're calculating something for the EPA. How we explain it to the public is a second issue. Once we calculated it for EPA, how to evaluate it and calculate it and give the public an understanding of what the issue is becomes a very different question.

Squire: From my experience having gone through "safety street" is that the local LEPC needs your help to explain and help them understand all the details, because they're nontechnical people and the technical aspects absolutely overwhelm them.

Hoffnagle: That's what I do, explain it to LEPC, and that is what has to be done. You're right.

Rudy Frey: *M.W. Kellogg*: I'm surprised with the correlation that you get with vapor ammonia clouds.

Hoffnagle: Well, Desert Tortise told us that the original experiment we're talking about was anhydrous ammonia released in a horizontal jet where clearly you had vapor plus aerosol liquid. The debate has always been how much aerosol liquid dropped out on the ground and how much aerosol liquid stayed with the cloud to make it dense. However, the reasoning or the concept out of that is most ammonia releases that aren't refrigerated down to the boiling point are going to be dense clouds. So, most of the modeling we're doing for ammonia facilities is with a dense cloud model, and once you are doing most of the work with a dense cloud model, you can use...vapor clouds.

Frey: I think the release model being issued by CCPS also addresses aerosols, and it's still somewhat controversial. With regard to your Figure 6, I'm always confused by these kind of dispersion programs where you have an end point of 4,000 m for your 200 parts per million for a 10 min puff release. What in fact do you mean by that? Do you mean that, during this 10 min interval, it might have gone up to 1,000 to 2,000 and for the next 50 min it has decayed down to a level of essentially zero and your overall average then becomes something like 200?

Hoffnagle: No.

Frey: The transient effect of this is very confusing, I believe, and I'm not even sure the dispersion program is correctly analyzed.

Hoffnagle: In my humble view as a meteorologist, we're doing pretty well at handling this, but this is a 10 min release which is only part of an hour. So, we are following this 10 min puff, as it goes to downtime step by time step. As it gets out to 4,000 m here, you can see that it gets flatter and flatter and that's because it's moving. It's disbursing in the downward direction. So, for instance, these are 10 min time steps, one, two, three, four, five. So, at 4,000 m downwind, the plume passes you in a lot longer time than 10 min. It takes almost 45 min from the time the first ammonia reaches you to the last ammonia trails off at the end.

Frey: Yeah, that's intuitively obvious. I'm trying to find out what this averaging really means. When you have 200, did it reach a peak of 2,000 or did it reach a peak of 2,050?

Hoffnagle: Well, that's why we wrote this post-processor which tells us what the distance is to the maximum for a half hour average of 238 ppm or a 5 min average of 690 ppm to make sure that no one is getting an average concentration over an hour of more than 200 parts per million even if they are closer.

Frey: I'll talk to you later about it.

Case Study of CO_2 Removal System Problems/Failures in Ammonia Plant

The article addresses the various problems/failures experienced in the CO_2 removal system of an ammonia plant in a short span operation of less than one year. Probable causes of failures and the corrective steps taken to avoid such failures in the future have also been discussed.

V. K. Bali and A. K. Maheshwari
Indian Farmers Fertilizer Cooperative Ltd. (IFFCO), Aonla Unit, Bareilly, Uttar Pradesh, India

Introduction

Indian Farmers Fertilizer Cooperative Ltd. operates two Ammonia plants, each with a name plate capacity of 1,350 MTPD of ammonia. Both of these plants have been designed based on Haldor Topsøe technology with steam reforming of natural gas and/or naphtha. Ammonia-1 is designed for natural gas feed stock and was commissioned in 1988. Ammonia-2 was commissioned in December, 1996 and is designed for both natural gas and naphtha feedstocks. The Benfield process was selected for the CO_2 removal system of Ammonia-1 which has been converted into the Giammarco-Vetrocoke (GV) dual activator system in April 1997 for achieving lower CO_2 slip and energy savings. For the Ammonia-2 plant, the GV dual activator low energy process has been selected for CO_2 removal system from the design stage. This article describes the problems/failures experienced in the CO_2 removal system of the Ammonia-2 plant during the very first year of its operation.

Process Technology Adopted For CO_2 Removal System

The CO_2 removal system of the ammonia Plant has a conventional design based on the GV dual activator process. The process is comprised of single-stage absorption and two-stage regeneration. Figure 1 shows the CO_2 removal system flowsheet.

Carbon dioxide is removed by absorption in hot aqueous potassium carbonate solution containing approximately 30 wt. % potash (K_2CO_3), partly converted into bicarbonate ($KHCO_3$). The solution further contains dual activators to effectively improve the overall performance of the system. Vanadium oxide is used as a corrosion inhibitor.

The process gas from the shift reactors is passed to the vetrocoke absorber, which contains stainless steel packing material distributed in five beds. The absorption is carried out in one stage. The major part of the circulating solution is fed without cooling to the middle of absorber at about 241°F (116°C). The remaining solution is fed to the top of the absorber after cooling to about 140°F (60°C). In the lower zone of the absorber, the bulk of the CO_2 is absorbed. In the upper zone, the reduced stream of cold solution is used to get low CO_2 slippage due to the low CO_2 vapor pressure of the dual activated solution.

The solution leaving the absorber bottom loaded with CO_2 is called the rich solution. The rich solution is transferred to a two-stage regeneration system operating at low pressures. The rich solution is depressurized through the hydraulic turbine and is sent to the top of the first regenerator operating at 14.2 psig (1.0 Kg/cm$_2$g) pressure. A stream of rich solution extracted from the top of first regenerator is depressurized through a control valve and enters the top of the second regenerator, working at a low pressure of 1.42 psig (0.1 Kg/cm$_2$g).

Corrosion Control In CO_2 Removal System

GV solution along with CO_2 at boiling temperature is very corrosive and would normally require stainless steel equipment. However, carbon steel equipment with passivation layers (oxidation layers) are being used successfully. The desired passivation layer is formed by controlled passivation in two phases called static passivation and dynamic passivation. The layer formed is tight, magnetic and tenacious and protects the carbon steel surfaces from corrosion. However, rubbing with hard sharp edges can scratch the layer. The GV system uses vanadium as a corrosion inhibitor. The recommended concentration of total vanadium is around 0.5% by weight as V205.

The hot potassium carbonate inhibited with vanadium can be safely operated, but is very sensitive for corrosion. In order to maintain the electrochemical potential required for the protection of the passivation layer of metallic surfaces, it is necessary to keep 30 to 40% of the total vanadium in the pentavalent form and never be allowed to be lower than 20%. This ratio is kept by means of the oxidation unit which treats a side stream solution with air.

Hence, formation of the proper passivation layer and its protection is very essential to avoid corrosion in the GV system equipment. Any damage to the passivation layer can cause very fast corrosion and subsequent leakages.

Problems/Failures Experienced

The various problems faced in the CO_2 removal system of the ammonia plant have been presented. Each problem/failure has been dealt with separately specifying the problem/failure faced, the cause of the problem, and various corrective steps undertaken to avoid such occurance in the future.

Second regenerator

The second regenerator is a carbon steel tower provided with stainless steel internals and operating at a pressure of 1.42 psig (0.1 Kg/cm$_2$g). It is a packed tower having 129 in. (3,230 mm) diameter and 1,643 in. (41,075 mm) height. It also contains stainless steel packing material distributed in two beds. Broad specifications and the general arrangement drawing of the second regenerator are shown in Figure 2.

Problem/failure description

Based on the failure history of this equipment in other plants, it was decided to measure the thickness of the second regenerator shell in the failure prone zones after about 9 months of operation. While the thickness measurement was in progress between A-2 and A-7 nozzles, a leak was observed on the other side of the shell on Nov. 9, 1997. This leak was located approximately 6 in. (150 mm) above the A-2 nozzle and 88 in. (2,200 mm) circumferentially towards the M-3 manhole. Initially, a hole of approximately 0.8 in. (20 mm) in diameter was observed which enlarged to a bigger size "eye shaped" hole within 2 h of the start of leakage as shown in Figure 3.

Thickness measurements were carried out around the leaking hole to ascertain the extent of thining in the shell. No thinning was found even around the hole, leading to the conclusion that the failure was localized as shown in Figure 4. The area of leakage was covered and welded with SS-304, 8 in. sch 10 pipe with a blind and a vent to arrest the leakage. The whole exercise

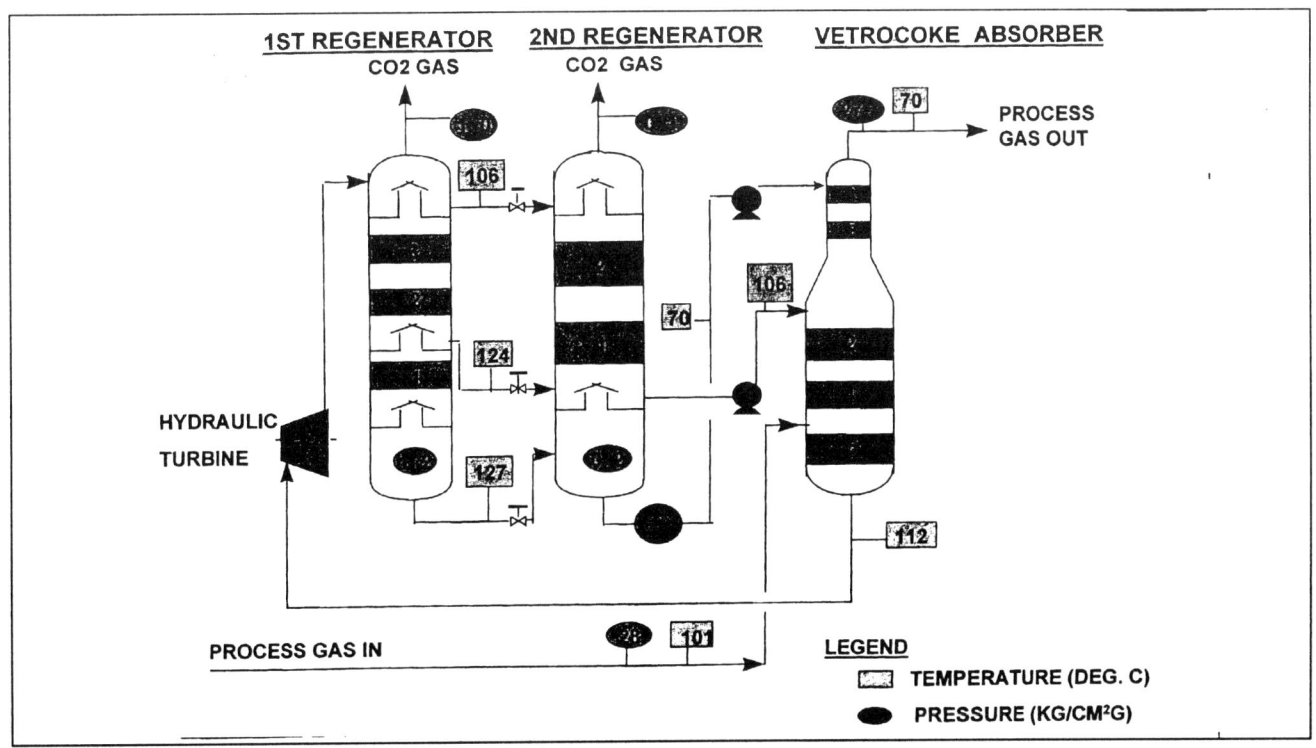

Figure 1. CO_2 removal system flowsheet.

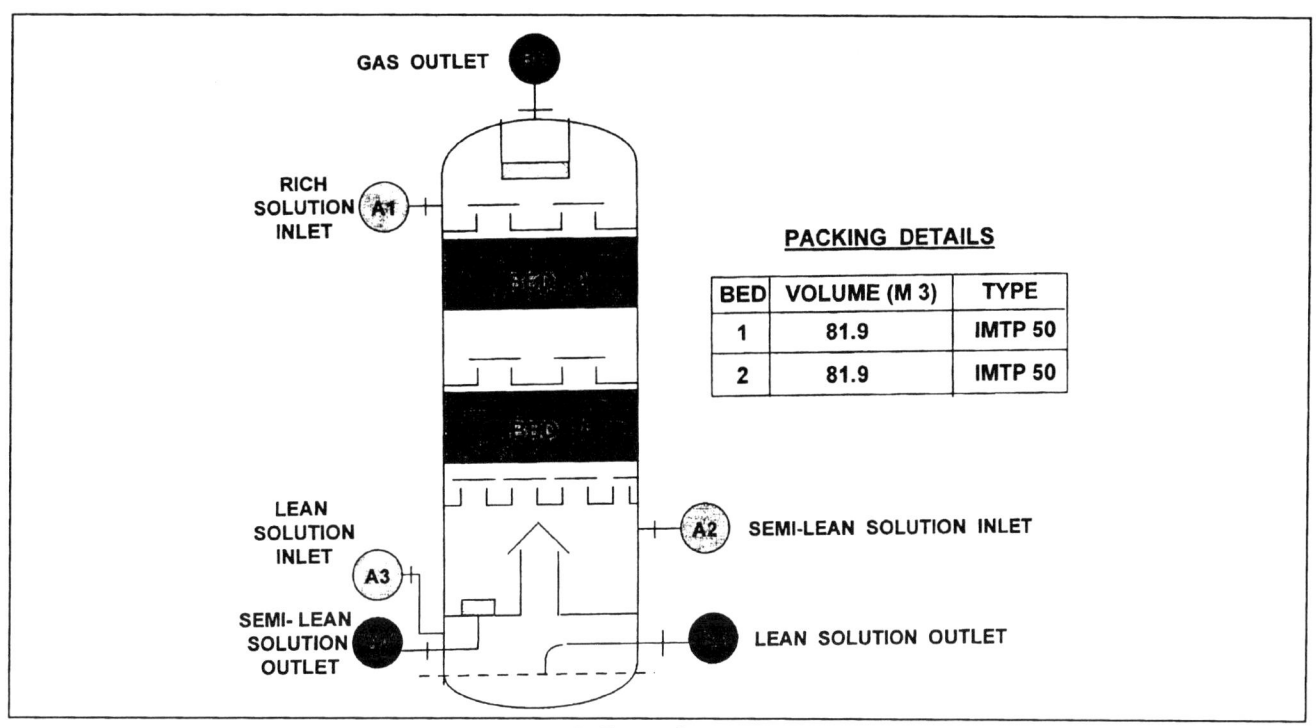

Figure 2. Second regenerator.

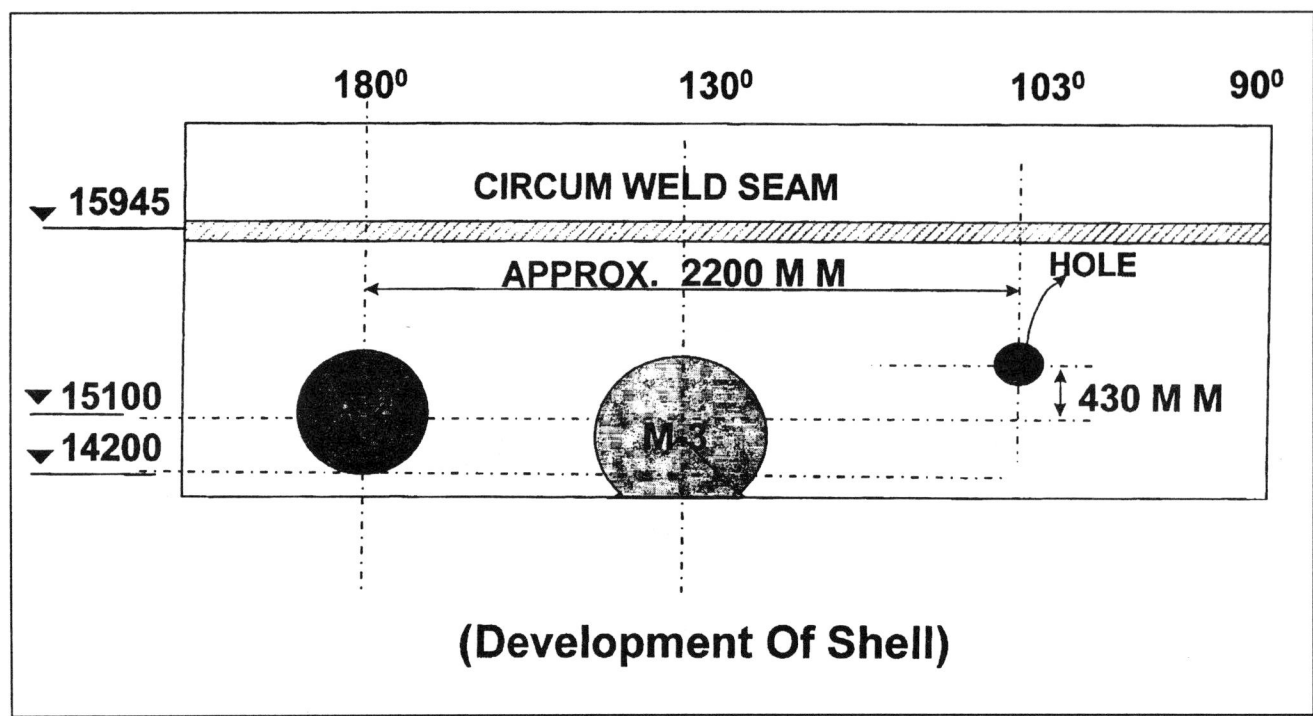

Figure 3. Second regenerator: leakage due to corrosion.

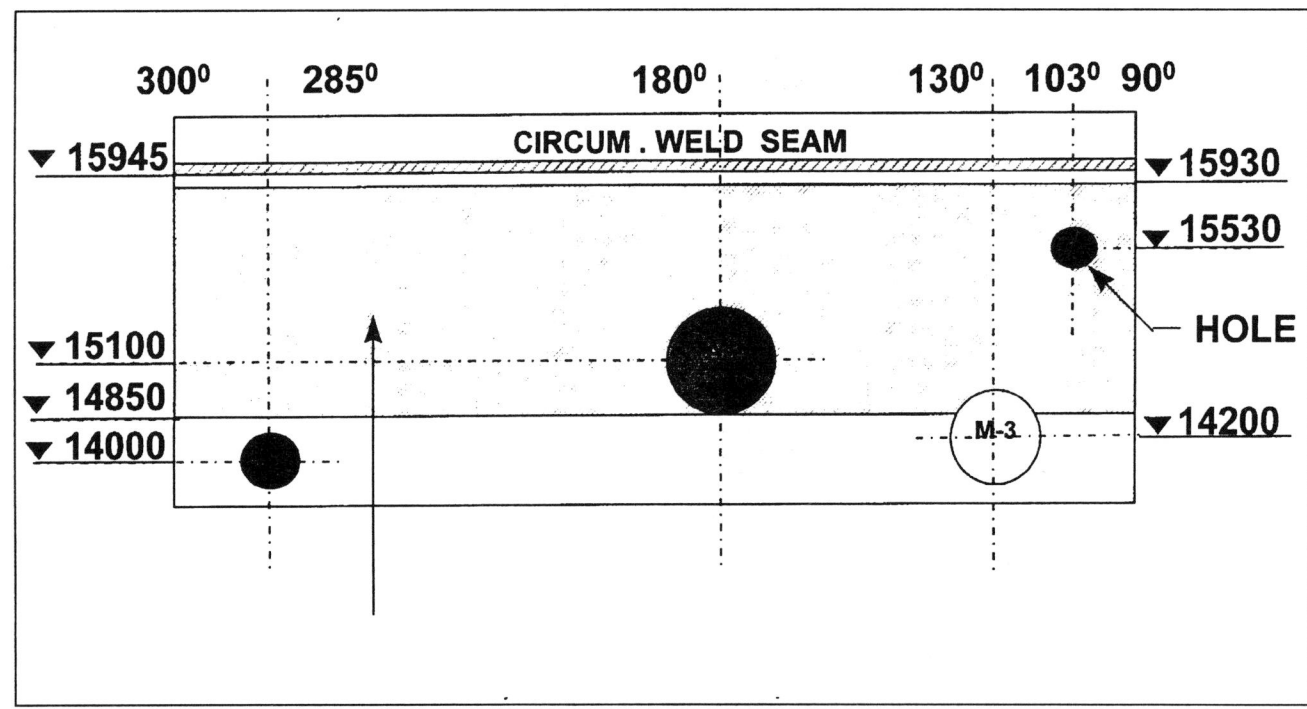

Figure 4. Second regenerator: leakage in shell due to corrosion.
Thickness measurement was carried out in the hatched area. No reduction, except failure, was observed.

was carried out while the plant was in operation and the equipment was in line.

As a preventive measure, it was decided to carry out thickness measurements around the affected area of the shell on a regular basis. It was observed that the thickness had been reduced to as little as 0.56 in. (14 mm) around the 8 in. stainless steel pipe, which had been welded to contain the leak. Stainless steel pads were welded around the 8 in. stainless steel pipe to strengthen the shell. The second regenerator was opened for inspection in October 1997. A hole of approximately 12 in. (300 mm) in diameter was observed from inside. Welding of the cleat between the SS 304 shroud and the vessel was also found broken. Photo 1 shows the damaged second regenerator shell.

In addition to the leakage in the shell, the following upset conditions of a minor nature were also observed in the second regenerator.
- Three segments of the steam distributor above the chimney at an elevation of 644 in. (16,100 mm) were found lifted from their support beam by about six to eight in. (150 to 200 mm), as shown in Figure 5. Photo 2 shows the disturbed internals of the steam distributor.
- The Bed No. 1 liquid re-distributor (Norten type) at an elevation of 26,750 mm was found lifted from its support ring by about 50 mm on one side, as shown in Figure 5.

Cause of the problem

The mechanical design of the fluid entry zone of second regenerator was found to be inadequate.

The support design of the stainless steel protective shroud was inadequate for the dynamic loads. This resulted in the shroud supports cracking and allowing the cleats of the shroud to hit the wall which broke the protective vanadium layer on the carbon steel shell. This allowed rapid corrosion of the shell and subsequent failure of pressure vessel (Figure 6).

Upon investigation it was also discovered that the material of the cleats was carbon steel rather than the stainless steel material specified on the approved drawings. Also, the number of cleats provided were at variance with the approved drawings.

It was further concluded that an internal annular passage should be avoided if possible to safeguard the installation against even the slightest possibility of any means of damaging the passivation layer. The welding of cleats to the shell to support the annular passage, which is likely to have some degree of vibration due to the process conditions, was identified to present such a possibility.

Various design options

Design problems in the fluid entry zone of the second regenerator were reported in 1993 in a 900 MTPD ammonia plant operating in the southern part of India. The flow was directly hitting the chimney which broke off and rubbed against the tower wall. This resulted in damage to the passivation layer and caused corrosion which, in a short period of time, resulted in equipment failure. Another operating problem experienced in this equipment was the difficulty in maintaining the proper solution level.

During the engineering and procurement phase of the IFFCO-Aonla Unit, it was decided to re-engineer the inlet arrangement using a Norton type inlet arrangement and providing an annular passage. The inlet arrangement was designed to divert the flow in two directions horizontally to avoid direct impingement on the chimney. A stainless steel annular passage was provided to prevent liquid impingement on the shell and possible disturbance of the passivation layer which could cause excessive corrosion. An annular passage was provided with an annular ring at the lower end to limit the disturbance of the surface, that is, to improve level measurement. The above arrangement is shown in Figure 7. Photo 3 shows the original inlet distribution arrangement.

However, ammonia plant of Aonla Unit and other similar plants based on above design have reported failures in this equipment in the fluid entry zone.

Yet another design option has been considered which has now been adopted and is described separately in this article.

Repairs/modifications

- The area of 24 in. x 24 in. (600 mm x 600 mm)

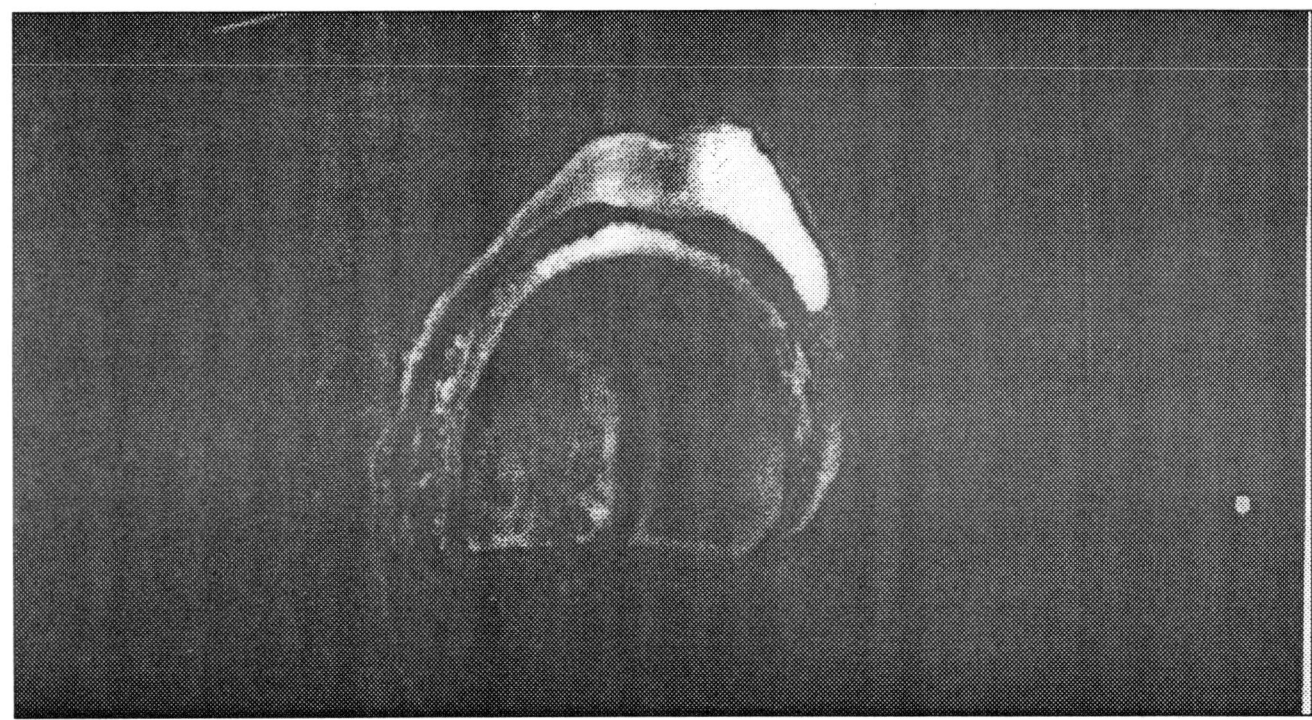

Photo 1. Damaged shell near solution inlet.

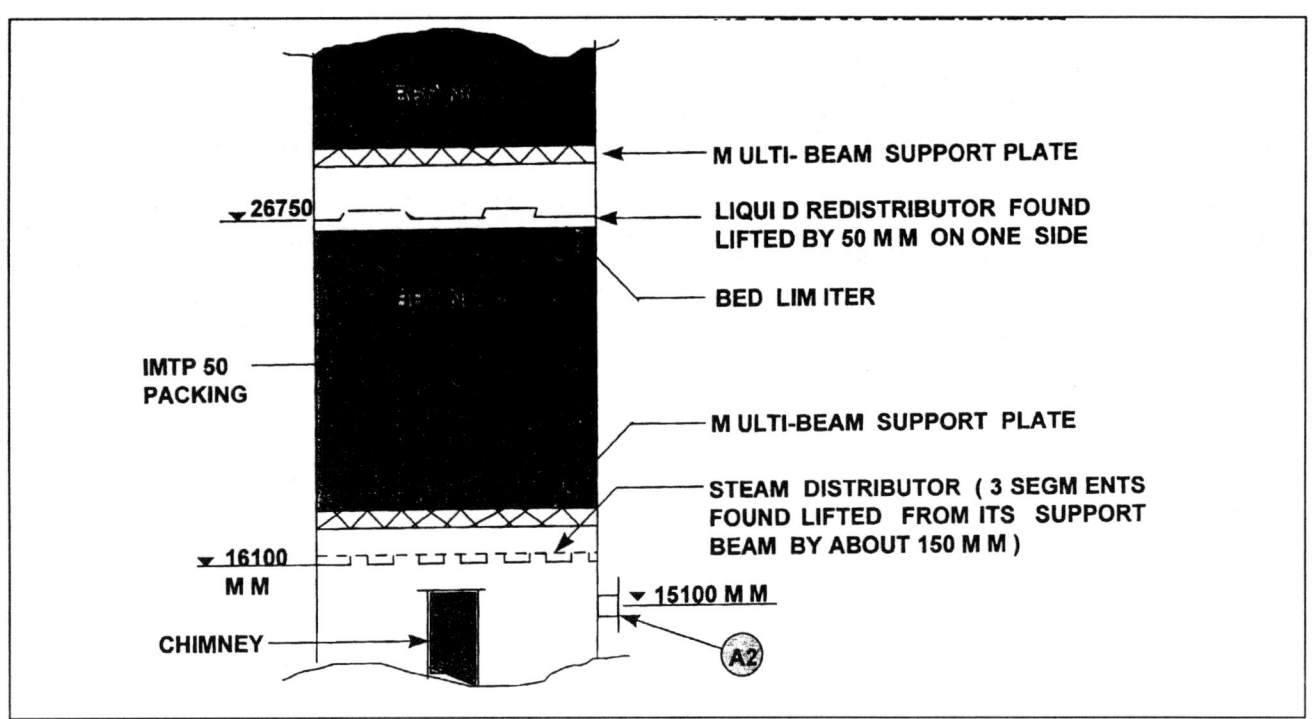

Figure 5. Second regenerator: failure of liquid redistributor and steam distributor.

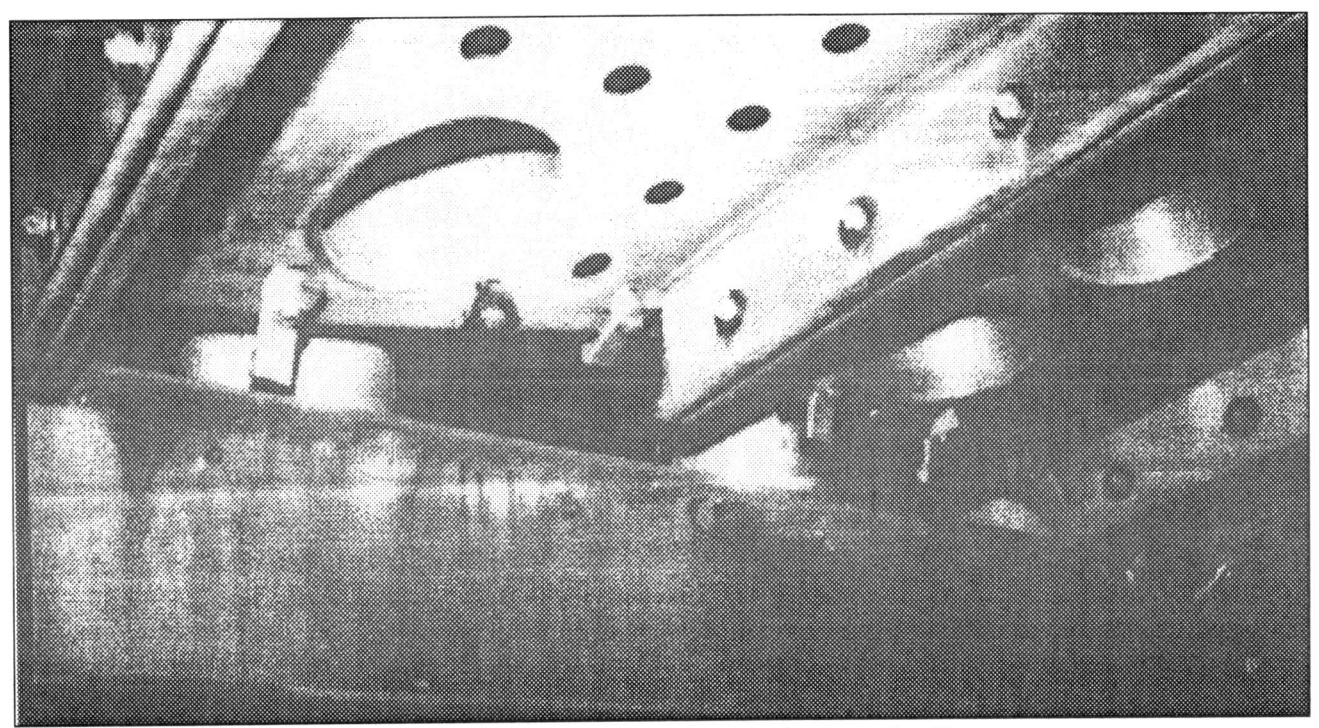
Photo 2. Second regenerator: disturbed steam distributor.

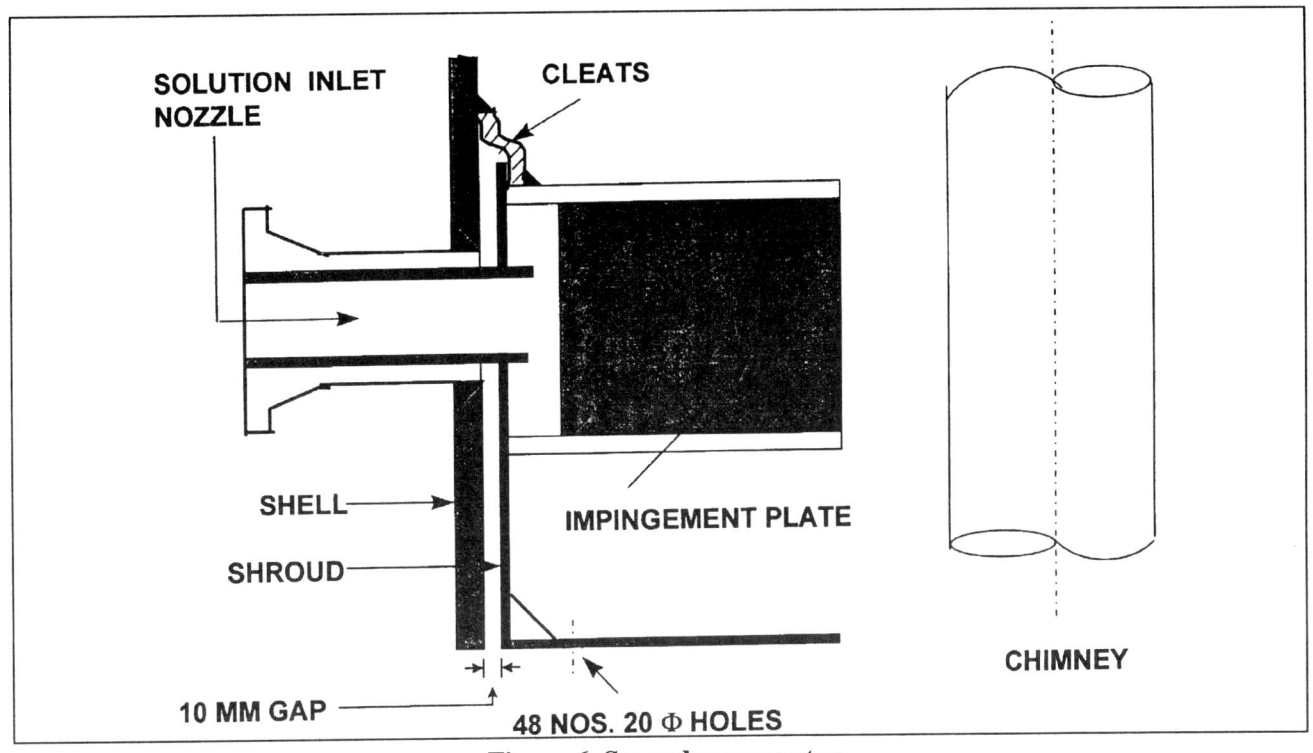
Figure 6. Second regenerator.

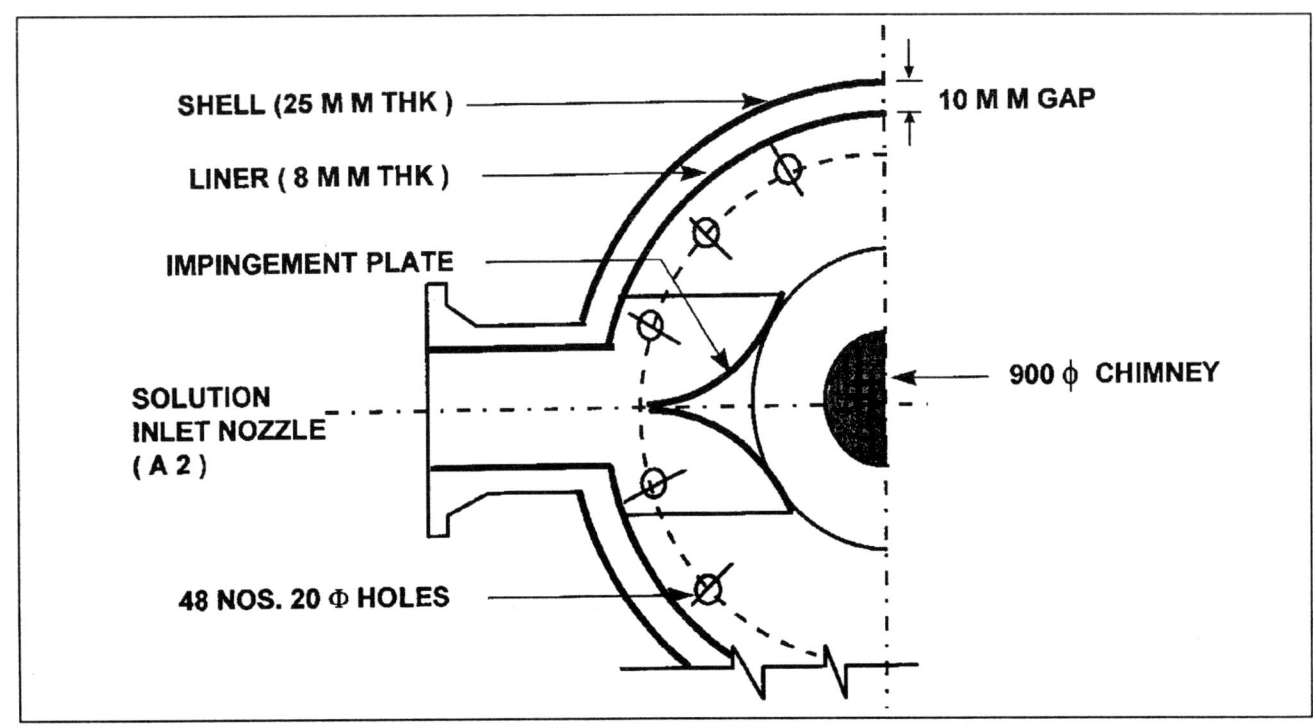

Figure 7. Second regenerator (original arrangement).

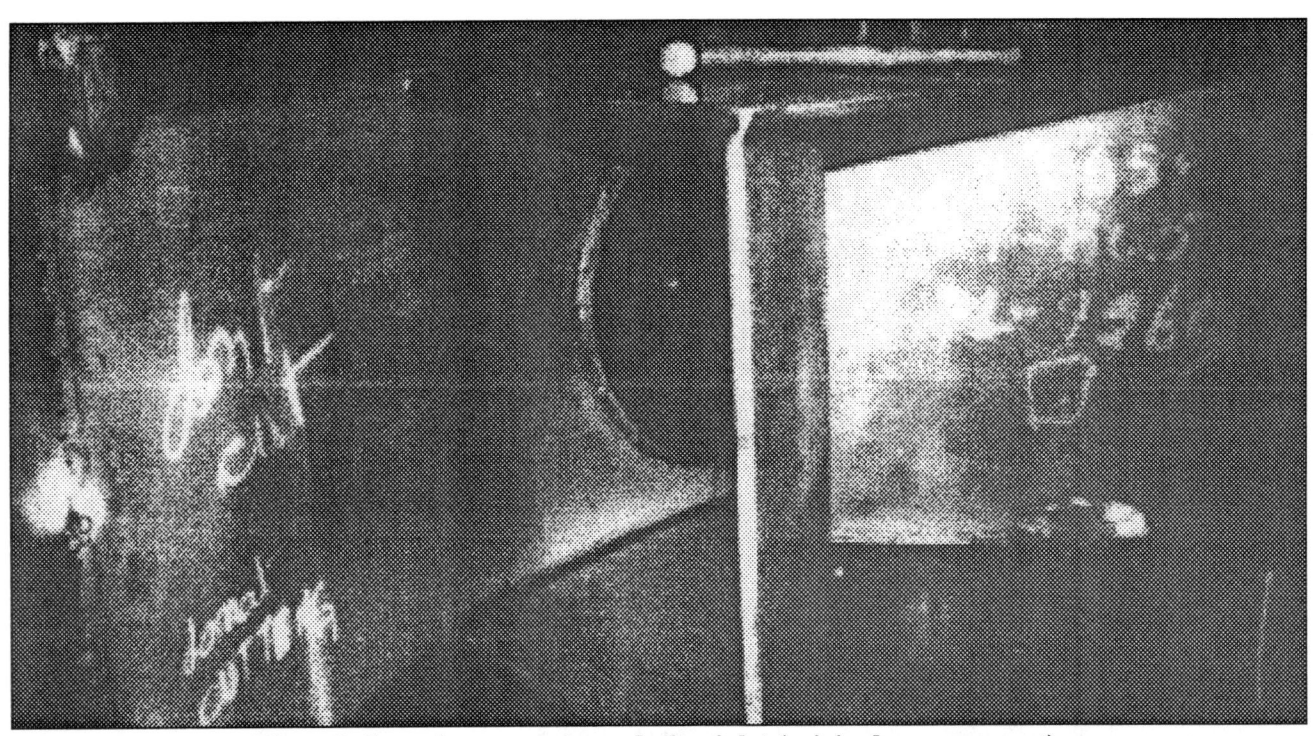

Photo 3. Second regenerator: solution inlet (original arrangement).

which was patched up from outside to arrest leakage while the plant was in operation, was removed. A new matching plate of the same size was welded into the shell.

- The existing nozzle entry configuration and stainless steel shroud was dismantled and removed.
- A new arrangement of the distributor and supporting arrangement as shown in Figure 8 was provided. This arrangement removes the possibility of damaging the passivation layer. At the same time, it ensures that the GV solution is uniformly distributed throughout the circumference of the second regenerator and does not hit the chimney and the surface of the tower.

Vetrocoke absorber

The Vetrocoke absorber is a carbon steel tower with stainless steel internals operating at 398 psig (28 Kg/cm$_2$g) pressure. It contains stainless steel packing material distributed in five beds. Liquid distributors and redistributors (called LRD) of stainless steel material have also been provided. Broad specifications and the general arrangement drg. of the Vetrocoke absorber is presented in Figure 9.

The overall performance of the Vetrocoke absorber was satisfactory as the CO_2 slip at the absorber exit was less than the design figure of 300 ppm. However, a detailed analysis of the performance of each bed conducted in August 1997 indicated that the second bed was not performing satisfactorily. Maldistribution of GV solution at top of the second bed was thought to be the probable cause and it was decided to open the absorber at the earliest opportunity.

Problem/Failure Description. The ammonia plant was shut down in October 1997 and this opportunity was utilized to open the various manholes of the absorber for inspection. The following failures were found as observed from various manholes.

Location: M3 Manhole. This manhole is located between Bed No. 2 and Bed No. 3 at an elevation of 1,056 in. (26,400 mm).

The Bed No. 2 liquid redistributor (Norton type) called LRD located at elevation of 1,030 in. (25,750 mm) was found buckled at the top and the J-bolts supporting the distributor had been sheared. The LRD was raised about 6 to 8 in. (150 to 200 mm) from its support ring.

Further upward movement of the LRD was restricted by the semi-lean solution distribution parting boxes placed above it, and by the semi-lean solution distribution pipes of nozzles A2 and A3 located at an elevation of 1,066 in. (26,640 mm).

The distance between the LRD and the Bed No. 3 containing IMTP 40 stainless steel packing had been reduced to around 24 in. (600 mm). A few loose rings were also found at the top of the LRD.

The parting box was also found damaged and buckled at the ends.

A few 8 in. NB nipples attached with the semilean solution distributor pipes of A2 and A3 nozzles were also found twisted.

The details indicating the above failures are given in Figure 10.

Location: M4 Manhole. This manhole is located between Bed No. 1 and 2 at an elevation of 631 in. (15,780 mm).

The Bed No. 1 liquid redistibutor (Norten type) called LRD located at an elevation of 610 in. (15,250 mm), which should have been below the M4 manhole level, was found hanging at approximately 20 in. (500 mm) above the top of the manhole. The LRD was found in distorted condition and its middle portion had taken convex shape viewed from bottom.

The whole Bed No. 1 had lifted up by approximately 64 in. (1,600 mm) from its original position. Loose IMTP 50 packing of Bed No. 1 were found all around the M4 manhole. Photos 4 and 5 indicate the disturbed beds of the absorber.

The support beam of bed No. 1 LRD with its ends distorted was found loose above manhole M4 and being stuck up at ring support of Bed No. 2.

The multibeam support plate of Bed No. 2 along with its supporting beam was not clearly visible. The beam had sheared from its support bracket and entered in the IMTP 40 packing of Bed No. 2 after leaving its original position.

Probable Causes. The possibility of construction defects and weak structure of the tower internals were considered to be probable causes of the failure. The plate thickness used for bed supports and clamps was only 2 mm. The extent of damage, however, suggested that forces of great magnitude acted in the upward

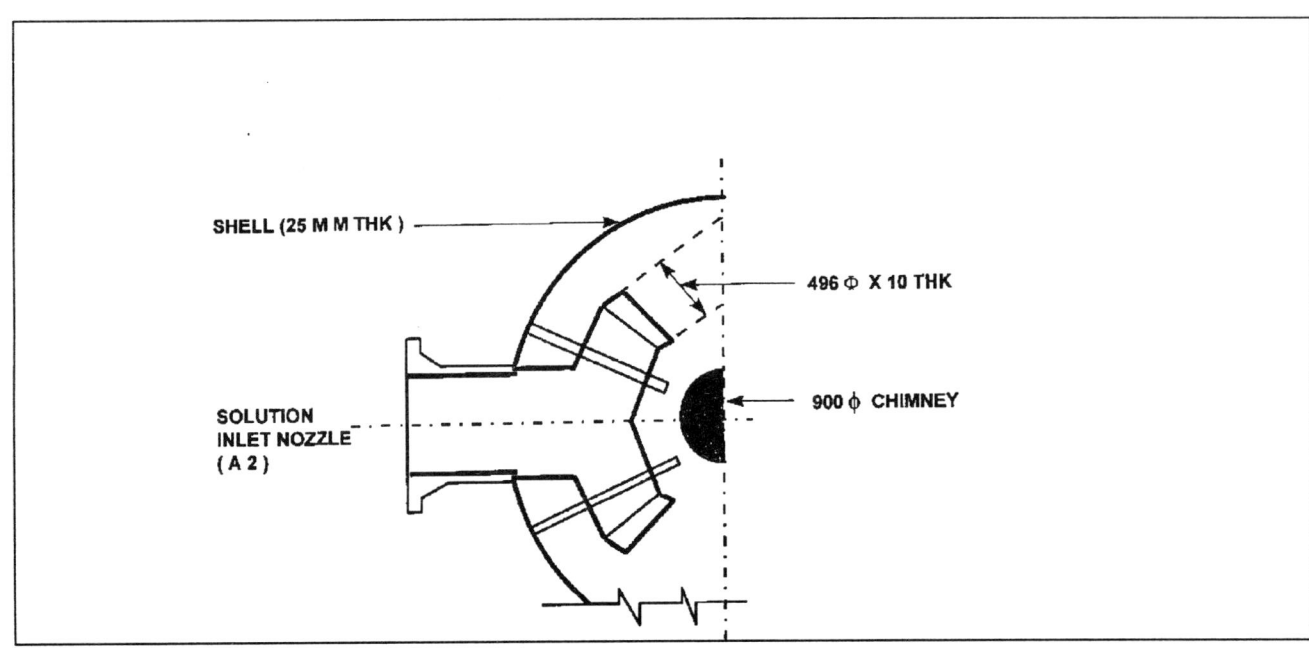

Figure 8. Second regenerator (modified arrangement).

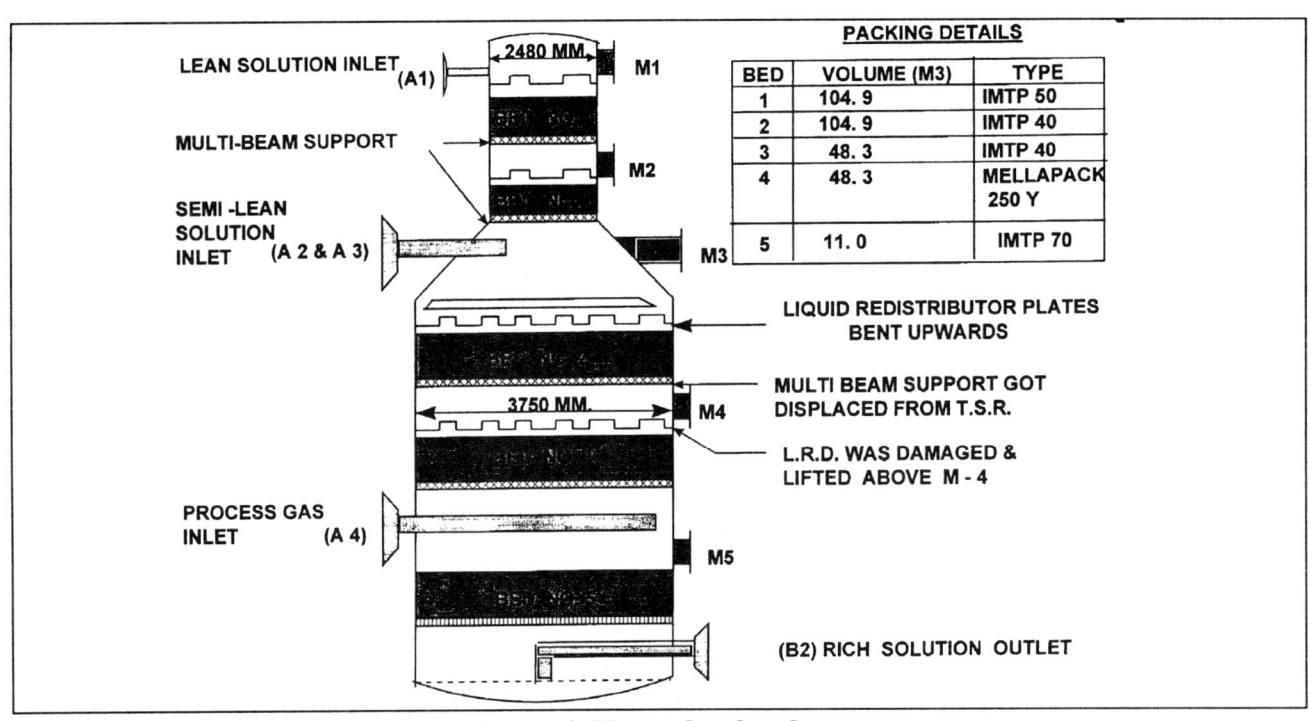

Figure 9. Vetrocoke absorber.

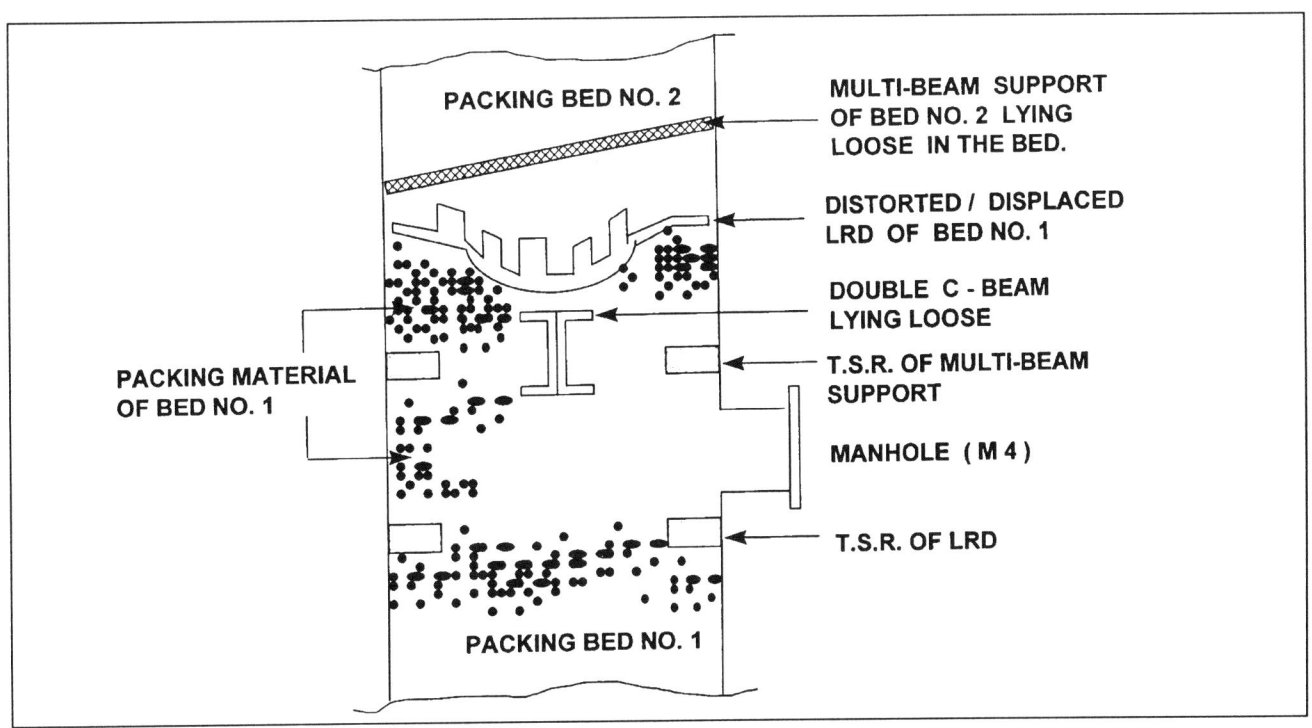

Figure 10. Vetrocoke absorber.

Photo 4. Vetrocoke absorber
Disturbed bed.

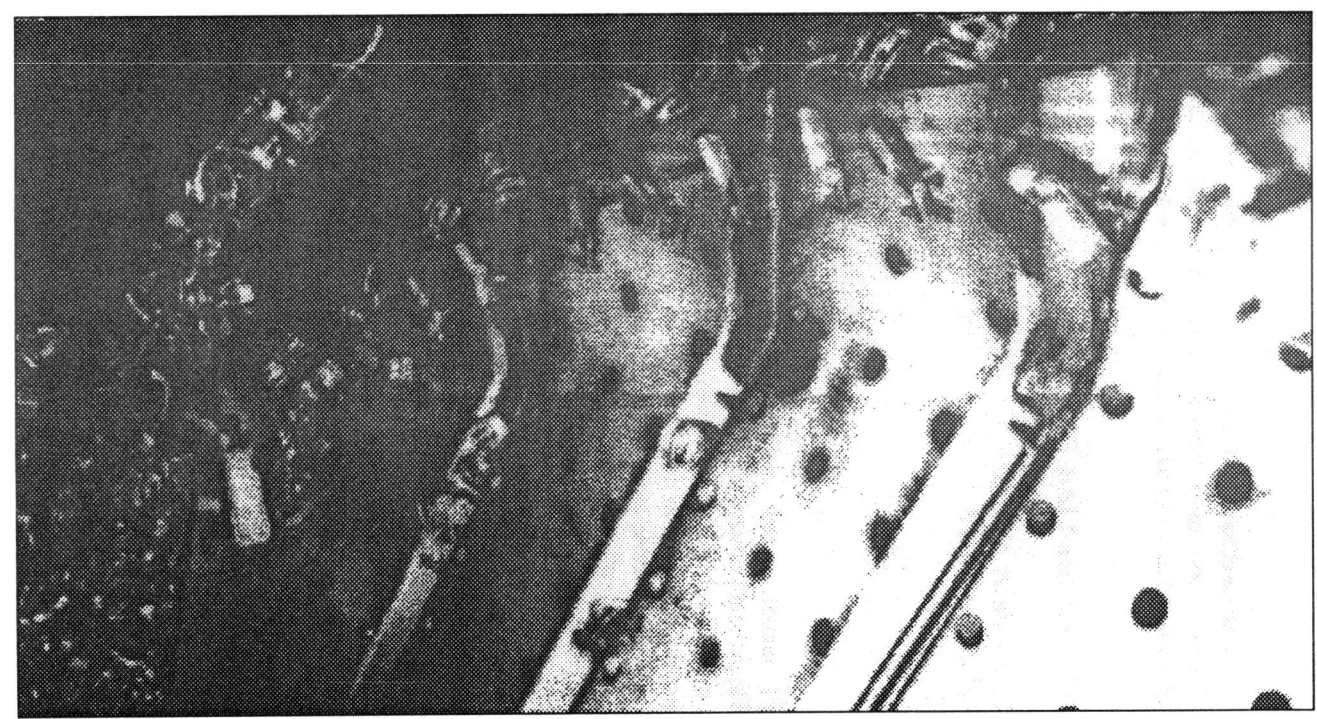

Photo 5. Vetrocoke absorber.
Disturbed first bed liquid redistributor.

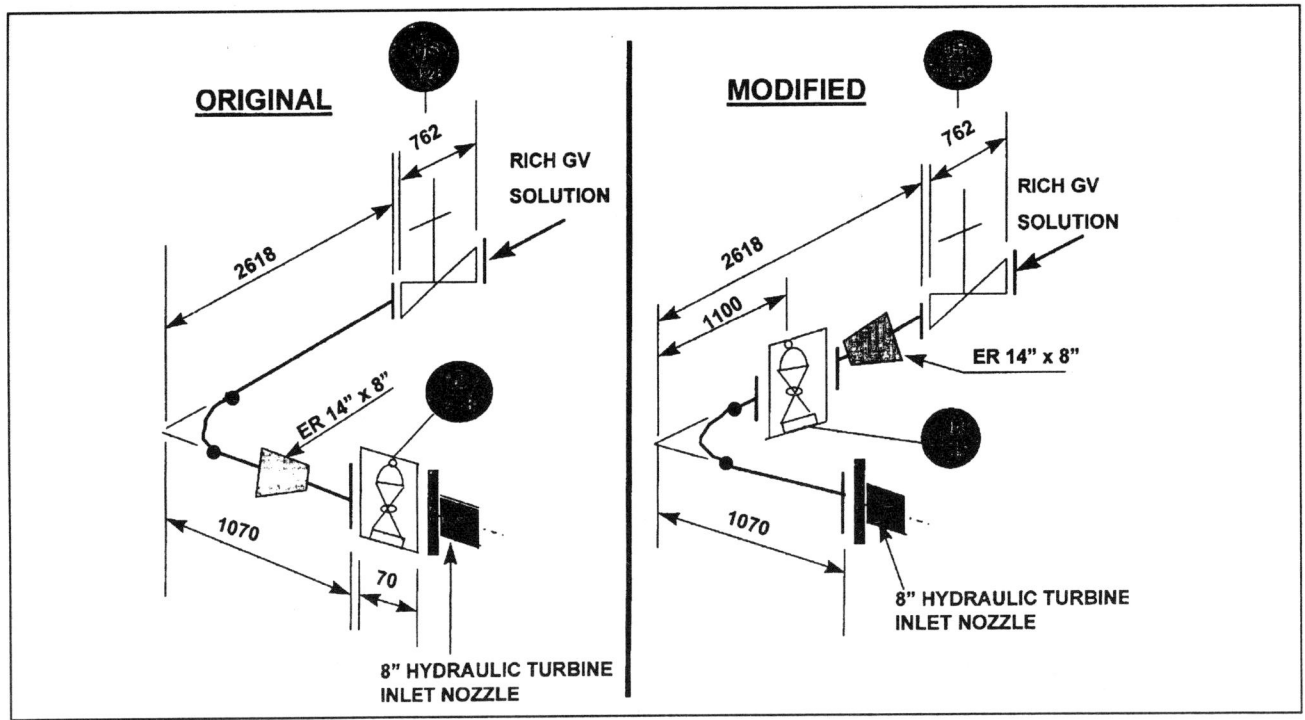

Figure 11. Hydraulic turbine inlet piping.

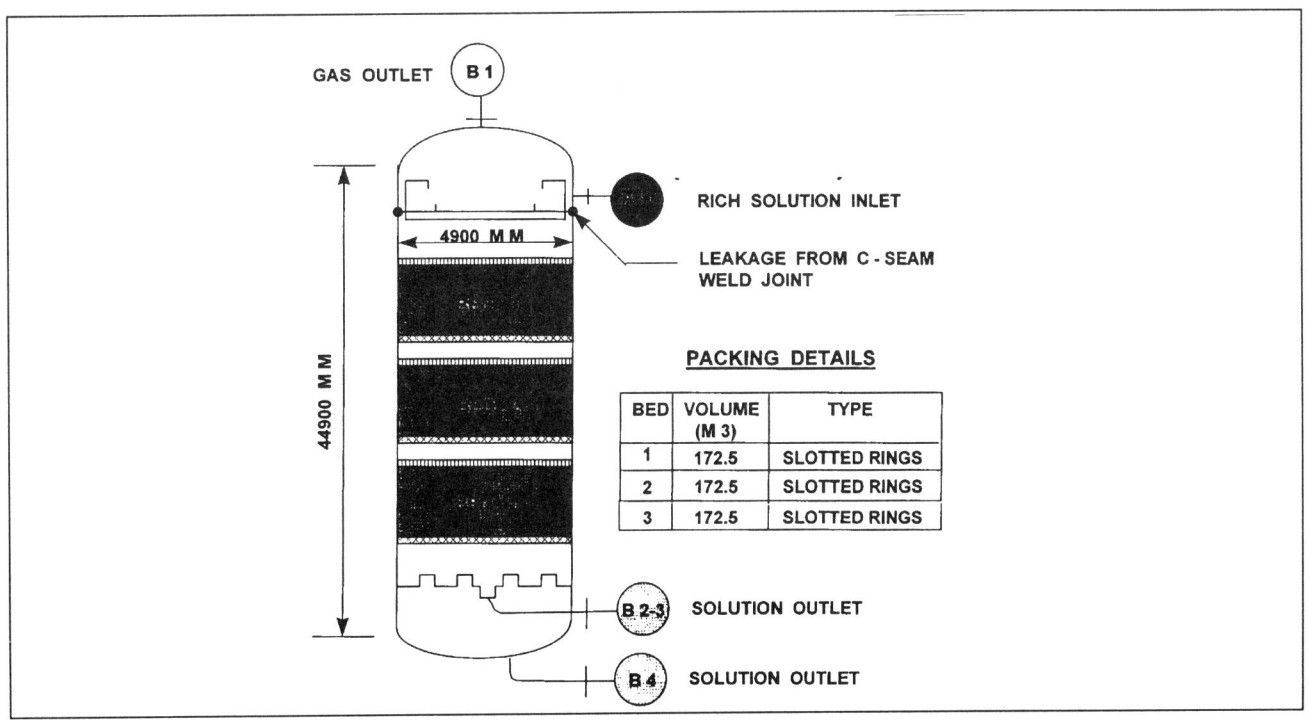

Figure 12. GV regenerator (old plant).

direction in the absorber. Hence, the cause of the failure cannot be attributed only to the weak design. Further, absorbers of the same design have been reported operating satisfactorily in other plants without any problems.

• The other possibility could be some sudden upward gas surge through the first and second bed of the absorber, which caused the upheaval of these beds and buckling of LRDs.

The process gas entering at the bottom of the absorber might have flowed backwards through the semi-lean inlet line via the ARC-NRV circulation line back to the solution draw-off tray in the second regenerator. The upward lifting of the steam distributor above the chimney in the semi-lean solution draw-off tray and also the uplifting of the first bed LRD of the second regenerator seems to support this view.

However, the above backflow could take place through this routine only when both ARC/NRV valves and the solenoid operated valves are not holding.

A study of the construction of the ARC/NRV valves indicated that a large quantity of gas passing backward through the NRV portion and then through the ARC portion could be possible only if the internals of the valve were severely damaged. These valves were opened to check their condition. The springs of these valves were found broken. The discs of these valves were also getting stuck up. The above conditions were creating possibilities of backflow.

Each semi-lean pump had a solenoid valve at the discharge which closes when the pump trips via the interlock I-301 A/B/C. For backflow to take place, these discharge valves must be in the open position. This can take place if the discharge valve does not close during the tripping of the pumps, due to failure of the interlock I-301. Malfunctioning of this interlock, however could not be confirmed.

Further, the backflow through the above route may result in reverse rotation of the semi-lean pumps. However, no damage to these pumps due to reverse rotation was observed. However, the absence of a reverse rotation of the semi-lean pump could be explained by the reasoning that the liquid passing in backflow through the valve was prevented by the motion of the decelerating machine.

• Another probable cause for the damage in the absorber could be the fast depressurization of the absorber by a sudden opening of the vent valve (PV-

60) located downstream of the absorber. This could have occurred during startup/shutdown of the plant. However, depressurization through this route could have resulted in the failures of the third and fourth bed as well. No failure in these beds, however, were found. This could be explained because the third and fourth beds are inherently stronger than the first and second beds, as the diameter there is 99 in. (2,480 mm) compared to 150 in. (3,750 mm) at the first and second beds, even though the internals and fittings are of the same thickness.

Repairs/Modifications. The following corrective actions have been suggested based on all probable causes of failures as discussed above.

- All damaged internals will be replaced with the next highest thickness.
- In order to prevent backflow, the solenoid valves should be interlocked with the low speed of turbines of the semi-lean solution pumps and lean solution pumps so that before all the liquid is drained off, the valve would have completely closed.
- The semi-lean flow control valve (FV-22) and the lean solution flow control valve (FV-23) should close shut on very low solution flow.
- Extreme care should be taken to ensure that the downstream vent valve (PV-60) is not opened suddenly under any circumstances.
- An additional NRV on each of the common headers of the semi-lean solution line and the lean solution line should be provided.

Hydraulic turbines

The rich GV solution at high pressure coming from the bottom of the absorber is let down and flashed in the upper portion of the first regenerator operating at low pressure. This let down in pressure is carried out through hydraulic turbines to supply power to turbine driven GV solution pumps and thus reduce the steam consumption of turbines.

Problem/Failure Description. It was observed that the hydraulic turbine was not developing power as per design and thus the steam consumption of steam turbines was high. On inspection, the casing vanes of hydraulic turbines were found to be eroded.

Probable Causes. Butterfly control valves have been provided at the inlet nozzle of the hydraulic turbine. The shaft pins of these valves have also been found to be broken probably due to flashing. The sudden increase in volume and the two-phase flow at the turbine inlet nozzle were the probable causes of the damage to the hydraulic casing vanes.

Repairs/Modifications. The butterfly control valves have now been shifted away from the inlet nozzle of hydraulic turbine to achieve laminar flow to the turbine inlet, as shown in Figure 11.

First regenerator

The first regenerator is a carbon steel tower provided with stainless steel internals with operating pressure of 14.2 psig (1 kg/cm$_2$g). It is a packed tower having 147 in. (3,680 mm) diameter, 1,855 in. (46,375 mm) height and contains stainless steel packing material distributed in three beds.

Problem/Failure Description. The rich solution line (20 in.) carrying rich solution from the bottom of the absorber to the top of the first regenerator through two inlet nozzles was vibrating heavily. Frequent leaks were observed at the welding joints at the upstream stub end of the butterfly valves provided in the inlet lines and these leakages were recurring frequently.

Probable Causes. The rich solution line is divided with two branches near the first regenerator and, hence, enters the vessel at two points. In both branches, butterfly valves have been provided near the first regenerator. These valves were causing restriction in the flow and hence the vibration in the lines. Vibration resulted in the increased load on the welding joints and the failure of the joints.

Repairs/Modifications. Both the 20 in. butterfly valves and flanges were removed and the gaps were filled by providing 20 in. SS-304 spool pieces.

GV regenerator

The GV regenerator in the CO$_2$ removal system of Ammonia-1 has been in operation since 1988. The plant was originally built with the Benfield process and was converted to the GV dual activator process in 1997.

The regenerator is a carbon steel tower provided with stainless steel internals and operating at 14.2 psig

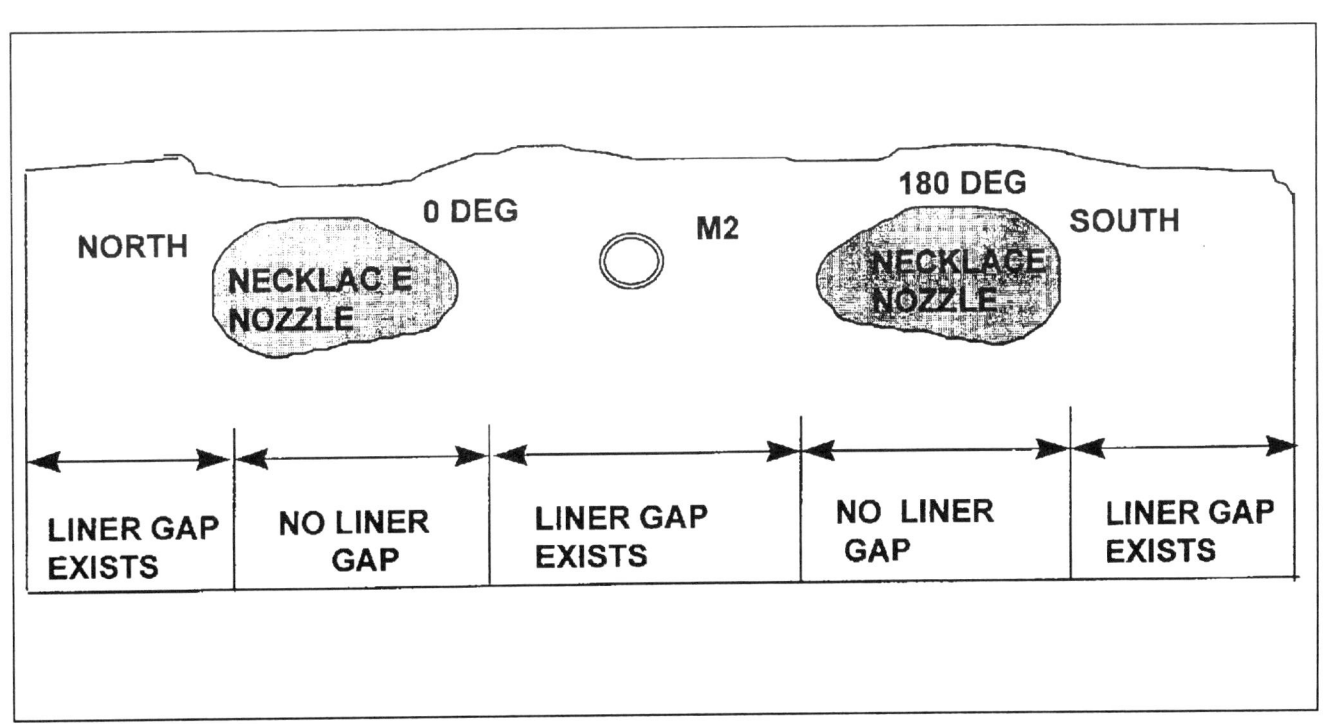

Figure 13. GV regenerator: development of liner.

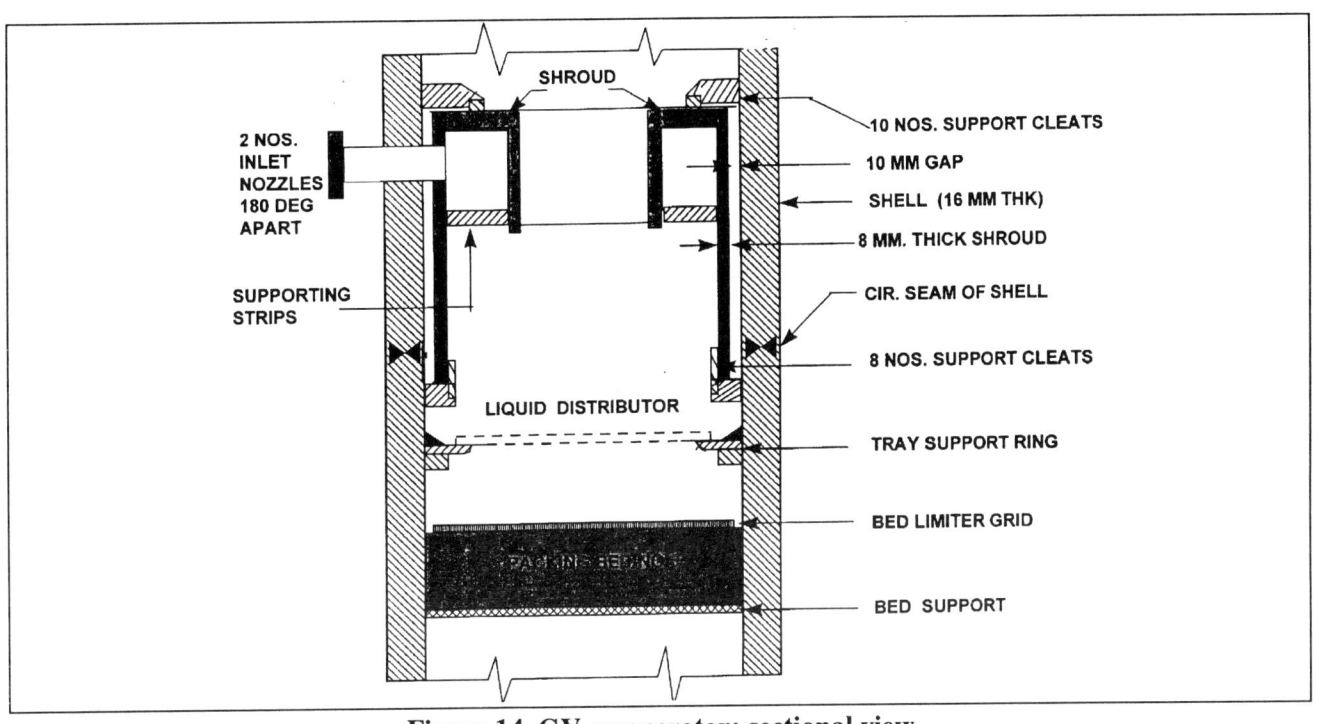

Figure 14. GV regenerator: sectional view.

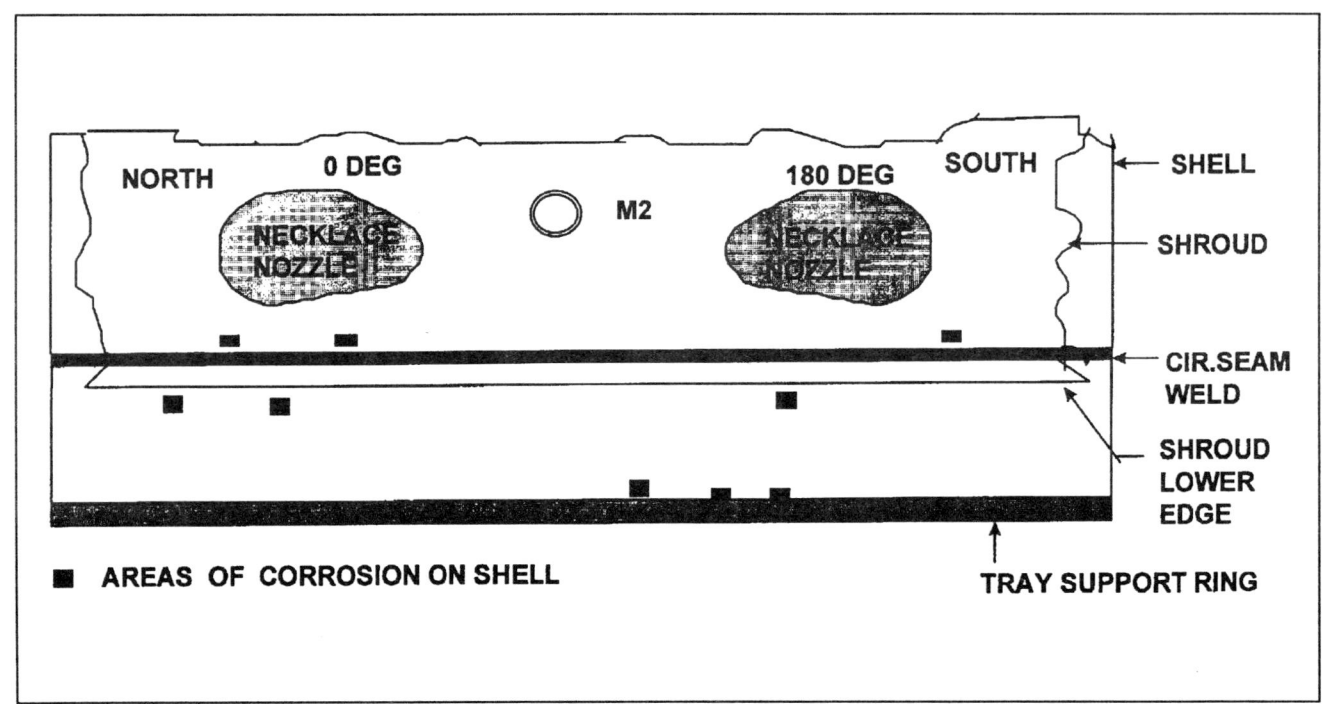

Figure 15. GV regenerator: development of shell and shroud showing areas of corrosion.

(1.0 Kg/cm$_2$g) pressure. It is a packed tower having 196 in. (4,900 mm) diameter, 1,796 in. (44,900 mm) height, and contains stainless steel packing material distributed in three beds. Rich solution to the regenerator is fed through two tangential entries (called a necklace), as shown in Figure 12. The above two inlet nozzles are welded to the 8 mm thick stainless steel liner provided to protect the carbon steel shell from severe inlet flow conditions. Broad specifications and the general arrangements drawing of the GV regenerator are presented in Figure 12.

The CO_2 removal system of ammonia plant has been operating normally except that the CO_2 slip was high at around 1,400–1,600 ppm. A consultants' expert in these systems was called in the last week of September 1997 to analyze the problem of high CO_2 slip. Following the recommendations of this consultant, various chemicals were added to the system to increase the concentration of chemicals in the solution.

Only marginal advantage in the reduction of CO_2 slip was observed. However, it was observed that the iron content in the solution was increasing. The iron content in the solution had increased from 67 PPM on September 30, 1997 to 127 ppm on October 15, 1997 in a very short span of two weeks time and was a clear-cut indication of heavy corrosion taking place in the system.

V+5 to total V ratio was being maintained at the same level of around 15% as was maintained with the Benfield system. KNO_2 and V_2O_5, however, were added to increase the ratio of V+5/V to stop further corrosion. However, iron level continued to increase in the solution.

It was observed on October 18, 1997 that the GV regenerator had started leaking from the top, resulting in continuous GV solution droplets falling down. The leaking zone was thoroughly inspected and the leak was arrested by welding. On October 21, 1997 another leak was observed about 180° opposite the previous leak. An attempt was made to arrest the leakage by providing a box around it. This was not possible as the vessel thickness had been reduced by corrosion to the extent that welding was impossible. Thickness measurements showed patches of reduced thickness. It was decided to shutdown the plant and carry out a thorough inspection and repair.

Problem/Failure Description. The Ammonia Plant was shut down in October 1997 to carry out a thorough inspection and repair of the GV regenerator. On opening the regenerator, the following observations were made.

- Black color deposition was found above the shroud.
- Both the necklace weld joints with liner were found cracked in two places in each joint. The length of the crack was about 12 in. (300 mm).
- The liner plate had gotten deformed and was touching the regenerator main shell at several places, as shown in Figure 13. The gap between the liner and the shell should be 10 mm as per design (Figure 14).
- Heavy corrosion on the shell near and including the tray support ring was found in three places, as shown in Figure 15.
- Corrosion of the shell at several places just below the lower edge of the SS liner were also found, as shown in Figure 15.
- Lower cleats welding with liner were found cracked.
- Some of the weldings of end plates of Omega trays were found broken. At two places, end plates were missing. A broken piece of 690 mm x 280 mm was also found loose.

Probable Causes. The following reasons put together can be attributed to the fast corrosion in the GV regenerator and its subsequent leakage:

- The stainless steel liner plate became deformed and was touching the shell at various places. Cleats of the liner were also found cracked. Also, there is two-phase flow at the inlet. This must have resulted in vibrations in the liner and damage to the passivation layer. The places where there was no gap between the shell and liner corrosion could be due to stagnated solution.
- The V+5/V ratio was slightly on the lower side at around 15% in comparison to the consultants recommendations of a minimum value of 20% and probably was not sufficient to give the desired protection to the carbon steel shell.
- Increase in concentration of chemicals further aggravated the situation for corrosion.

Repairs/Modifications.
- It was decided to cut the liner by about 8 in. (200 mm) from the bottom at the places where there were no gaps between the liner and the shell to check for further damage to the shell. After cutting, it was discovered that some of the shell areas and the circumferential seal welds behind the liner were found badly corroded. Another 18 in. wide by 80 in. long (450 mm x 2,000 mm) section of the liner was removed to inspect the condition of the shell. No further corrosion was observed on the shell behind the liner.
- A total of about 35 stainless steel cleats were welded behind liner to maintain a uniform gap between the liner and the shell throughout the periphery. This was done to ensure no further contact between the liner and the shell in the future to avoid damage to the passivation layer and to avoid stagnation of the solution.
- Some of the welding of end plates of omega trays, which was found to be broken, were rewelded.
- All the corroded areas of the shell were repaired by filling material with welding. All the repair welds were ground finish and DP tested. The 18 in. (450 mm) width liner was rewelded in position.
- The cracks on both necklace to liner joints were also repaired and DP tested.

Conclusion

The failures presented in the article were caused by a number of different factors including design deficiencies, defects introduced during manufacturing or fabrication, service related deterioration, upsets during plant operation, and so on.

The cases presented do not indicate that a particular system or design is more prone to failures than others. Instead, these examples must be carefully analyzed to prevent their occurance in other plants.

The awareness of the conditions which produce failures helps the plant personnel to reduce the potential for failures. This also helps in purchasing the most suitable equipment for a given operation and ensuring proper design and fabrication of the equipment.

Safe Work Procedures for Profitability

Companies have historically prepared safe work procedures to meet regulatory requirements. These regulations have been written over a long period of time, without regard for creating a consistent, integrated system. Resulting documentation and approach was often fragmented with no consideration of profitability issues. A fully integrated Safe Work Permit Procedure system is discussed.

Margaret M. R. Eastman and James R. Sawers
Knowledge Technologies, Charleston, SC 29401

Introduction

Traditional plant safe work documentation has been written under the guidance of the safety officer. The result was usually plant-wide generic procedures outlining the principles of equipment lockout, vessel entry, hot work, and so on. The procedure writers did not write equipment-specific procedures nor did they always enforce the principles outlined in the regulations that formed the basis for the principles as many incidents and citations indicate. Different permits were created as the need arose, each with regulatory compliance the driving factor in the design of the permit and its administration. Rarely was profitability a consideration in the documentation that was created.

Discussion

Knowledge Technologies has developed a new approach for documenting safe practices for equipment lockout, confined space entry, hot work, excavation, and so on. This documentation approach is called a Safe Work Permit Procedure system. It is designed to comply with the U.S. regulations written for the chemical process industries. Specifically, this includes Lock/Tag/Try (LTT), Confined Space Entry (CSE), Hot Work, Excavation, and Vehicle Entry. These hazards are currently covered under separate regulations. Since interpretations given by regulatory authorities indicate that some of these regulations require equipment-specific checklists, KT uses that standard throughout the Safe Work Procedure system.

In addition, KT has designed a Return-to-Work Procedure to prevent costly events. Examples of omissions that have had extremely costly consequences include failure to refill the oil sumps, leaving oily rags inside equipment. Such omissions have significantly affected product quality, equipment performance, and unnecessarily increased unscheduled downtime due to a failure to get equipment back in service after turnarounds.

The Safe Work Permit system consists of a permit card to which endorsements for all relevant hazards such as LTT, CSE, and so on, are attached. A comprehensive Safe Work Permit Procedure consisting of instructions for using the Safe Work Permit and for preparing and using all endorsements can be created.

The Safe Work Permit Procedure includes:

• A site-wide procedure for implementing the use of Safe Work Permits including all appropriate endorsements.

• A set of site-wide procedures for writing equipment-specific checklists. Blank equipment-specific checklists for each type of endorsement and return-to-service are included for use in preparing the initial library of equipment-specific checklists.

• An Approval Procedure including "fast-track" system to cover those situations when a checklist is needed immediately and does not already exist. Specifically, the "fast-track" approval can be used when the full formal approval process is not practical; it is not meant to replace the full approval process but to serve as a precursor to it. (This approach permits the Safe Work Permit system to be enforced, although it is not economically desirable to write all checklists now; for example, some conceivable CSE checklists may not be needed for years.)

• A System Maintenance and Updating Procedure.

• Collections of equipment-specific checklists.

Lock/Tag/Try Endorsement

The LTT Checklists are designed to protect personnel against accidents caused by *all* forms of energy including but not limited to:

• Electricity including static charge.

• Springs that are compressed, elongated, flexed, torqued, and so on.

• Weights above ground level.

• Compressed flammable, toxic, corrosive, or explosive gases or fluids including steam.

• Flywheels and rotating equipment.

• Caustic or reactive chemicals.

• Hydraulics.

• High temperatures or cryogenic (extremely low temperature) materials.

• Air, pneumatic, and vacuum systems.

• Magnetic fields and stored electrical charge.

These forms of energy include both those traditionally covered by the former "Lockout/Tagout" Procedures designed for electrical isolation, as well as the "Line Breaking" procedures designed to protect personnel from dangerous substances contained in piping and vessels.

If an Equipment-Specific Lock/Tag/Try Procedure does not exist for the particular piece of equipment, it must be written before that equipment can be locked/tagged/tried. A template to prepare an initial draft of the Equipment-Specific LTT Checklist has been prepared and will be illustrated at the conclusion of this article. The preparer must obtain the required approvals and authorization, complete the LTT, edit the equipment-specific procedure, and submit it to the individual assigned to maintain documentation for inclusion the plant documentation system.

Confined Space Entry Endorsement

The CSE Checklists are designed to protect personnel against accidents caused by any deprivation of adequate breathing air. The CSE endorsement follows the conditions outlined in current U.S. regulations.

A confined space is defined as a space that has adequate size and configuration for employee entry, has limited means of access or egress, and is not designed for continuous employee occupancy.

A "permit-required confined space" is defined as a confined space that presents or has a potential to present one or more of the following: an atmospheric hazard, an engulfment hazard, a configuration hazard, or any other recognized serious hazard.

Engulfment is defined as the surrounding and effective capture of a person by a liquid or finely divided (flowable) solid substance that can be aspirated to cause death by filling or plugging the respiratory system or that can exert enough force on the body to cause death by strangulation, constriction, or crushing.

A hazardous atmosphere is defined as an atmosphere that may expose employees to the risk of death, incapacitation, impairment of ability to self-rescue (escape unaided from a permit space), injury, or acute illness from one or more of the following causes:

- Flammable gas, vapor, or mist in excess of 10% of its lower flammability limit (LFL).

- Airborne combustible dust at a concentration that meets or exceeds its LFL. (Note: This concentration may be approximated as a condition in which the dust obscures vision at a distance of 5 ft. [1.52 m] or less.)

- Atmospheric oxygen concentration below 19.5% or above 23.5%.

- Atmospheric concentration of any substance for which a dose or a permissible exposure limit is published in Subpart G, *Occupational Health and Environmental Control*, or in Subpart Z, *Toxic and Hazardous Substances*.

- Any other atmospheric condition that is immediately dangerous to life or health.

Roles of Individuals in Confined Space Entry follow those outlined in the new regulations. They are:

Entrant: Employee who is authorized by employer to enter a permit space.

Attendant: Stationed outside one or more permit spaces and keeps a constant watch on the worker(s) inside the confined space.

Advises the worker(s) inside the confined space to stop work and leave the confined space if any condition outside might endanger the worker(s) inside. This includes ensuring that no work is taking place above the confined space and that no equipment or tools fall or are dropped into the confined space.

Orders the worker(s) to leave the confined space immediately if any of them shows signs of sickness or acts abnormally.

Obtains help before attempting to rescue anyone and uses the proper equipment to do so.

Entry Supervisor: Ensures all check-off items on the CSE Checklist have been completed and ensures all entrants have currently valid training as entrants.

Ensures all entrants are wearing specified PPE and have both normal and escape breathing equipment.

Records the names of all persons entering the confined space on the CSE Personnel Endorsement Sheet.

Attaches a copy of the CSE Checklist and Personnel Endorsement Sheet to the Safe Work Permit and posts near the point of entry or at a convenient point of ascent to the confined space.

Ensures that no spark-generating work is done unless a Hot Work endorsement has been completed and attached to the Safe Work Permit.

Provides an attendant who has currently valid training and has been briefed on the specific Confined Space Procedure now being started.

Verifies that the Maintenance Supervisor has completed all of his/her assigned tasks.

Remains in the immediate area.

Maintenance Supervisor: Provides PPE to all entrants as specified in the CSE Checklist.

Ensures that all entrants have been trained in correct use of PPE and that they use it.

Provides ladders as a means of access/egress where necessary.

If an Equipment-Specific Confined Space Procedure does not exist for a particular piece of equipment, it must be written before that equipment can be entered. The preparer can use a template to prepare an initial draft and obtain the required approvals and authorization, complete the CSE and return to service, edit and submit it to the Job Documentation Coordinator for inclusion in the plant documentation process. A template to prepare an initial draft of the Equipment-Specific CSE Checklist has been prepared and will be illustrated at the conclusion of this article.

Hot Work Endorsement

The Hot Work Checklists are designed to protect personnel on-site against accidents caused by explosions or fire resulting from that work. Hot work endorsements are required if any of the following conditions are met:

- Sparks or flames generated by:
  - Welding/cutting.
  - Grinding.
  - Drilling.
  - Chiseling.
- Electrical work on energized equipment.
- Systems under pressure when:
  - Hot tapping.
  - Furmaniting.
  - Pigtail crimping.
- Other situations that in the judgment of the author should require a Hot Work endorsement.

A template to prepare an initial draft of a Hot Work Checklist has been prepared and will be illustrated at the conclusion of this article.

Excavation Endorsement

Evacuation Checklists are designed to protect personnel on-site against accidents caused by underground cables and piping or personnel injury or death due to possible cave-ins. The preparer can prepare excavation and return to service checklists using the same method as described above.

Vehicle Entry Checklists

Vehicle Entry Checklists are designed to protect personnel and to avoid damage to plant and equipment. It covers all motor vehicles operated on the plant site and not on designated roadways, parking lots, and regularly permitted and posted functions and areas (such as forklift trucks in the warehouse). The preparer can prepare vehicle entry and return to service checklists using the same method as described above.

Atmospheric Test Checklist

An atmospheric checklist is designed to prevent a worker from working around flammable, toxic, or explosive gasses without taking the necessary precautions. The preparer can prepare atmospheric test checklists using the same method as described above.

A template to prepare an initial draft of an Atmospheric Test Checklist has been prepared and will be illustrated at the conclusion of this article.

Approval Procedure

Procedures must be prepared for both types of situations:

- Approval of Originally Written Checklists.
- Approval of one-time Checklists for Inclusion in the Safe Work Permit Documentation System.

System Updating

A system for maintaining and updating documents must be established as part of the plant Management of Change program.

Demonstration of Safe Work Permits

The documents that can be converted to a Safe Work permit system for the purposes of demonstration in this article include Lock/Tag/Try, Confined Space Entry, Hot Work Endorsement, Excavation Endorsement, Vehicle Entry Endorsement, Atmospheric Test Checklist and others.

Although not required by U.S. regulations, Return-to-Service procedures should be prepared for use after any of the Safe Work Permit procedures or endorsements have been completed. This vital information is needed to prevent the costly errors and omissions, which on occasion cause significantly delayed startups and avoidable, unscheduled downtime.

Conclusions

Using a Safe Work Permit Procedure system will

help companies avoid costly repairs, extended unscheduled downtime, and degraded quality product.

Regulatory compliance will be enhanced; writing the necessary documentation will be simplified by following a systematic, well-conceived approach.

Management Tool For Undertaking Quantified Risk Assessment Studies

VRJ has developed a series of management tools that systematically approach risk evaluation for new and existing plants. The main tool is VRJ Quantitative Risk Assessment (VRJQRA), a program designed for undertaking quantified risk assessment studies in order to calculate risk exposures to the community and the plant workforce. This article discusses VRJQRA and other management tools that are used to focus the results of the risk assessment on ongoing management of risks and hazards in the ammonia industry.

Katherine Filippin and Mark Jarman
VRJ Risk Engineers Pty Ltd., Melbourne, Australia

Introduction

Quantified risk assessment (QRA) describes a technique used to systematically calculate the cumulative likelihood of different consequences for all possible hazards associated with a plant or facility. The primary focus of a QRA in this context is the assessment of accidental hazardous events that may cause injury or death, with particular focus on people outside the plant boundary.

Quantified risk assessment studies are required in Australia whenever a new plant is built, a major expansion is planned, or (as with the recent introduction of the Major Hazard Facilities Regulations) every 5 years. Risk engineering is becoming an increasingly important tool in the planning, commissioning, and operations of businesses.

The ammonia industry can benefit from QRA. QRA can be used to identify areas of particular concern and evaluate control strategies that are in place or have been proposed.

VRJ undertakes QRAs using the program VRJQRA. The program is computer-based and was developed in-house to provide a system for calculating, modeling, and recording all the information associated with a QRA. The program and analysis follow the traditional steps for a QRA (Figure 1).

Hazard Identification

The first step of QRA is hazard identification. Hazard identification can be undertaken in many ways. One of the most useful ways is to conduct a workshop, which can use the experience and knowledge of the project team and allows the plant personnel to be involved in an existing facility. In the case of a new facility the hazard identification is based primarily on the information contained in process and instrumentation diagrams (P&IDs) and process flow diagrams (PFDs).

The model VRJQRA provides a pictorial hazard identification method. The electronic versions of the process diagrams are imported into the computer model. An overlay of intelligent objects representing

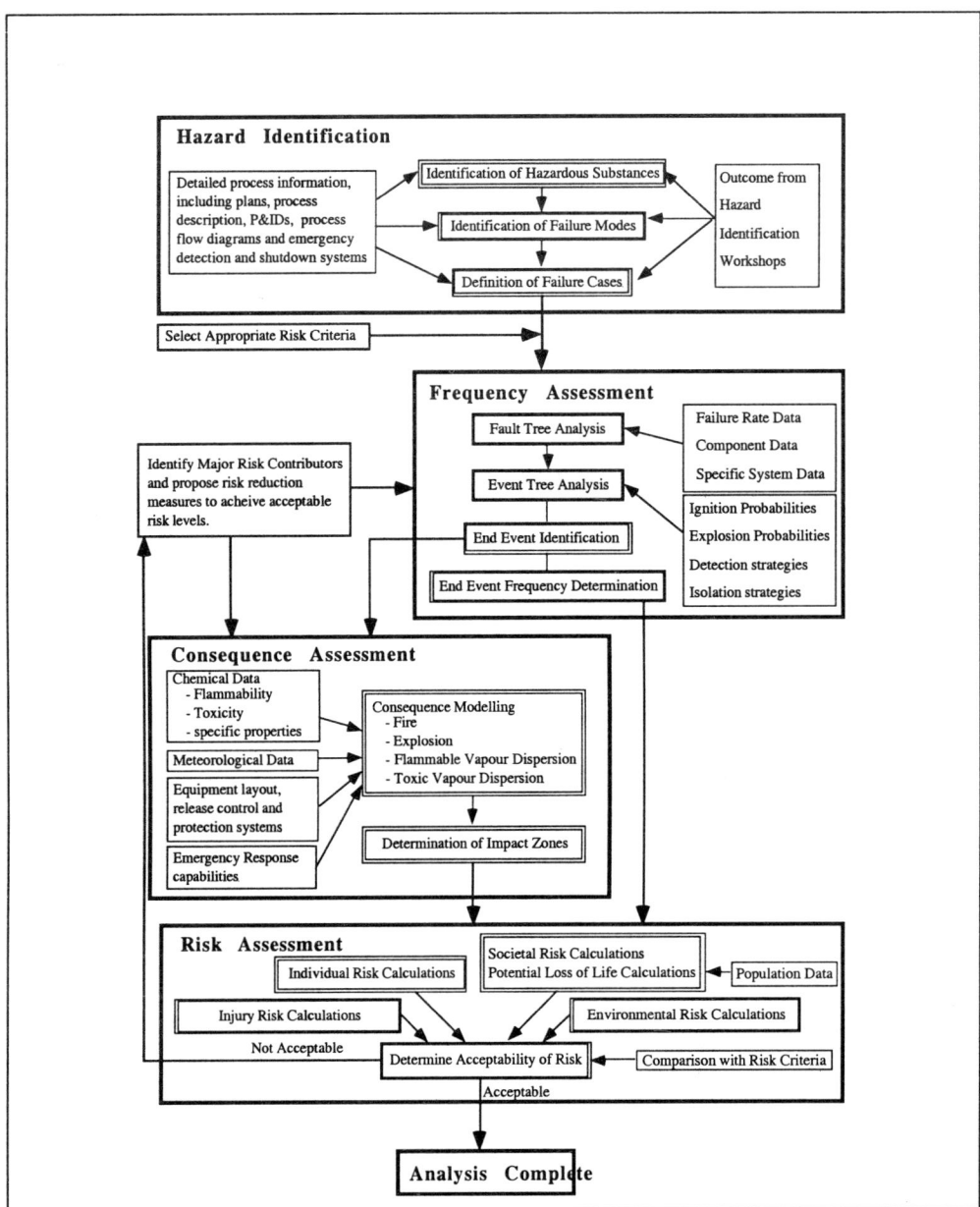

Figure 1. Quantitative risk assessment methodology.

failure items, such as process vessels, valves, flanges, storage tanks, pumps, and piping are placed on the diagrams. These items are grouped into sections that can be isolated by emergency shutdown valves throughout the operation. The sections, referred to as "hazard sections," are pipes, vessels, and storage tanks and the equipment associated with them. Figure 2 shows a sample section of a process diagram with the failure items overlaid on it.

Frequency Assessment

Once the hazard identification has been completed, frequency assessment and consequence assessments can be undertaken.

Frequency assessment is used to determine the likelihood of a failure. Failure frequency data for piping, valves, vessels (process and storage) and other process items are determined using information from databases and available literature. Site-specific data are used directly for assigning failure frequencies to specific hazardous events if such data are available.

The potential failures for each hazard section identified are reduced to a representative selection of hole sizes. The frequency for each hole size is calculated by summing the failure frequency data for all the items associated with the particular hazard section.

The information recorded about each hazardous section is carried through to the release event calculator. The release event calculator is also used to record information about the product and the process conditions for each hazard section. Figure 3 shows a printout of the release event calculator.

The release event calculator is divided into four sections. The first section displays information about the hazardous item, including the name of the item, the type of item, the upstream and downstream hazard items and the analyst.

The second part of the calculator shows information about the process conditions. This section indicates the product being analyzed, the temperature and pressure, a liquid level in a vessel, and any flow into the vessel. A chemical database provides the properties for all the products being analyzed. The liquid level is used to divide possible failures into liquid releases and vapor releases.

The third part of the calculator describes the release summary. This information concerns the representative hole sizes that have been selected. The information is divided into a liquid release summary and a vapor release summary. The frequency of occurrence for each hole size is recorded, along with the calculated release rate. A button links the events to an event tree set up specifically for that event.

Release rates are determined by VRJQRA based on the specific properties and process conditions of the substance discharged. This ensures that all releases are modeled as accurately as possible. For liquid releases, the Bernoulli equation is used to determine the release rate. For vapor releases, relationships for both choked and nonchoked flow are used. The release rates are assumed to remain constant for the duration of the release. This is a conservative assumption because depressurizing effects reduce the release rate over time. If the calculated release rate exceeds the process flow rate for a particular hazard section, the process flow rate is used so that overestimation of the flow rate will not produce significantly higher results for longer releases.

The last section on the calculator is the contributing item summary. This section lists all the items that contribute to the failure frequency of the hazardous section under examination. The list includes all valves, flanges, fittings, filters and other items in the section.

The recording method within VRJQRA allows the analyst easy access to information about a hazard. As part of the reporting process the hazards identified are shown pictorially (Figure 2) with each section identified by a tag number as well as in a table format that lists all the contributors to the event (Table 1). A summary of the process information can also be provided for every hazard section.

This initial stage of frequency assessment leads into the consequence assessment.

Consequence Assessment

The possible consequences for a failure in a particular hazard section are determined once the representative hole sizes have been identified. These are examined using the event trees that form part of the VRJQRA program.

Event-tree analysis is utilized to determine the possible release sizes, the resulting event and corresponding frequency. VRJQRA has a series of different event trees that can be utilized, to determine the possible consequences of a release. These include different event trees for liquids and vapors, continuous process releases and storage releases.

The event tree is developed from left to right through a series of branches. The branches represent factors which influence the final outcome. Possible influencing factors are operator intervention, ignition probability (either early or delayed),

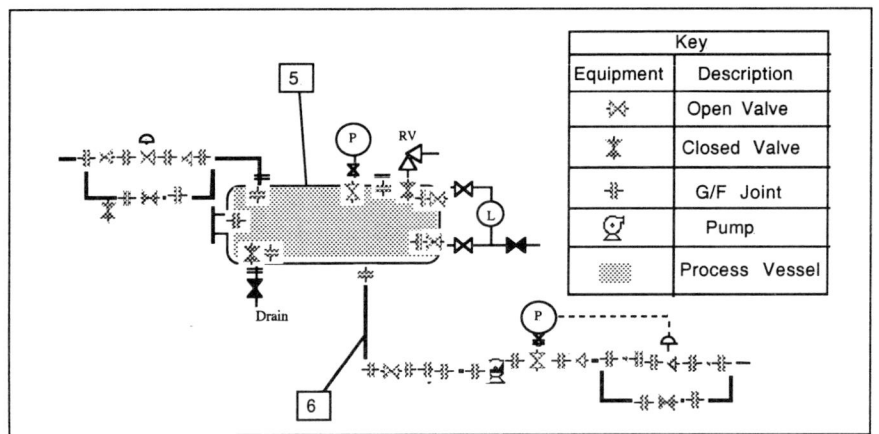

Figure 2. Sample P&ID showing failure items.

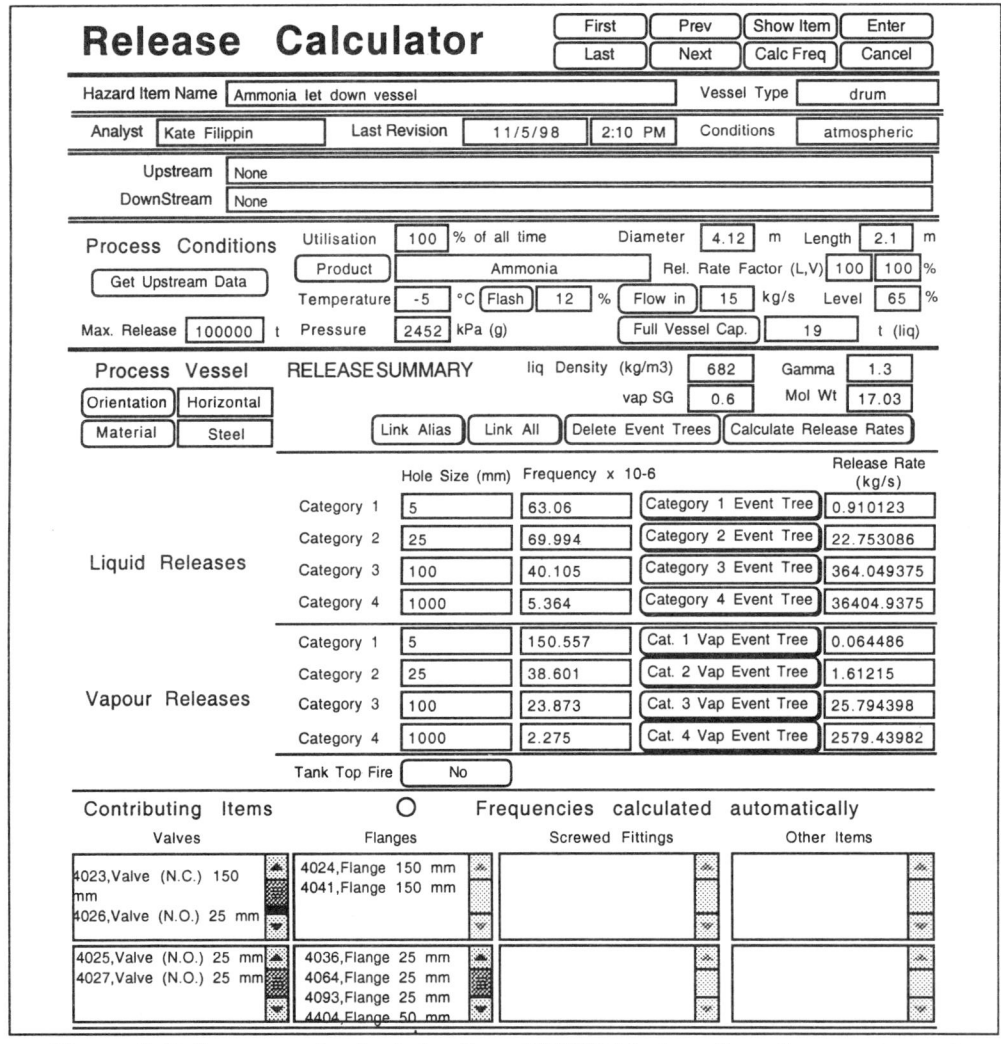

Figure 3. Release event calculator from VRJQRA. (continued on next page)

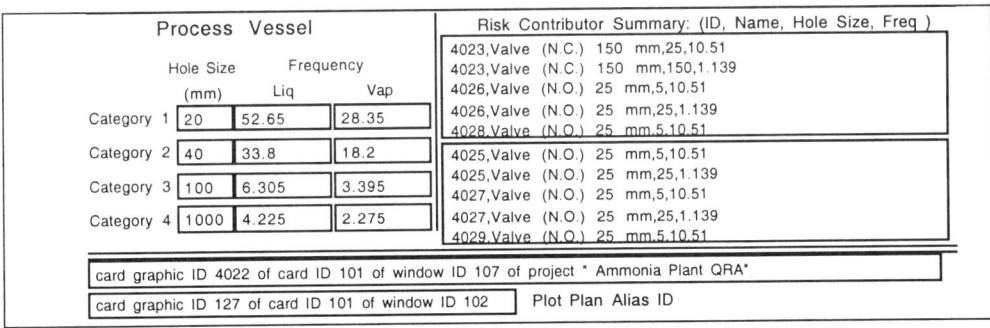

Figure 3. (continued from previous page) Release event calculator from VRJQRA.

Table 1. Example of Item Summary from VRJQRA

Process Item Name	Tag No.	Type	Length	Diameter	Valves	Flanges	Other
D103-Ammonia Drum	5	Proc. Vessel	8 m	2.1 m	5	6	0
Line from Drum to Storage	6	Proc. Pipe	25 m	150 mm	7	15	1 (Pump-Centrifugal double-seal)

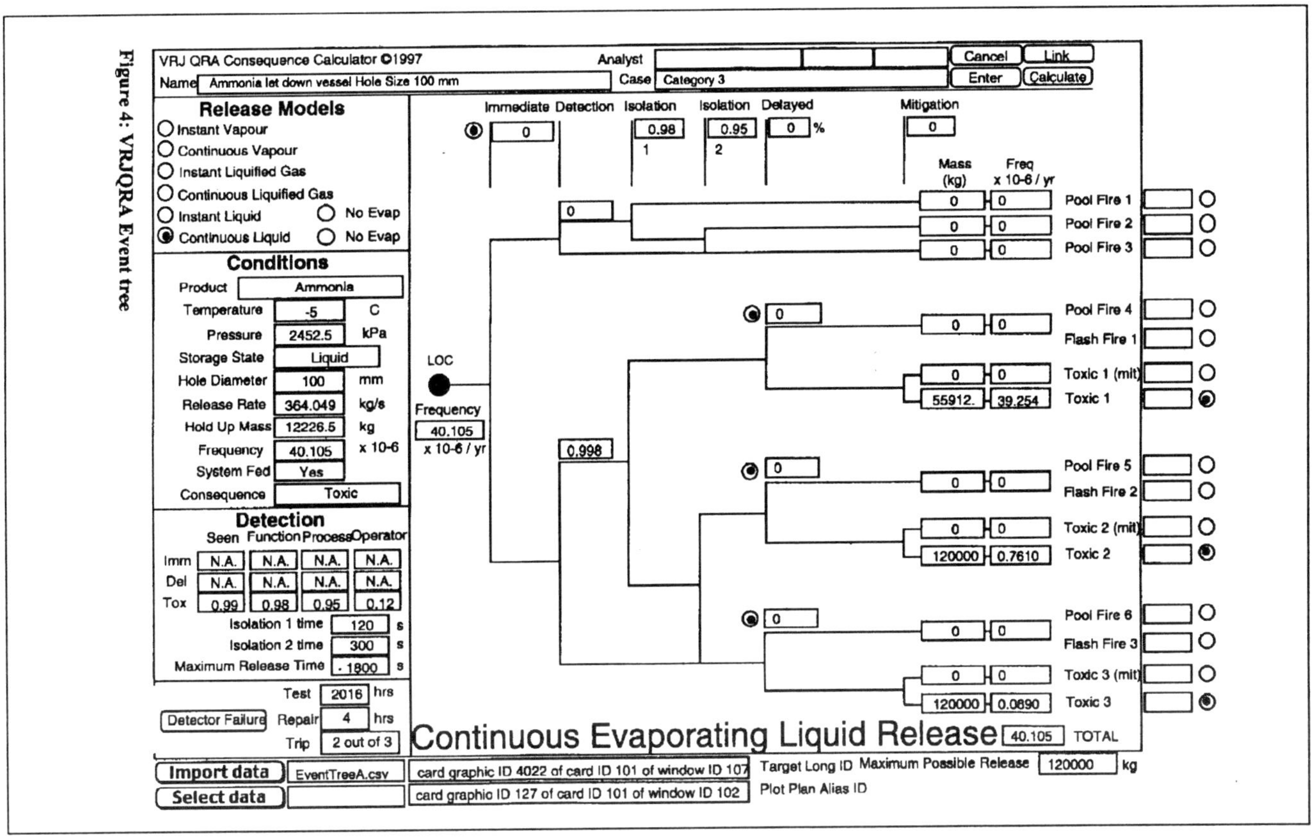

Figure 4. VRJQRA event tree.

and process detection (such as gas detectors) and isolation systems. A representative time for each possibility along with information about release rates and holdup volumes allows calculation of the release size.

The event tree outcomes include such consequences as fireballs, explosions, flash fires, pool fires, jet fires and toxic releases.

Each release scenario is assigned a frequency based on detection and isolation probabilities and the likelihood of ignition, if applicable.

Figure 4 shows an example of an event tree for a toxic release of liquid ammonia.

Once the end event is established the appropriate calculation can be undertaken to determine impact distances from fireballs, vapor cloud explosion, flash fire, pool fire, jet fire, or toxic dose scenarios. Meteorological data is incorporated in the modeling at this stage. Different weather stability categories with a corresponding frequency of occurrence allow various consequences to be considered.

Toxic releases

In order to simulate the impact of potential accidental releases of toxic products, VRJ uses the computer model AUSTOX (Centre for Applied Mathematical Modeling, Monash University, Melbourne, Australia). The model has the ability to simulate the source emissions for a particular accident scenario and estimate the impact of these emissions. Dose integration calculations have been included in the AUSTOX model, enabling more accurate prediction of the impacts of toxic releases than is possible by using concentration impacts alone.

AUSTOX has been configured to determine the cumulative dose at points surrounding the release in order to produce contours of iso-dose. The dosage levels the contours represent are determined from a probit equation by selecting levels of impact. As dose is integrated over time it is not necessary to know exposure times. Final concentration footprints are produced as contours that can be imported into VRJQRA for use in risk analysis.

Fireballs

A fireball can occur as a result of a large instantaneous release of vapor. Fireball impacts are calculated using the TNO model. This allows the calculation of the fireball dimensions and duration, which enables evaluation of the heat impacts on the surroundings.

Vapor cloud explosions

A vapor cloud explosion is an explosive deflagration of a dispersed flammable vapor air mixture. Explosions can result in fatalities due to the direct impact of overpressure on a person as well as indirect impacts from debris, missiles, and building collapse.

There are two models used in the QRA for the calculation of vapor cloud explosions. The first is the HSE model (CCPS, 1994), which relates the overpressure to the equivalent mass of TNT. This model is used for plants where there is limited confinement.

The second model is the multienergy damage model developed by TNO (TNO, 1979). The flammable mass is determined from the release rate and the resultant gas dispersion. The gas dispersion used to determine the flammable mass and the location of the flammable vapor cloud after dispersion has occurred is the computer model HGSYSTEM (developed by Shell Research, Thornton Research Centre, The United Kingdom).

Flash fires

If a flammable cloud encounters an ignition source and a vapor cloud explosion does not occur, the result will be a flash fire. A flash fire may also result from a liquid release if there is a significant amount of flashing or evaporation. In a flash fire the flame front moves through the vapor cloud, consuming the portions of the cloud in which the concentration is above the lower flammability limit. The dispersion of the released vapor is dependent on atmospheric conditions.

Flash fire impacts are determined by calculating the distance to flammable concentrations using the

HGSYSTEM gas dispersion model.

Pool fires

A pool fire is the result of a spill of flammable liquid that forms a pool on the ground that ignites. The heat radiation that pool fires emit presents a risk to human life. The pool-fire model considers the influence of wind on flame tilt and flame drag and the effect of absorption of heat by water vapor in the air (transmissivity).

The diameter of the pool depends on the release rate and any confinement due to bunding. The algorithm used for the radiation calculations is chosen based on the particular properties of the product being released.

Jet fires

A jet fire can result from a prolonged pressurized release of a gas. The vapor forms a turbulent jet that forms a jet flame when ignited. Personnel are at risk from both the flame front and the radiation levels emitted. The jet-fire model used for calculations is based on the Chamberlain Flare model (Chamberlain, 1987). This model approximates the flame shape to a frustum of a cone. The flame shape is used in calculating the view factor. The jet-fire model is sensitive to hole orientation and wind strength and is therefore able to model jet fires very realistically.

Risk Assessment

Individual risk

"Individual fatality risk" is the risk of death to a person at a particular point and assumes that the person will be present at that point at all times.

Quantification of risk (frequency of a given consequence level) is developed by a method in which sets of overlapping risk events, each with a frequency value and consequence radius, are combined to show risk contours.

VRJQRA is used to develop the risk contours. This method of calculation involves plotting each of the events calculated on a site plan that has been electronically entered into the program. The site plan is set up with a geographical coordinate system. Every identified event is represented by an intelligent object that knows its identity, location, and risk value. This information makes it possible to choose any point on the site plan and calculate the risk at that point.

The calculation for risk at any point from a particular event, will be

$$IR = F \times PWD \times PWC \times PF \qquad (1)$$

where IR is individual risk (chances per million per year), F is event frequency (chances per million per year), P_{WD} is probability of wind direction, P_{WC} is probability of the weather condition/stability, and P_F is probability of fatality. The probability of fatality is multiplied by the frequency when determining individual risk, as it is a measure of the likelihood of events impacting on an individual person at a given point. The total risk at a point is the sum of all the individual risks from all the events that impact on the point under consideration.

After calculation of the risk values for a sufficient number of points around the site an iso-risk contour can be plotted to show the different risk levels at the site. Figure 5 shows an example of individual risk contours for a 1,350 metric t/d ammonia plant. The contours for this plant are quite skewed. This is due to the higher frequency of one particular wind direction for the planned location.

Societal risk

Societal risk reflects the likelihood of accidents involving multiple fatalities. A societal risk analysis combines the consequences and likelihood of an incident occurring with population information to evaluate the risk of fatalities to a group of people. It can be used for land use planning off site or for determining the risk to personnel onsite.

Societal risk is most effectively presented in the form of a graph, combining the cumulative frequency (F) of one or more fatalities (N). This is generally known as an *F-N* curve.

Societal risk is determined using the same consequence calculations used to determine individual risk. However, instead of presenting the risk summed for all

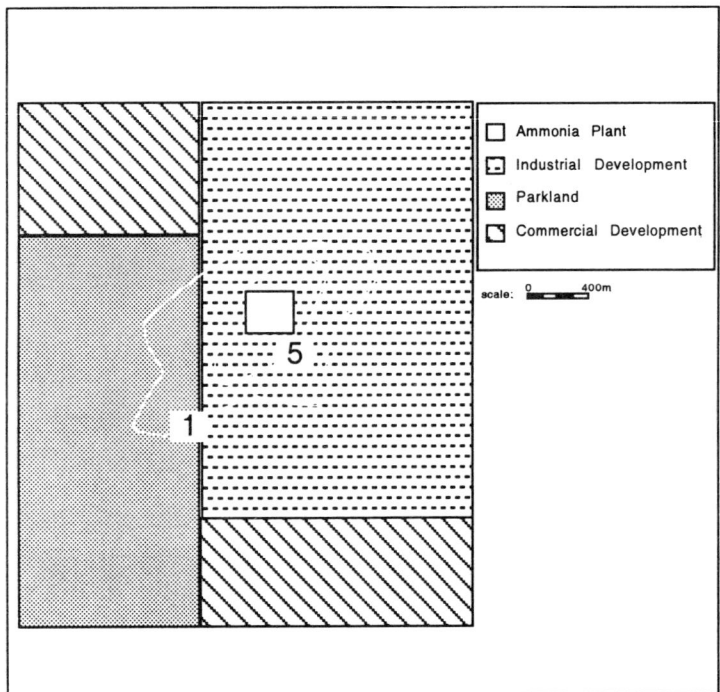

Figure 5. Individual risk contours (chances per million per year) for an ammonia plant.

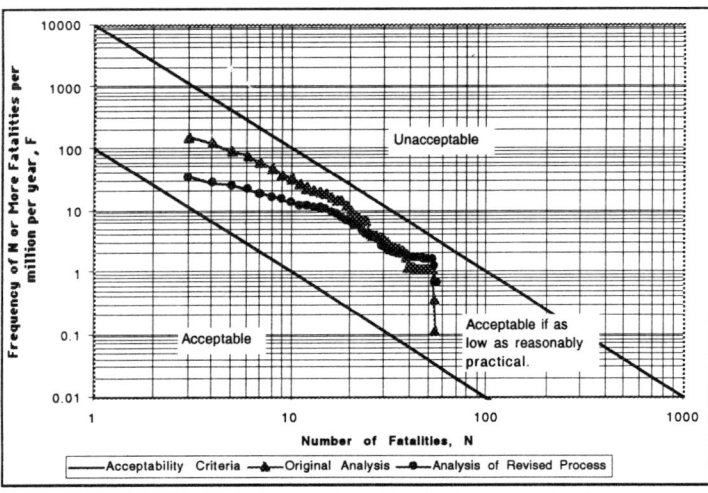

Figure 6. Comparison of societal risk results for two ammonia urea plants.

lated areas around the site. Each area is assigned a population density based on the number of hours the area is populated. The calculation of expected fatalities includes consideration for the predicted percentage of persons outdoors and the predicted percentage of fatality for both those persons indoors and outdoors should an incident occur.

The program evaluates the expected number of fatalities in the identified areas for each of the hazardous events that are determined for the individual risk. A series of F-N pairs are produced and these are used to determine a cumulative frequency for different values of N. The results are plotted to provide the societal risk curve. Part one of Figure 6 shows an example of a societal risk curve.

Risk Minimization and Acceptability

Once the risk has been calculated the results are compared with published risk criteria. Figure 7 shows a summary of some individual risk criteria and the acceptable land use at different risk levels.

The results of the QRA are compared to the risk criteria applicable to the location of the plant under examination. The comparison can be used in a number of ways. The risk contours can be used to determine possible land use around the site. This ensures that sensitive populations such as schools are not located too close to an existing or proposed facility.

The results for a new or existing plant can also be used to identify required risk reduction. The QRA report outlines all the information relating to major risk contributors. This includes the information about existing process conditions, detection and intervention strategies, and other contributing factors. The results can be used to identify the need for additional isolation points in the process, gas detectors, or an emergency shutdown system.

VRJQRA allows the risk-reduction measures to be fed back into the model so that new risk contours can be plotted for comparison. This provides flexibility so that risk-reduction options can be compared.

incidents independent of the density of the population, as is the case for individual risk, each possible incident outcome is considered distinctly and its frequency (F) and the number of fatalities (N) that could occur are recorded as an F-N pair.

VRJQRA is used to produce the F-N pairs required to plot the societal risk curve. A site plan is used with an overlay of intelligent objects to represent the popu-

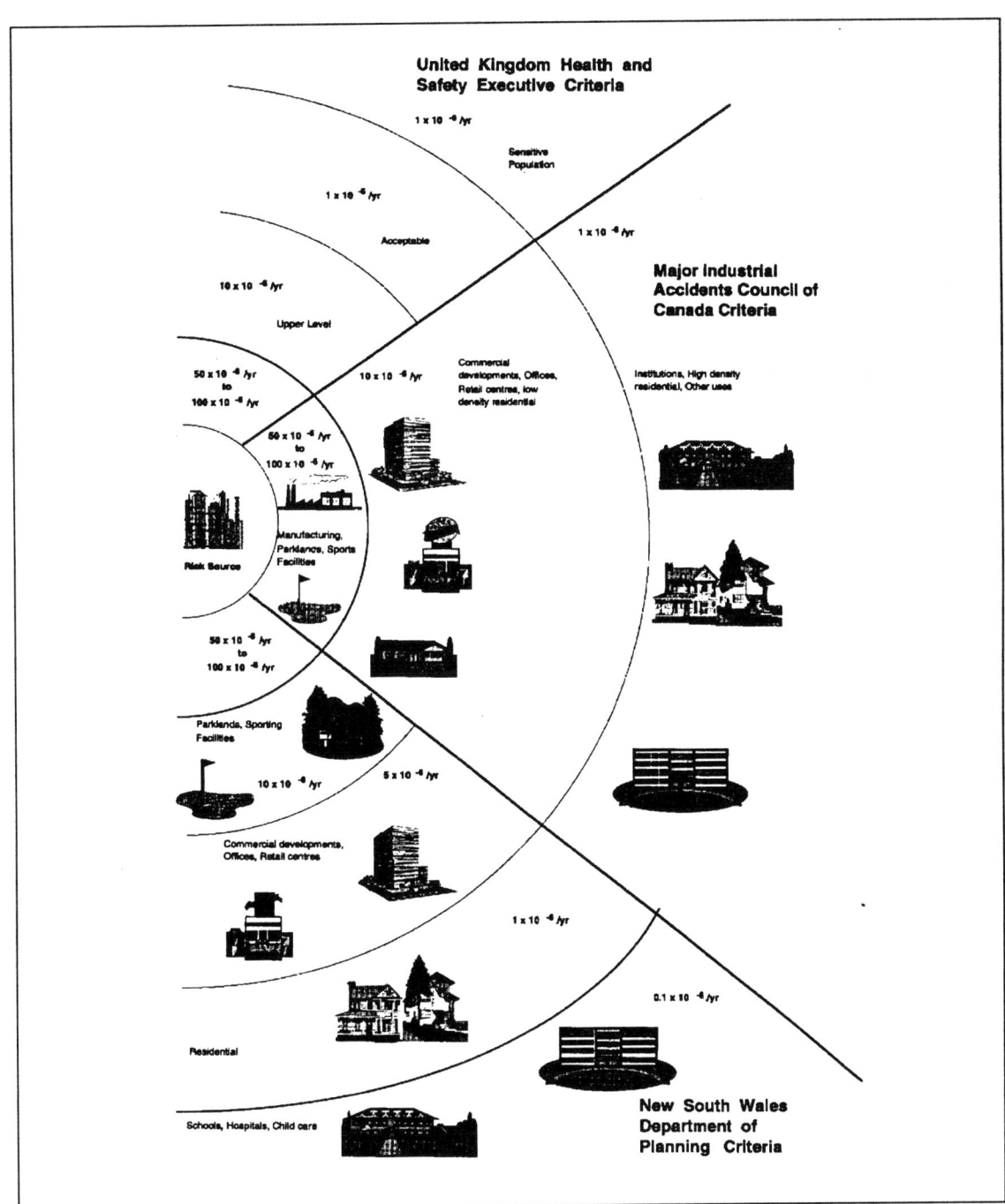

Figure 7. Individual risk criteria for critical exposed group.

Application of Risk Reduction Using VRJQRA

Example 1

Figure 8 shows the risk contours for a proposed ammonia urea plant. The plant was designed to produce 2,200 ton/d of urea. An ammonia storage and handling system was also included in the design. A QRA was undertaken for the plant using the VRJQRA program. The initial risk study produced large individual risk contours that extended significant distances offsite and did not meet the criteria for risk acceptance for the chosen location.

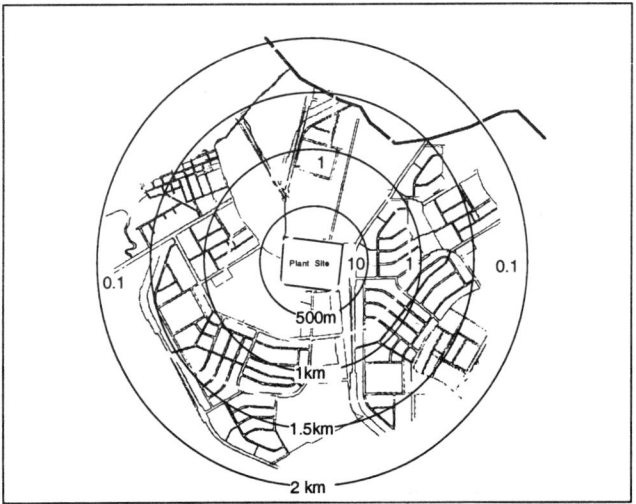

Figure 8. Individual risk contours (chances per million per year) for an ammonia urea plant.

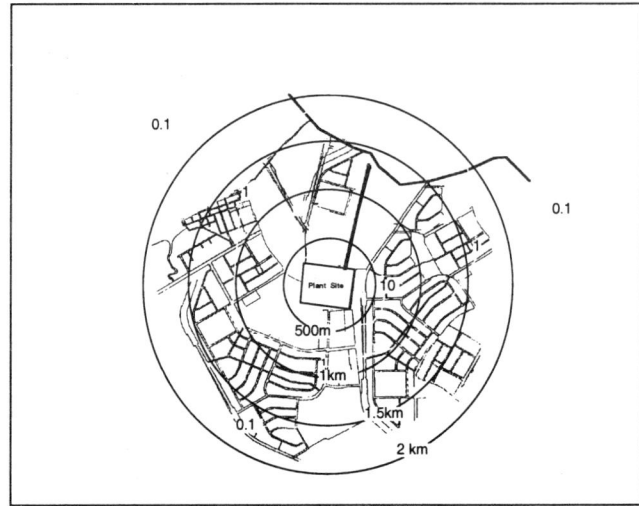

Figure 9. Revised individual risk contours (chance per million per year) for ammonia urea plant with changes to urea process.

The first step in risk reduction is to identify the major risk contributors. A summary of risk contributors was extracted from the QRA model. The examination of the risk contributors for the plant found that a large contributor to the risk was a release of liquefied ammonia from the recycle stripping process.

A reduction in the risk can be achieved by reducing either the frequency or the consequences of the event.

In this case a reduction in the frequency was not a feasible option as the facility was already designed using high integrity process items (with expected low failure frequency) and high detection and isolation probabilities.

A reduction in consequence would require a reduction in product mass released or better containment. As the process to be utilized had not been finalized, the client decided to examine an alternative urea process for comparison. The alternative process utilized carbon dioxide as the stripping agent. The risk results (Figure 9) produced for the alternative process showed some improvement. This was due to a reduction in potential for an ammonia release and the lower consequences for a carbon dioxide release.

The change in process was a feasible option for the proposed plant and allowed the individual risk to be reduced to satisfy the criteria.

Figure 6 shows the societal risk results for the same two plants. The risk criteria used for determining

Table 2. Comparison of Consequence Distances for Contained and Uncontained Liquid Ammonia Releases

Scenario	Gas Dispersion Distance (m) for 10% Fatality (No Bunding)	Gas Dispersion Distance (m) for 10% Fatality (Bunding)
Continuous Release of Liquid Ammonia from a 25-mm Hole	1,555	1,010
Continuous Release of Ammonia from a 100-mm Hole	2,880	1,025
Instantaneous Release of Liquid Ammonia	1,020	790

acceptability for the societal risk is that suggested by an Australian state government. The societal risk for the initial process was found to be in the region where risk reduction is still desirable. The societal risk for the changed plant process found that the risk was reduced, but not to an acceptable level. The main factor driving the societal risk levels up was found to be the close proximity of the plant to populated areas. At the time of writing of this paper the risk was being assessed for a new site location, which was in a less populated area.

Example 2

In a ammonia/urea plant of similar size, the individual risk results were also found to be too high for the risk criteria applicable to the location. Examination of the risk contributors found that there was insufficient detection and shutdown systems were insufficient. In order to minimize the size and probability of a possible release, additional ammonia detectors were included. The use of alarms in the double seal pumps in liquid ammonia service were also included as a means of detecting releases and minimizing their size. The possibility of bunding for the major liquid ammonia releases was also analyzed. The gas dispersion calculations for the possible releases from one particular vessel are presented in Table 2. The calculations are for a 10% chance of fatality at a time when the weather conditions are calm. The releases are for 26.5 metric t of liquid ammonia. The first set of calculations are for releases which have no containment, while the second set of results are for releases contained by a 200-m^2 bund.

The results show that providing a bunded area for the vessel being examined significantly reduces the consequence distances for a release from the vessel.

A decision was made to provide bunding for all areas of the plant where a major liquid ammonia release was possible. The inclusion of this risk reduction measure brought the individual risk to a level that was more acceptable.

Ongoing Risk Management Systems

As part of the ongoing risk management for a plant, the risks identified in the QRA study can be logged in a hazard register. The VRJQRA program completely summarizes the hazards identified so they can be imported into a hazard register and managed throughout the life of the plant. The hazard register system used provides live tracking with the ability to sign off hazards that have been eliminated. The use of a hazard register can also assist in tracking incidents, so that the failure rate data used in future updates of the plant QRA are more specific for the particular process being examined.

The VRJQRA program also provides the basis for a fire safety study and emergency response procedure. The information from VRJQRA is also used for developing safety management systems for the plant. The information used to identify particular hazards in the QRA can be used to examine specific scenarios and the impact they will have on surrounding equipment. It is useful for identifying emergency access routes and fire fighting requirements. This information is the basis for developing an emergency response plan. The QRA system is extremely useful for being able to evaluate the effects of changing management strategies such as testing strategies for critical trip systems and undertaking sensitivity studies.

Conclusion

The VRJQRA model provides an effective method of undertaking QRA. It enables a precise calculation procedure to be undertaken with a uniform approach over an entire facility. The storage of information within the program allows a complete auditing of the calculations and results at any time.

The use of this tool is valuable for identifying risks at the planning stage of a project, when the cost of risk reduction measures will be minimized. The ability to use the information resulting from the study provides a systematic approach to developing hazard registers, safety management systems and emergency procedure plans.

In the ammonia industry it is particularly important to identify and manage risks. The use of a tool such as VRJQRA provides an effective means of identifying the major areas of risk in the plant and identifying how effective the proposed risk reduction techniques will be. The identification of risks at an early stage in the design process should allow for better planning and management of risk.

Literature Cited

Center for Chemical Process Safety (American Institute of Chemical Engineers), Guidelines for Evaluating the Characteristics of Vapor Cloud Explosions, Flash Fires and BLEVES, pp. 117–118 (1994).

Chamberlain, G. A., "Developments in Design Methods for Predicting Thermal Radiation from Flares," *Chem. Eng. Res. Des.*, **65**, 299 (1987).

TNO Yellow Books, *Methods for the Calculation of the Physical Effects of the Escape of Dangerous Materials*, (1979).

DISCUSSION

R.A. McConnell, *ICI Chemicals and Polymers Ltd.*: In the case you showed where a second option taken is optimal from the risk point of view to the local population, was the inverse going to apply such that the state involved would prevent hospitals and houses being built right up against the gate to the new plant preventing any future development?

Filippin: We hope so.

McConnell: Is it not enshrined in the law of the state concerned?

Filippin: At the moment, in the state where the plant is located, the planning criteria is based on criteria set in other states. It is still in the development stage and we hope that this will take into account future development around the plant.

Syngas Purification in Gasification-Based Ammonia/Urea Plants

A new syngas purification process for gasification-based ammonia/urea plants using a UOP Selexol/PSA process sequence is presented. This new process can be used in place of the conventional Rectisol/Nitrogen Wash process. Capital and operating costs and plant performance for a typical coal-based ammonia plant using these two different processes are compared. These results also apply to syngas production from other feedstocks such as coke and residual oil, and other plant sizes.

John Y. Mak and David Heaven
Fluor-Daniel, Irvine, CA 92698

Introduction

Feasibility studies of a coal-to-ammonia/urea fertilizer complex demonstrated that the use of the UOP Selexol/PSA process technology for synthesis gas purification is superior to the conventional route of Rectisol/Nitrogen Wash. To completely evaluate these two gas processing routes, the differences in capital and operating costs, the consequences of varying operating parameters, and the impacts on surrounding units were investigated. Four cases were evaluated using Texaco Quench Gasification technology to gasify coal, considering the additional CO_2 co-production for urea manufacture. Material balances were developed for each case on the basis of producing 2,000 MTD of ammonia and, as an option, 1,475 MTD of CO_2 for urea manufacture. Capital and operating costs were developed for the entire plant so that the overall effects of the two different synthesis gas purification technology routes could be fully evaluated. The results show that the Selexol/PSA process is lower in both capital and operating costs than the Rectisol/Nitrogen Wash process. The results of this study apply to various feedstocks (coal, petroleum coke, and refinery heavy residues) and to different plant sizes and gasification technologies.

Independent of this work, a 1,000 MTD ammonia/urea fertilizer based upon Texaco gasification of refinery petroleum coke using the UOP Selexol/PSA technology for synthesis gas purification is currently under engineering/construction in North America. The low value of refinery residue feedstocks and this gasification/syngas purification route to fertilizer can be competitive with the conventional natural gas reforming processes.

Case Studies

This section describes the fertilizer complex configurations for the four cases of syngas production. Overall block flow diagrams, heat and material balances and process flow diagrams are shown in the subsequent sections. The gasification complex is designed to produce a syngas that is further processed to produce 2,000 MTD ammonia. Illinois No. 6 coal was used as the feed basis for these cases.

Case 1: Selexol/PSA without CO_2 Production

Case 1 is the base case to which all other configurations are compared. This case uses Selexol/PSA for acid gas removal and hydrogen purification. The coal feed to the plant is 2,593 MTD to produce 2,000 MTD of ammonia. The overall block flow diagram for this case is depicted in Figure 1. The overall heat and material balance is summarized in Table 1. The process units are briefly described below.

A single train Air Separation Unit (2,340 MTD) is used to supply high-pressure oxygen at 99.5% purity to the gasification unit plus low-pressure oxygen to the oxygen blown Claus Unit. The use of oxygen in the Claus Unit reduces the sulfur plant size and improves its operating efficiency. The Air Separation Unit also supplies high-pressure gaseous nitrogen which is mixed with the hydrogen from the Hydrogen Recovery Unit (PSA) to make up a stoichiometric feed to the ammonia plant. The PSA unit, designed by UOP, uses low-pressure nitrogen for purging, resulting in a substantial improvement in hydrogen recovery.

Coal from storage is conveyed to the coal grinding and slurry preparation system at a rate of 2,593 MTD. Coal and recycle carbon is wet ground with water to produce a coal-slurry. The slurry and a 99.5% purity oxygen stream from the air separation plant are fed to the Texaco Quench Gasifier operating at 1,000 psig for the production of a raw syngas stream. The particulate-laden quench and scrubber water stream is further treated for soot removal and recovery of unconverted carbon.

The CO in the syngas from the gasification unit is converted to hydrogen using two stages of sour shift reactors via the water-shift reaction.

$$CO + H_2O \rightarrow H_2 + CO_2$$

Since the raw syngas has been saturated with water in the scrubbing process, it contains sufficient steam for the shift reaction and no additional steam is required. The heat contained in the reactor effluent is used to generate various levels of steam.

A side reaction of the CO shift catalyst converts most of the COS to CO_2 and H_2S by the catalytic hydrolysis reaction, which also occurs in the presence of steam.

$$COS + H_2O \rightarrow H_2S + CO_2$$

The syngas is then fed to the UOP Selexol unit that is designed to remove 99.5% of the H_2S and to produce a CO_2 stream suitable for urea manufacture.

The process flow diagram for the Selexol unit is depicted in Figure 2.

Selexol Unit

The HP Absorber is designed with an inter-cooler for controlling the absorption temperature and maximizing the acid gas loading of the rich solvent. This exchanger results in reducing the lean solvent circulation. The Selexol solvent reduces the H_2S content of the syngas from 0.92 mol. % to 18 ppmv while also removing about 50% of COS.

Table 1. Case 1: Material Balance for Selexol/PSA Without CO_2 Production

Stream Number	1	2	3	4	5	6	7	8	9	10	11	12	13
Description	Coal Feed	Coal Slurry To Gasifiers	Oxygen To Gasifiers	Raw Syngas To Syngas Scrubbing	Scrubbed Syngas	Shifted Syngas To Selexol	Sulfur From Oxygen Claus Unit	Treated Syngas To PSA	Total Purge Gas	LP Nitrogen From ASU	HP Nitrogen From ASU	Hydrogen From PSA	Syngas To Ammonia Synthesis
Component	lb/hr	lb/hr	Lbmol/hr	Lbmol/hr	Lbmol/hr	Lbmol/hr	Lbmol/hr	Lbmol/hr	Lbmol/hr	Lbmol/hr	Lbmol/hr	Lbmol/hr	Lbmol/hr
CO				9,722	9,722	445		444	445				
Hydrogen				8,065	8,065	17,330		17,315	1,053			16,276	16,276
CO2				3,751	3,751	12,941		9,522	12,941				
Methane				22	22	22		22	19			3	3
Argon			25	33	33	33		33	22			12	12
Nitrogen			9	97	97	97		97	4,168	4,240	5,216	169	5,385
H2S				286	286	293		1	1				
COS				7	7	0		0	0				
H2O		145,973		5,777	34,169	65		64	64				
Oxygen			6,685										
Sulfur							293						
Coal	238,167	238,167											
Total	238,167	384,140	6,718	27,761	56,153	31,226	293	27,498	18,714	4,240	5,216	16,460	21,676
Mol Wt			32	21	19	20	32	17	36	28	28	2	9

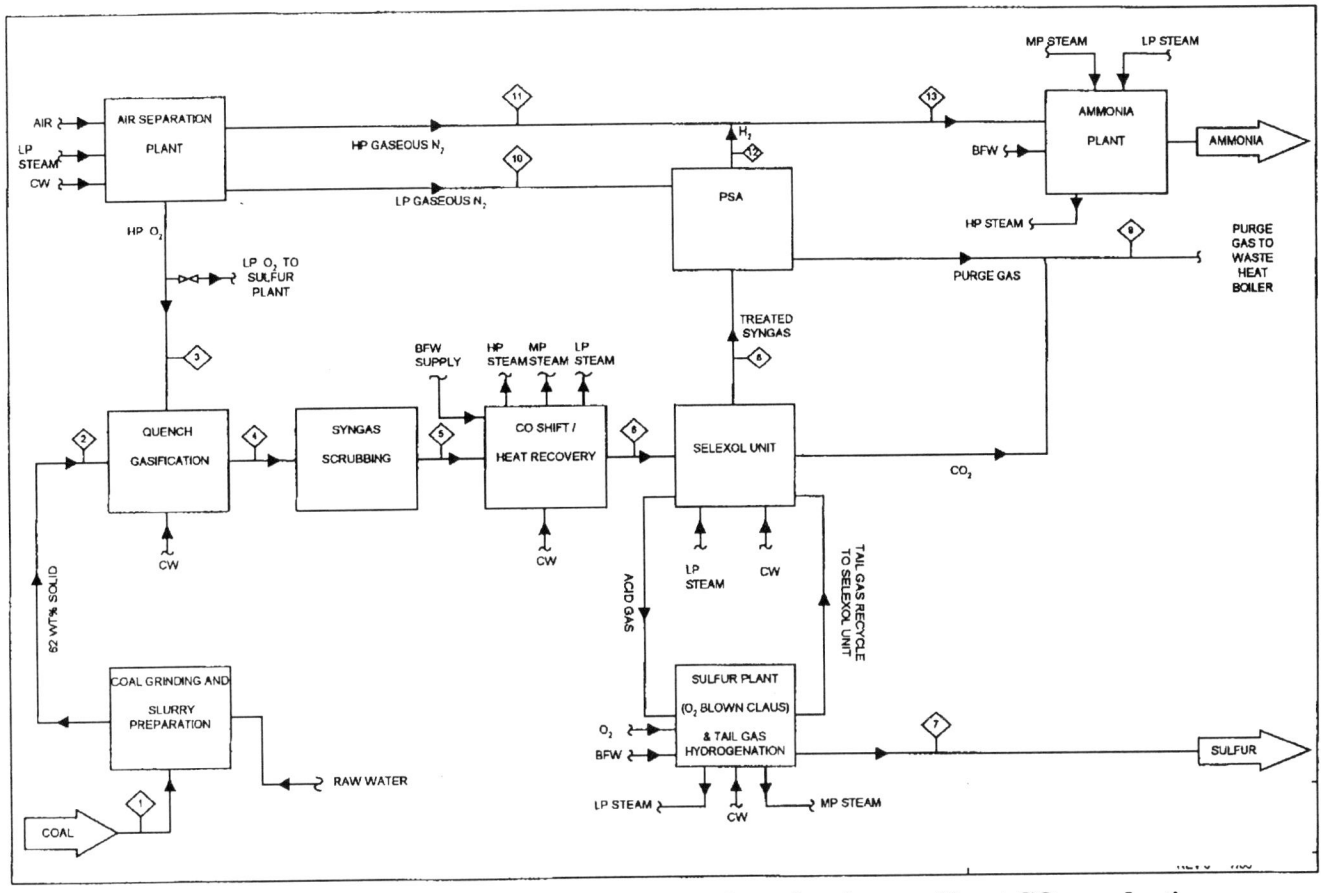

Figure 1. Case 1 (base case): coal to ammonia complex selexol case without CO_2 production.

The rich Selexol solvent from the HP Absorber is let down to 450 psia, and the flash gas is recycled back to the absorber so that its hydrogen content is recovered. The solvent is then heated in the lean/rich exchangers to 244°F and is then flashed to 200 psia. The LP flash vapor is sent to the H_2S Concentrator, which is designed to reject the bulk of the CO_2 thereby concentrating the H_2S content of the acid gas feed to the Claus unit.

The H_2S Concentrator uses a pre-saturator to produce a CO_2 stream with a very low H_2S content. The lean Selexol is first mixed with the overhead vapor and is cooled to 0°F in an overhead heat exchanger. Most of the CO_2 absorption occurs in this exchanger and the solvent becomes saturated with CO_2. The CO_2 saturated liquid is pumped back to the column for H_2S absorption.

PSA Unit

The PSA unit is designed to purify the treated gas from the Selexol unit and produce 99% purity hydrogen for the ammonia plant. This PSA unit is a new design offered by UOP that improves the overall hydrogen recovery in the gasification-to-ammonia plants.

The PSA unit uses conventional multibed adsorption/desorption cycle design, with nitrogen being used for purging the bed in the desorption cycle. Conventional design uses hydrogen for purging during the desorption cycle, which results in loss of hydrogen in the purge gas. In a gasification/ammonia plant, very pure nitrogen is available from the Air Separation Plant and purging the beds using nitrogen improves the overall hydrogen recovery, typically from 88%–90% to 92%–94%. The tail gas from the PSA unit contains a substantial amount of nitrogen and

Table 2. Case 2: Material Balance for Selexol/PSA With CO_2 Production

Stream Number	1	2	3	4	5	6	7	8	9	10	11	12	13	14	15
Description	Coal Feed	Coal Slurry To Gasifiers	Oxygen To Gasifiers	Raw Syngas To Syngas Scrubbing	Scrubbed Syngas	Shifted Syngas To Selexol	Sulfur From Oxygen Claus Unit	Treated Syngas To PSA	Total Purge Gas	LP Nitrogen From ASU	HP Nitrogen From ASU	Hydrogen From PSA	Syngas To Ammonia Synthesis	LP (Gas) CO_2 To Urea Plant	HP (Liquid) CO_2 To Urea Plant
COMPONENT	lb/hr	lb/hr	Lbmol/hr	Lbmol/hr	Lbmol/hr	Lbmol/hr	Lbmol/hr	Lbmol/hr	Lbmol/hr	Lbmol/hr	Lbmol/hr	Lbmol/hr	Lbmol/hr	Lbmol/hr	Lbmol/hr
CO				9,722	9,722	445		444	444						
Hydrogen				8,065	8,065	17,330		17,315	1,039			16,276	16,276		
CO2				3,751	3,751	12,941		9,522	9,861					1,232	1,848
Methane				22	22	22		22	19			3	3		
Argon			25	33	33	33		33	22			12	12	0	0
Nitrogen			9	97	97	97		97	4,168	4,240	5,216	169	5,385	0	0
H2S				286	286	293		1	1						
COS				7	7	0		0	0						
H2O		145,973		5,777	34,169	65		64	64						
Oxygen			6,685												
Sulfur							293								
Coal	238,167	238,167													
TOTAL	238,167	384,140	6,718	27,761	56,153	31,226	293	27,498	15,634	4,240	5,216	16,460	21,676	1,232	1,848
Mol Wt			32	21	19	20	32	17	36	28	28	2	9	44	44

requires supplementary natural gas for combustion.

Case 2: Selexol/PSA With CO_2 production

Case 2 is similar to Case 1 with the exception that 1,475 MTD CO_2 is recovered and purified for the urea plant. In this case, the CO_2 waste stream is further compressed from 185 psia to the CO_2 Compression/Liquefaction unit. The overall block diagram for this case is depicted in Figure 3 and the mass balance is shown in Table 2.

The CO_2 Compression/Liquefaction unit is shown in Figure 4. The CO_2 stream from the Selexol unit is compressed to 425 psia, and steam is injected into the CO_2 to hydrolyze the residual COS. The effluent is sent to a ZnO bed that removes its sulfur content down to 1 ppm to meet the CO_2 specification for the urea plant.

The sulfur-free CO_2 stream is cooled by cooling water and sent to a knockout drum where the condensed liquid, mostly water, is removed. The CO_2 stream is further dried using a molecular sieve dryer in order to avoid water freezing in the downstream exchangers. A feed/effluent first cools the CO_2, and a

Table 3. Case 3: Material Balance for Rectisol/N_2 Wash Without CO_2 Production

Stream Number	1	2	3	4	5	6	7	8	9	10	11	12	13
Description	Coal Feed To Slurry Preparation	Coal Slurry To Gasifiers	Oxygen To Gasifiers	Raw Syngas To Syngas Scrubbing	Scrubbed Syngas	Shifted Syngas To Rectisol	Sulfur From Oxygen Claus Unit	Treated Syngas To Nitrogen Wash	Total Purge Gas	Stripping Nitrogen	HP Nitrogen From ASU	Recycle Gas From N2 Wash To Rectisol Unit	Syngas To Ammonia Synthesis
COMPONENT	lb/hr	lb/hr	Lbmol/hr	Lbmol/hr	Lbmol/hr	Lbmol/hr	Lbmol/hr	Lbmol/hr	Lbmol/hr	Lbmol/hr	Lbmol/hr	Lbmol/hr	Lbmol/hr
CO				7,574	7,574	347		339	0			3	
Hydrogen				7,621	7,621	16,375		16,387	99			74	16,276
CO2				3,319	3,319	11,456			11,456				
Methane			9	78	78	78		71	78				
Argon			23	22	22	22		21	22			1	0	1
Nitrogen				157	157	157		174	1,541	1,082	5,688	5	5,385
H2S				247	247	254							
COS				7	7	0							
H2O		137,954		5,452	25,012	47		47					
Oxygen			6,326										
Sulfur							254						
Coal	225,083	225,083											
Total	225,083	363,038	6,358	24,477	44,037	28,735	254	16,993	13,589	1,082	5,688	82	21,662
MOL WT			32	19	19	20	32	3	42	28	28	5	9

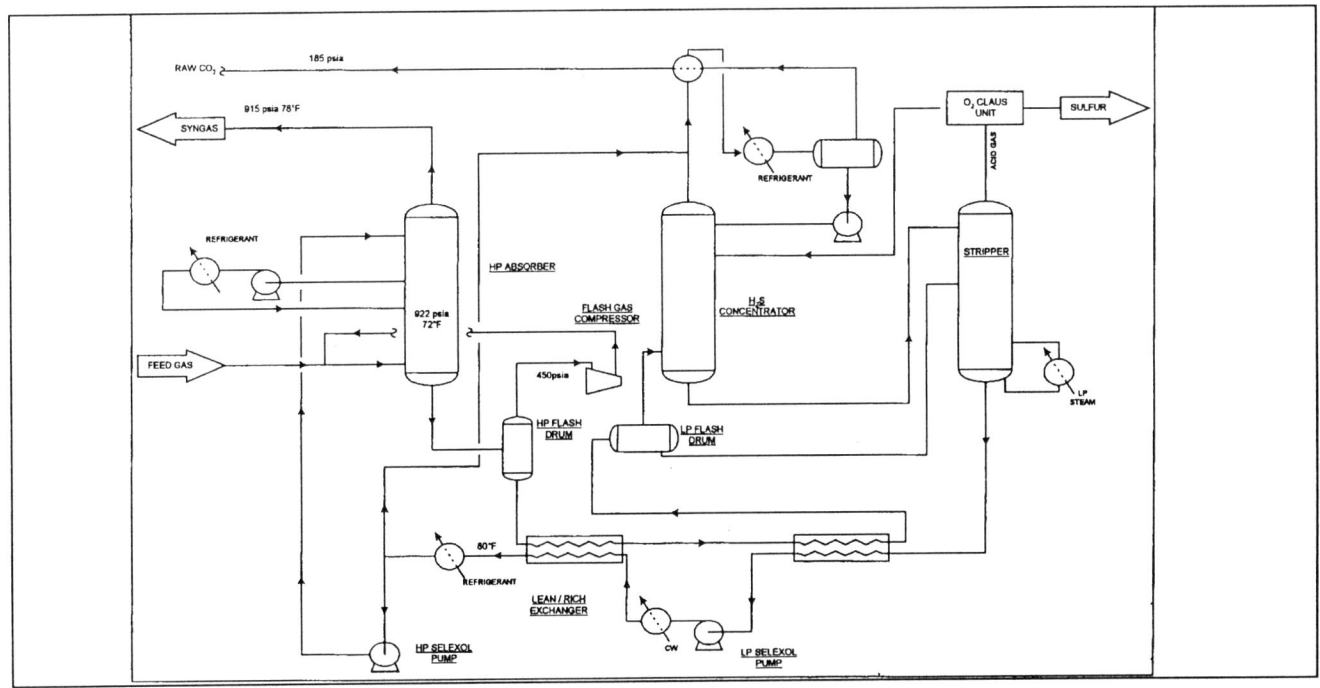

Figure 2. Typical selexol unit.

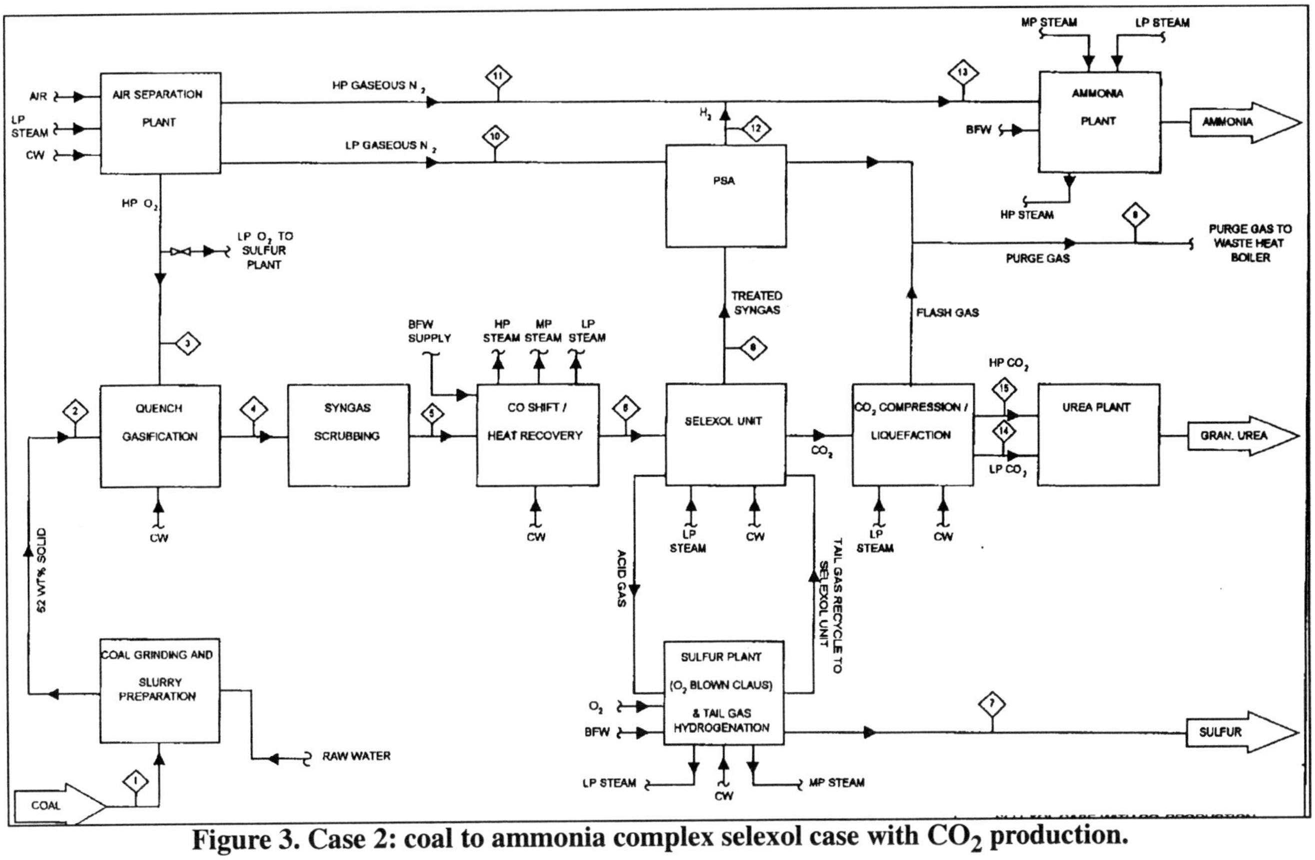

Figure 3. Case 2: coal to ammonia complex selexol case with CO_2 production.

Table 4. Case 4: Material Balance for Rectisol/N₂ Wash With CO₂ Production

Stream Number	1	2	3	4	5	6	7	8	9	10	11	12	13	14	15
Description	Coal Feed	Coal Slurry To Gasifiers	Oxygen To Gasifiers	Raw Syngas To Syngas Scrubbing	Scrubbed Syngas	Shifted Syngas To Rectisol	Sulfur From Oxygen Claus Unit	Treated Syngas To Nitrogen Wash	Total Purge Gas	Stripping Nitrogen From ASU	HP Nitrogen From ASU	Recycle Gas From N2 Wash To Rectisol Unit	Syngas To Ammonia Synthesis	LP (Gas) CO2 To Urea Plant	HP (Liquid) CO2 To Urea Plant
COMPONENT	lb/hr	lb/hr	Lbmol/hr	Lbmol/hr	Lbmol/hr	Lbmol/hr	Lbmol/hr	Lbmol/hr	Lbmol/hr	Lbmol/hr	Lbmol/hr	Lbmol/hr	Lbmol/hr	Lbmol/hr	Lbmol/hr
CO				7,574	7,574	347		339	0			3			
Hydrogen				7,621	7,621	16,375		16,387	99			74	16,276		
CO2				3,319	3,319	11,456		8,376						1,232	1,848
Methane			9	78	78	78		71	78						
Argon			23	22	22	22		21	22			1	0	1	
Nitrogen				157	157	157		174	1,541	1,082	5,688	5	5,385		
H2S				247	247	254									
COS				7	7	1									
H2O		137,954		5,452	25,012	47		47							
Oxygen			6,326				254								
Sulfur															
Coal	225,083	225,083													
Total	225,083	363,038	6,358	24,477	44,037	28,735	254	16,993	10,509	1,082	5,688	82	21,662	1232	4848
Mol Wt			32	19	19	20	32	3	42	28	28	5	9	44	44

chiller using propane refrigeration at 5°F liquefies most of the CO₂. The cooled stream is flashed in the CO₂ Flash Drum, and is sent to a stripper where the residual hydrogen and CO are removed down to a very low level. The liquid CO₂ product is then pumped to the pressure required by the urea plant.

Case 3: Rectiso/N₂ wash without CO₂ production

The Case 3 process flow scheme is similar to the base case except that the Rectisol/N₂ Wash process is used as the acid gas removal /hydrogen purification step. Because of the higher hydrogen recovery in this case, the coal feed to the plant is lower (2,450 MTD vs. 2,593 MTD in the base case). The overall block diagram for this case is depicted in Figure 5. The overall heat and material balance is shown in Table 3.

The Rectisol process uses methanol as the solvent for acid gas removal and has been used for acid gas removal in conventional "Coal to Ammonia" plant applications. The process operates at cryogenic temperatures and when used in combination with the N₂ Wash Unit, produces syngas with a very low levels of impurities.

The N₂ Wash unit is designed as an integrated unit with the Rectisol unit, which improves its overall energy consumption and hydrogen recovery. The flash gas from the N₂ Flash Drum downstream of the N₂ Wash Column is recycled back to the Rectisol absorber, which eliminates hydrogen loss from the Wash Unit.

Similar to the UOP Selexol unit design, the acid gas from the Rectisol unit is sent to the oxygen-blown Claus unit, and the tail gas from the Claus unit is hydrogenated and recycled back to the acid gas removal unit so that SO$_x$ emissions are eliminated.

The main differences between the Rectisol unit and the Selexol unit are summarized as follows:

- The Rectisol unit removes all the CO₂ and H₂S from the syngas, while Selexol selectively removes H₂S, and the PSA unit removes the residual CO₂.

- The Rectisol unit operates under cryogenic temperatures (typical, -80°F), while Selexol operates at warmer temperatures (0 to 40°F).

- Low-pressure nitrogen is required by the Rectisol unit for CO₂ stripping in producing a concentrated H₂S feed to the Claus unit and in the regeneration of the solvent. The Selexol solvent is regenerated by steam and H₂S is concentrated by rejecting CO₂ in the Pre-saturator.

- The Rectisol/N₂ Wash design produces a very pure syngas feed which results in a very minimal purge from the ammonia synthesis loop. The PSA design allows a residual amount of argon and methane (200 to 700 ppmv) which requires a slightly higher

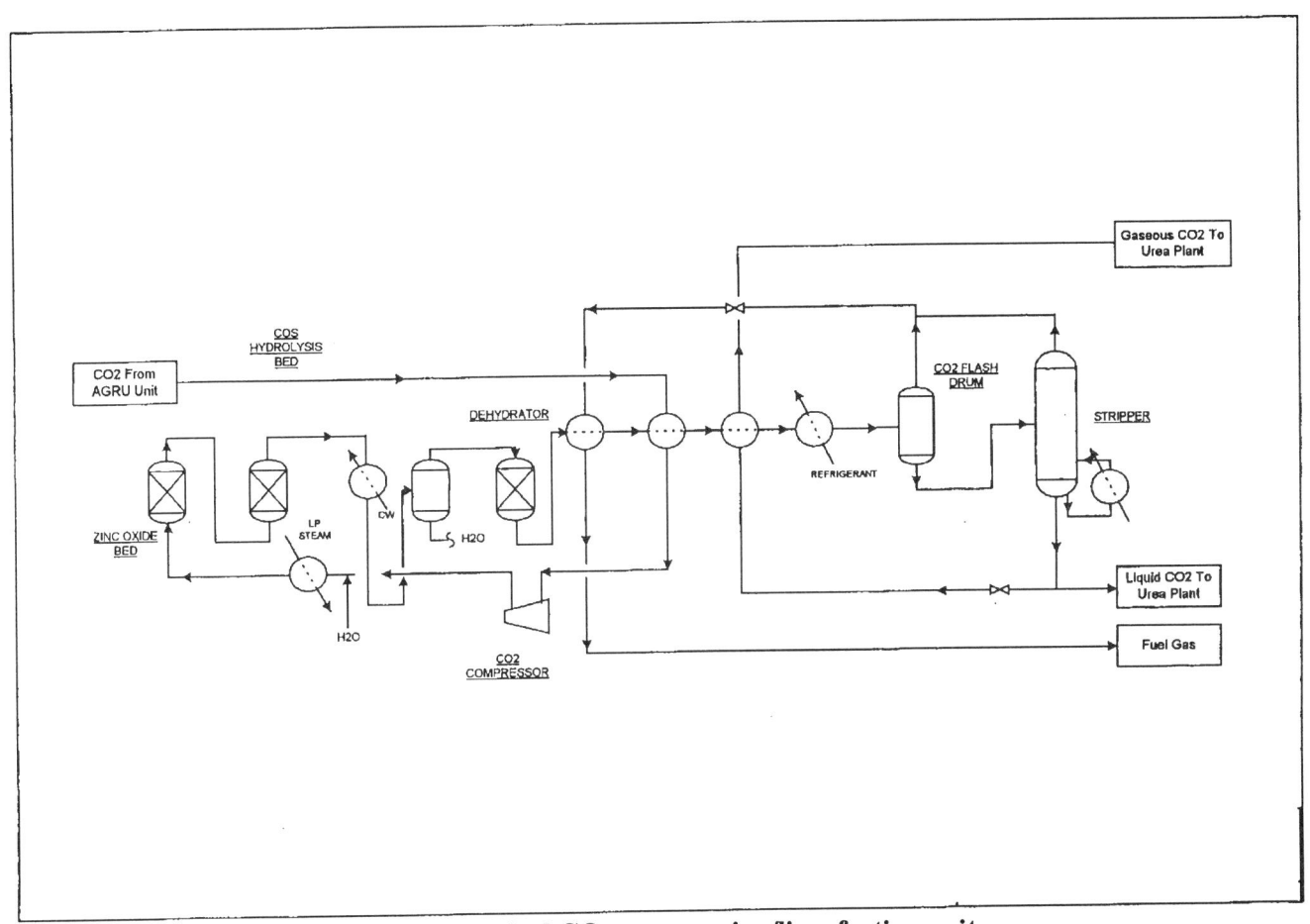

Figure 4. Typical CO_2 compression/liquefaction unit.

purge from the ammonia plant. In addition, if the oxygen content of the nitrogen from the air separation is greater than the limit specified by the ammonia converter design, an oxygen removal bed is required upstream of the ammonia plant.

The process flow diagram of a typical Rectisol unit is depicted in Figure 6. The methanol scrubber consists of an upper bulk CO_2 removal section and a lower H_2S removal section. A portion of the CO_2 rich methanol is drawn from the CO_2 removal section and letdown in pressure to the CO_2 flash drum. The H_2S saturated methanol from the bottom of the H_2S removal section is let down to the H_2S flash drum. The flash gas from both drums is recompressed and recycled to the inlet of the unit. The gas stream from the top of this column is warmed by the incoming syngas and is then sent to the CO_2 methanol adsorption unit.

The methanol from the bottom of the CO_2 Stripper is sent to the H_2S Concentrator, which uses low-pressure nitrogen for stripping the bulk of CO_2 from the methanol, thereby concentrating its H_2S content. The solvent is finally regenerated in the methanol regenerator, which produces a concentrated H_2S feed to the Claus unit.

The N_2 Wash Unit consists of a cold box, a N_2 Wash column, and a flash drum. The process is depicted in Figure 7.

The feed gas originates at the top of the methanol scrubber column and is sent to the CO_2/methanol adsorber. Traces of CO_2 and methanol are removed here to avoid the formation of solids inside the cryogenic section of the unit. The treated gas from the adsorber is cooled down against product streams and is then sent to the N_2 Wash column. The impurities still remaining in the hydrogen syngas (Argon, CO and methane) are removed by means of liquid nitrogen. Cooling is accomplished by the Joule-Thompson effect. Purified gas leaves the top of the column and is sent to ammonia synthesis. The liquid from the bottom of the column is expanded into a flash drum. The flash gas is recycled back as feed to the Rectisol unit, while the bottoms is vaporized against incoming feed streams, providing cooling for the process before it is sent to the waste fuel boiler.

The nitrogen required for the wash process and for making up the stoichiometric feed to the ammonia plant enters the cold box at an ambient temperature and is cooled against product streams. The nitrogen is split into two streams. One stream is further cooled and sent to the top of the column to perform the absorption. The other stream is added to the purified hydrogen to maintain the required stoichiometric ratio for the ammonia plant.

Case 4: Rectisol/N_2 Wash With CO_2 Production

This case is the same as Case 3 with the exception that a CO_2 stream is produced for the Urea Plant. A portion of the CO_2 stream from the top of the CO_2 stripper is sent to the CO_2 compression/liquefaction unit while the remainder is sent to the waste fuel burner. The overall block diagram for this case is depicted in Figure 8. The heat and material balance is shown in Table 4.

The Rectisol process produces CO_2 at a lower pressure (33 psia) than the Selexol process (185 psia) and, consequently, the CO_2 Compression/liquefaction unit requires additional compression stages.

Cost Evaluation

Capital cost

Capital cost estimates were prepared for each of the four cases using equipment and unit capacity factoring techniques and licensor quotations. The capital costs are based on a U.S. location and third-quarter 1996 time frame. The capital cost estimation results are summarized in Table 5.

The capital costs for the units surrounding the Selexol/PSA Unit (or Rectisol/N_2 Wash process), that is, the balance of plant (BOP), are determined using the overall unit capacity-factoring method. The balance of plant represents the sum of the front-end units in the plant and does not reflect the cost of any unit downstream of the purification unit, as the costs of the downstream units are the same for all the cases.

As seen in Table 5, the capital cost for the gas purification unit is lower for the UOP Selexol/PSA case because of less processing equipment and less expensive material of construction. On the other hand, the Rectisol/N_2 Wash unit is lower in the BOP cost, because of its higher hydrogen recovery. The net overall cost savings for selecting Selexol/PSA over Rectisol/N_2 Wash amount to approximately $18 MM.

Table 5. Selexol/PSA vs. Rectisol/N_2 Wash Capital Cost Comparison

	Case 1 (Selexol/ PSA)	Case 3 (Rectisol/ N_2 Wash)	Case 2 (Selexol/ PSA)	Case 4 (Rectisol/ N_2 Wash)
	No CO_2 Production		With CO_2 Production	
Capital Cost ($1,000)				
Selexol Unit	16,330	N/A	16,330	N/A
PSA Unit	24,530	N/A	24,530	N/A
Rectisol/N_2 Wash Units	N/A	72,000	N/A	72,000
Balance of Plant (Front End Units) (BOP)	307,200	294,300	310,010	297,850
Incremental Gas Purification Capital Costs	Base	31,140	0	31,140
Incremental BOP Costs	Base	(12,900)	2,810	(9,350)
Incremental Overall Plant Capital Costs	Base	18,240	2,180	21,790

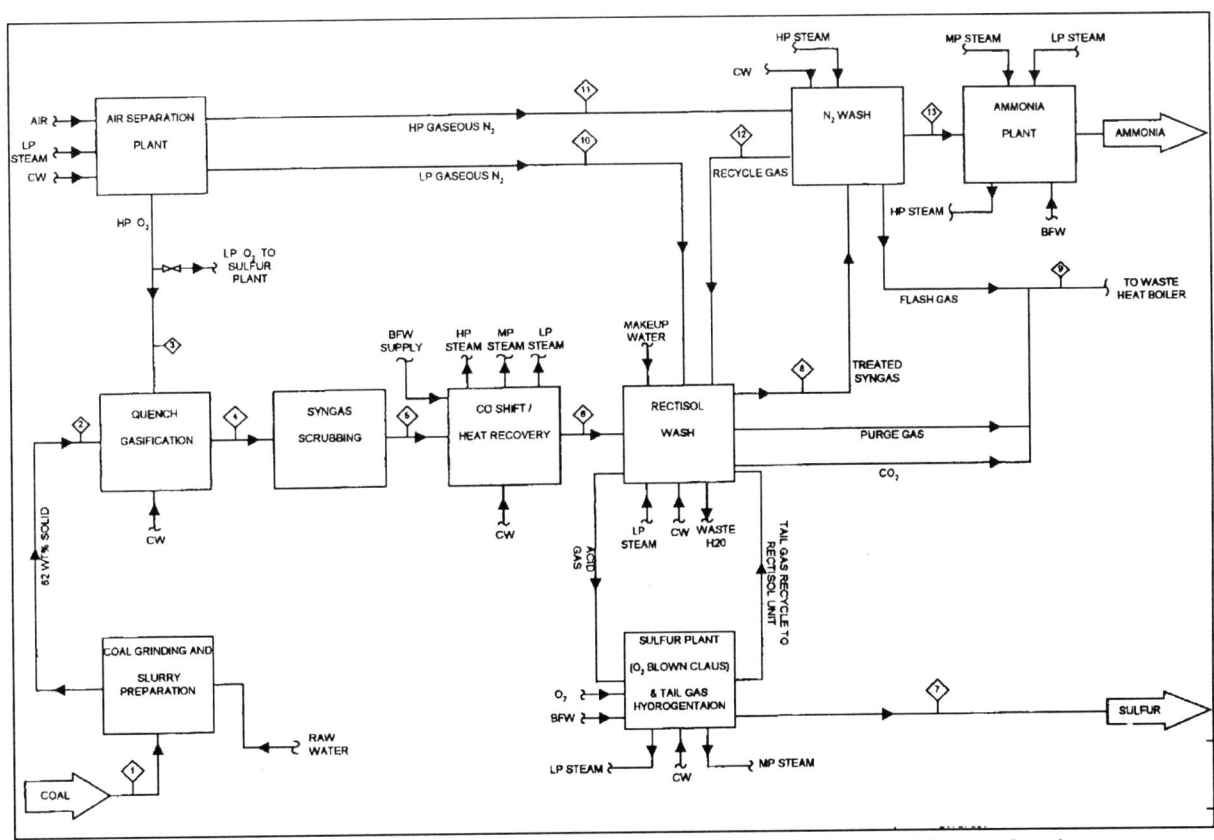

Figure 5. Case 3: coal to ammonia complex rectisol case without CO_2 production.

Table 6. Selexol/PSA vs. Rectisol/N_2 Wash Annual Operating Cost Comparison

	Case 1 (Selexol/ PSA)	Case 3 (Rectisol/ N_2 Wash)	Case 2 (Selexol/ PSA)	Case 4 (Rectisol/ N_2 Wash)
	No CO_2 Production		With CO_2 Production	
Annual Operating Cost ($1,000)				
Annual Fixed Operating Costs	14,200	14,540	14,260	14,740
Annual Utilities and Feed Costs	25,700	26,540	27,350	28,780
Annual Catalyst and Chemical Costs	1,280	1,430	1,570	1,710
Total Annual Operating Costs	41,180	42,510	43,180	45,230
Incremental Annual Fixed Op. Costs	Base	340	60	540
Incremental Annual Utilities and Feed Costs	Base	840	1,650	3,080
Incremental Annual Cat. and Chem. Costs	Base	150	290	430
Total Incremental Annual Operating Costs	Base	1,330	2,000	4,050

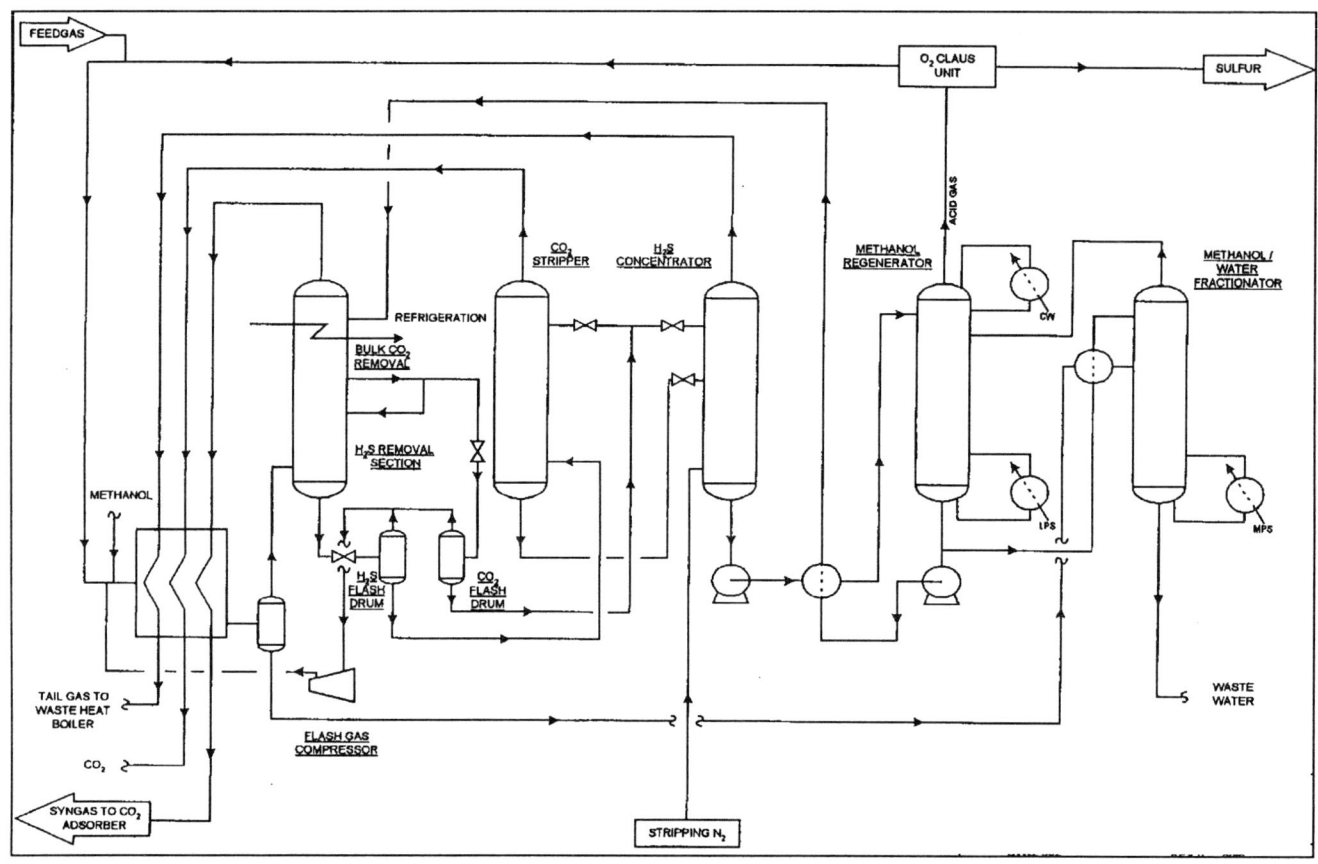

Figure 6. Typical rectisol wash unit.

Table 7. Selexol/PSA vs. Rectisol/N_2 Wash Required Feedstock Comparison to Produce 2,000 MTD Ammonia

	Required Tar Feedstock (MTD)		
	Coal	Coke	Residual Oil (tar)
Purification Units			
UOP Selexol /PSA	2,600	2,230	1,610
Rectisol/Nitrogen Wash	2,450	2,100	1,520
Incremental Feedstock Requirements			
UOP Selexol/PSA	Base	(370)	(990)
Rectisol/Nitrogen Wash	Base	(350)	(930)

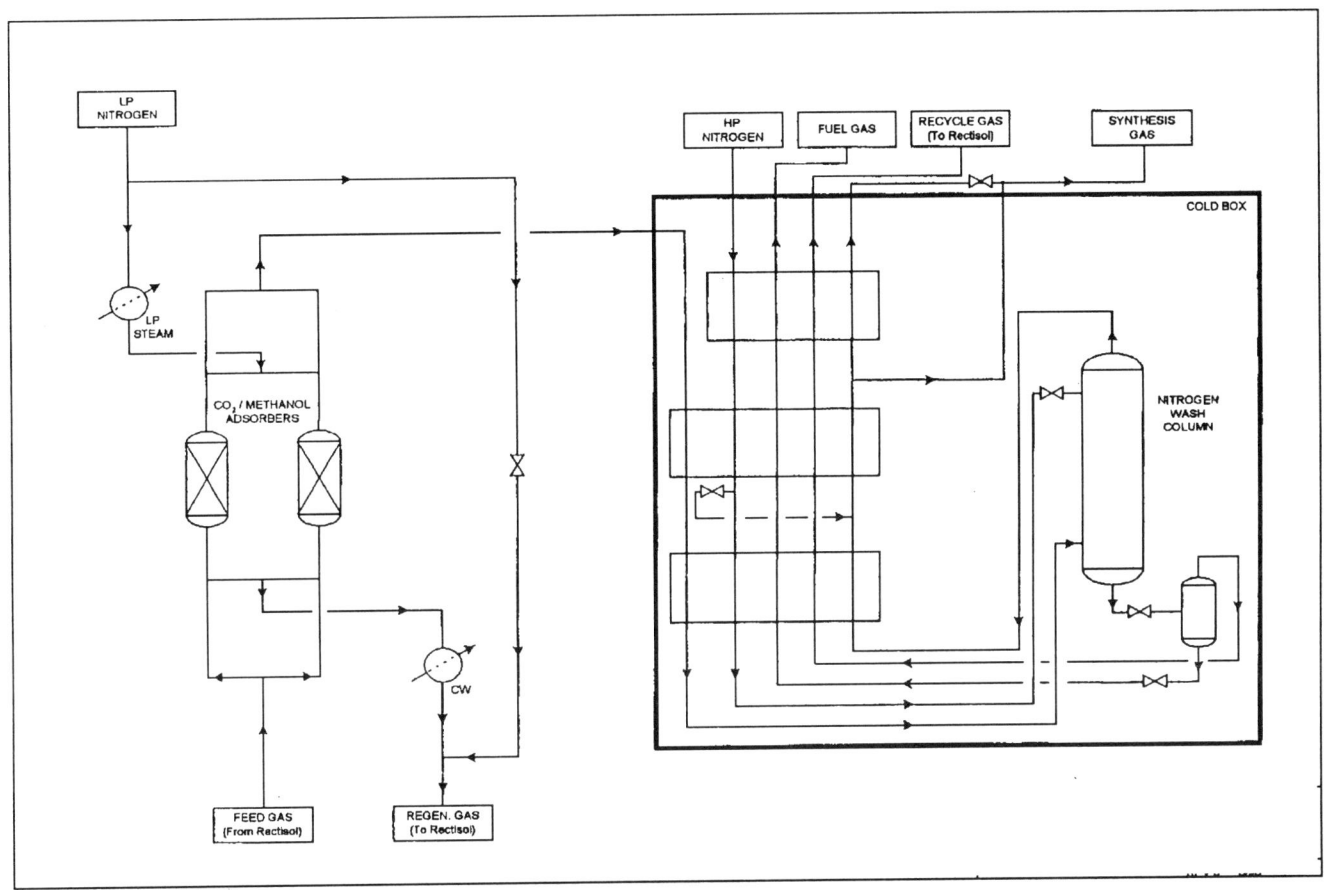

Figure 7. Typical nitrogen wash unit.

CO_2 co-production adds to the savings of the Selexol case because of its less compression requirement.

Operating Costs

The operating costs for the overall plant are divided into fixed and variable components. The fixed costs are composed of operating labor, maintenance labor, and administration, support labor and maintenance materials. The variable operating costs depend upon the capacity and the configuration of the plant and are composed of utility costs, feed costs, and catalyst and chemical costs.

A summary of the annual operating costs can be found in Table 6. Selexol/PSA requires less utility to operate because less refrigeration is required and its purge gas is used to generate power/steam to supply the process. The net savings by the UOP Selexol/PSA case are about $1.3 MM per year. CO_2 coproduction further adds to the operating cost savings because of the less CO_2 compression requirement by the Selexol unit.

Effect of Operating Parameters on Plant Performance and Costs

The following parameters have been studied to evaluate their effects on the acid gas removal and hydrogen recovery design. These parameters and their effects are discussed qualitatively below.

Evaluation of different feedstocks

Heat and material balances were developed and analyzed for the Selexol/PSA and the Rectisol/N_2 Wash processes for two other feedstocks: coke and residual oil. The comparison results are discussed below.

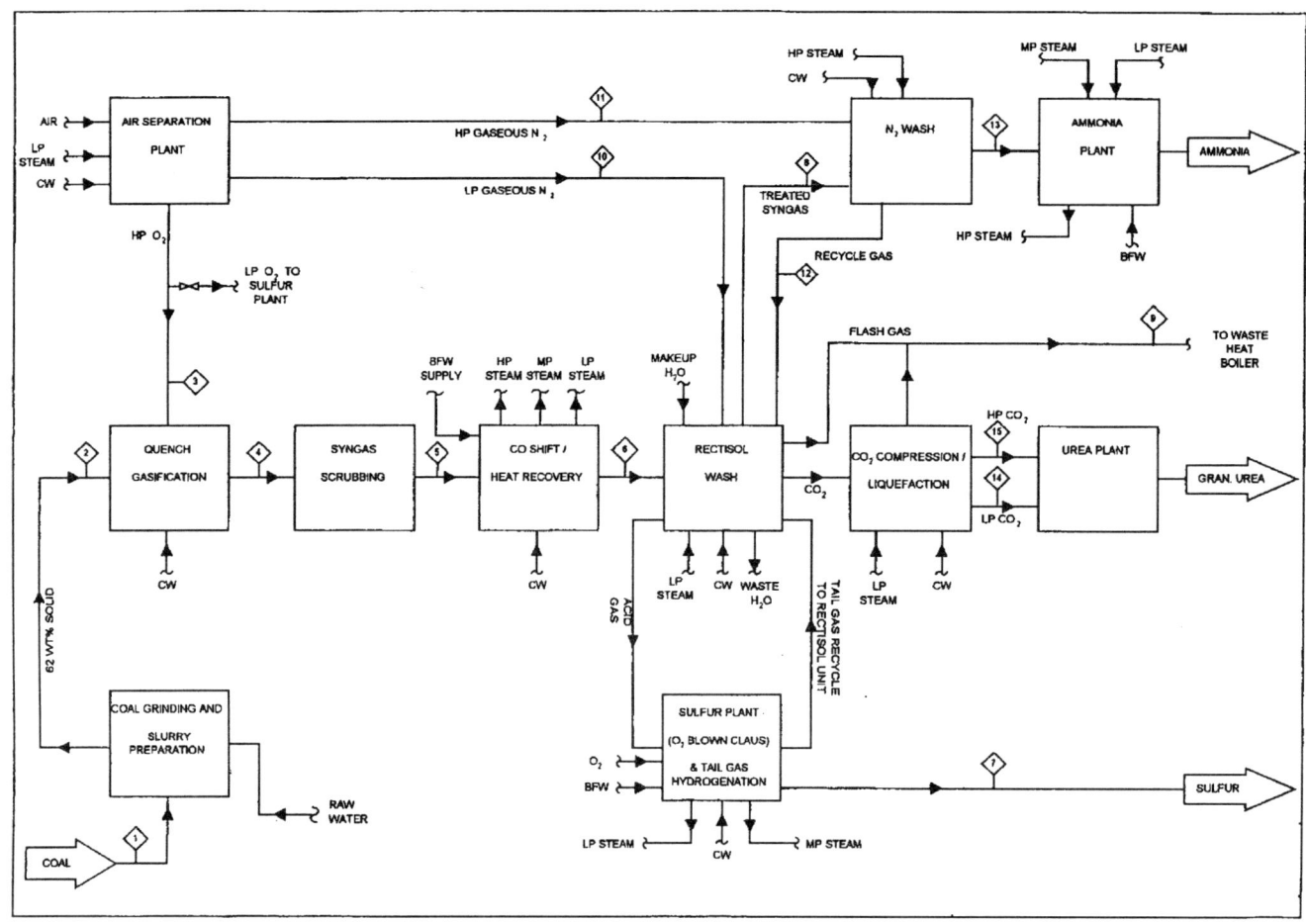

Figure 8. Case 4: coal to ammonia complex rectisol case with CO_2 production.

Evaluation of coke as a feedstock

Petroleum coke (usually from a delayed coker unit) can be used as a feedstock to the gasification unit for hydrogen production. The overall plant requires less coke than coal as feed to produce the same amount of ammonia. This is mainly due to the higher heating value of petroleum coke as compared to that of coal. Coal contains inert materials such as sulfur and ash, and is usually at least 15% lower in heating value than coke. As a result, the front-end of the plant is proportionally reduced in size. Once again, it is observed that less petroleum coke (about 6%) is required for the Rectisol/N_2 Wash case than for the Selexol/PSA case due to its higher hydrogen recovery efficiency.

Evaluation of residual oil as a feedstock

Residual oils such as vacuum residue, tar, and asphalt can also be used as a gasifier feedstock. Tar feed is used in this sensitivity analysis. When other types of residual oil are used, the amount of feedstock and utility consumption will vary, and, in general, the feedstock requirement increases in the order of vacuum residue, tar, and asphalt.

The overall plant requires less residual oil (38%) than coal feed to produce the same amount of ammonia. The front-end solid handling equipment in a coal/coke gasification plant is not required in an oil-fed plant, and this results in a considerable size reduction. Once again, it is observed that less residual oil (about 6%) is required for the Rectisol/N_2 Wash case than for the Selexol/PSA case due to the higher hydrogen recovery efficiency. The results of this evaluation are summarized in Table 7.

Evaluation of a Different Gasification Technology

The Shell Gasification technology was evaluated using coal as the feedstock to determine the applicability of the Selexol/PSA process to this technology. The Shell Coal or Coke Gasification technology is different than the Texaco process in that the Shell process uses nitrogen as a "carrier," or transport fluid, rather than water as in the Texaco Quench process. When gasification is considered by itself, the Shell process produces a syngas higher in CO and H_2 than the Texaco process because less CO shifting occurs in the gasifier. The UOP Selexol/PSA process is applicable to this gasification technology, as the only difference is the higher nitrogen content in the syngas.

Evaluation of Varying Plant Sizes

The reduction in plant costs for small plant sizes can be estimated by capacity factoring from the base case. The exponent factor is expected to be slightly lower in the Rectisol/N_2 Wash case than the Selexol/PSA case because of the complexity and more equipment counts in the Rectisol/N_2 Wash design. The operating costs can be estimated by proportioning based on the base plant capacity. The conclusions of this study will not be affected by the variation in plant sizes.

Conclusion

This feasibility shows that the UOP Selexol/PSA technology is clearly an attractive alternate to the conventional Rectisol/N_2 cryogenic route, both in capital and operating costs. With the advent of UOP's new PSA technology and with its improvement in hydrogen recovery, the UOP Selexol/PSA technology offers an economic and a simple plant in terms of design and operation when compared to the conventional Rectisol/N_2 Wash design.

Acknowledgments

Flour-Daniel wishes to express their gratitude to UOP/Equipment and Systems Dept. for their sponsorship and support of this project.

Literature Cited

Kohl, A, and R. Nielsen, *Gas Purification*, 5th Ed., Gulf Publishing Company, Houston, TX, pp. 1202-1223 (1997).

Mak, Y. J., "Gasification to Ammonia Feasibility Studies" Fluor-Daniel internal reports (1995).

Mak, Y. J., "Study on Selexol/ PSA vs Rectisol/Nitrogen Wash in Coal to Ammonia Plant", Fluor Daniel/UOP Report (Dec. 1996).